Silicon Photonics

Wiley Series in Materials for Electronic and Optoelectronic Applications

www.wiley.com/go/meoa

Series Editors

Professor Arthur Willoughby, University of Southampton, Southampton, UK
Dr Peter Capper, SELEX Galileo Infrared Ltd, Southampton, UK
Professor Safa Kasap, University of Saskatchewan, Saskatoon, Canada

Published Titles

Bulk Crystal Growth of Electronic, Optical and Optoelectronic Materials, Edited by P. Capper
Properties of Group-IV, III – V and II – VI Semiconductors, S. Adachi
Charge Transport in Disordered Solids with Applications in Electronics, Edited by S. Baranovski
Optical Properties of Condensed Matter and Applications, Edited by J. Singh
Thin Film Solar Cells: Fabrication, Characterization and Applications, Edited by J. Poortmans and V. Arkhipov
Dielectric Films for Advanced Microelectronics, Edited by M. R. Baklanov, M. Green and K. Maex
Liquid Phase Epitaxy of Electronic, Optical and Optoelectronic Materials, Edited by P. Capper and M. Mauk
Molecular Electronics: From Principles to Practice, M. Petty
CVD Diamond for Electronic Devices and Sensors, Edited by R. S. Sussmann
Properties of Semiconductor Alloys: Group-IV, III – V and II – VI Semiconductors, S. Adachi
Mercury Cadmium Telluride, Edited by P. Capper and J. Garland
Zinc Oxide Materials for Electronic and Optoelectronic Device Applications, Edited by C. Litton, D.C. Reynolds and T. C. Collins

Forthcoming Titles

Lead-Free Solders: Materials Reliability for Electronics, Edited by K. N. Subramanian
Nanostructured Waveguides: Fundamentals and Applications, M. Skorobogatiy

Silicon Photonics

Fundamentals and Devices

M. JAMAL DEEN

Department of Electrical and Computer Engineering, McMaster University, Canada

P. K. BASU

Institute of Radiophysics and Electronics, University of Calcutta, India

WILEY

A John Wiley & Sons, Ltd., Publication

This edition rst published 2012

Registered office
John Wiley & Sons Ltd, The Atrium, Southern Gate, Chichester, West Sussex, PO19 8SQ, United Kingdom

For details of our global editorial of ces, for customer services and for information about how to apply for permission to reuse the copyright material in this book please see our website at www.wiley.com.

Library of Congress Cataloging-in-Publication Data

Deen, M. Jamal.
Silicon photonics : fundamentals and devices / M.J. Deen, P.K. Basu.
p. cm.
Includes bibliographical references and index.
ISBN 978-0-470-51750-5 (cloth)
1. Silicon–Optical properties. 2. Optoelectronic devices. 3. Photonics.
I. Basu, P. K. (Prasanta Kumar) II. Title.
TA1750.D44 2012
621.3815′2–dc23

2011050047

A catalogue record for this book is available from the British Library.

Print ISBN: 9780470517505

Set in 10/12pt Times Roman by Thomson Digital, Noida, India

Printed in the UK

Dedication

PKB
To Chitrani (wife), Rikmantra (son), and Kaberi (sister)

MJD
To Meena (wife), and Arif, Imran, and Tariq (sons)

Contents

Series Preface

Wiley Series in Materials for Electronic and Optoelectronic Applications

This book series is devoted to the rapidly developing class of materials used for electronic and optoelectronic applications. It is designed to provide much-needed information on the fundamental scientific principles of these materials, together with how these are employed in technological applications. The books are aimed at (postgraduate) students, researchers and technologists, engaged in research, development and the study of materials in electronics and photonics, and industrial scientists developing new materials, devices and circuits for the electronic, optoelectronic and communications industries.

The development of new electronic and optoelectronic materials depends not only on materials engineering at a practical level, but also on a clear understanding of the properties of materials, and the fundamental science behind these properties. It is the properties of a material that eventually determine its usefulness in an application. The series therefore also includes such titles as electrical conduction in solids, optical properties, thermal properties and so on, all with applications and examples of materials in electronics and optoelectronics. The characterization of materials is also covered within the series in as much as it is impossible to develop new materials without the proper characterization of their structure and properties. Structure–property relationships have always been fundamentally and intrinsically important to materials science and engineering.

Materials science is well known for being one of the most interdisciplinary sciences. It is the interdisciplinary aspect of materials science that has led to many exciting discoveries, new materials and new applications. It is not unusual to find scientists with a chemical engineering background working on materials projects with applications in electronics. In selecting titles for the series, we have tried to maintain the interdisciplinary aspect of the field, and hence its excitement to researchers in this field.

Arthur Willoughby
Peter Capper
Safa Kasap

Preface

Semiconductor research made a giant leap forward with the invention of the transistor in 1947. Since then, a great many researchers engaged in the study of fundamental physical processes of semiconductors: mostly silicon (Si) or germanium (Ge), their growth technology, and device fabrication methods. Initially the emphasis was on the study of electronic processes with a view to optimizing the device's performance. Some efforts were made to study optical processes in semiconductors, primarily to understand the band structure of the materials. The idea to use semiconductors and their junctions for optoelectronic systems applications was not well defined in the early stages of the development of semiconductor technology.

The next decade (i.e., the 1950s) saw the emergence of Si as the leading electronic material. In fact, silicon is the material par excellence. The decade starting from 1960 marked the dominance of Si in electronics, thanks to the development of planar technology, and, later on, integrated circuits. This dominance of Si is still in force today; almost more than 95% of all electronic circuits are now grown on a Si platform. This dominance is expected to continue following the trend over the last few decades where the number of transistors in a chip doubles about every 18 months and the size of the individual transistors is shrinking at the same pace. The length of the channel in a field effect transistor has shrunk to a few tens of nanometers, and the downsizing is expected to continue in the near foreseeable future.

Coming back to photonics, the subject of semiconductor optoelectronics owes its origin to the announcement of semiconductor lasers almost simultaneously by four groups in 1962. The proposal to use optical fiber as a communication medium came in 1966. The semiconductor laser was soon vastly improved by using heterojunctions grown by liquid phase epitaxy (LPE), molecular beam epitaxy (MBE), or metal–organic vapor phase epitaxy (MOVPE). These developments led to remarkable progress in the area of optoelectronics or photonics, in which optical communications, among many other applications, form an ever-growing industry.

Unfortunately, silicon, the most widely used material in electronics, does not find the same niche in optoelectronics. The indirect nature of its band gap prevents the realization of efficient light-emitting diodes (LEDs) or laser diodes; the absence of linear electro-optic effect stands in the way of achieving high-speed modulators using Si-based materials; and, in addition, the band gap of Si does not match the standard wavelengths of 1.3 μm or 1.55 μm in present fiber-optic telecommunications systems. If, however, the problems could be solved, the integration of electronic and photonic devices on Si platform will be possible with all the benefits of Si technology that include low cost, high packing density, compact size, and high reliability. This dream had been cherished by many people over the last four decades, with little relative successes compared to silicon electronics.

Currently, there is one niche area in optoelectronics in which Si nds its place. This is the planar lightwave circuits (PLCs) used in optical ber communications. These include primitive devices like planar waveguides with silica-on-Si (SOS), silicon –on-insulator (SOI), directional couplers, Y-junctions, as well as more complex passive circuits like arrayed waveguide gratings, add-drop multiplexers and hybrid lightwave circuits embodying light sources, light detectors, multiplexers and demultiplexers, and various other components. These circuits rely on standard microelectronics technology and are therefore cheap and reliable. The Si-based PLC came into usage during the mid-1980s.

Yet, the main challenge to realize sources and modulators could not be overcome. The motivation for Si photonics came from the IC industry itself in the form of interconnect bottleneck. Electrical connectors in integrated circuits introduce larger time delays than the gate delay, and multilayered interconnections need insulators in between. Optical interconnects can solve both these problems.

There appeared several novel concepts and demonstrations of light emission from Si which include porous Si, Si nanocrystals, and erbium-doped Si. But the true solution is yet to be found. However, in view of the importance of the issue, almost all important players in the integrated circuits manufacturing business have started extensive research in silicon or group IV photonics. Governments of countries with major research investments such as the United States, Canada, Japan, countries of the European Union, China, India, and Brazil are providing substantial funds to nd useful solutions. And the Institute of Electrical and Electronic Engineers (IEEE) is now holding an annual conference titled "Group IV Photonics."

A number of review articles and at least two books on this subject have appeared. One book on silicon photonics by Reed and Knights is the rst textbook in the subject. It emphasizes the waveguiding and modulation properties of Si-based structures, and brie y discusses their light emission properties. The volume on silicon photonics edited by Pavesi and Lockwood is a compilation of articles written by experts, and the chapters are excellent source of references for both new and experienced researchers in the eld.

The need for a textbook dealing rst with the fundamental aspects of silicon and then covering the application areas was articulated by the present authors as early as 2001, when Professor Basu made his rst visit to McMaster University at the invitation of Professor Deen. Initially, the composition of the book was not clearly de ned because, in some areas, there were very few published reports, for example on sources and modulators, and their performance was far below that of III–V semiconductor counterparts. The contents evolved to become better de ned during Professor Basu's subsequent visits to McMaster University and in close collaborations with Professor Deen. It was decided then to cover the basic optoelectronic processes in bulk and quantum nanostructures, sources, detectors, modulators, wave propagation in guided structures, and components for dense wavelength division multiplexing (DWDM) optical communications systems. And a chapter on growth and fabrication was also planned. The present text covers all these topics.

This book is an outcome of nearly a decade's efforts by the authors. Care has been exercised to include the latest developments in the eld as much as possible. However, since the eld is growing and evolving at a rapid pace, it has not been possible to do justice to all the recent developments. The book is primarily a text with the targeted audience of senior undergraduates, graduate students, practicing engineers and technologists, and beginners in the eld, for whom introduction to basic principles and overall development in the eld are more important. Interested readers may nd a vast amount of contemporary literature once they get acquainted with the subject to help develop their understanding of the subject a number of examples within each chapter and also problems at the end of each chapter have been included.

Most of the chapters of the book were written during the periods when Professor Basu (PKB) visited McMaster University and in collaboration with Professor Deen. Professor Basu acknowledges the support, hospitality, and very congenial academic and non-academic atmosphere created by Professor Deen and his team, Dr. Ognian Marinov, Dr. Nikhil R. Das, and Mrs. Saswati Das, during his stay in Hamilton. In addition, PKB's students and colleagues, Dr. (Mrs.) Sumitra Ghosh, Dr. (Mrs.) Bratati Mukhopadhyay, Dr. (Mrs.) Gopa Sen, and Dr. Abhijit Biswas, merit special mention for their help in drawing gures, making available a number of papers, and providing useful collaboration. PKB also acknowledges partial nancial support from the Department of Science and Technology, Government of India, through project No. SR/S2/CMP-34/2006, as well as from the University Grants Commission through the UGC–Basic Scienti c Research Faculty Fellowship Programme. Finally, he records his indebtedness to his late father Jitendranath, late mother Amita, wife Chitrani, son Rikmantra, and sister Kaberi for their encouragement and for giving him complete freedom to pursue academic work without caring for household duties.

An undertaking such as this is possible with the help of many collaborators, students, and colleagues to all of them, we express our sincerest thanks and heartfelt gratitude. Professor Deen MJD and PKB is particularly indebted to Dr. Ognian Marinov for his continued assistance during this project and to many of his former students and collaborators, especially Professor Yaser El-Batawy (Cairo University), Dr. Anirban Bandyopadhyay (Intel), Dr. Yegao Xiao (Cross-Light Software Inc.), Professor Nikhil Das (University of Calcutta) and Professor Javier de la Hidalga Wade (INAOEP). He is also deeply indebted to his family – Meena, Arif, Imran, and Tariq – for their care, love, and support, and to his late parents who taught him the value of education.

MJD
PKB

1

Introduction to Silicon Photonics

1.1 Introduction

Silicon is the material par excellence. It is the most widely studied material in the history of civilization. In fact, the present-day information age has dawned with an electronics revolution brought about by the maturity of silicon-based microelectronics.

The growth of the silicon industry follows the now-famous Moore's law, which states that the number of transistors in an integrated circuit chip doubles every 12 months (since revised to every 18 months). However, during the last few years there has been indication of the decline of Moore's law. There are doubts whether in future silicon-based integrated circuits (ICs) will deliver the same advantages and functionalities as shown today.

The weakest point of silicon is that proper light emitters and modulators cannot be realized by using it due to the indirect nature of its band gap. On the other hand, there is a steady increase in the area of photonics, in the form of optical communication and networking, optical information processing, and consumer electronics based on light. Present-day photonics relies on compound semiconductors and their alloys. Although discrete devices using these materials show very good performance, when it comes to integration of these devices, preferably on the same substrate, the levels of integration and performance are far below what has already been achieved in electronic integration. It is natural to expect that monolithic optoelectronic integrated circuits (OEICs) will provide the same advantages, that is, low cost due to batch fabrication, high functionality, scaling for denser integration, and so on, as provided by silicon ICs.

If, however, it is possible to grow OEICs on silicon and integrate with electronic ICs by using the same production facilities, the bene ts to be accrued need no further elaboration. Si-based systems will then be used in all elds of electronics, computers, and communication. This is the dream cherished by many workers over the last few decades, though that dream is yet to materialize. In spite of this, Si-based photonics remained an active area of research and over the last 10–15 years some signi cant milestones have been achieved.

Silicon Photonics: Fundamentals and Devices, First Edition. M. Jamal Deen and P. K. Basu.

Another important area of application of silicon photonics is in the very large-scale integrated (VLSI) circuit itself. The complexity of present-day ICs has reached such a high level that the interconnects within it are formed on a number of levels. At present the number is six, but within a few years it will be doubled. The metallic interconnects, mainly Cu, provide delay due to resistor–capacitor (RC) time constants, which far exceed the transit time delay associated with the individual transistors. If the increase in speed is to be maintained at the same rate for the next-generation ICs, the interconnect bottleneck must be properly addressed. Optics is believed to be the right solution to the problem.

In the present chapter, we shall give an overview of the developments in silicon electronics, the present status, and the problems faced to achieve the goals discussed in this section. We rst give a very short history of the development of silicon-based microelectronics, the present status, and the international roadmap for future development. The two most important areas in which presence of silicon is needed, that is, communication and interconnect in chips, will then be described, giving the reader an idea of the present scenario and the problems faced. In this connection, the alternative to monolithic integration, that is, the hybrid integration technology, followed at present and the related problems will be pointed out. Finally, the scope of the present book will be outlined.

A number of text, research monographs, and reviews have already appeared dealing with silicon photonics. The reader is referred to such sources, all of which contain a large number of useful references [1–13].

1.2 VLSI: Past, Present, and Future Roadmap

The announcement of the rst point contact transistor on December 23, 1947, marks the birth of the electronics era. Although the material used in this device and its improved version, the junction transistor, was Ge, it was soon felt that single-crystal silicon would be a better alternative. The rst silicon bipolar transistor came in 1954. The concept of the IC was rst explored in 1958, and its working was demonstrated by using discrete components. A few months later, an IC using planar technology was developed. The bipolar transistor technology was developed earlier and was applied to the rst IC memory in the 1960s. Although bipolar transistors are the fastest at the individual circuit level, their large power dissipation and very low integration level ($\sim 10^4$ circuits per chip), compared to today's VLSI standard, do not promote their use.

The control of conductivity in the surface of a semiconductor by an external electric eld was proposed in the early 1930s. Attempts for conductivity modulation during the early part of the 1950s were not very successful. The rst metal oxide semiconductor eld effect transistor (MOSFET) using SiO_2 as the gate insulator on silicon substrate was fabricated in the 1960s. The complementary MOS (CMOS) transistor was fabricated in 1963, and its advantage of lower power consumption was rmly established. The one-transistor dynamic random-access memory (DRAM) cell was announced in 1968, and the rst microprocessor was marketed in 1971 [14].

The advantages of silicon as an electronic material are too many to recount here. Silicon is available in nature abundantly. It can be puri ed to a very high level. Native oxide silica is a very good insulator, is stable, and can withstand a large eld across it; and above all, the interface charge between silica and silicon can be reduced to a minimal level. SiO_2 can be easily patterned by photolithography [15].

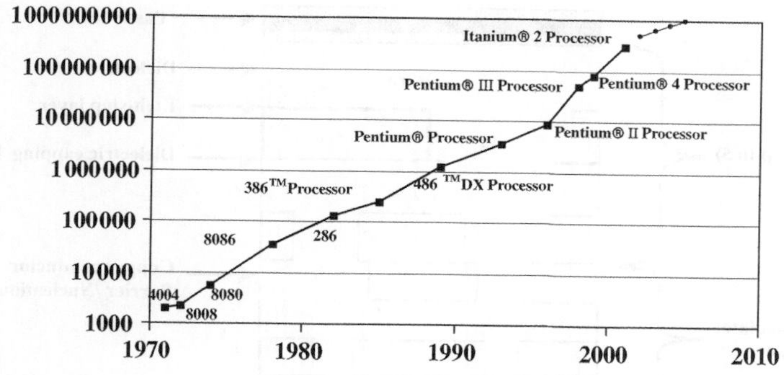

Figure 1.1 Graph showing the number of transistors in a chip since 1970.

Although bipolar transistors are faster than MOSFETs at the individual level, the low power consumption in CMOS, adaptability to planar processing, reduced size of the transistor, and larger packing density and ease of fabrication with reduced number of masks have made the CMOS technology on silicon substrate the sole technology followed by industries [16]. Figure 1.1 illustrates the developments of VLSIs over the last three decades and the technology roadmap for the coming decade [17].

At the rate shown in Figure 1.1, there were 1 billion transistors on a single die before 2007. With increasing numbers of transistors per die, the minimum feature size, or roughly the channel length in a single transistor, was around 70 nm in 2008. For almost the past 30 years, the feature size in IC lithography has been reduced at a rate of 0.7 × every three years. It is predicted that the feature size will reach 35 nm in 2014.

1.3 The Interconnect Problem in VLSI

On-chip interconnect is nothing but electrical wiring. According to the International Technology Roadmap for Semiconductors (ITRS), an interconnect is electrical wiring that distributes clock and other signals, and provides power and ground to and among the various circuits or systems functions on a chip.

The process devoted to metallization and interconnect involves deposition of metals, interlevel dielectrics deposition, and etching steps. A typical interconnect structure is shown in Figure 1.2.

The earlier approach was aluminum deposition and dry-etch de nition. Currently, copper wires are introduced. The global wires connect different functional units, distributing clock signals and power among them. The length of the global wires scales down with chip size. On the other hand, local wires connect the gates, sources, and drains of close MOSFETs of the same functional units and their length scales with gate size.

Although the downscaling of transistor increases the speed, the same is not true for the downsizing of interconnect. In earlier ICs in use around the 1980s, the delay in interconnects was propagation limited, that is, it was limited by the time of propagation of the electromagnetic waves associated to SiO_2, rather than the RC time constant, which was ~1 ps. It is predicted that for 35 nm technology generation, the interconnect response time of 0.1 mm copper line with a low-k dielectric (k = 2) will be about 250 ps, about two orders of magnitude higher, and will account for delays related to the RC time constant of the wire.

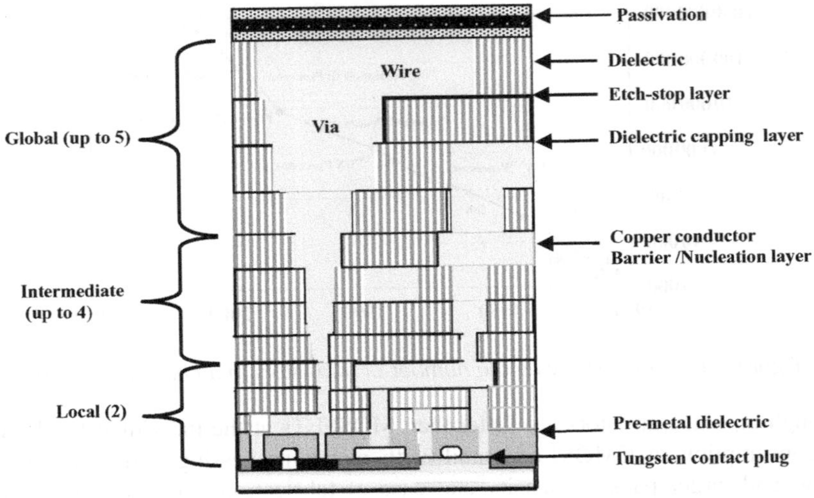

Figure 1.2 *Schematic view of electrical interconnect in very large-scale integrated (VLSI) circuit.*

The second issue with interconnect is power consumption. With scaling of transistors, the power consumption by the interconnect exceeds that by a transistor. For example, for 1.0 μm devices, the switching energies in the transistor and 1 mm long interconnect were, respectively, 300 and 400 fJ. The predicted values for 35 nm technology are 0.1 and 3 fJ, respectively, indicating that the ratio between dissipation in the interconnect and in the transistor is about 30.

Figure 1.3 shows a trend of interconnect propagation delay for 1 cm length with feature size and year for aluminum metal and silica insulators, copper and low-k dielectrics, and projected optical waveguide technology. It indicates that Al-based technology has reached the performance limit at 0.55 μm; the reduced resistance due to Cu and reduced capacitance due to low-k dielectric ensure a performance improvement limit up to 0.18 μm technology.

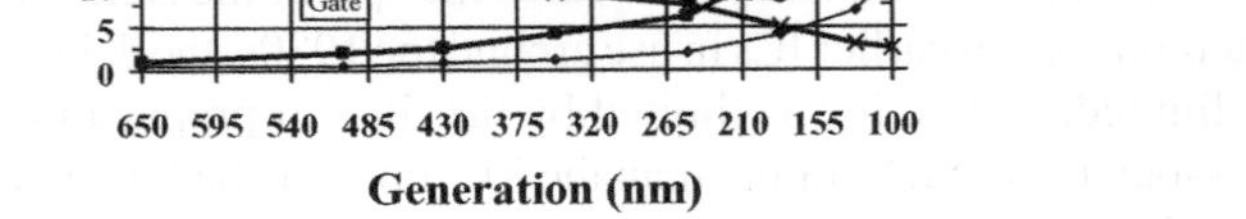

Figure 1.3 *Variation of interconnects propagation delay with year and feature size for Al–SiO_2, Cu–low-k dielectric, and projected optical interconnects technology.*

The optical interconnect using silicon microphotonics technology offers a potential solution to the RC time delay associated with traditional metal interconnects. Using photons as bits of information, instead of electrons, a speedier performance of the devices is expected. Use of photons also solves the power dissipation problem. Photons propagate in transparent media with less heat dissipation and almost no cross-talk. Unlike electrical current beams, light beams can cross one another without using any insulator. The multilevel interconnection scheme shown in Figure 1.2 is not needed when light beams are used for interconnects within the chip.

Further discussion of the use of light waves for chip-to-chip or board-to-board connections will follow in Section 1.6.

1.4 The Long-Haul Optical Communication Link

Fiber-optic communication links have spread today over the whole globe like a spider's net. A still larger number of links is being added. Today's ber-optic links employ dense wavelength division multiplexing (DWDM), in which huge amounts of data carried by hundreds of carrier wavelengths, each modulated at a high bit rate (~10 Gb/s or more), are transmitted by a single strand of a ber. In Section 1.4.1, we shall rst discuss the basic link, and the components used [18]. In Section 1.4.2, the materials used to grow the devices and the methods of integration of the devices will be pointed out.

1.4.1 Basic Link and Components

Figure 1.4 shows a block schematic of the WDM communication link. Voice, picture, or computer data, in digital format, are impressed on each laser emitting at a particular wavelength (e.g., λ_1). Either the laser may be directly modulated, or an external modulator may be used to impress the signal on the laser beam. A multiplexer combines the modulated signals coming from the bank of lasers, and the combination is transmitted by an optical ber. After traversing a distance of a few hundred kilometers, the signal becomes attenuated

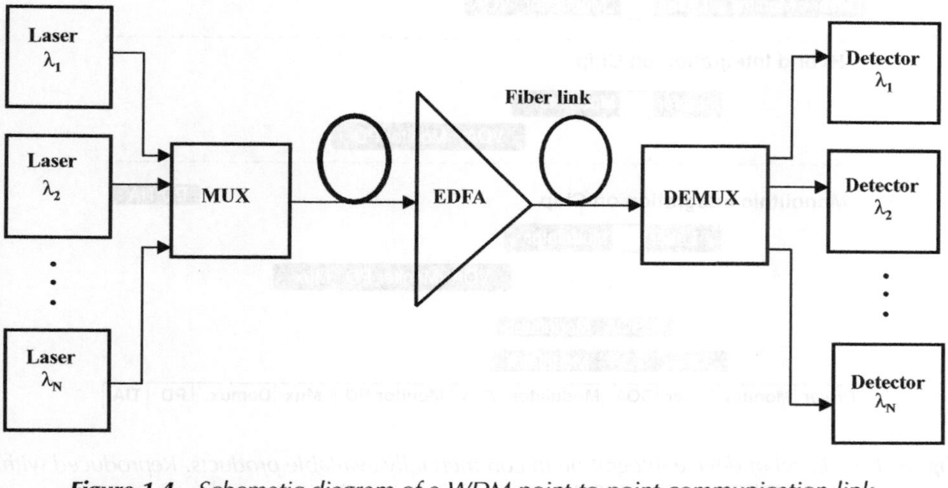

Figure 1.4 *Schematic diagram of a WDM point-to-point communication link.*

and the digital pulses considerably spread due to material dispersion of the ber. The combined signal is then regenerated by a regenerator (not shown in Figure 1.4). A photodetector rst converts the weak and distorted optical signal into a stream of electrical pulses. These pulses are then reshaped and retimed by a decision circuit. The cleaned electrical pulses are then converted to optical pulses by a laser, and the stream of pulses propagates through another long section of the ber. At present, a number of optical ampli ers are inserted at regular intervals in the link to boost up the intensity of optical signals, adding noise at the same time. A repeater or regenerator, which includes a detector, a laser, and different electronic circuits as described in this chapter, is then employed to reshape the pulses.

The transmitter and receiver units of the link need additional sub-units. The basic device in the transmitter, that is, the laser, is to be properly biased by a driver and the light output power should be accurately controlled by a monitor circuit. An optical ampli er, usually a semiconductor optical ampli er (SOA), may boost the laser power up. A variable optical attenuator (VOA) is sometimes necessary to reduce or control the intensity. In the receiver unit, the optical signal is detected by a photodetector. The weak electrical signal is then ampli ed rst by a low-noise preampli er and then by power ampli ers. Further processing systems are needed before these signals are converted back to the original format, at which time they are transmitted. The bottom part of Figure 1.5 indicates the occurrence of various devices in the order they appear in an optical link. Further discussion of Figure 1.5 will be made in connection with the integration of devices, which is discussed in Section 1.4.2.

The present-day communication links work around 1.55 μm, the wavelength at which optical bers have minimum attenuation. The optical ampli ers, the Er-doped ber ampli ers (EDFAs), also work around this wavelength. Earlier systems used a 1.3 μm

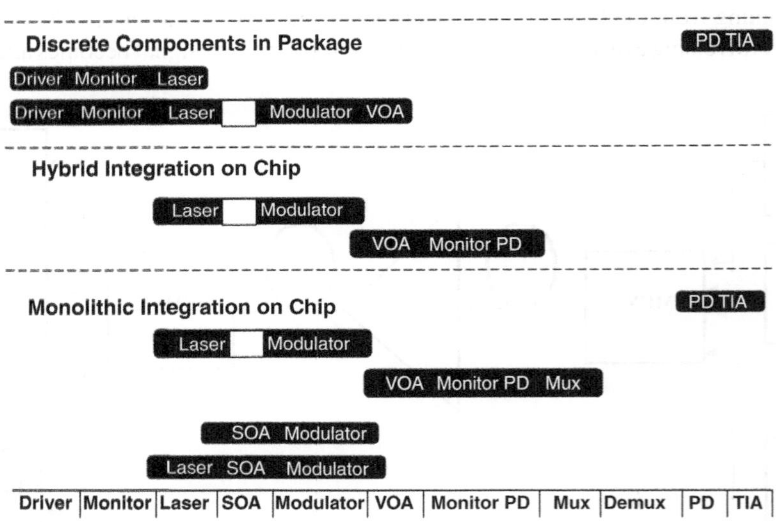

Figure 1.5 *Level of device integration in commercially available products. Reproduced with permission from [2] Copyright (2004) Springer Science + Business Media*

transmission window of the ber, at which the material dispersion of the ber is minimum. The DWDM system employs hundreds of wavelengths, separated approximately from each other by about 100 GHz (0.8 nm), covering both the 1.3 and 1.55 μm windows.

The requirement that all the optoelectronic and electronic components must work at the two wavelengths completely rules out the use of silicon-based active devices in the telecommunications network, since the cutoff wavelength of Si is about 1.1 μm. The materials of choice are quaternary alloy InGaAsP and ternary alloy InGaAs grown on InP substrate.

1.4.2 Materials and Integration

There are many differences between today's microelectronics and photonics industries. While silicon is the only material in the former, a variety of materials are used in photonics. InP substrate is used for source and detector development, silica as the ber material, different semiconductors and even insulators like $LiNbO_3$ for modulators, an Si platform for passive lightwave circuits used in DWDM, and Si-based ICs for driver and controller circuits. No single material or single technology is leading the market. The production technology is still primitive, and the level of integration is far below the level achieved in microelectronics.

In a truly monolithic IC, all components, that is, electronic circuits, light sources, photodetectors, modulators, waveguides, multiplexers, and so on, are grown on the same piece of semiconductor substrate. Since good sources and modulators have not yet been realized using silicon, and since ef cient silicon photodetectors do not exist at 1.55 μm, monolithic integration on a silicon platform is at present ruled out. It is possible, in principle, to use InP substrate for integration. However, the small wafer size, high cost, and other factors limiting the manufacturability have hindered progress in this direction.

Hybrid integration, in which disparate parts are assembled onto one common platform, has been pursued for quite some time. In common hybrid optical components, III–V compound light sources and detectors are attached onto silicon on insulator (SOI), silica, or polymer platforms. These components are commercially available. Figure 1.5 gives examples of discrete components in package, hybrid integration, and monolithic integration. The order of the devices on the axis agrees with the order of appearance of the devices in an optical link. As mentioned already, monolithic integration is almost exclusively on an InP platform; however, photodetectors and transimpedance ampli ers (TIAs) have been grown on silicon.

The main interest in hybrid technologies lies in the combination of III–V semiconductor laser diodes with Si integrated circuits for optical ber communication or optical interconnects. For this purpose, GaAs or InP is grown on Si and then processed, or, alternatively, laser devices are detached from their substrate by an epitaxial liftoff process and then bonded to Si substrate.

For growth of GaAs and InP on Si, the large lattice mismatch (4% for GaAs and 8% for InP), different thermal expansion coef cients, and fast diffusion of Si as impurities all create dif culties in maintaining low defect densities in compounds for laser production. Although several new techniques have been developed to overcome these dif culties, it is too early to predict the long-term success of the techniques.

In the epitaxial liftoff technique, wet chemical etching is performed and then the III–V heterostructure is oated off and transferred to a planar Si substrate. The bonding occurs due to the van der Waals force. In the wafer fusion process, the two materials (of high quality) are brought into intimate contact under hydrogen ambient at around 450 °C. Under uniform direct pressure, the substrates form robust chemical bonds. One of the substrates, for example compound semiconductor, is selectively etched and photonic devices grown onto it are integrated with silicon electronic circuits. Once again, it is dif cult to predict how far these technologies will be successful in commercial production.

It appears, therefore, that the most satisfactory solution to the above-mentioned problems would be achieved when all the optoelectronic and photonic components could be grown on a single substrate, for which silicon seems to be the best choice. The extensive experience in Si fabrication and processing could then be put to maximum use. Unfortunately, however, the lack of suitable emitter and especially a laser based on Si, as well as of a fast modulator, stands in the way of achieving the coveted goal.

1.5 Data Network

While long-haul optical communication systems work at 1.55 μm, to exploit the minimum attenuation in the ber, local area networks span smaller distance and area. In this case, working at other wavelengths, at the cost of higher attenuation of signals, may be of advantage due to the availability of cheap components. A possible system employs GaAs-based lasers at around 800 nm, at which wavelength Si photodetectors and other electronic circuits would offer a low-cost solution. If, in addition, Si emitters are available, further reduction in cost is highly expected.

A large part of data communication network is anchored to servers and desktop computers that utilize Si devices. The large potential volume of the market and the competition with copper cables will necessitate more use of inexpensive optical bers. Si-based photonic components will offer the cheapest solution to the network.

1.6 Conclusions

From the discussions in the above sections, the following points emerge:

- The interconnect problem within a chip is taking an alarming shape. Optical interconnect based on silicon technology may offer a viable solution.
- The long-haul optical communication link employs at present a number of different devices (viz., lasers, modulators, power monitor and control, ampli ers, photodetectors, photoreceivers, multiplexers, demultiplexers, lters and other passive lightwave circuits, and active network components like wavelength converters, etc.). Apart from passive components, most of the active components are fabricated on the InP platform. A truly monolithic OEIC on silicon may offer all the advantages of integration including cost reduction.
- Si-based photonic devices may offer lower cost in the sector of data networks covering shorter distances.
- Si microphotonics seem to be an attractive solution for next-generation optical interconnects for chip-to-chip or board-to-board interconnects.
- Discrete silicon photonic devices like light-emitting diodes (LEDs) and lasers are in demand for consumer electronics, display, and mobile communication, and as mid-infrared or THz emitters.

Si-based passive lightwave circuits are well developed and nd use in commercial optical communication and networking systems. Si-based photodetectors technology is also mature, and the devices are already in use. There are attempts to extend the operation to the important telecommunication window at 1.55 μm. On the other hand, the indirect band gap in silicon makes it dif cult to realize ef cient light emitter and to achieve laser action. This is also the reason why the intrinsic modulation bandwidth of silicon-based modulators is substantially lower than the compound semiconductor-based counterpart.

The challenge in the area of Si photonics is therefore to develop suitable emitter and modulator and then to integrate all the active and passive devices on a single silicon chip. The next challenge would obviously be to increase the level of integration, functionality, and yield coupled with reduction of cost.

1.7 Scope of the Book

In the earlier sections, an attempt has been made to identify the areas where Si photonics may be useful and advantageous. A few photonic devices are in the matured state of development, while a few others, notably emitters and modulators, are in the stage of early research or at most at the development stage in the laboratory. The book aims at providing the basic principles of operation of the devices, the structures of the devices, and an idea of the state-of-the art developments. The following is a brief description of the chapter-wise coverage of different topics.

Chapter 2 describes the fundamental electronic properties of silicon and its alloys, and of heterostructures made with Si-based materials. The band structure, density-of-states in bulk silicon, and Ge are discussed rst followed by similar discussions on Si and Ge-based alloys. A general introduction of heterostructures and band line-up then follows. The special features of Si-based heterostructures, the pseudomorphiv growth, are then introduced. The band structure modi cations arising out of strain and band offsets and band line-ups are then introduced. Recently, direct band gap has been achieved in tensile-strained Ge layers grown on Sn-based Si alloys. The theory and results obtained are presented.

Chapter 3 is devoted to quantum nanostructures. In the beginning, a simpli ed picture of quantum con nement and calculation of energy levels, density-of-states in quantum wells (QWs) using well-studied GaAs–GaAlAs material system is given, followed by a brief description of re ned theories. Quantum wires (QWRs) and quantum dots (QDs) are then introduced. Similar discussions using Si–SiGe systems are then included, mentioning the recent results on direct band gap systems. Finally, the effect of electric eld on the subband structures is discussed.

Chapter 4 gives an introduction to the optical processes in bulk semiconductors. The semiclassical theory of absorption is rst introduced. Both direct gap and indirect gap materials are covered. The relationship between absorption and gain is then established. The chapter then discusses other types of absorption and different forms of radiative and nonradiative recombination. The basic idea of exciton formation and a simpli ed theory of excitonic absorption are then presented.

Chapter 5 is similar in structure as in Chapter 4; however, here the basic theory of optical processes as modi ed for QW, QWR, and QD is presented. The theory of absorption, gain, and recombination in direct and indirect gap materials and in different types of band alignments is presented. Excitonic processes and the effect of electric elds on the properties of excitons, particularly in QWs, are discussed.

The remaining chapters are devoted to silicon photonic devices, covering description of the structures, operating principles, performance, and the application areas. These chapters may be grouped into two categories. Active devices including light emitters, modulators, photodetectors, and recently reported Raman lasers form the rst group to which Chapters 5–9 belong. Passive lightwave circuits belong to the second group, and Chapters 10–12 discuss various aspects related to the circuits. The topics covered are the propagation of electromagnetic waves, waveguiding action, loss mechanisms, coupling between waveguides, various passive devices used in dense wavelength division multiplexed ber-optic communication systems, and the application areas. The device fabrication processes are described in Chapter 13.

Chapter 6 discusses the phenomena of light emission in silicon, related materials, and nanostructures formed using Si, Ge, and their alloys. The devices reported are still at the primitive stage. The chapter makes a list of all the methods followed so far to achieve ef cient light emission, the structures of the devices, the basic principle of light emission in each structure, and the performance achieved so far.

Chapter 7 deals with Si-based light modulators. This device also is in the early stage of development. The different approaches followed by different authors, the results obtained by them, and possible ways of improving the performance form the subject of discussion in this chapter. Some recent breakthroughs in the area are also reported.

Chapter 8 is devoted to silicon photodetectors. It starts with a general discussion of optical receiver systems, and it discusses optical detection from an engineering perspective and the most important performance measures of photodetectors. Then various examples of silicon-based–pn photodiodes, p*in* photodetectors, Schottky barrier photodiodes, avalanche photodiodes, and bipolar and MOS phototransistors are presented and discussed. Various examples of photodiodes and of phototransistors in standard silicon CMOS and BiCMOS technologies are described. Important emerging technologies such as silicon-on-insulator, and various types of photodetectors that could be fabricated using heteroepitaxial techniques such as silicon–silicon germanium multiple quantum wells and germanium-on-silicon, are presented and discussed. The chapter concludes with related theoretical discussion from a practical perspective so that time constants, operating frequencies, and signal-to-noise ratios can be calculated. Throughout the chapter, numerical examples are provided to add numerical details to the discussions.

Chapter 9 is devoted to stimulated Raman scattering and other nonlinear effects in silicon. After giving a brief idea of Raman scattering, recent results of stimulated Raman gain in silicon and the structure and characteristics of Raman lasers are described. The chapter ends with a brief discussion of other nonlinear effects in silicon that play important roles in silicon-based photonic devices.

Chapter 10 gives the introduction to the waveguides that form the basic unit of all passive lightwave circuits to be discussed in later chapters. This chapter explains the principle of waveguiding in dielectric waveguides from both the ray-optic theory and the electromagnetic theory. The simple forms of slab and three-dimensional waveguides are presented and the propagation characteristics in these structures are then discussed. This chapter then introduces various sources of loss in a waveguide and underlines the principle of calculating the loss. It also discusses how waveguides are coupled to sources, detectors, and other elements.

Chapter 11 discusses the principle of operation of several planar waveguide devices used in actual optical communication systems and networking. Coupling between different waveguides occurs in many such devices. The general coupled mode theory is presented rst, followed by the application of the theory to two important classes of devices: directional coupler and Bragg grating. The operating principles of splitters, directional couplers, Mach–Zehnder interferometers, Fabry–Perot resonators, Bragg gratings, mirrors, ring resonators, and resonant cavities are then presented.

Chapter 12 embodies the description of important devices used in DWDM systems. At rst, the arrayed wave guides (AWGs), extensively used in present systems, are described in detail including the structure, principle of operation, performance, and so on. The different material systems on silicon platform are then discussed, and some comparative gures are quoted. The chapter includes two more devices, the Bragg grating in planar form and the Eschelle grating.

Chapter 13 is devoted to description of device fabrication processes. The different growth processes, like epitaxy and chemical vapor deposition (CVD), are rst discussed. The planar technology, lithography, and etching processes are then described. Different methods used for the fabrication of waveguides are also presented in this chapter.

References

1. Reed, G.T. and Knights, A.P. (2004) *Silicon Photonics: An Introduction*, Wiley Interscience, New York.
2. Pavesi, L. and Lockwood, D.J. (eds) (2004) *Silicon Photonics: Topics in Applied Physics*, vol. **94**, Springer, Berlin.
3. Soref, R.A. (2008) The advances and challenges of silicon photonics. *Advances in Opt. Technol.*, Art. ID 472305.
4. Fitzgerald, E.A. and Kimerling, L.C. (1998) *Silicon-Based Microphotonics and Integrated Optoelectronics*, vol. **23**, MRS Bulletin, pp. 39–47.
5. Lockwod, D.J. (ed) (1997) *Light Emission in Silicon: From Physics to Devices, Semiconductors and Semimetals*, vol. **49**, Academic Press, San Diego.
6. Pavesi, L., Gaponenko, S., and Dal Negro, L. (eds) (2003) *Towards the First Silicon Laser, NATO Science Series*, Kluwer Academic, Dordrecht.
7. Bisi, O., Campisano, U., Pavesi, L., and Priolo, F. (eds) (1999) *Silicon-Based Microphotonics: From Basics to Applications*, IOS Press, Amsterdam.
8. Ossicini, S., Pavesi, L., and Priolo, F. (2003) *Light-Emitting Silicon for Microphotonics*, Springer, Berlin.
9. Pavesi, L. (2003) Will silicon be the photonic material of the third millenium? *J. Phys. Condens Matter*, **15**, R1169–R1196.
10. Salib, M. *et al.* (2004) Silicon photonics. *Intel. Technol. J.*, **08**(02), 142–160.
11. See papers in the special issue on silicon photonics: Fauchet, P.M. and Shin, J.H. (eds) (2006) *IEEE J. Sel. Top. Quantum. Electron.*, **12**(6), Part 2, 1327–1755. See also the invited papers on Si photonics, (2010) *Nat. Photon.*, **4**, 491–544.
12. Jalali, B. (2008) Can silicon change photonics? *Phys. Stat. Sol. (a)*, **205**, 213–224.
13. Jalali, B., Paniccia, M., and Read, G.T. (2006) Silicon photonics. *IEEE Microwave Mag.*, **7**, 58–68.

14. For history of development of devices, the reader is referred to a special issue: Pearson, G.L.(guest ed) and Haitz, R.H. (ed) (1976) *IEEE Trans. Electron. Dev.*, **ED 23**(7), 595–785.
15. Plummer, J.D., Deal, M.D., and Grif n, P.B. (2001) *Silicon VLSI Technology: Fundamentals, Practice and Modeling*, Prentice Hall, Upper Saddle River, NJ.
16. Taur, Y. and Ning, T.H. (1999) *Fundamentals of Modern VLSI Devices*, Cambridge University Press, Cambridge, UK.
17. Semiconductor Industry Association (2003) *The International Technology Roadmap for Semiconductors*, International SEMATECH, Austin, TX, http://public.itrs.net.
18. Ramaswami, R. and Sivarajan, K.N. (2002) *Optical Networks: A Practical Perspective*, 2nd edn, Morgan Kaufmann, San Francisco.

2

Basic Properties of Silicon

2.1 Introduction

There are many textbooks, introductory and advanced, that deal with fundamental properties of solid-state materials including semiconductors. Research monograms are also available that emphasize one or two focused and important aspects of material silicon or a particular device using silicon. Since silicon and different structures using Si are invariably used in microelectronics, specialized textbooks discuss growth, materials and device characterization, the electronic properties of materials, device behavior, and so on.

The aim of the present book is to discuss the use of silicon and related materials in photonic devices. Therefore the optoelectronic properties of Si will be the subject of interest. In the present chapter, we plan to mention briefly the basic properties that may be useful in understanding the operation of Si-based photonic devices. It is not possible to cover an in-depth survey of all the important properties. Therefore, only the introductory ideas will be given, and the reader will be referred to proper sources for finer details.

The chapter begins with a brief introduction to the band diagram; specifically, the energy–wave vector (E–$\mathbf{k}$) relationship will be described. The most important bands participating in electronic and optoelectronic processes and some of the useful parameters needed in further studies will be introduced.

Most of the modern optoelectronic and photonic devices rely on heterojunctions, that is, junctions between two dissimilar semiconductors. Often, nanostructures like quantum wells, wires, and dots prepared by using heterostructures are used to realize present photonic devices. The formation of heterojunctions by using silicon and its alloys will then be introduced. The basic concepts related to heterojunction band line-up, band offsets, and strained layer epitaxy using Si, its alloys, and other group IV elements will be presented in this chapter along with some basic electronic properties like band structure modification under strain, effective mass, and so on. Finally, the recently introduced concept of realizing direct gap type I structure in tensile-strained Ge will be introduced.

Silicon Photonics: Fundamentals and Devices, First Edition. M. Jamal Deen and P. K. Basu.

2.2 Band Structure

2.2.1 E–k Diagram: General Considerations

Semiconductors comprising only Group IV elements like Si or Ge have a diamond crystal structure. The cubic cell consists of two interpenetrating face-centered cubic (fcc) lattices displaced with respect to each other along the (111) direction by one quarter of the body diagonal of the elemental cube. A primitive unit cell has two Si atoms, and therefore there is a center-of-inversion midway between the two nearest neighbors.

On the other hand, semiconductors made of Group III and V elements, like InSb or GaAs, have a zinc-blende crystal structure, which is derived from the diamond structure by making the two interpenetrating fcc lattices different. In GaAs, for example, one fcc lattice contains the Ga atom, while the other fcc lattice contains the As atom. Since the two nearest neighbors are different, there is no inversion symmetry. This difference from Si as well as the different bonding mechanisms is responsible for the changes in band structure of the two materials.

To understand the transport and optical processes in a semiconductor, it is sufficient to know the behavior of outermost valence electrons in a crystal. A perfect crystal consists of many valence electrons and ions. The many-electron problem is reduced to a single-electron problem by using various levels of approximation, and finally the one particle Schrödinger equation is solved. The wave function of an electron is written as

$$\Psi_{\mathbf{k}}(\mathbf{r}) = U_{\mathbf{k}}(\mathbf{r}) \exp(i\mathbf{k} \cdot \mathbf{r}), \tag{2.1}$$

where Ψ is the wave function, $\mathbf{k}$ is the wave vector of the electron in the crystal, $\mathbf{r}$ is the position coordinate of the electron, and U is an envelope function having the periodicity of the lattice, that is, $\mathbf{U}(\mathbf{r}) = \mathbf{U}(\mathbf{r}+\mathbf{a})$, with $\mathbf{a}$ being the lattice vector. All textbooks on solid-state physics discuss the solution of the wave equation under simplified approximations. The concept of effective mass in a crystal is introduced from the E– $\mathbf{k}$ diagram so obtained. The quantity $\hbar\mathbf{k}$ is called the *crystal momentum*.

It is difficult to get full idea of the band structure in any semiconductor by a single experiment. Different experimental methods give the knowledge of the structure over only a very limited range of energy. Therefore, in most situations, band structure obtained from theoretical analysis forms the starting point for all subsequent research. Calculation of the band structure of any semiconductor is fairly complicated. The results obtained by Chelikowsky and Cohen [1] by empirical pseudopotential method are often quoted. Figure 2.1 gives the band diagram for Si. The same for GaAs is also included for comparison.

In the diagram, plots of energy (E) as a function of wave vector **k**, known as the *dispersion relation*, are given for **k** lying within the first Brillouin zone. The electron energy increases along the vertical axis, while the hole energy increases downward. The dispersion relation is plotted for different high-symmetry directions in the crystals.

The upper part of Figure 2.1 gives the band structure of Si for the principal directions in the Brillouin zone. The maximum of the uppermost filled band, called the *valence band*, occurs at the Γ point ($\mathbf{k}=0$), and the minimum of the lowest empty (conduction) band occurs along the [100] direction about 0.85 away to the zone boundary. Due to crystal symmetry, there are six such minima (called the Δ *valleys*) in the conduction band, as we shall discuss in more detail in this chapter. This is an example of an indirect gap semiconductor whose conduction and valence band extremas occur at different points in

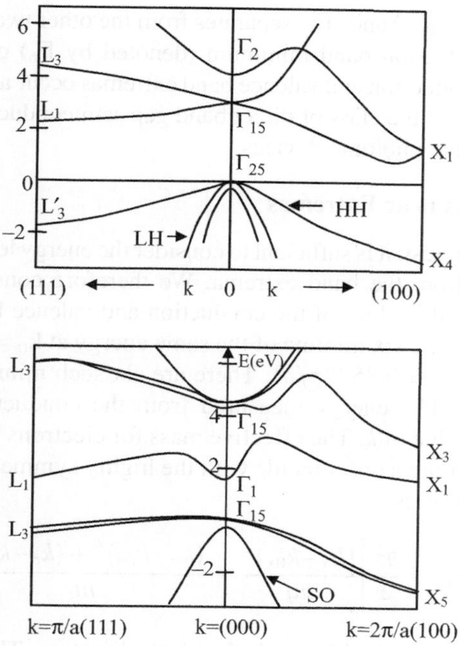

Figure 2.1 *Band structure for Si (top) and GaAs (below).*

the Brillouin zone. The magnitude of the indirect energy gap is 1.12 eV at room temperature. The direct gap at Γ point ($\Gamma_{15} - \Gamma_{25}'$) is 3.45 eV at 5 K.

The valence band at Γ_{25}' is degenerate, as seen from Figure 2.1. This band is derived from the atomic 3*p* states. They are split by spin–orbit interaction, but the splitting is too small to be detected in the scale of Figure 2.1. As one moves from the Γ point, the degeneracy is lost for finite **k** and the two bands are called heavy-hole (HH) and light-hole (LH) bands according to the values of the effective masses defined as (see the "Suggested Readings" section at the end of this chapter)

$$\left(\frac{1}{m}\right)_{ij} = \frac{1}{\hbar^2}\frac{\partial^2 E}{\partial k_i \partial k_j}, \tag{2.2}$$

where m_{ij} is a component of the effective mass tensor, and i and j are the directions along the crystal.

Example 2.1

The band, for which energy decreases more slowly, offers larger mass according to Eq. (2.2) and hence is called the *HH band.*

The calculated band structure of GaAs shown in Figure 2.1 shows some similarities and basic differences from that for Si. First, the two degenerate HH and LH valence bands have maxima at $\mathbf{k} = 0$ (Γ_8). However in this case, the spin–orbit splitting is appreciable (0.34 eV) and the split-off (SO) valence band (Γ_7) separates from the other two even at $\mathbf{k}=0$. Unlike in Si, the principal conduction band minimum (denoted by Γ_6) occurs at $\mathbf{k}=0$ in this material. Because the conduction and valence band extremas occur at the same point, GaAs is a well-known member of the class of direct band gap semiconductors, which are almost invariably used to produce photonic devices.

2.2.2 Band Properties near Extremas

In most of the cases of interest, it is sufficient to consider the energy levels within a few times of the thermal energy from the band extrema. We therefore consider the properties of electrons and holes near the edges of the conduction and valence bands.

In the case of Si, there are six minima of the same energy at $\mathbf{k}_0=(\pm k_0,0,0)$, $(0, \pm k_0,0)$, and $(0, 0, \pm k_0)$, where $k_0 \approx 0.85\,(2\pi/a)$. There are six such minima or valleys. This is depicted in Figure 2.2. The energy measured from the conduction band edge varies parabolically with the value of k. The effective mass for electrons is not isotropic but is a tensor for each valley. If the axes coincide with the highly symmetric axes, then the E–$\mathbf{k}$ relation may be expressed as

$$E_{\mathbf{k}} = E_{\mathbf{k_0}} + \frac{\hbar^2}{2}\left[\frac{(k_x-k_{0x})^2}{m_l} + \frac{(k_y-k_{0y})^2+(k_z-k_{0z})^2}{m_t}\right]. \tag{2.3}$$

In this equation, the x-axis coincides with the [100] direction. The surfaces of constant energy are ellipsoids of revolution with major and minor axes proportional to $\sqrt{m_l}$ and $\sqrt{m_t}$, respectively. The set of constant energy surfaces is shown in Figure 2.2.

Example 2.2

For the two valleys having longitudinal axis along the [100] direction, the mass along this direction is m_l, while the masses along the [010] and [001] directions are the transverse mass.

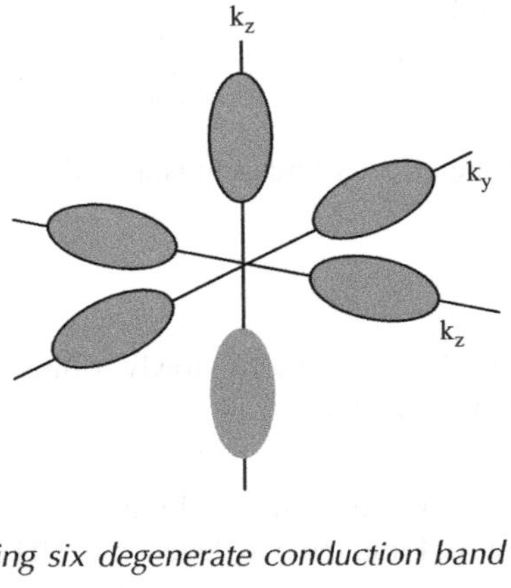

Figure 2.2 *Diagram showing six degenerate conduction band valleys and constant energy surfaces for electrons in silicon.*

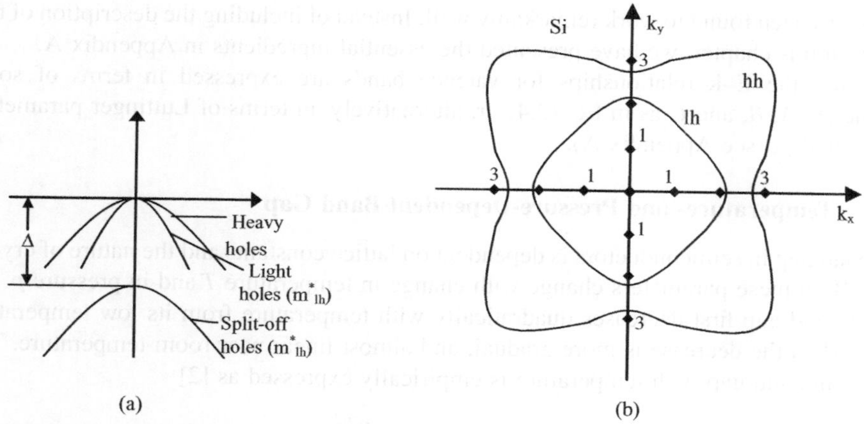

Figure 2.3 *(a) The valence band structure in silicon; and (b) constant energy contours in the (100) plane in* **k***-space for the two warped energy surfaces of the valence band of Si.*

The values of the effective masses are $m_l = 0.915\, m_0$ and $m_t = 0.19\, m_0$, where m_0 is the free electron mass.

Consider now the three valence bands shown in Figure 2.3a; they are the heavy-hole (HH), light-hole (LH), and split-off (SO) bands. The E–**k** relationship near the band extrema for the HH and LH bands has been calculated by using degenerate perturbation theory, usually known as **k.p** perturbation theory (see Appendix A). The following is the dispersion relation for the valence bands:

$$E(\mathbf{k}) = Ak^2 \pm \left[B^2k^4 + C^2(k_x^2k_y^2 + k_y^2k_z^2 + k_z^2k_x^2\right]^{1/2}, \tag{2.4}$$

where A, B, and C are various constants related to the momentum matrix elements to be defined in Appendix A and Chapter 3. The + and − signs are related to the LH and HH bands, respectively. The constant energy surfaces described by Eq. (2.4) are nonspherical for $C \neq 0$, and are referred to as *warped surfaces*. The intersection of these surfaces with the (100) plane in **k** space are plotted in Figure 2.3b for silicon.

The E–**k** relationship for the SO band is given by

$$E(\mathbf{k}) = -\Delta_{so} + Ak^2 \tag{2.5}$$

where the constant A is the same as in Eq. (2.4). The constant energy surfaces are spherical, and the SO band is shifted from the two other bands by an amount Δ_{so}.

2.2.3 Refined Theory for Band Structures

The simple theory for band structure, as discussed in this chapter, is in some cases inadequate to describe optical processes in bulk materials as well as in quantum heterostructures. In particular, the complicated nature of the valence band structure calls for further refinements in the theory. Usually the states near the band extrema are of importance for the study of low-field transport and optical processes. In that respect, the **k.p** perturbation theory has been found to work remarkably well. Instead of including the description of this theory in this chapter, we have presented the essential ingredients in Appendix A.

Usually the E–**k** relationships for valence bands are expressed in terms of some parameters A, B, and C as in Eq. (2.4) or, alternatively, in terms of Luttinger parameters γ_1, γ_2, and γ_3 (see Appendix A).

2.2.4 Temperature- and Pressure-Dependent Band Gap

The band gap in semiconductors is dependent on lattice constants and the nature of crystal bond. Both these parameters change with change in temperature T and in pressure p.

The band gap first decreases quadratically with temperature from its low temperature value. Then the decrease is more gradual, and almost linear near room temperature. The change in band gap with temperature is empirically expressed as [2]

$$E_g(T) = E_g(0) - \frac{\alpha T^2}{T + \beta}, \tag{2.6}$$

where α and β are constants that vary from material to material. The values for Si are $\alpha = 7.021 \times 10^{-4}$, and $\beta = 1108\,\text{K}$ [3].

The dependence of band gap on pressure is complicated and depends on the direction along which the pressure is applied. If the pressure is applied along the [111] direction, then all the six conduction band valleys behave identically. The actual gap depends on a physical constant, known as the *deformation potential constant* of the conduction band and valence band. The sixfold degeneracy is lifted when the pressure is applied along the [100] direction. The nature of band gap changes will be discussed in this chapter when dealing with strained Si–SiGe heterostructures.

2.2.5 Band Structure in Ge

In most of the early work in semiconductors, Ge received the attention of workers. However, as the superiority of Si in VLSI technology was established, the study of electronic and optical properties in Ge was conducted mainly for academic interest.

With increased interest in Si-based hetero-epitaxy, Ge and its alloy with Si (i.e., $Si_{1-x}Ge_x$) enjoyed renewed interest from the workers. The properties of alloys and heterojunctions will be treated in this chapter. In this subsection, we mention the essential features of the band diagram and E–**k** relationship for electrons and holes.

Like Si, Ge is also an indirect gap semiconductor. The E–**k** diagram for different crystallographic directions is shown in Figure 2.4.

The valence band structure is quite similar to that in Si, with HH and LH bands degenerate at $\mathbf{k} = 0$. The split-off band is shifted by an amount Δ_{so}. Both the HH and LH bands show warped nature.

The lowest conduction band minima in Ge occur at the L point along the [111] crystallographic direction, and the corresponding band gap is nearly 0.66 eV. The next higher conduction band occurs at the zone center and the direct gap is nearly 0.8 eV, thus making the difference between L and Γ point minima about 140 meV only. This proximity leads to indirect-to-direct gap crossover under favorable situations, as we shall see in this chapter.

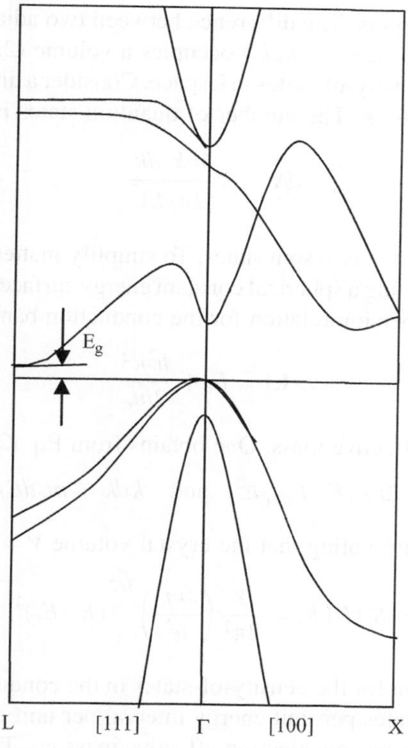

Figure 2.4 *Band diagram of Ge.*

The L-point minima occur at the edge of the Brillouin zone. From symmetry conditions, the number of such minima should be eight; however, as each of the minima is shared by two adjacent zones, the degeneracy of the conduction band is taken to be four.

There are six equivalent conduction band valleys along the [100] directions: X-valleys, which lie higher than the Γ valleys.

Considering the L-minima, the constant energy surfaces are ellipsoidal, with elongation along the [111] directions. The longitudinal mass is higher than the transverse mass.

2.3 Density-of-States Function

The knowledge of the number of available energy states in the conduction and valence bands in a semiconductor is needed to obtain the expressions for many physical parameters. Let us assume that the crystal is a cube of dimension L in each of the x, y, and z directions. The electron wave function given by Eq. (2.1) must vanish at the crystal boundaries. This condition is fulfilled if

$$\begin{aligned} k_x L &= 2\pi n_x \\ k_y L &= 2\pi n_y \, , \\ k_z L &= 2\pi n_z \end{aligned} \tag{2.7}$$

where n_x, n_y, and n_z are integers. The difference between two adjacent k_x values is $(2\pi/L)$. Thus, each allowed value of $\mathbf{k}$ $(= k_x, k_y, k_z)$ occupies a volume $(2\pi/L)^3$ in $\mathbf{k}$-space.

We now determine the density-of-states in $\mathbf{k}$ space. Consider a differential volume $4\pi k^2 dk$ bounded by radii k and $k + dk$. The number of quantum states is then

$$dN = 2\frac{4\pi k^2 dk}{(2\pi/L)^3} \tag{2.8}$$

Factor 2 takes into account the two spin states. To simplify matters, let us first consider an isotropic semiconductor having a spherical constant energy surface for electrons. A common example is GaAs. The dispersion relation for the conduction band electrons is given by

$$E(\mathbf{k}) = E_c + \frac{\hbar^2 k^2}{2m_c}, \tag{2.9}$$

where m_c is the electron effective mass. One obtains from Eq. (2.9)

$$k^2 = 2m_c(E - E_c)/\hbar^2 \quad \text{and} \quad kdk = m_c dE/\hbar^2.$$

Using these in Eq. (2.8) and noting that the crystal volume $V = L^3$, one readily obtains

$$dN = S_c(E)dE = \frac{1}{2\pi^2}\left(\frac{2m_c}{\hbar^2}\right)^{3/2}(E - E_c)^{1/2}dE \tag{2.10}$$

The above is the expression for the density-of-states in the conduction band, $S_c(E)$, that is, the number of available states per unit energy interval per unit crystal volume valid for a semiconductor having an isotropic electron effective mass m_c. The expression is modified for Si and Ge in which the effective mass is anisotropic. Consider a valley of Si having longitudinal mass m_l along [100] direction for which the E– $\mathbf{k}$ relation is

$$E(\mathbf{k}) = E_c + \frac{\hbar^2}{2}\left(\frac{(k_x - k_{x0})^2}{m_l} + \frac{k_y^2}{m_t} + \frac{k_z^2}{m_t}\right) \tag{2.11}$$

Using the transformation

$$k_x^* = \left(\frac{m_0}{m_l}\right)^{1/2}(k_x - k_{x0}), \quad k_y^* = \left(\frac{m_0}{m_t}\right)^{1/2}k_y, \quad k_z^* = \left(\frac{m_0}{m_t}\right)^{1/2}k_z,$$

the above equation can be reduced to the equation with isotropic mass m_O, that is, $E(\mathbf{k}^*) = (\hbar^2 k^{*2}/2m_0)$.

The DOS in $\mathbf{k}^*$ space is obtained from Eq. (2.10) by replacing m_c with m_O. To obtain the DOS, we note that the volume element

$$dk_x^* dk_y^* dk_z^* = (m_0^3/m_l m_t^2)\, dk_x dk_y dk_z$$

Therefore, transforming back into $\mathbf{k}$ space and noting that the conduction band degeneracy is n_c, the required DOS function becomes

$$S_c(E)dE = \frac{n_c}{2\pi^2}\left(\frac{2m_d}{\hbar^2}\right)^{3/2}(E - E_c)^{1/2}dE, \tag{2.12}$$

where $m_d = (m_l m_t^2)^{1/3}$ is the density-of-states effective mass.

The carrier density in a semiconductor can be calculated by using the density-of-states function given by Eqs. (2.10) and (2.12). In addition, we need consider the probability that a state of energy E is occupied, expressed in terms of Fermi function,

$$f(E) = \frac{1}{\exp\left[(E - E_F)/k_BT\right] + 1}, \tag{2.13}$$

where E_F is the Fermi level. In most situations the Boltzmann approximation

$$f(E) = \exp\left[-(E - E_F)/k_BT\right] \tag{2.14}$$

is valid. The electron density may therefore be written as

$$n = \int_{E_c}^{\infty} S_c(E) f(E) dE \tag{2.15}$$

Using Eqs. (2.12) and (2.14) and introducing a variable $x = (E - E_c)/k_BT$, Eq. (2.15) can be expressed as

$$n = \frac{n_c}{2\pi^2}\left(\frac{2m_d k_B T}{\hbar^2}\right)^{3/2} e^{-(E_c - E_F)/k_BT} \int_0^{\infty} x^{1/2} e^{-x} dx$$

The integral is in the form of Gamma function and equals $\sqrt{\pi}/2$. The number of electrons per unit volume is then expressed as

$$\begin{aligned} n &= 2n_c\left(\frac{m_d k_B T}{2\pi\hbar^2}\right)^{3/2} e^{-(E_c - E_F)/k_BT} \\ &= N_c e^{-(E_c - E_F)/k_BT} \end{aligned} \tag{2.16}$$

Example 2.3

The prefactor N_c in Eq. (2.16) is called the effective density of states. Let $m_l = 0.92\, m_O$, $m_t = 0.19\, m_O$, and the valley degeneracy factor $n_c = 6$. The value of N_c becomes $3.2 \times 10^{19}\ \text{cm}^{-3}$.

The density-of-states effective mass for holes in all cubic semiconductors is expressed as

$$m_{dh} = (m_{hh}^{3/2} + m_{lh}^{3/2})^{2/3} \tag{2.17}$$

where m_{hh} and m_{lh} are, respectively, the heavy-hole and light-hole effective masses. The hole density can easily be expressed by replacing m_{de} by m_{dh}, n_c by n_h, and N_c by N_v in Eq. (2.16).

2.4 Impurities

It is well known that the introduction of impurities drastically changes the properties of semiconductors, especially the electrical conductivity of the materials. Impurities can also affect the optical properties in semiconductors. In this section, we shall consider mainly the energy levels of the impurity ions in a semiconductor. The effect of impurities in optoelectronic processes will be presented in a later section.

2.4.1 Donors and Acceptors

Consider first that a pentavalent impurity, such as P, is introduced into a semiconductor crystal having tetravalent Si atoms. Si is bonded with its four nearest neighbors by sharing its four valence electrons. When a P atom replaces a Si atom, four of the outermost electrons of the P atom participate in the bonding process. The fifth electron is loosely bound to its parent nucleus, which possesses a charge of $+e$. The combination of a positively charged P nucleus and an electron revolving around it is equivalent to the hydrogen atom picture. Bohr's theory of the H-atom may readily be applied, taking into account the fact that the electron, revolving in a material of relative permittivity ε_r, possesses an effective mass, m^*, characteristic in a crystal. The ionization energy is written simply as

$$E_D = \frac{\hbar^2}{2m^* a_B^2} = \frac{m^* e^4}{2(4\pi\varepsilon_0)^2 \hbar^2 \varepsilon_r^2}, \tag{2.18}$$

where a_B is the effective Bohr radius of the impurity atom.

Example 2.4

Let us assume $\varepsilon_r = 12$ and $m^* = 0.26\, m_0$, which correspond approximately to Si. The ionization energy as calculated from Eq. (2.18) is nearly 25 meV.

It appears from the above discussion that the electron is loosely bound to the parent atom. If an amount of energy E_D is provided, the electron will detach itself from the influence of the nucleus and will move to the conduction band, where it is free. The symbol E_D is termed the *donor-binding energy*. The pentavalent atom P acts as a donor in Si, since the atom donates an electron to the conduction band.

By a similar argument, a trivalent impurity, like B, Al, and so on, will act as an acceptor to receive an electron from the otherwise filled valence band, thereby creating a hole there. An acceptor binding energy may also be defined and calculated in a similar way.

In the above description and in Eq. (2.9), the donor-binding energy has been expressed for the ground state of the electron in the impurity potential. The excited states (corresponding to the Bohr orbit number, $n > 1$) also exist and the ionization energies are modified.

The energy eigenvalues and wavefunctions of the impurity atoms were determined from the effective mass theory developed by Luttinger and Kohn [4]. Indeed, the theory developed an expression as given by Eq. (2.18) for the binding energy. A complication develops for donor impurities in Si or Ge, in which the constant energy surfaces are not spheres, but ellipsoids of revolution. Approximate methods have been developed to address the problem, and a value of 31.3 meV has been obtained for donor-binding energy in Si [5].

Table 2.1 *Ionization energy of donors (in meV) in Si. Reproduced from Semiconductor Physics and Applications by M. Balkanski and R. F. Wallis. Copyright (2000) Oxford University Press.*

State	Theory	P	As	Sb
$1s(A_1)$	40.65	45.5	53.7	42.7
$1s(T_2)$	30.05	33.9	32.6	32.9
$1s(E)$	28.95	32.6	31.2	30.5
$2p_0$	11.51	11.5	11.5	11.6
$2p_\pm$	6.40	6.4	6.3	6.3
$3p_0$	5.48	5.5	5.5	5.3
$3p_\pm$	3.12	3.1	3.1	3.0

The anisotropy in the effective mass causes a splitting of excited states like p states of the impurities. As the number of equivalent conduction band valleys in Si is six, there is a set of donor levels associated with each valley. A given donor level therefore possesses a sixfold degeneracy. Another additional consideration in the calculation of donor-binding energy is the wave vector (**k**) dependence of the permittivity, which leads to coupling between impurity states associated with different extrema. Baldareshi [6] showed that the sixfold ground state for donors in Si is split into a singlet (A_1) ground state, a doublet excited state (E), and a triplet (T_2) excited state. The ionization energies of donors in Si are given in Table 2.1.

2.4.2 Isoelectronic Impurities

Isoelectronic impurities in several semiconductors are responsible for efficient light emission. The common example is N -doped GaP, a p–n junction which is used to produce light-emitting diodes (LEDs) at visible wavelengths. Introduction of isoelectronic impurities in Si and its alloy SiGe has led to the observation of efficient luminescence. In the following, we shall point out the basic difference between the ionized impurities, like donors and acceptors, and the isoelectronic impurities, leaving further discussion to a later section related to excitons.

An isoelectronic impurity is one having the same outer electron configuration as the atom of the host crystal which it replaces. Examples are germanium in silicon, or arsenic in gallium phosphide. A useful distinction can be made between charged centers (i.e., donor or acceptor) and neutral (or "isoelectronic") centers. A single donor has one excess electron (such as Si : P), meaning that the defect region has a net positive core charge. Similarly, a single acceptor (such as Si : B) has a net negative localized charge. An isoelectronic center (such as Si : C or Si : Cu) has no net charge in the local bonding region. A hole or an electron can be localized at such a center by a local (core) potential; subsequently, the secondary particle can be captured by the Coulomb field, so that a shallow localized state is formed within the forbidden gap. The main difference between neutral and charge centers is the absence of the long-range Coulomb potential. This simple picture is adequate for single donors, acceptors, and isoelectronic centers. The origin of the core potential may be ascribed to a change in electronegativity in the impurity site from that of the host crystal, a localized strain due to change in bond length, and so on.

2.5 Alloys of Silicon and Other Group IV Elements

2.5.1 Different Alloy Systems

Electronic and optical properties of binary alloy of Si and Ge: $Si_{1-x}Ge_x$ were studied as early as 1958 by Braunstein *et al.* [7]. The widespread use of heterojunctions made of III–V compounds and their binary, ternary, and quaternary alloys in electronic and photonic devices prompted workers to undertake studies of the properties of alloys of group IV elements and to make devices using these alloys grown on Si substrate. Electronic devices like heterojunction bipolar transistors and heterostructure field effect transistors using SiGe–Si combinations have already found commercial applications. Recently it has been found that the addition of C in SiGe alloy introduces a number of beneficial effects. A lot of work has recently been undertaken for tailoring the band gap or even for obtaining direct-gap Group IV material on Si substrate. The materials studied so far are binary alloys of Si, Ge, C, and Sn as well as ternary alloys of Si, Ge, C, and Sn.

In Sections 2.5.2 and 2.5.3, the basic properties of these alloys will be discussed.

2.5.2 Lattice Constants

Solid solutions of Si and Ge can be formed over the complete range of compositions, and the alloy $Si_{1-x}Ge_x$ is thermodynamically stable. Ordered compound SiGe (analogous to SiC) has not been reported. The lattice constant of the alloy SiGe obeys the relationship

$$a_x = (1-x)a_{Si} + xa_{Ge} - x(1-x)c, \tag{2.19}$$

where c is the bowing parameter $= 25.6 \times 10^{-3}$ Å. In most of the calculations of the parameters for the alloy including lattice constant, however, Vegard's law (linear interpolation: $c = 0$) is employed.

It is not clear how to apply Vegard's lay for the alloy $Si_{1-y}C_y$ which has been grown for all compositions. One may use the constant either for diamond or for the ordered compound SiC. The following interpolation for unstrained alloy $Si_{1-y}C_y$ has been suggested:

$$a_y = 5.4310 - 2.439y + 0.5705y^2 \text{ Å}. \tag{2.20}$$

This reproduces the lattice constants of Si ($y = 0$), diamond ($y = 1$), and ordered cubic compound SiC ($y = 0.5$).

Introduction of C into Si or SiGe alloy affects the lattice constant in both ways. If substitutional, the lattice constant is reduced in accordance with the lower value of the lattice constant of C. On the other hand, interstitial C increases the lattice constant. In most cases, the lattice constant of the unstrained alloy $Si_{1-x-y}Ge_xC_y$ is given by Vegard's law as

$$a_{xy} = (1-x-y)a_{Si} + xa_{Ge} + ya_C \tag{2.21}$$

Most of the parameter values of α-Sn are unknown. The lattice constant of the alloy $Si_xGe_ySn_{1-x-y}$ will be assumed to obey Vegard's law.

Table B.1 in Appendix B lists most of the parameter values of the group IV elements, Si, Ge, C, and Sn.

2.5.3 Band Structures of Unstrained Alloys

Binary alloy $Si_{1-x}Ge_x$, formed by mixing two elemental semiconductors Si and Ge, is used in many modern electronic and photonic devices. Alloying changes the band gap of Si in two different ways. First, the disposition of band extrema in **k**-space is altered. A number of investigations confirm that the valleys remain Si-like (*X*-valleys) as long as the composition *x* remains less than about 0.85. With further increase in Ge concentration, the conduction band valleys assume Ge-like character. Elemental Ge is also an indirect gap material, in which the lowest conduction band valleys occur at the edge of the Brillouin zone along the [111] direction (referred to as *L* valleys). These valleys are fourfold degenerate. The valence band extrema in Ge lies at $\mathbf{k}=0$, and HH and LH bands degenerate at the center of the zone, and in addition there is the SO band.

Next comes the question of how the value of the fundamental band gap is altered due to alloying. Assuming linear variation, the expression for the gap in a binary alloy $A_{1-x}B_x$ is simply expressed by Vegard's law as

$$E_g(x) = E_g(A) + x[E_g(B) - E_g(A)]. \tag{2.22}$$

Deviation from this linear relation has been observed, however. Braunstein *et al.* [7] first studied the band gap of the alloy from optical absorption measurement, and the experimental data were subsequently found to be in agreement with theoretical values. Weber and Alonso [8] presented the following expression for the band gap of the alloy by fitting their data obtained from photoluminescence measurement at 4.2 K.

$$\begin{aligned} E_g^X(x) &= 1.155 - 0.43x + 0.206x^2, \quad 0 < x < 0.85 \\ E_g^L(x) &= 2.010 - 1.270x, \quad x > 0.85. \end{aligned} \tag{2.23}$$

The data obtained by various workers [7–9] are plotted in Figure 2.5. Although the trends of the curves by Braunstein *et al.* and by Weber and Alonso are similar, deviations of up to 40 meV between the data by the two groups are observed. Taking the above expression as standard at 4.2 K, the following empirical relation expresses the temperature dependence of the band gap for the unstrained alloy:

$$E_g(x,T) = E_g(x,0) - \frac{4.73 \times 10^{-4} T^2}{T - 636} \tag{2.24}$$

The effect of incorporation of a small amount of substitutional C in SiGe (Ge content < 50%) alloy on the band gap has been modeled by assuming Si-like indirect band structure, that is, valence band maxima at Γ point and a conduction band minimum at Δ. The expression for the band gap is

$$E_g(x,y) = 1.155 - 0.43x + 0.206x^2 - 1.7y\,\text{eV}, \tag{2.25}$$

indicating that the gap decreases at the rate 17 meV/%C, at low temperatures.

The atomic percent of substitutional C in the alloy remains below 3% in almost all reports published so far.

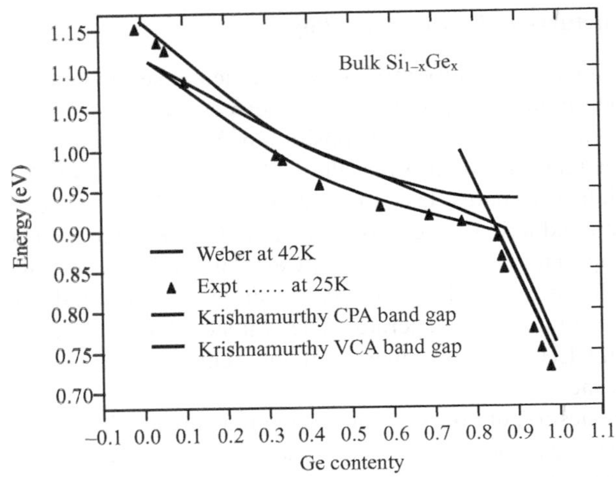

Figure 2.5 *Band gap energy of unstrained $Si_{1-x}Ge_x$ alloy. Reproduced with permission from [9]. Copyright (2004) Institute of Physics.*

2.6 Heterojunctions and Band Lineup

Heterojunctions are junctions formed by two dissimilar semiconductors having different band gaps, permittivities, electron affinities, and so on. While the focus in this text is on heterojunctions made of Si and other group IV materials, we present in this section a few general concepts and definitions by using the well-studied materials combination of GaAs and its alloy $Ga_{1-x}Al_xAs$.

The lattice constants of GaAs and AlAs are respectively 5.6533 and 5.6612 Å. The 0.14% lattice mismatch between the two is still further reduced between GaAs and its alloy GaAlAs. This combination is an example of lattice-matched heteroepitaxy, in which the alloy is grown on GaAs substrate. The band gap of the alloy is expressed as

$$E_{g\Gamma} = 1.424 + 1.247x\,\text{eV},\ (0 < x < 0.45). \tag{2.26}$$

When a heterojunction is made by growing the alloy on GaAs, the difference in the band gap $\Delta E_g = 1.247x$ is consumed by two steps ΔE_C and ΔE_V, occurring respectively at the conduction and valence band edges at the heterointerface as indicated in Figure 2.6a.

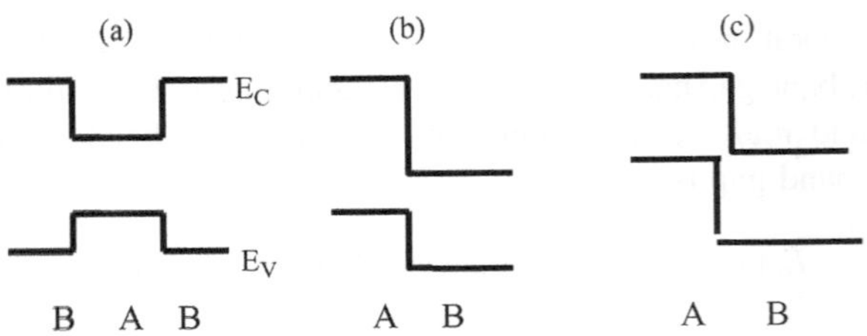

Figure 2.6 *Different types of band alignment: (a) type I, (b) type II, and (c) broken-gap type II band alignments.*

There are different theories to calculate the band offsets ΔE_C and ΔE_V; however, in actual calculations the experimentally determined ratio $\Delta E_C : \Delta E_V \approx 60:40$ is considered.

Example 2.5

Consider the $Ga_{0.7}Al_{0.3}As$–GaAs heterojunction. Since $x = 0.3$, $\Delta E_g = 0.3745$ eV and ΔE_C and ΔE_V are, respectively, 0.2245 and 0.1497 eV.

It appears from Figure 2.6a, showing the heterojunction band alignment, that both the conduction and valence band extremas occur at the GaAs region: the layer having lower band gap (layer A, in the figure sandwiched between higher gap material B). Such an alignment is termed as a *type I alignment*. Other combinations, such as InGaAs–InP, GaN–AlGaN, and so on, also have a type I lineup.

Figure 2.6b shows a different kind of band lineup, in which the lowest conduction band state is in one material (B) and the highest valence band state is in the other material (A). However, the lowest conduction band state is higher than the highest valence band state. Such a lineup is characteristic of a type II heterostructure. The electrons and holes are then confined in separate regions, and the effective band gap is somewhat reduced. A widely studied combination belonging to this type is GaInAs–GaAsSb.

Finally there exists a situation, as shown in Figure 2.6c, in which the conduction band edge in one layer is below the valence band edge in the other layer or the difference is negligibly small. This is the situation in a broken-gap type II alignment. The lineup is found in a InAs–GaSb system.

2.7 Si-Based Heterostructures

2.7.1 Lattice-Mismatched Heteroepitaxy

Most of the discussions in Section 2.6 were based on the direct-gap material GaAs and its alloy GaAlAs in which almost perfect lattice matching exists between the two materials.

Heterojunctions made by using lattice-mismatched pairs have been investigated over the last two or three decades, and their novel properties have been exploited to fabricate a large number of electronic and optoelectronic devices. Usually lattice mismatch between the two constituent semiconductors gives rise to strain. The strained layer epitaxy and strained heterojunctions merit an extensive discussion.

The heterostructures involving silicon and other group IV elements and their alloys invariably lead to strained heterostructures. We consider at present heterojunctions made of Si and its alloy $Si_{1-x}Ge_x$.

The lattice constant of Si is 5.4311 Å and of Ge is 5.6579 Å. This large lattice mismatch, ~4%, between the two allows the growth of Ge or the alloy SiGe on Si under restricted conditions. The layers are usually strained, and strain alters the band gap, heterojunctiion band alignment, and band structure in a very complicated way.

2.7.2 Pseudomorphic Growth and Critical Thickness

When an epitaxial layer having a larger lattice constant is deposited on a substrate having a lower lattice constant, usually misfit dislocations develop in the interface between the two. The growth of Ge or alloy SiGe on Si will therefore introduce such dislocations. However, when the misfit between the growing layer and the substrate is sufficiently small, the epitaxial film may grow on the substrate with a lattice constant equal to that of the substrate. The result is that the film is under strain and also that misfit dislocation is now absent. The condition to be maintained for such a pseudomorphic growth is that the layer thickness should be below a critical value. If the film thickness exceeds this, then growth with misfit dislocation will be preferred. The argument put forward is as follows. The strain energy increases linearly with film thickness; on the other hand, the energy of dislocation increases superlinearly first, but saturates at larger thickness. For thickness lower than the critical value, the strain energy is lower and therefore pseudomorphic growth is preferred.

There are several theories that relate critical layer thickness to material parameters and those related to dislocations. The expressions will be given in other subsections of Section 2.7.

2.7.3 Elasticity Theory: Stress and Strain

The theory presented below follows closely the presentation by Zollner [10]. The epitaxial layers are usually grown on (001)-oriented Si substrate. The overlayers therefore experience forces along the [100] and [010] directions. The resulting stress, **X**, is described by the following stress tensor

$$\mathbf{X}_{\text{biaxial}} = \begin{pmatrix} X & 0 & 0 \\ 0 & X & 0 \\ 0 & 0 & 0 \end{pmatrix}. \tag{2.27}$$

where $X > 0$ for tensile stress ($Si_{1-y}C_y$ on Si) and $X < 0$ for compressive stress ($Si_{1-x}Ge_x$ on Si). The effect of stress is to produce strain and to change lattice constants along different directions. Taking a cubic crystal of length a_0 in all directions, the changed dimension of the crystal will be described by the following:

$$\vec{a}' = (1+\varepsilon)\vec{a}_0 = \begin{pmatrix} 1+\varepsilon_{xx} & \varepsilon_{xy} & \varepsilon_{xz} \\ \varepsilon_{yx} & 1+\varepsilon_{yy} & \varepsilon_{yz} \\ \varepsilon_{zx} & \varepsilon_{zy} & 1+\varepsilon_{zz} \end{pmatrix} \begin{pmatrix} a_0 \\ a_0 \\ a_0 \end{pmatrix}. \tag{2.28}$$

The dimensionless quantity ε is called the strain tensor. Its trace is the relative volume change. When the deformation is small, strain is proportional to stress by Hooke's law and the proportionality is expressed by the compliance tensor **S**. The inverse of **S** is the elastic stiffness tensor **c**, which has components c_{ij}, called the *elastic stiffness constants*. For cubic crystals, the number of independent components in each of S and **c** is only three. The relationships are given in Eq. (2.29).

$$\begin{aligned} c_{11} &= \frac{S_{11}+S_{12}}{(S_{11}-S_{12})\,(S_{11}+2S_{12})} \\ c_{12} &= \frac{-S_{12}}{(S_{11}-S_{12})\,(S_{11}+2S_{12})} \\ S_{11} &= \frac{c_{11}+c_{12}}{(c_{11}-c_{12})\,(c_{11}+2c_{12})} \\ S_{12} &= \frac{-c_{12}}{(c_{11}-c_{12})\,(c_{11}+2c_{12})} \\ S_{44} &= \frac{1}{c_{44}} \qquad c_{44} = \frac{1}{S_{44}}s \end{aligned} \tag{2.29}$$

Consider the growth of alloys on Si layer oriented along the (001) plane. The relationship between strain and stress is given by

$$\varepsilon = \begin{pmatrix} (S_{11}+S_{12})X & 0 & 0 \\ 0 & (S_{11}+S_{12})X & 0 \\ 0 & 0 & 2S_{12}X \end{pmatrix}. \tag{2.30}$$

The in-plane strain (perpendicular to the growth direction) is $\varepsilon_{\parallel} = \varepsilon_{xx} = \varepsilon_{yy}$, and the strain along the vertical direction is $\varepsilon_{\perp} = \varepsilon_{zz}$.

The above strain tensor can be written as the sum of two diagonal tensors representing hydrostatic strain and shear strains as follows:

$$\varepsilon = \varepsilon_H \begin{pmatrix} 1 & 0 & 0 \\ 0 & 1 & 0 \\ 0 & 0 & 1 \end{pmatrix} + \varepsilon_S \begin{pmatrix} -1 & 0 & 0 \\ 0 & -1 & 0 \\ 0 & 0 & 2 \end{pmatrix} \tag{2.31}$$

$$\varepsilon_H = 2(S_{11}+2S_{12})X/3, \quad \varepsilon_S = (S_{12}-S_{11})X/3.$$

The hydrostatic (ε_H) and uniaxial shear (ε_S) strains for (001) biaxial stress are therefore

$$\varepsilon_H = (\varepsilon_{\perp}+2\varepsilon_{//})/3, \quad \text{and} \quad \varepsilon_S = (\varepsilon_{\perp}-\varepsilon_{//})/3.$$

Now, equating the lattice constant a_{xy} of the freestanding unstrained alloy with a_0, we may write the strain components as

$$\varepsilon_{//} = \frac{a_{Si}}{a_{xy}} - 1, \tag{2.32a}$$

$$\varepsilon_{\perp} = \frac{2S_{12}}{S_{11}+S_{12}}\varepsilon_{//} = -2\frac{c_{12}}{c_{11}}\varepsilon_{//} = -2\frac{v}{1-v}\varepsilon_{//}. \tag{2.32b}$$

The Poisson ratio v is defined as

$$v = \frac{c_{12}}{c_{11} + c_{12}} = -\frac{S_{12}}{S_{11}} \tag{2.33}$$

The elastic energy density for a strain arising out of a (001) biaxial stress is given by

$$u = \frac{1}{2}\sum_{ij} X_{ij}\varepsilon_{ij} = \left(c_{11} + c_{12} - 2\frac{c_{12}^2}{c_{12}}\right)\varepsilon_{ll}^2. \tag{2.34}$$

The different strain components may now be calculated as a function of Ge content in the alloy $Si_{1-x}Ge_x$ or of C content in $Si_{1-y}C_y$ by using the expressions given here. The following example will illustrate the use of the expressions.

Example 2.6

The values of in-plane, perpendicular, hydrostatic, and shear components of strain will be calculated for $Si_{0.6}Ge_{0.4}$ grown on a Si substrate.

The lattice constant a_{xy} of the alloy is 5.5220 Å, using Vegard's law. The in-plane strain is $\varepsilon_{\|} = -0.0165 = -1.65\%$; $c_{11} = 1.51$, and $c_{12} = 0.5766$. Therefore the perpendicular strain $\varepsilon_{\perp} = 0.0126 = 1.26\%$, $\varepsilon_H = -0.0068 = -0.68\%$, and $\varepsilon_S = +0.0097 = 0.97\%$.

2.7.4 Expressions for Critical Thickness

When the thickness of the epilayer or the lattice mismatch is too high, the energy density becomes large and strain is relieved by the generation of misfit dislocation. Complete strain relaxation is rare. Below critical thickness, pseudomorphic growth is possible. The critical layer thickness depends on the growth temperature, the misfit between the thermal expansion coefficients of the substrate and epilayer, and several other parameters. The following expression for critical thickness is due to People and Bean [11],

$$h_c = \frac{1}{16\pi\sqrt{2}}\frac{1-v}{1+v}\frac{b^2}{a_{xy}}\frac{1}{\varepsilon_{ll}^2}\ln\left(\frac{h_c}{b}\right) \tag{2.35}$$

where $b \approx a_{xy}/\sqrt{2}$ is the Burger vector and all other symbols are previously defined. There exists another approximate expression, $h_c \approx a_{Si}/2\varepsilon_{ll}$.

Example 2.7

Approximate value of critical thickness: the strain between the film (F) and the substrate (S) is expressed as $\varepsilon = (a_F - a_S)/a_S$, and the approximate expression for critical thickness is $h_C \approx a_S/2\,|\varepsilon|$.

Consider growth of $Si_{0.6}Ge_{0.4}$ on Si. Using the lattice constants given above, $a_{SiGe} =$ 5.52172, $\varepsilon = 0.0167$, and $h_C = 163$ Å.

Example 2.8

The critical thickness of ternary alloy $Si_{0.53}Ge_{0.44}C_{0.03}$ will be evaluated by using more accurate expression. Using the lattice constants given here, $a_{xy} = 5.474$ Å, the Burger vector $b = a_{xy}/\sqrt{2} = 3.87$ Å, and the Poisson ratio $\nu = 0.2706$. Solving the above, $h_C = 250$ nm.

2.7.5 Strain Symmetric Structures and Virtual Substrates

The growth of lattice-mismatched epilayer on a substrate is illustrated in Figure 2.7. In Figure 2.7a, the epitaxial layer and the substrate are shown isolated and the epilayer has a larger lattice constant. This is the situation for a $Si_{1-x}Ge_x$ layer on a Si substrate or, more generally, for a thin $Si_{1-x}Ge_x$ layer grown on top of a $Si_{1-y}Ge_y$ film, for $x > y$. In this case, the top layer is compressively strained as shown in Figure 2.7a. On the other hand, for $x < y$, the two layers are shown isolated in the right side of Figure 2.7a, and when the growth is complete (as shown in the right of Figure 2.7b) the top layer is tensile strained. It follows from elasticity theory that an in-plane strain also produces a perpendicular strain. It is also easy to understand that if the two layers A and B as shown in Figure 2.7 are grown alternately, then the in-plane strain of layer A, $\varepsilon_{\parallel}{}^{A}$, is related to the strain in layer B, $\varepsilon_{\parallel}{}^{B}$, by the relation

$$\varepsilon^A = -\left(\frac{G_B h_B}{G_A h_A}\right)\varepsilon^B, \qquad (2.36)$$

where h represents the height of the film and G is the shear modulus of each layer. It therefore follows that the average strain of the two layers may be zero, and pseudomorphic growth of structure may be made by the successive growth of compressive and tensile strained layers. The strain-symmetrized superlattice requires a substrate of lattice-constant intermediate between the two extremes.

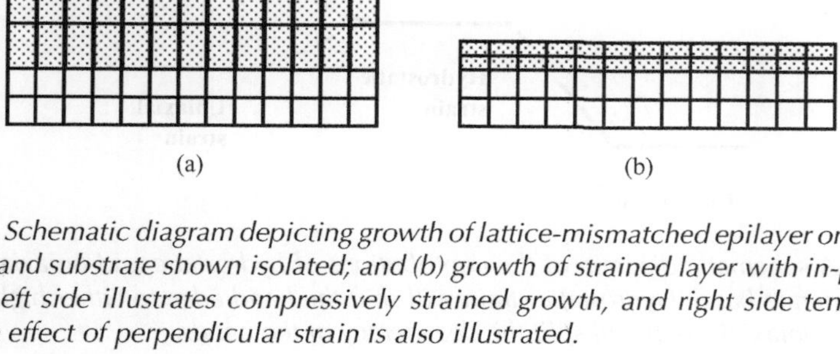

Figure 2.7 *Schematic diagram depicting growth of lattice-mismatched epilayer on a substrate. (a) Epilayer and substrate shown isolated; and (b) growth of strained layer with in-plane lattice matching. Left side illustrates compressively strained growth, and right side tensile-strained growth. The effect of perpendicular strain is also illustrated.*

In many heterostructures, only one or two strained layers are grown on top of either a bulk silicon wafer or a relaxed $Si_{1-y}Ge_y$ virtual substrate. Virtual substrates allow tensile strained layers along with strain-symmetrized superlattices (SLs) to be grown. In many cases, a thick $Si_{1-y}Ge_y$ layer with thickness well above the critical thickness is grown directly on a silicon wafer and acts as a virtual substrate. However, a large number of dislocations are created at the virtual substrate–substrate interface, deteriorating device performance. One of the techniques used to avoid the presence of dislocation is to slowly grade the Ge content from 0% up to the required Ge composition.

Example 2.9

As an example of growth of a strain-symmetric SL structure, we consider the growth of Si and $Si_{1-x}Ge_x$ layers on a $Si_{0.6}Ge_{0.4}$ buffer substrate on a Si substrate. If Si and Ge layers are grown alternately on this thick substrate, then Si will be tensile strained and Ge will be compressed. The critical thickness values are h_C (Si) = 166 Å and h_C (Ge) = 113Å.

2.7.6 Band Offsets and Band Lineup

As noted already, when two semiconductors join up to form a heterojunction, discontinuities occur at the conduction and valence bands at the heterointerface. If one of the semiconductors is coherently strained on the other, the band structure is modified by three basic effects: an alloy-only effect for the unstrained material ΔE_a, the shift due to hydrostatic strain ΔE_h, and the change due to uniaxial or biaxial strain ΔE_s. The total change in band is expressed as

$$\Delta E = \Delta E_a + \Delta E_h + \Delta E_s \tag{2.37}$$

The effect is schematically illustrated in Figure 2.8.

The following discussion is mostly based on the work by Yang *et al.* [12], who made a systematic study of the available experimental data and theoretical results.

When the Ge content in the alloy $Si_{1-x}Ge_x$ is less than 85%, the alloy exhibits Si-like band structure, in which the conduction band is sixfold degenerate and the valence band is twofold degenerate at the band extrema. Let us consider pseudomorphic growth of the $Si_{1-x}Ge_x$ layer on a relaxed $Si_{1-y}Ge_y$ substrate. If $x > y$ the strain is compressive, or it is

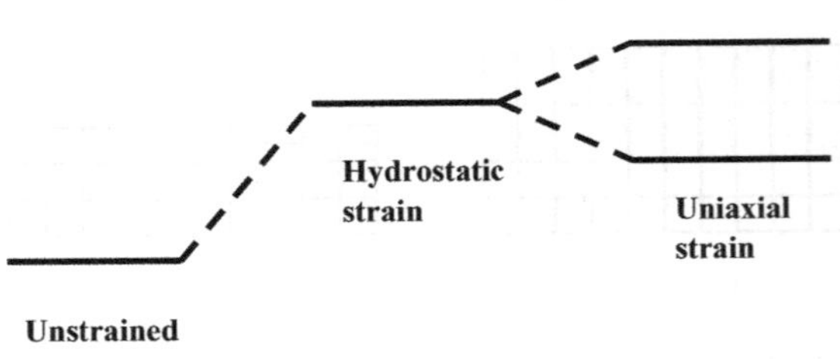

Figure 2.8 *A schematic diagram of the contributions of hydrostatic and uniaxial strain to a doubly degenerate band. Hydrostatic strain will shift the band edge to either higher or lower energy, while uniaxial strain will split the degeneracy of the bands.*

tensile if $x < y$. The strain has two components; the hydrostatic component shifts the average band energy level, while the uniaxial component splits the degenerate bands, as illustrated in Figure 2.8.

The lattice constant of the relaxed substrate $Si_{1-y}Ge_y$ as a function of Ge content, y, is given by

$$a_0(y) = a_0(Si) + 0.200326y(1-y) + [a_0(Ge) - a_0(Si)]y^2 \text{ Å} \tag{2.38}$$

Denoting the in-plane and perpendicular lattice constants by $a_{//}$ and $a_{\perp}$, respectively, in the strained $Si_{1-x}Ge_x$ layer,

$$\begin{aligned} a_{//} &= a_0(y) \\ a_{\perp} &= a_0(x)\left[1 - 2\frac{c_{12}(x)}{c_{11}(x)}\frac{a_{//} - a_0(x)}{a_0(x)}\right]. \end{aligned} \tag{2.39}$$

where the elastic constants $c_{11}(x)$ and $c_{12}(x)$ are obtained from the linear interpolation of the constants for Si and Ge.

The hydrostatic component that results from the change in volume of the material changes the position of the conduction and valence band edges by

$$\Delta E_c = a_c \frac{\Delta V}{V} = a_c(2\varepsilon_{ii} + \varepsilon_{\perp}) = 2\varepsilon_{\shortparallel}\frac{1-2v}{1-v}a_c, \tag{2.40}$$

$$\Delta E_v = a_v \frac{\Delta V}{V} = a_v(2\varepsilon_{ii} + \varepsilon_{\perp}) = 2\varepsilon_{\shortparallel}\frac{1-2v}{1-v}a_v. \tag{2.41}$$

where a_c and a_v are the hydrostatic deformation potential constants, and v is the Poisson ratio, that relates in-plane and perpendicular strain components by $\varepsilon = (-2v\varepsilon)/(1-v)$.

The other component of strain is the uniaxial strain which splits the degeneracy of both the conduction and valence bands, as shown in Figure 2.8.

When uniaxial strain is applied along the highly symmetric direction [111], there is no effect on the Δ minimum. Strain along the [100] direction produces a splitting in the six Δ minima with changes as follows:

$$\Delta E_c^{001} = \frac{2}{3}\Xi_u(\varepsilon_{\perp} - \varepsilon_{//}), \tag{2.42a}$$

$$\Delta E_c^{100,010} = -\frac{1}{3}\Xi_u(\varepsilon_{\perp} - \varepsilon_{//}). \tag{2.42b}$$

where Ξ_u is the deformation potential constant for uniaxial strain.

The uniaxial strain–induced splitting of the conduction band is therefore

$$\Delta E_{c,s} = \Xi_u(\varepsilon_{\perp} - \varepsilon_{//}) = \Xi_u \frac{1+v}{1-v} \cdot m, \quad m = \frac{a_0(y) - a_0(Si)}{a_0(Si)}$$

The term m represents the mismatch parameter.

Example 2.10

The uniaxial strain–induced splitting of the conduction band is calculated by assuming Vegard's law for the lattice constants. Noting that the uniaxial deformation potential constant $\Xi_u = 9.16\,\text{eV}$ and Poisson ratio $v = 0.278$, the strain-induced splitting is $\Delta E_{c,s} = 0.677\ y$.

For perpendicular strains along the [001] and [111] directions, these splittings can be expressed as a shift from the band edge (ΔE) as

$$\Delta E_{LH} = -\frac{1}{6}\Delta_0 + \frac{1}{4}\delta E + \frac{1}{2}\sqrt{\Delta_0^2 + \Delta_0\delta E + \frac{9}{4}(\delta E)^2}, \quad (2.43a)$$

$$\Delta E_{HH} = \frac{1}{3}\Delta_0 - \frac{1}{2}\delta E, \quad (2.43b)$$

$$\Delta E_{HH} = \frac{1}{3}\Delta_0 - \frac{1}{2}\delta E, \quad (2.43b)$$

$$\Delta E_{SO} = \frac{1}{6}\Delta_0 - \frac{1}{4}\delta E + \frac{1}{2}\sqrt{\Delta_0^2 + \Delta_0\delta E + \frac{9}{4}(\delta E)^2}. \quad (2.43c)$$

The value of δE in these equations is evaluated according to the direction of application of the strain and is expressed as

$$\delta E_{001} = 2b(\varepsilon_\perp - \varepsilon_{//}) \quad \text{for [001] strain}, \quad (2.44a)$$

$$\delta E_{111} = \frac{2}{\sqrt{3}} d(\varepsilon_\perp - \varepsilon_{//}), \quad \text{for [111] strain}. \quad (2.44b)$$

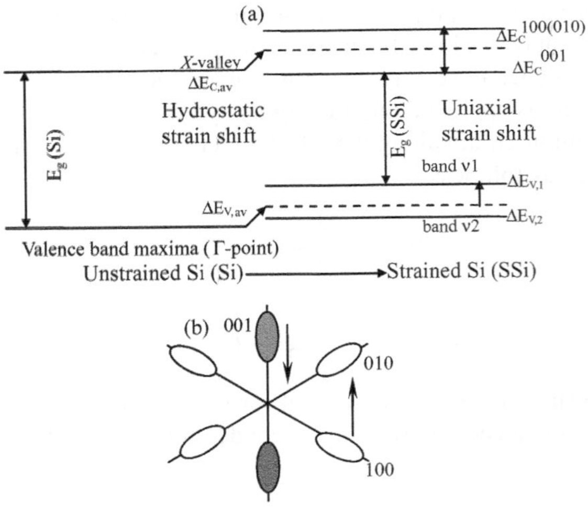

Figure 2.9 How strain affects the band structure of silicon: (a) hydrodynamic and uniaxial strain components; and (b) uniaxial strain effect on the sixfold degenerate ellipsoids in the conduction band. Reproduced with permission from [12]. Copyright (2004) Institute of Physics.

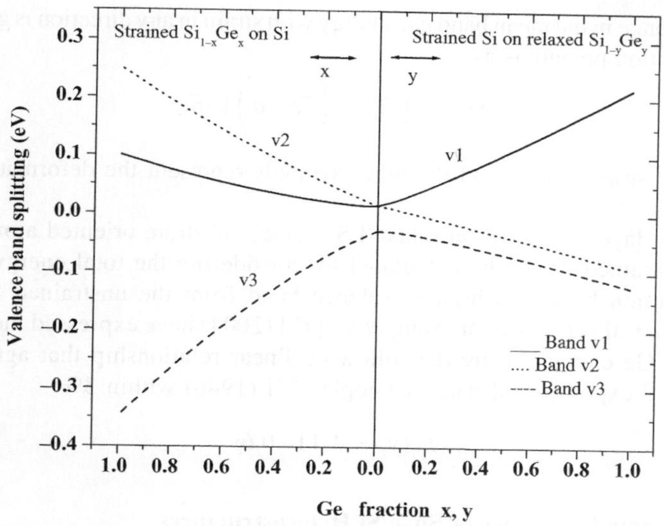

Figure 2.10 *The valence band splitting (taking the average valence band as the reference level. Reproduced with permission from [12]. Copyright (2004) Institute of Physics.*

where b and d are the uniaxial deformation potential constants for valence band for tetragonal strain.

Figure 2.9 shows schematic diagrams of the band shifts and splitting of the conduction and valence bands of strained $Si_{1-x}Ge_x$ alloys on relaxed $Si_{1-y}Ge_y$ ($x < y$) under various strain conditions.

Figure 2.9 depicts how the energy levels shift and split due to tensile strain. As illustrated, the hydrostatic component raises both the conduction and valence band edges, and the uniaxial component splits the degeneracy. The twofold degenerate conduction bands (shaded) are lowered, while the other four valleys are raised in energy.

The strain also shifts and splits the degenerate valence bands $v1$ and $v2$ at the Γ points.

The calculated splitting of the valence bands into levels v1, v2, and v3 is shown in Figure 2.10 for both compressive and tensile strains. The split VB v2 is a pure $|3/3, 3/2\rangle$ state, whereas bands v1 and v3 are mixtures of states $|3/2, 1/2\rangle$ and $|1/2, 1/2\rangle$.

Example 2.11

Consider the growth of strained $Si_{0.3}Ge_{0.7}$ on Si. Using linear interpolation of all parameters, $a_0(x) = 5.58976$ A, $\varepsilon_{//} = -0.028402$, $a_\perp = 1.0216 a_0(x)$, $\varepsilon_\perp = 0.0216$, $b(x) = -1.99$, $\Delta_0(x) = 0.2204$, $\delta E = -0.199$ eV, $\delta E_{v2} = 0.173$ eV, $\delta E_{v1} = 0.0672$ eV, $\delta E_{v3} = -0.2406$ eV.

The values may be checked with the values given in Figure 2.10.

The total change in the mean band gap energy with strain in any direction is given in terms of the deformation potentials as

$$\Delta E_g = \left(\Xi_d + \frac{1}{3}\Xi_u - a\right) 1 \cdot \vec{\varepsilon}, \tag{2.45}$$

where $\vec{\varepsilon}$ is the strain tensor, and the other symbols represent the deformation potential constants.

When the Si layer is grown on relaxed $Si_{1-y}Ge_y$ substrate oriented along [100], the band gap of strained Si can be calculated by considering the total energy shift of the lowest conduction band and highest valence band from the unstrained Si band gap. Figure 2.9 makes this point clear. Yang *et al.* [12] (2004) have expressed the band gap as a function of Ge content, y, by the following linear relationship that agrees with the calculation and experimental data of People [13] (1986) within 3%:

$$E_g(y) = 1.11 - 0.6y \tag{2.46}$$

2.7.7 Electronic Properties of SiGe/Si Heterostructures

The presence of strain in the heterostructure alters the conduction and valence band location as well as the curvature of the bands to alter the effective masses [14].

As mentioned already, the valence band in both Si and Ge consists of HH, LH, and SO bands. However, it is the conduction band where the major differences occur. Ge has four conduction band minima at L points along the <111> directions in the Brillouin zone. Si has six valleys along the Δ direction at the X point (at a distance of about 85% from the center to the zone boundary). The conduction band valleys of Ge and Si are shown in Figure 2.11.

The band gap for unstrained (100)-oriented $Si_{1-x}Ge_x$ material is shown in Figure 2.12. The abrupt discontinuity in the curve at $x = 0.85$ is due to change of the conduction band

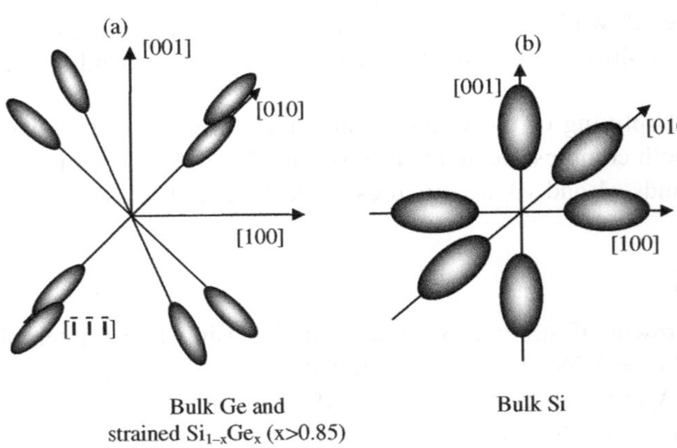

Figure 2.11 *(a) The conduction band valleys of Ge along the [111] directions; and (b) the conduction band valleys of silicon along the [100] directions.*

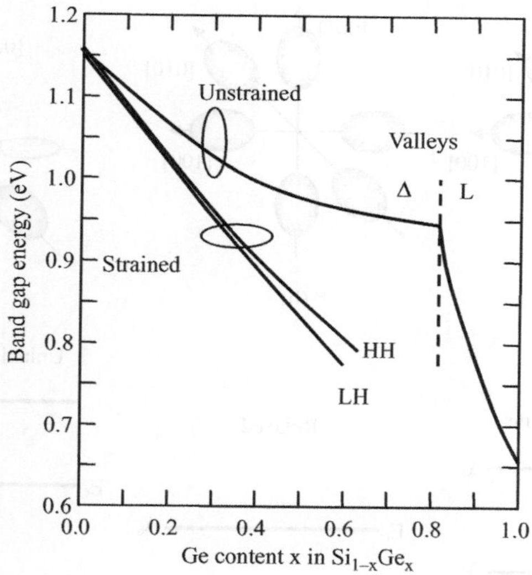

Figure 2.12 *The band gap for strained $Si_{1-x}Ge_x$ grown on bulk-silicon substrates and for unstrained $Si_{1-x}Ge_x$. Reproduced with permission from [15]. Copyright (2004) Institute of Physics.*

from Si-like minima to Ge-like minima (*L* valleys). In Figure 2.12, the band gap for strained $Si_{1-x}Ge_x$ grown on relaxed (100)-oriented Si is also shown. The degeneracy of HH and LH is seen to be lifted by strain.

In addition to the band gap, knowledge about band alignment in strained SiGe–Si systems is also important. The valence band of the strained $Si_{1-x}Ge_x$ layer grown on $Si_{1-y}Ge_y$ virtual substrate has a valence band discontinuity ΔE_v, obtained from pseudopotential theory, as

$$\Delta E_v = (0.74-0.06y)(x-y)(\text{eV}). \tag{2.47}$$

The conduction band discontinuity is somewhat complicated. As Figure 2.13 shows, for uniaxial compressive strain the Δ_4 valleys are lower in energy, while for uniaxial tensile strain the Δ_2 valleys are lower. For a uniaxially tensile-strained Si layer on a $Si_{1-y}Ge_y$ virtual substrate, the conduction band discontinuity produces a quantum well (QW) in the strained Si layer. The magnitude of this discontinuity is given by

$$\Delta E_c = 0.6y\ (\text{eV})\ \text{for}\ x < 0.6. \tag{2.48}$$

The discontinuities expressed by Eqs. (2.47) and (2.48) are shown in Figure 2.14. When the Ge concentration in the heterolayer exceeds 0.85, there is a small region where the L valleys are lowest in energy. For all other Ge concentrations, the minima of the conduction band are Δ valleys. For tensile strain, a QW is produced with only two Δ valleys; in the presence of compressive strain, a small barrier is produced.

The band alignment is type II in character for the vast majority of Ge compositions. When the Ge concentration is around 0.85, a type I structure is obtained.

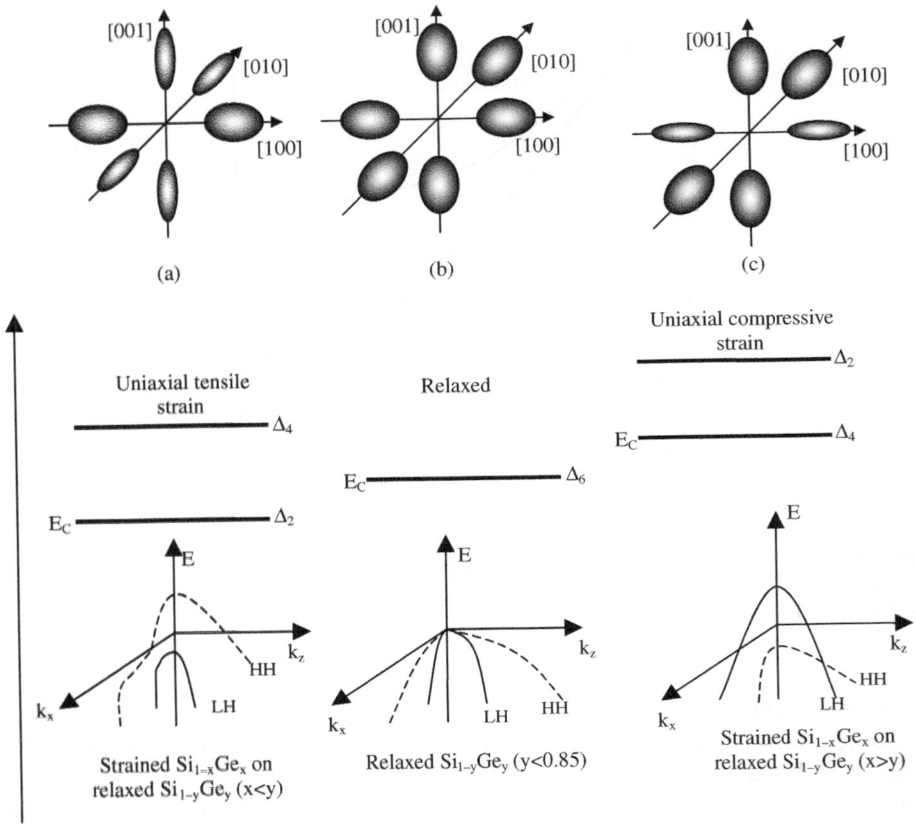

Figure 2.13 *The conduction band, valley structure, and valence band splitting of (a) uniaxial tensile strained silicon or Si1-xGex grown on a (100) Si1-yGey virtual substrate with x < y; (b) bulk silicon and relaxed (100) Si1-xGex with x < 0.85; and (c) uniaxial compressively strained (100) Si1-xGex grown on a Si1-yGey virtual substrate with x > y. Reproduced with permission from [15]. Copyright (2004) Institute of Physics.*

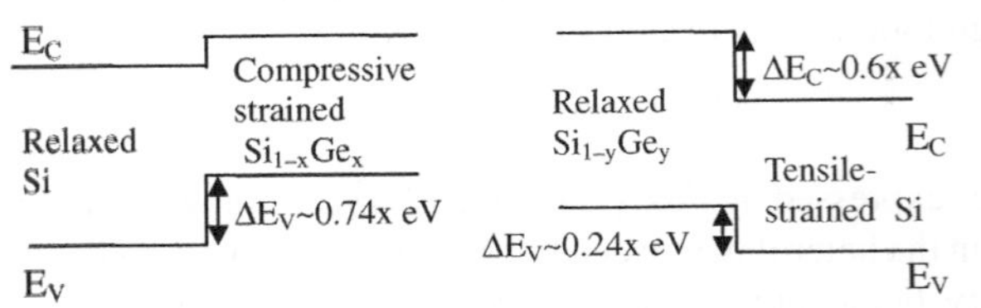

Figure 2.14 *The band alignments and discontinuities for (a) a compressively strained-$Si_{1-x}Ge_x$ heterolayer grown on relaxed silicon; and (b) a tensile-strained-Si layer grown on relaxed $Si_{1-y}Ge_y$.*

Effective masses: The effective masses are different for different compositions of the layer and virtual substrate, which change the character of the lowest conduction band. When a strained Si layer is grown on relaxed (100)-oriented $Si_{1-y}Ge_y$ virtual substrate, the lowest two valleys have in-plane effective mass nearly equal to $0.195\,m_0$, while the calculated longitudinal effective mass is $0.918\,m_0$, which changes little with tensile strain.

The effective mass of holes is more complicated to describe. For strained Si (100) substrate, the effective masses are given by Luttinger parameters which are obtained as linear interpolation of the values for bulk Si and Ge.

2.8 Direct Gap: Ge/SiGeSn Heterojunctions

In this section, some of the recent developments in heterostructures made of group IV elements and their alloys will be mentioned. The most interesting features of the heterostructures are that direct gap can be obtained in the active material, which is mainly Ge, and that the band alignments are type I. Further, the multilayered structures may be grown on silicon platform.

2.8.1 Structures

The conventional heterostructures combining Ge, $Si_{1-x}Ge_x$, and $Si_{1-x-y}Ge_xC_y$ have the limitations that the active layers are indirect gap in nature, and in addition the Si–$Si_{1-x}Ge_x$ heterostructures mostly show type II band alignment. These heterostructures are not useful as light emitters. SiGeC grown on Si is believed to show type I alignment [10]; however, the band gap is still indirect.

The band gap variation in SiGe alloys with incorporation of Sn has been investigated by Soref and coworkers [14, 16], who predicted that the direct band gap may be achieved in SiGeSn ternary alloy for some compositions of Sn.

Recently, two significant achievements have been made in the quest for direct-gap group IV materials. The first is the growth of device-quality $Ge_{1-y}Sn_y$ alloy, in which an indirect-to-direct gap crossover is expected for $y \sim 0.2$ [17]. The second is the development of $Ge_{1-x-y}Si_xSn_y$ ternary alloy [18], the band gap of which has been theoretically studied by Soref and Perry.

Menéndez and Kouvetakis [19] theoretically predicted that the direct-gap nature of the band gap may be observed in the tensile-strained Ge layer. The lowest direct gap in Ge (L point) is only 140 meV above the direct gap (Γ point). With the application of tensile strain, the energy of both edges decreases, but the Γ conduction band edge decreases more rapidly than the L valley to effect indirect-to-direct crossover. This idea cannot be applied to the Ge-$Si_{1-x}Ge_x$ system, since the smaller lattice constant of Si makes the Ge layer compressively strained. The lattice constant of Sn is, on the other hand, larger and hence growth of the Ge layer on $Ge_{1-y}Sn_y$ is expected to solve the problem. However, the direct gap in GeSn decreases very rapidly and therefore the direct gap in Ge may not be the lowest gap. Incorporation of Si in GeSn raises the band gap, but at the same time reduces the tensile strain. Menéndez and Kouvekatis predicted that under suitable conditions, the direct band gap in Ge is also the lowest band gap in Ge–$Ge_{1-x-y}Si_xSn_y$ multilayers, and further the band lineup is type I.

We have in Section 2.8.2 essentially reproduced the theory of band structure and band lineup presented by Menéndez and Kouvetakis.

2.8.2 Band Edges and Band Lineup

The method is a generalization of van de Walle's approach [20, 21]. The average of the three valence bands is taken as the reference level. In the work, the strain-dependent spin–orbit Hamiltonian has been included, in addition. Considering growth of the (001)-oriented Ge layer, lattice-matched to a relaxed $Ge_{1-x-y}Si_xSn_y$ layer, the following expressions for the band edges of Ge are derived:

$$E_{v\Gamma}(\text{Ge}) = -\frac{\Delta_0(x,y)}{3} - \Delta E_{v,av}(x,y) - \frac{\Delta_0(\text{Ge})}{6} + \delta E_h^0 + \frac{1}{4}\delta E_{001} + \frac{1}{2}\sqrt{\left(\Delta_0(Ge) + \frac{1}{2}\delta E_{001}\right)^2 + 2(\delta E'_{001})^2}, \tag{2.49a}$$

$$E_{c\Gamma}(\text{Ge}) = E_{v\Gamma}(\text{Ge}) + E_0(\text{Ge}) + \delta E_h^{c\Gamma} \tag{2.49b}$$

$$E_{cL}(\text{Ge}) = E_{v\Gamma}(\text{Ge}) + E_{ind}(\text{Ge}) + \delta E_h^{cL}. \tag{2.49c}$$

All the energies are measured using the valence band edge of the alloy SiGeSn as the reference. Here $E_{v\Gamma}, E_{c\Gamma}$ and E_{cL} have the usual meanings. $E_0(\text{Ge}) = E_{c\Gamma} - E_{v\Gamma} = 0.805$ eV is the direct gap in Ge, and $E_{ind}(\text{Ge}) = E_{cL} - E_{v\Gamma} = 0.664$ eV is the indirect gap in Ge. Using linear interpolation, the spin–orbit splitting in the alloy is expressed as $\Delta_0(x,y) = \Delta_0(\text{Ge}) - 0.26x + 0.47y$, with $\Delta_0(\text{Ge}) = 0.30$ eV. Also, $\Delta E_{v,av}(x,y) = 0.69y - 0.48x$. The strain shifts are given by

$$\delta E_{001} = -2(b_1 + 2b_2)\left[\frac{2C_{12}}{C_{11}} + 1\right]e_{//},\ \delta E'_{001} = -2(b_1 - b_2)\left[\frac{2C_{12}}{C_{11}} + 1\right]e_{//},$$

$$\delta E_h^v = 2a_v\left[1 - \frac{C_{12}}{C_{11}}\right]e_{//},\ \delta E_c^{c\Gamma} = 2a_c\left[1 - \frac{C_{12}}{C_{11}}\right]e_{//},\ \delta E_c^{cL} = 2\left[\Xi_d + \frac{1}{3}\Xi_u\right]\left[1 - \frac{C_{12}}{C_{11}}\right]e_{//}.$$

The in-plane strain is calculated from the relation

$$e_{//} = \left[a(Ge_{1-x-y}Si_xSn_y) - a(\text{Ge})\right]/a(\text{Ge}),$$

where a is the lattice constant. The different strain components are evaluated by using Vegard's law, and the parameters used in the calculation are included in Appendix B.

The band edges in the ternary are calculated by using linear interpolation between the constituents. The expressions are

$$E_{c\Gamma}(x,y) = E_c(\text{Ge}) + 3.14x - 1.225y \text{ and } E_{cL}(x,y) = E_{cL}(s) + 1.27x - 0.524y.$$

Example 2.12

The values of band edge energies will be calculated for $x = 0.15$ and $y = 0.25$. Using the parameters given in Table B.1 (in Appendix B), we obtain the following values $E_{v\Gamma}(\text{Ge}) = 0.1\,\text{eV}$, $E_{c\Gamma}(\text{Ge}) = 0.6252\,\text{eV}$, and $E_{cL}(\text{Ge}) = 0.7162\,\text{eV}$. The corresponding values of $E_{c\Gamma}(x, y) = 0.9692\,\text{eV}$ and $E_{cL}(x, y) = 0.7235\,\text{eV}$. This example indicates that the band structure in strained Ge is indeed direct in nature. These values belong to the family of curves shown in Figure 2.15.

It has been found that Ge becomes a direct-gap semiconductor for a wide range of concentration and also the band alignment is type I, with both the valence and conduction band edges localized in the Ge layer.

The values of different band edges calculated by the procedure discussed in this example are shown in Figure 2.15 for a fixed value of Si concentration: $x = 0.15$. It appears that the indirect-to-direct crossover occurs for $y \approx 0.17$. The zero of energy is chosen at the top of the valence band in the ternary barrier layer.

Recently, a self-consistent pseudo-potential plane wave method, within the mixed-alloy supercell model of alloys, has been employed in [22]. This method is more accurate than virtual-crystal approximation (VCA). The calculations suggest that good-sized (up to ~0.5 eV) direct-gap materials are achievable in tensile-strained Ge or $Ge_{1-x}Sn_x$ alloy layers. An unstrained $Ge_{1-x}Sn_x$ bulk layer becomes a direct-gap material for $x > 0.17$, but offers lower gap $\sim < 0.2\,\text{eV}$.

There are several recent works, mostly theoretical, predicting the type I direct-gap structure [23] and performance of lasers, and exploiting the direct-gap nature of strained Ge layers using relaxed SiGeSn barriers [24].

It appears from the discussions here that direct gap in tensile-strained Ge grown on a relaxed SiGeSn barrier cannot be used as an emitter at the 1550 nm wavelength range for telecommunication applications. In a recent proposal [25], it has been pointed out that a small amount of C in Ge raises the direct band gap to about 0.8 eV. The results are shown in

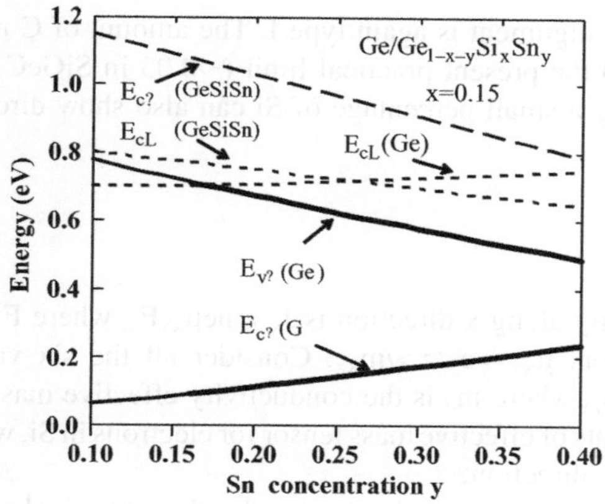

Figure 2.15 *Band edge variation of Γ, L points in tensile-strained Ge–SiGeSn layers. Reproduced with permission from [19]. Copyright (2004) American Institute of Physics.*

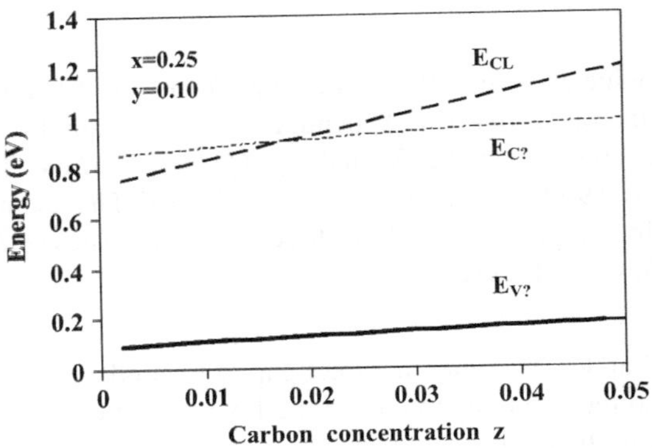

Figure 2.16 Band edges in GeC–SiGeSn.

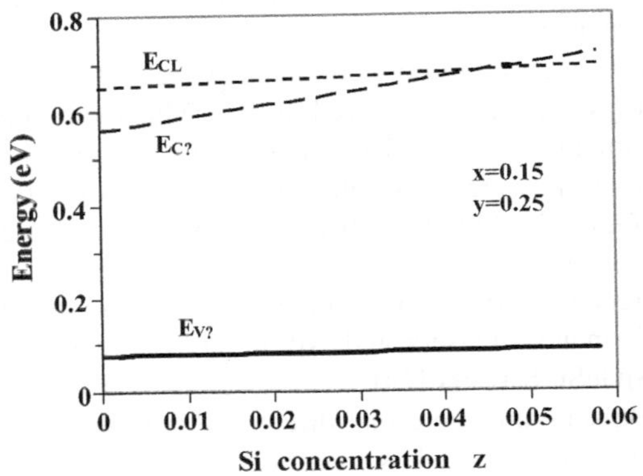

Figure 2.17 Band edges in SiGe–SiGeSn.

Figure 2.16. The band alignment is again type I. The amount of C needed is only 0.015, which does not exceed the present practical limit (~0.03 in SiGeC). It is also predicted that SiGe layers having a small percentage of Si can also show direct gap, as illustrated in Figure 2.17.

Problems

2.1 The current density along x direction is $J_x = ne\mu_{xx}F_x$, where F is the field along the x = (100) direction. $\mu_{xx} = e<\tau>/m_{xx}$. Consider all the six valleys and show that $J_x = ne^2<t>F_x/m_c$, where m_c is the conductivity effective mass.

2.2 Obtain the elements of effective mass tensor for electrons in Si, when one of the axes is along the <111> direction.

2.3 The longitudinal mass in the conduction band valley of Ge is along the [111] direction. Use a suitable coordinate system and suitable coordinate transformation to obtain the elements of the effective mass tensor in the [100] set of the coordinate system.

2.4 Calculate the total number of states in silicon between E_c and $E_c + k_BT$ at 300 K.

2.5 Write the *E*–**k** relationship for the conduction band of silicon in terms of a general Cartesian coordinate system.

2.6 Using the expression for a temperature-dependent band gap, calculate the gap at 300 and 1200 K. Take $E_g(0) = 1.17$ eV, $\alpha = 4.73 \times 10^{-4}$ eV/K, and $\beta = 636$ K.

2.7 Apply Bohr's theory of the H atom to obtain Eq. (2.18): the expression for the donor-binding energy.

2.8 Under a high electric field, the electrons become "hot" with an effective temperature $T_e > T_L$, with T_L being the lattice temperature. The energy balance equation $e\mu F^2 = (3/2)\, k_B(T_e - T_L)/\tau_e$ is used to determine the electron temperature. Prove that when the electric field is along the <100> direction, the two conduction band valleys having longitudinal mass along the field direction will be colder than the remaining four valleys.

2.9 Using the same argument as above, prove that when the field is along the <111> direction, all the six valleys will be equally hot.

2.10 At 0 K, the Fermi function is box-like. Using this, express the Fermi energy in terms of the carrier density.

2.11 Obtain the expression for density-of-states effective mass for holes as given by Eq. (2.17).

2.12 Obtain the expression for hole density by considering heavy and light holes.

2.13 Express the Fermi wave vector in terms of the electron density at absolute zero.

2.14 Obtain the general expression for electron density given as $n = N_c F_{1/2}(\eta)$, where

$$F_{1/2}(\eta) = (2/\sqrt{\pi}) \int_0^\infty \frac{x^{1/2} dx}{[1 + \exp(x - \eta)]},\ \eta = (E_F - E_c)/k_BT.$$

2.15 Show that when $\eta \ll -1$, $F_{1/2}(\eta) = \exp(\eta)$ and the electron density is expressed by Eq. (2.15), valid for nondegenerate semiconductors.

2.16 When $\eta \gg 1$, $F_{1/2}(\eta) \approx [4\eta^{3/2}/3\sqrt{\pi}]$. Using this, express Fermi energy in terms of electron density. Verify that the results agree with those derived in Problem 2.14.

2.17 The lattice constants of Si, Ge, and C are respectively 5.431, 5.658, and 2.7 A. Determine the composition of $Si_{1\text{-}x\text{-}y}Ge_xC_y$ for perfect lattice matching with Si. Use Vegard's law.

2.18 Find the combination of *x* and *y*: the Ge and C contents, that will yield constant values of $h_C = 250$, 500, 1000, 2000, and 4000 Å. See the plots in Zollner [8], fig. 12.5.

2.19 Calculate the values of in-plane, perpendicular, hydrostatic, and shear strains as a function of composition x and (x,y) of $Si_{1\text{-}x}Ge_x$ and $Si_{1\text{-}x\text{-}y}Ge_xC_y$ grown on Si substrate. Use the values of parameters given in Appendix B. Plot the values of different quantities against x and (x,y).

2.20 Calculate the band gap of unstrained SiGe alloy at different temperatures using the expressions given in the text, and compare the values of experimental data given by Braunstein *et al.* [7].

2.21 Calculate the values of valence band splitting for all three bands as a function of Ge fraction x and y for strained $Si_{1\text{-}x}Ge_x$ on Si and strained Si on relaxed $Si_{1\text{-}y}Ge_y$ following Example 2.11. Compare your results with the curves in Figure 2.10, taking the average valence band as the reference level.

References

1. Cohen, M.L. and Chelikowsky, J.R. (1989) Electronic structure and optical properties of semiconductors, in *Springer Series in Solid-State Science 75*(ed. M. Cardona), Springer, Berlin.
2. Varshni, Y.P. (1967) Band-to-band radiative recombination in groups IV, VI, and III–V semiconductors (I). *Phys. Stat. Sol.*, **19**, 459–514.
3. Bucher, K., Bruns, J., and Wagemann, H.G. (1994) Absorption coefficient of silicon: an assessment of measurements and the simulation of temperature variation. *J. Appl. Phys.*, **75**, 1127–1132.
4. Luttinger, J.M. and Kohn, W. (1955) Motion of electrons and holes in perturbed periodic fields. *Phys. Rev.*, **97**, 869–883.
5. Faulkner, R.A. (1969) Higher donor excited states for prolate-spheroid conduction bands: a reevaluation of silicon and germanium. *Phys. Rev.*, **184**, 713–721.
6. Baldareschi, A. (1970) Valley-orbit interactions in semiconductors. *Phys Rev. B*, **1**, 4673–4677.
7. Braunstein, R., Moore, A.R., and Herman, F. (1958) Intrinsic optical absorption in germanium-silicon alloys. *Phys. Rev.*, **109**, 695–710.
8. Weber, J. and Alonso, M.J. (1989) Near band gap photoluminescence of Si-Ge alloys. *Phys. Rev. B*, **40**, 5683–5693.
9. Krishnamurthy, S., Sher, A., and Chen, A.B. (1986) Band structures of $Si_xGe_{1-(x}$ alloys. *Phys. Rev. B*, **33**, 1026–1035.
10. Zollner, S., Kouvetakis, J., and Mayer, J.W. (2002) in *Silicon–Germanium Carbon Alloys: Growth, Properties and Applications*, vol. 15 (eds S.T. Pantelidesand S. Zollner), in Series Optoelectronic Properties of Semiconductors and Super-Lattices, Series editor M.O. Manasreh, Taylor & Francis Books, Inc., New York, London, Chapter 2.
11. People, R. and Bean, J.C. (1985) Calculation of critical layer thickness versus lattice mismatch for Ge_xSi_{1-x} strained layer heterostructures. *Appl. Phys. Lett*, **47**, 322–325.
12. Yang, L., Watling, J.R., Wilkins, R.C.W. *et al.* (2004) Si/SiGe heterostructures parameters for device simulation. *Semicond. Sci. Technol.*, **19**, 1174–1182.
13. People, R. (1986) Physics and applications of Ge_xSi_{1-x}/Si strained-layer heterostructures. *IEEE J. Quantum Electron.*, **22**, 1696–1710.
14. Soref, R.A. and Friedman, L. (1993) Direct-gap Ge/GeSn/Si and GeSn/Ge/Si heterostructures. *Superlatt. Microstruct.*, **14**, 189–194.
15. Paul, D.J. (2004) Si/SiGe heterostructures: from material and physics to devices and circuits. *Semicond. Sci. Technol.*, **19**, R75–R108.
16. Soref, R.A. and Perry, C.H. (1991) Predicted bandgap of the new semiconductor SiGeSn. *J. Appl. Phys.*, **69**, 539–541.
17. Taraci, J., Tolle, J., Kouvetakis, J. *et al.* (2001) Simple chemical routes to diamond cubic germanium tin alloys. *Appl. Phys. Lett.*, **78**, 3607–3609.
18. Bauer, M., Ritter, C., Crozier, P.A. *et al.* (2003) Synthesis of ternary SiGeSn semiconductors on Si (100) via Sn Ge buffer layers. *Appl. Phys. Lett.*, **83**, 2163–2165.
19. Menéndez, J. and Kouvetakis, J. (2004) Type-I Ge/Ge Si Sn strained layer heterostructures with a direct Ge bandgap. *Appl. Phys. Lett.*, **85**, 1175–1177.
20. Van de Walle, C.G. (1989) Band lineups and deformation potentials in the model-solid theory. *Phys. Rev. B*, **39**, 1871–1883.
21. Van de Walle, C.G. and Martin, R.M. (1986) Theoretical calculations of heterojunction discontinuities in the Si/Ge system. *Phys. Rev. B*, **34**, 5621–5634.
22. Moontragoon, P., Ikonic, Z., and Harrison, P. (2007) Band structure calculations of Si-Ge-Sn alloys: achieving direct band gap materials. *Semicond. Sci. Technol.*, **22**, 742–748.
23. Virgilio, M. and Grosso, G. (2006) Type-I alignment and direct fundamental gap in SiGe based heterostructures. *J. Phys.: Condens. Matter*, **18**, 1021–1031.

24. Chang, S.-W. and Chuang, S.L. (2007) Theory of optical gain of Ge–Si_x Ge_y Sn_{1-x-y} quantum-well lasers. *IEEE J. Quantum Electron.*, **43**, 249–256.
25. Ghosh, S. and Basu, P.K. (2010) The calculated composition of $Ge_{1-z}C_z/Ge_{1-x-y}Si_xSn_y$ heterostructure grown on Si for direct gap emission from $Ge_{1-z}C_z$ at 1.55 μm. *Solid State Comm.*, **150**, 844–847.

Suggested Readings

Balkanski, M. and Wallis, R.F. (2000) *Semiconductor Physics and Applications*, Oxford Univ Press, Oxford, UK.

Bastard, G. (1988) *Wave Mechanics Applied to Semiconductor Heterostructures*, Les Editions de Physique, Les Ulis, France.

Basu, P.K. (2003) *Theory of Optical Processes in Semiconductors: Bulk and Microstructures*, Clarendon Press, Oxford, UK.

Davies, J.H. (1998) *The Physics of Low-Dimensional Semiconductors: An Introduction*, Cambridge Univ Press, Cambridge, UK.

Harrison, P. (2000) *Quantum Wells, Wires and Dots: Theoretical and Computational Physics*, John Wiley & Sons, Ltd., Chichester, UK.

Manasreh, O. (2005) *Semiconductor Heterojunctions and Nanostructures*, McGraw Hill, New York.

Mitin, V., Strocio, M.A., and Kochelap (1999) *Quantum Heterostructures*, John Wiley, New York.

Neamen, D.A. (1992) *Semiconductor Physics and Devices: Basic Principles*, McGraw-Hill, New York.

Ridley, B.K. (2000) *Quantum Processes in Semiconductors*, 5th edn, Clarendon Press, Oxford, UK.

Singh, J. (2003) *Electronic and Optoelectronic Properties of Semiconductor Structures*, Cambridge Univ. Press, Cambridge, UK.

Smith, R.A. (1964) *Semiconductors*, Cambridge Univ Press, Cambridge, UK.

Weisbuch, C. and Vinter, B. (1991) *Quantum Semiconductor Structures*, Academic, San Diego.

3

Quantum Structures

3.1 Introduction

Most of the modern optoelectronic and photonic devices rely on heterojunctions, that is, junctions between two dissimilar semiconductors. Over the last two or three decades, nanostructures like quantum wells (QWs), quantum wires (QWRs), and quantum dots (QDs) prepared by using heterostructures are finding more and more use in realizing photonic devices. Most of the recent developments in the area of Si photonics also are made with quantum heterostructures. This chapter gives first an introduction to quantum nanostructures in general. The basic concept of quantum confinement in quantum wells, wires, and dots is first introduced and illustrated by using the conventional compound semiconductors and their alloys. Some basic electronic properties of low-dimensional systems formed in these quantum nanostructures are then discussed. The peculiarities in quantum structures made of silicon and its alloys with other group IV elements are finally pointed out.

3.2 Quantum Wells

Most of the present electronic and optoelectronic devices are based on nanostructures in which at least one dimension of the layer of interest is only a few nanometers thick. The behavior of particles in the nanostructure differs grossly from the same in bulk materials and the motion of particles is quantum mechanically confined along the concerned direction. The confinement gives rise to electron or hole gas of lower dimensions, as will be explained in this chapter. The confinement and existence of low-dimensional electron systems have also been observed in silicon and its alloys. Before we discuss the specific properties related to silicon, we first consider the conditions necessary for quantum confinement, the structures that support the low-dimensional systems, and the changes in electronic properties from the bulk materials. To start with, we first consider a quantum well (QW) that supports two-dimensional electron gases. We shall consider the well-studied GaAs–AlGaAs material system for the sake of illustration.

Silicon Photonics: Fundamentals and Devices, First Edition. M. Jamal Deen and P. K. Basu.

3.2.1 Condition for Quantum Confinement

The electrons in a bulk semiconductor have free motion in all the three dimensions, as indicated by the E–$\mathbf{k}$ relationship expressed by

$$E = \frac{\hbar^2 k^2}{2m_e}, k \equiv k_x, k_y, k_z. \tag{3.1}$$

Here the electron mass m_e is isotropic as in direct-gap materials like GaAs. For electrons in Si, Eq. (2.3) applies. If the electron is confined in a one-dimensional (1D) potential well having width comparable to its de Broglie wavelength, then the free motion of the electron is inhibited in that direction. The electron waves are then standing waves in nature and the energy becomes quantized. The electron is, however, free to move along the other two dimensions, and accordingly, a two-dimensional electron gas results. The electron wavelength $\lambda = 2\pi/k$ is related to its energy E by

$$\lambda = \frac{2\pi\hbar}{\sqrt{2m_e E}}. \tag{3.2}$$

Example 3.1

Let us assume that the energy of the electron is 25 meV and the effective mass of the electron is $m_e = m_l^* = 0.91\, m_0$: the longitudinal mass in the Si conduction band. Then $\lambda \approx 8$ nm. The value for GaAs will be ≈ 30 nm using $m_e = 0.07\, m_0$.

Example 3.1 gives us an idea of the length scale of the semiconductor layer needed for observation of quantum confinement. There exist other length scales like mean free path and so on.

3.2.2 A Representative Structure

The 1D potential well described in Section 3.2.1 may be realized in a double heterojunction. A heterojunction is formed by two dissimilar semiconductors having differences in band gap, permittivity, and other parameters as explained in Chapter 2. Consider the simplest case of a double heterojunction made by GaAs and its alloy $Ga_{1-x}Al_xAs$, as shown in Figure 3.1. In the simplest approximation, the difference in the band gap of the two semiconductors, $E_g^{AlGaAs} - E_g^{GaAs}$, is consumed by discontinuities (steps) in the conduction and valence bands in the two heterointerfaces. The heterointerfaces are assumed to be abrupt, which condition is more or less satisfied in present-day structures grown by MBE or MOCVD. It appears from the figure that rectangular potential wells with abrupt walls at the two heterointerfaces exist in the valence and conduction bands of the lower gap GaAs sandwiched between two AlGaAs layers. If the thickness d of GaAs along the growth direction is comparable to the de Broglie wavelength, then the quantum confinement mentioned in Section 3.2.1 is expected. The structure shown in Figure 3.1 is called a square quantum well (QW) structure. In this example of QW structure, the GaAs layer is called the *well material* and the two AlGaAs layers surrounding the well are called the *barrier layers*.

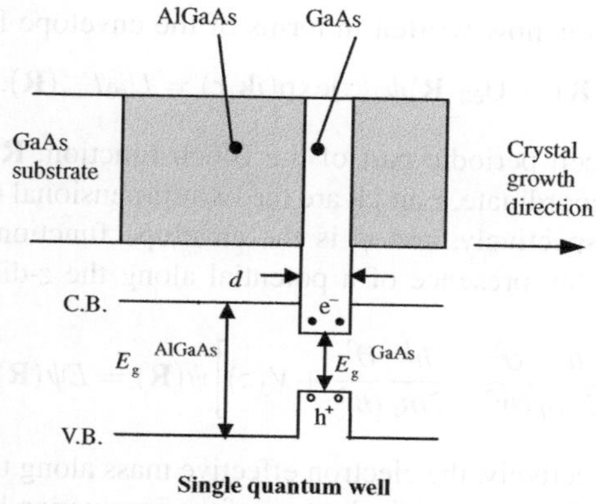

Figure 3.1 *The quantum well structure formed by sandwiching the GaAs layer in between two AlGaAs layers.*

3.2.3 Simplified Energy Levels

The calculation of the energy levels in a square potential well is presented in all elementary textbooks on quantum mechanics [1]. The energy levels in the conduction and valence bands shown in Figure 3.2 are calculated by using the effective mass theory. To simplify matters, we assume that the lattice constant and the effective mass of electrons in the two materials are equal. Since the motion of electrons is confined along the z-direction, electrons are free along the (x–y) plane. The conduction band wavefunction in

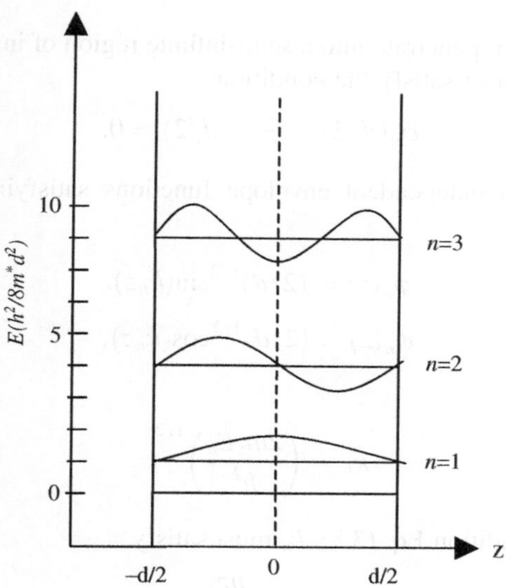

Figure 3.2 *Energy levels and envelope functions for the infinite square well.*

the well material (GaAs) is now written in terms of the envelope function $F_{cn}(\mathbf{R})$ as

$$\Psi_c(\mathbf{R}) = U_{c0}(\mathbf{R})\phi_n(z)\exp(i\mathbf{k}.\mathbf{r}) = U_{c0}F_{cn}(\mathbf{R}). \tag{3.3}$$

where U represents the cell periodic part of the Bloch function, $\mathbf{R}$ ($\equiv \mathbf{r}$, z) is the three-dimensional (3D) space coordinate, $\mathbf{r}$ and $\mathbf{k}$ are the two-dimensional (2D) space coordinate and 2D wave vector, respectively, and ϕ is the envelope function. The effective mass Schrödinger equation in the presence of a potential along the z-direction, $V(z)$, may be written as

$$\left[-\frac{\hbar^2}{2m_{II}}\frac{\partial^2}{\partial \mathbf{r}^2} - \frac{\hbar^2}{2m_z}\frac{\partial^2}{\partial z^2} + V(z)\right]\psi(\mathbf{R}) = E\psi(\mathbf{R}). \tag{3.4}$$

Here m_{II} and m_z are, respectively, the electron effective mass along the layer and along the z-direction, $V(z)$ denotes the energy at the bottom of the conduction band, and E is the total energy of the electron. The total wavefunction is written in product form as

$$\psi(\mathbf{R}) = \chi(\mathbf{r})\phi_n(z) \tag{3.5}$$

Using this in Eq. (3.4) we may obtain the following equation for $\phi_n(z)$:

$$\left[-\frac{\hbar^2}{2m_z}\frac{d^2}{dz^2} + V(z)\right]\phi_n(z) = E_n\phi_n(z). \tag{3.6}$$

where n stands for different quantized energy levels.

The solution of Eq. (3.6) is greatly simplified under the infinite well approximation, having $V = 0$ within the well and $V = \infty$ outside the well. Specifically, for the well of width d,

$$\begin{aligned} V(z) &= 0, |z| < d/2, \\ &= \infty, |z| > d/2. \end{aligned} \tag{3.7}$$

Since the electron cannot penetrate into a semi-infinite region of infinite potential energy, the envelope function must satisfy the condition

$$\phi_n(d/2) = \phi_n(-d/2) = 0. \tag{3.8}$$

The following are two independent envelope functions satisfying the effective mass equations:

$$\phi_n(z) = (2/d)^{1/2}\sin(k_n z), \tag{3.9a}$$

$$\phi_n(z) = (2/d)^{1/2}\cos(k_n z), \tag{3.9b}$$

where

$$k_n = \left(\frac{2m_z E_n}{\hbar^2}\right)^{1/2}. \tag{3.10}$$

From the boundary condition Eq. (3.8), k_n must satisfy

$$k_n = \frac{n\pi}{d}, \tag{3.11}$$

where n is a positive integer, with even values applying to Eq. (3.9a) and odd values to Eq. (3.9b). The energy eigenvalues are given by

$$E_n = \frac{\hbar^2}{2m_z}\left(\frac{\pi n}{d}\right)^2 n = 1, 2, 3 \ldots \tag{3.12}$$

A diagram of the first few energy levels (or *subbands*, as they are called) and envelope functions is given in Figure 3.2. The total energy including the translational energy parallel to the interface is now expressed as

$$E(\mathbf{k}, n) = E_n + \frac{\hbar^2 k^2}{2m_{II}}. \tag{3.13}$$

Example 3.2

Consider two semiconductors GaAs and $Ga_{0.7}Al_{0.3}As$. The band gaps of the two are, respectively, 1.43 eV and 1.8 eV. If it is assumed that the band gap difference is partitioned as ΔE_c: $\Delta E_v \equiv$ 60:40, where ΔE_c and ΔE_v are the steps in conduction and valence bands at the heterointerface, then $\Delta E_c = 222$ meV and $\Delta E_v = 148$ meV.

Example 3.3

Let the width of GaAs QW d = 10 nm. Then the energy of the first subband in the conduction band is $E_1 = 56.1$ meV and of the second subband is $E_2 = 225$ meV, by using $m_e = 0.067\, m_0$.

The energy dispersion relation is shown in Figure 3.3. The dispersion relation for holes is expressed identically as Eq. (3.13) by using the appropriate hole mass.

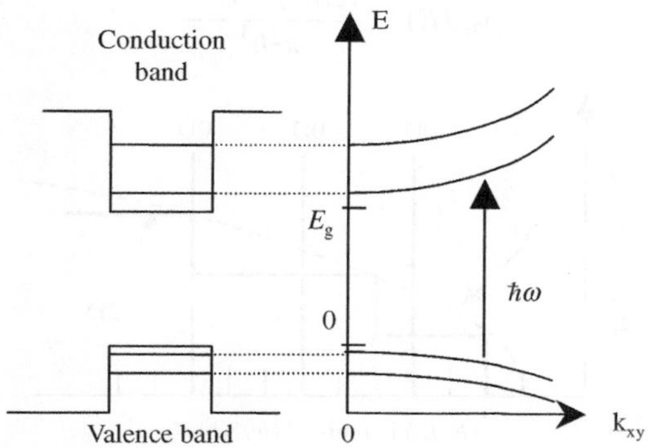

Figure 3.3 AnE–$\mathbf{k}$ diagram of two-dimensional electron and hole gas. Two subbands for each of the conduction and valence bands are shown.

3.2.4 Density-of-States in Two Dimensions

The density-of-states (DOS) is calculated by assuming that the lengths of the sample are L_x and L_y, respectively, along the x- and y-directions. Assuming periodic boundary conditions, along x and y, the allowed values of wave vectors are

$$k_x = \frac{2\pi m_x}{L_x}, k_y = \frac{2\pi m_y}{L_y}, \tag{3.14}$$

where m_x and m_y are integers and $k = (k_x^2 + k_y^2)^{1/2}$. The area in the (k_x, k_y) plane that contains one allowed value of k is $(2\pi)^2/L_xL_y$. Using Eq. (3.14) the area between two circles corresponding to E and $E + dE$ is $2\pi kdk$. The number of states associated with this area is

$$dN = \frac{2\pi kdk}{(2\pi)^2} L_xL_y = \frac{kdk}{2\pi} L_xL_y. \tag{3.15}$$

The change in energy is as obtained from Eq. (3.13)

$$dE = \frac{\hbar^2 kdk}{m_{II}}. \tag{3.16}$$

The density-of-states is therefore

$$\frac{dN}{dE} = \frac{m_{II}}{2\pi\hbar^2} L_xL_y. \tag{3.17}$$

Introducing a factor 2 for spin and normalizing to unit area, one obtains for the two-dimensional density-of-states function

$$\rho_{2D} = \frac{m_{II}}{\pi\hbar^2}, \tag{3.18}$$

which is independent of energy. In a QW, each subband contributes this amount to the total density-of-states. The profile is like a staircase, as shown in Figure 3.4. It will be seen later that absorption in a QW follows the same pattern.

The DOS for bulk material (3D case) is given by (and see Eq. (2.10))

$$\rho_{3D}(E) = \frac{(2m^*)^{3/2} E^{1/2}}{\pi^2\hbar^3}. \tag{3.19}$$

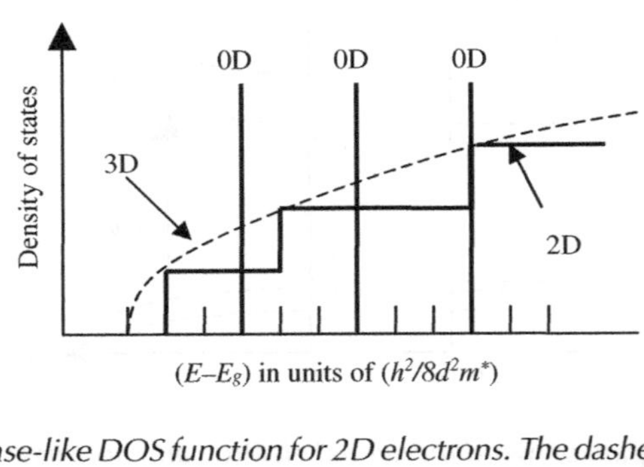

Figure 3.4 The staircase-like DOS function for 2D electrons. The dashed curve represents a 3D DOS function. The DOS for quantum dots (0D) structures is also included.

The variation of the 3D DOS function is also shown in Figure 3.4. It appears that the 3D DOS approaches zero at the band edge. On the other hand, the 2D DOS function is finite at each subband edge and remains constant over a particular subband.

3.2.5 Finite Quantum Well

In a real QW structure, the height of the potential well at the heterointerfaces is never infinitely high (see Example 3.2). The finite QW is characterized by a potential V_0 shown in Figure 3.5, at the heterointerface.

The potential $V(z)$ may now be defined as

$$\begin{aligned} V(z) &= 0, |z| < d/2, \\ &= V_0, |z| > d/2 \end{aligned} \tag{3.20}$$

The envelope functions in the barriers are no longer zero; rather, they penetrate into the barrier due to finiteness of the barrier height V_0. The solutions of the effective mass equation take the form

$$\phi_n(z) = C_1 \cos k_w z, \quad |z| < d/2 \tag{3.21a}$$

$$= C_2 \exp[-k_b(z - d/2)], \quad z > d/2, \tag{3.21b}$$

$$= C_2 \exp[k_b(z + d/2)], \quad z < -d/2. \tag{3.21c}$$

The solutions given above are even functions. Cs are constants and the subscripts w and b refer, respectively, to the well and barrier. The odd solutions can also be written similarly by replacing cos by sin in Eq. (3.21a). The exponential functions in the other two are

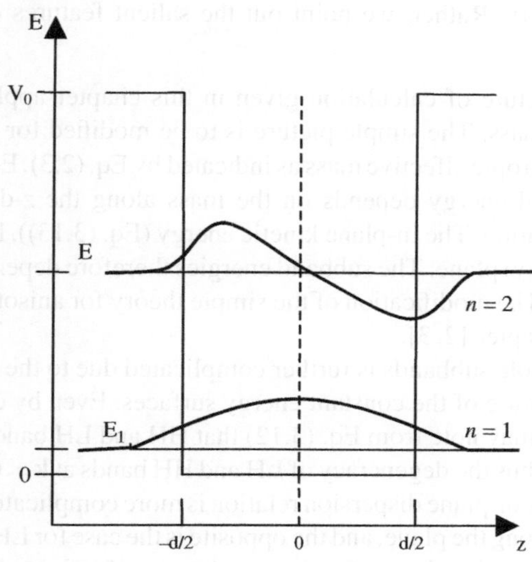

Figure 3.5 *Potential profile for a finite QW structure. The envelope functions penetrate into the barrier materials.*

unchanged. The energy eigenvalues for both the cases may be expressed as

$$E_n = \frac{\hbar^2 k_w^2}{2m_w^*}, \text{ or } E_n = V_0 - \frac{\hbar^2 k_b^2}{2m_b^*}, \tag{3.22}$$

The values of k_w and k_b are determined by boundary conditions at $\pm d/2$ that ϕ_n and $(1/m^*)d\phi_n/dz$ are continuous. For even solutions, one obtains

$$C_1 \cos(k_w d/2) = C_2, \tag{3.23a}$$

$$(k_w/m_w^*)C_1 \sin(k_w d/2) = k_b C_2/m_b^*. \tag{3.23b}$$

The nontrivial solution of Eq. (3.23a) is

$$(k_w/m_w^*)\tan(k_w d/2) = (k_b/m_b^*). \tag{3.24}$$

From Eq. (3.22) one obtains by eliminating E_n

$$\frac{\hbar^2 k_w^2}{2m_w^*} + \frac{\hbar^2 k_b^2}{2m_b^*} = V_0. \tag{3.25}$$

Equations (3.24) and (3.25) may be solved by graphically or numerically for k_w and k_b, and their values are then substituted in Eq. (3.22) to give the energy eigenvalues.

3.2.6 Refined Methods

The above simple model for the calculation of energy eigenvalues and envelope functions must be drastically modified in real situations. We make no attempt to present the refinements in the theory. Rather, we point out the salient features of the modifications introduced.

1. First, the simple picture of calculation given in this chapter applies to bands having isotropic effective mass. The simple picture is to be modified for semiconductors like silicon having anisotropic effective mass as indicated by Eq. (2.3). Equation (3.12) points out that the subband energy depends on the mass along the z-direction, that is, the direction of quantization. The in-plane kinetic energy (Eq. (3.13)), however, is governed by the mass in the (x,y) plane. The subband energies therefore depend on the direction of growth of the well. The modification of the simple theory for anisotropic masses will be presented in this chapter [2, 3].
2. The calculation of hole subbands is further complicated due to the anisotropic effective mass and warped nature of the constant energy surfaces. Even by considering isotropic effective mass, one may note from Eq. (3.12) that HH and LH bands will have different subband energies. Thus the degeneracy of LH and HH bands at $\mathbf{k} = 0$ in the bulk material is lifted in a QW. The in-plane dispersion relation is more complicated. The HH subbands show lighter mass along the plane, and the opposite is the case for LH subbands. The band mixing should be considered even for the simple case of infinite potential well. In real situations, the band structure is calculated by using **k.p** perturbation theory. These refinements will be given in Appendix. A.

3. The potential in the calculation given in this chapter has been taken to be constant even in the finite well problem. In actual situations, one or both semiconductors forming the heterojunctions may be doped and there is a transfer of carriers from one to the other. As a result, space charges develop at the heterojunctions. The shape of the potential becomes fairly complicated then, and the energy eigenvalues and envelope functions of different subbands are totally dependent on the shape of $V(z)$. The carriers distribute themselves among different subbands according to the carrier statistics, and this distribution also governs the shape of the potential $V(z)$. The determination of the energies and eigenfunctions therefore reduces to the solution of coupled Schrödinger and Poisson equations self-consistently. When the electron density is large, the effect of electron–electron interaction is to be included in determining the self-consistent potentials and eigenvalues. The reader is referred to different texts suggested and to papers and monographs discussing these problems [2–4].

3.2.7 Different Band Alignments

The different types of band alignments have been presented in Chapter 2 (see Section 2.6 as well as Figure 2.6). As may be seen from the band diagram of GaAs–AlGaAs QW, shown in Figures 2.6 and 3.3 the subbands in both conduction and valence bands of the system occur in the GaAs layer. The type of band alignment is known as *type I*. There exist other types of band alignments in heterojunctions. We shall mention here only one type: the type II alignment, which is illustrated in Figure 3.6 by using the heterojunction between GaAs and InAs. It is found from the diagram that the conduction band minimum occurs at InAs; on the other han,d the valence band maximum occurs in GaAs. That means that electrons and holes are confined in different layers, unlike the case of type I alignment discussed so far. The type II alignment is found in Si–SiGe systems also, as we shall see in this chapter. A special point to note in this connection is that electron–hole recombination is indirect in real space, whereas the recombination is spatially direct in type I structures.

3.3 Quantum Wires and Dots

In Section 3.2, we have seen that confinement of electrons or holes in a narrow 1D potential well, along the z-direction, say, leads to 2D electron and hole gas. The structure is called a QW. In this section, we shall consider the effect of introducing further confinement to the motion of electrons and study the relevant subband structures and a few electronic properties briefly. The simple theory with infinite potential barriers will be presented.

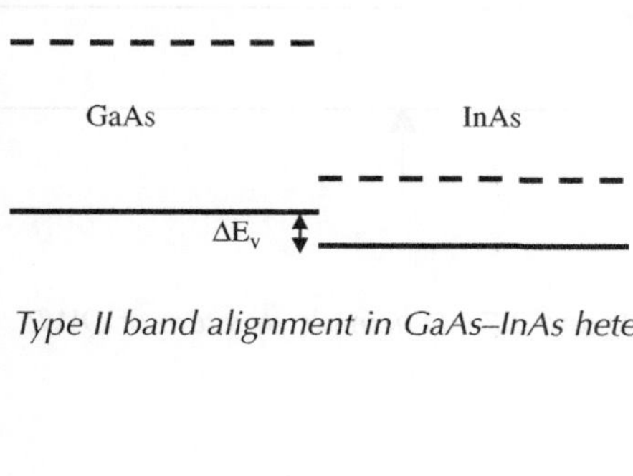

Figure 3.6 Type II band alignment in GaAs–InAs heterojunctions.

3.3.1 Subbands and DOS in Quantum Wires

Assume first that the confinement is extended into two dimensions, that is, along the y and z directions. Extending the earlier arguments, the electron motion is now confined in two directions, y and z; the only direction the electrons are free to move along is the x-direction. We now encounter a one-dimensional electron gas (1 DEG). Because of the nature of the potential, the structure supporting 1D electrons (or holes) is called the quantum wire (QWR) structure. A schematic diagram of the structure is shown in Figure 3.7.

Following the arguments given in Section 3.2.3, the envelope function for electrons in a 2D confining potential $V(y,z)$ along y and z should satisfy the effective mass equation

$$\left[-\frac{\hbar^2}{2m^*}\left(\frac{d^2}{dy^2}+\frac{d^2}{dz^2}\right)+V(y,z)\right]\phi(y,z)=E_{mn}\phi(y,z). \tag{3.26}$$

The effective mass has been assumed to be isotropic. In the simplest situation of infinite barriers along both the dimensions, $\phi(y,z)=\phi(y)\phi(z)$, and as noted earlier both of them may be expressed by sinusoidal functions. The subband energies may be expressed as

$$E_{mn}=\frac{\hbar^2}{2m^*}\left[\left(\frac{m\pi}{d_y}\right)^2+\left(\frac{n\pi}{d_z}\right)^2\right]. \tag{3.27}$$

Here d's are the widths of the well in respective directions, and m and n are integers including zero, but both cannot be zero. The dispersion relation is now

$$E(k_x)=E_{mn}+\frac{\hbar^2k_x^2}{2m^*}, \tag{3.28}$$

indicating free motion along the x-direction: the axis of the QWR.

The expression for the DOS function for 1D electrons may be worked out along the same lines and it takes the form

$$\rho(1D)=\frac{2d_x}{\pi\hbar}\left[\frac{m^*}{2(E-E_{mn})}\right]^{1/2}, \tag{3.29}$$

where d_x is the length of the sample. It follows therefore that the DOS for 1D shows singularities at the subband edges.

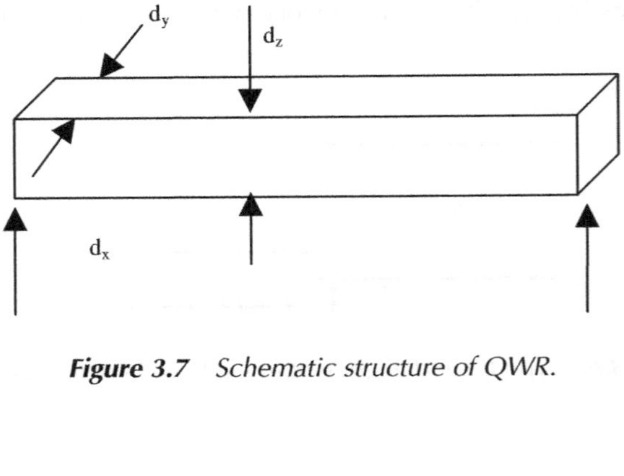

Figure 3.7 *Schematic structure of QWR.*

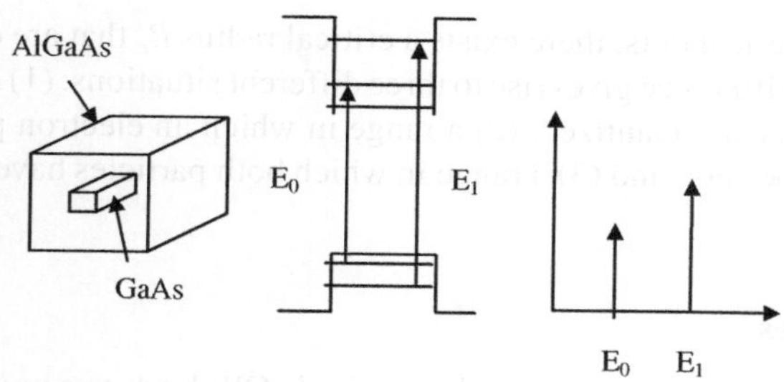

Figure 3.8 *Schematic diagram of a quantum box and the DOS function.*

3.3.2 Quantum Dots

Suppose now that barriers are created along all the three directions of the sample. To visualize this, let us consider a cubic box of GaAs, having widths of a few nanometers in all directions, is surrounded by AlGaAs as shown in Figure 3.8.

There exists now a 3D potential well for both electrons and holes. The motion of the carriers will now be confined in all three dimensions, and a zero-dimensional (0D) systems results. Such a system is referred to as a *quantum box*, a quantum dot (QD), or a *superatom* (if the shape is spherical). Considering the barrier height to be infinite, the energy levels can easily be expressed as

$$E_{n_x,n_y,n_z} = \frac{\hbar^2\pi^2}{2m^*}\left(\frac{n_x^2}{d_x^2} + \frac{n_y^2}{d_y^2} + \frac{n_z^2}{d_z^2}\right). \tag{3.30}$$

where, as usual, d's are the dimensions in the respective directions. For a spherical dot, the subband energies may be expressed in terms of three parameters describing levels for hydrogen atoms. The energy levels are now discrete. The DOS in the ideal case consists of delta functions centered on each discrete level. The nature of DOS for both QWR and QD is shown in Figure 3.9.

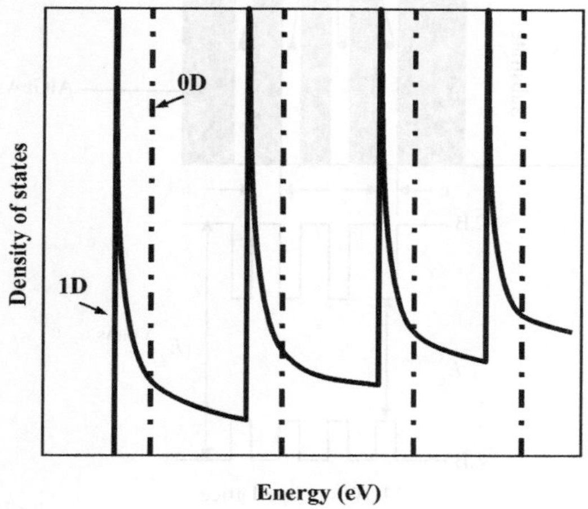

Figure 3.9 *Density-of-states function of a QWR (1D) and a QD (0D) structure.*

In the case of spherical dots, there exists a critical radius R_c that are different for holes and electrons. This difference gives rise to three different situations: (1) a range of radius R in which no particles are quantized, (2) a range in which an electron possesses a bound state while a hole does not, and (3) a range in which both particles have bound states (see Problem 3.22).

3.4 Superlattices

A repetition of the layer sequence to produce a single QW leads to a multiple QW (MQW) structure, as shown in Figure 3.10. The individual QWs in MQWs are uncoupled and therefore the energy levels remain discrete, with values given by Eq. (3.10) for infinite barrier.

Coupling between adjacent wells may take place in a number of ways. If the barrier layer thickness is small or the barrier height is low or if both of them are effective, then the exponential tails of the envelope functions as shown in Figure 3.5 in two wells overlap, giving rise to coupling. The situation is analogous to two mutually coupled tuned circuits. When the coupling is strong, the resonant frequency breaks into two well-separated peaks. For two coupled wells, the coupling leads to two closely spaced energy levels centered about the original degenerate energy level. The presence of N number of coupled wells gives rise to a similar number of closely spaced energy levels. When the spacing between the levels is small, the levels are almost continuously distributed and an energy band, called the *miniband*, is formed.

The periodic potential in a superlattice is shown in Figure 3.10. The simplest method of obtaining the eigenvalues and envelope functions for this case is to employ the well-known Kronig–Penney model or the tight-binding model treated in almost all textbooks on

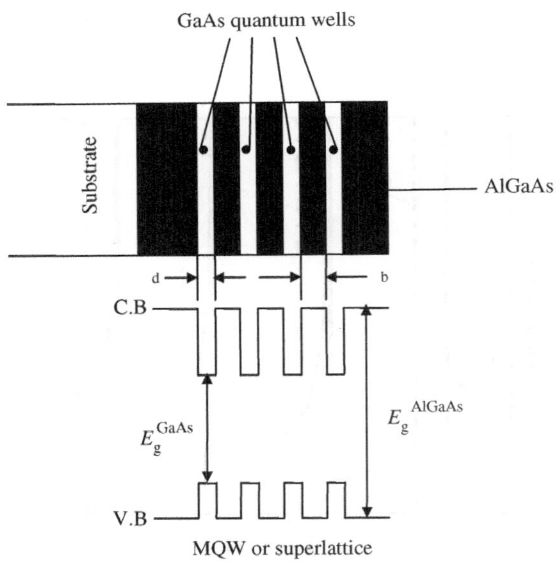

Figure 3.10 Schematic diagram of a semiconductor multiple QW (MQW) or a superlattice.

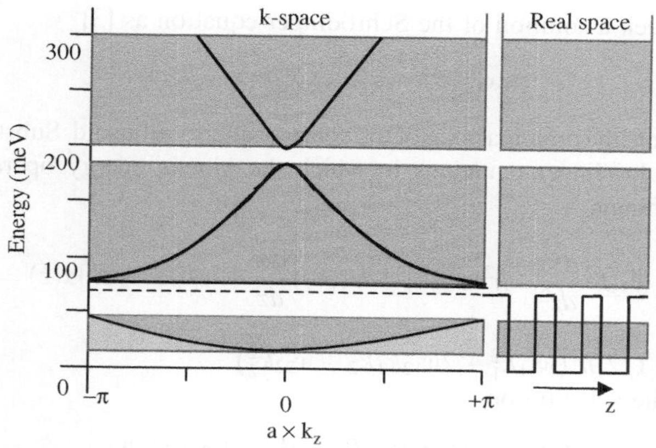

Figure 3.11 *Miniband (shaded region) and minigap (clear region) in a superlattice, shown in both real and k-space.*

solid-state physics. The models consider rectangular potentials shown as in this chapter with different atoms spaced at a distance of a_0, the interatomic distance. Both the models prove the existence of energy bands separated by gaps occurring at wave vectors $k_z = \pm \pi/a_0$.

In the present situation, the periodicity is $a = d + b$, where d and b are, respectively, the well and barrier widths. As the lattice spacing $a >> a_0$, the structure is rightly called the *superlattice*. The E–k_z diagram obtained from the analysis is shown in Figure 3.11. The energy bands now exist in the minizone ranging from $-\pi/a$ to $+\pi/a$. Note that $a \gg a_0$ and therefore the bands are termed as minibands. The band diagram shows gaps, called *minigaps*, at the edges of the minizone.

3.5 Si-Based Quantum Structures

Sections 3.2 through 3.4 presented the basic ideas of quantum confinement; simplified expressions for subband energies in QWs, QWRs, and QDs; and expressions for DOS functions in the structures by considering isotropic effective masses for electrons and holes.

In the present section the modifications to be introduced for materials having anisotropic mass will be presented, taking Si as the example.

3.5.1 Electron Subband Structure

Since Si and Ge have ellipsoidal conduction band valleys, the effective mass is expressed as a tensor. The kinetic energy operator for the Schrödinger equation is

$$T = -\frac{\hbar^2}{2}\sum_{i,j} w_{ij}\frac{\partial^2}{\partial x_i \partial x_j} \qquad (3.31)$$

where w_{ij}'s are the elements of the reciprocal effective mass tensor for the particular conduction band minimum being considered. Since the potential energy $V(z)$ is a function of

z only, we may seek a solution of the Schrödinger equation as [3]

$$\phi(x,y,z) = \alpha(z)\exp(jk_1x + jk_2y) \tag{3.32}$$

where 1, . . . 3 denote the principal axes of the constant energy ellipsoid. Substituting this in the effective mass Schrödinger equations in which the kinetic energy operator is given by Eq. (3.31), one obtains

$$-\frac{1}{2}w_{33}\hbar^2\frac{d^2\alpha}{dz^2} - \hbar^2(w_{13}k_1 + w_{33}k_2)\frac{d\alpha}{dz} + [eV(z) + E']\alpha(z) = 0, \tag{3.33}$$

where $E = E' + (1/2)\hbar^2(w_{11}k_1^2 + 2w_{12}k_1k_2 + w_{22}k_2^2)$.

Making now the substitution

$$\alpha(z) = \chi(z)\exp[-jz(w_{13}k_1 + w_{23}k_2)/w_{33}] \tag{3.34}$$

to eliminate the first derivative with respect to z, we find that

$$\frac{d^2\phi_i}{dz^2} + \frac{2m_3}{\hbar^2}\left[E_i'' + eV(z)\right]\phi_i(z) = 0 \tag{3.35}$$

where $m_3 = w_{33}^{-1}$ and a subscript i has been introduced to label the solutions. Thus the total energy becomes

$$E_i(k_1,k_2) = E_i'' + \frac{1}{2}\hbar^2\left[\left(w_{11} - \frac{w_{13}^2}{w_{33}}\right)k_1^2 + 2\left(w_{12} - \frac{w_{13}w_{23}}{w_{33}}\right)k_1k_2 + \left(w_{22} - \frac{w_{23}^2}{w_{33}}\right)k_2^2\right] \tag{3.36}$$

In the above, E_i'' denotes the subband energy and the other term is the kinetic energy in the xy plane.

Three effective masses enter in the energy of the electrons. One is $m_3 = w_{33}^{-1}$, which determines the subband energy. The other two masses, m_1 and m_2, are the principal effective masses of the constant energy ellipses associated with motion in the xy plane. A suitable transformation yields the principal effective masses m_x and m_y in the plane.

The subband structures may be calculated for high-symmetry surfaces, since the effective masses take the simpler forms. As an illustration we consider the subband structure for the (100) orientation of Si [5]. As shown in Figure 3.12(a), the valleys 1 and 2 have their longitudinal mass m_l ($=0.916\,m_0$) along the growth direction or normal to the surface. Since $m_l > m_t$, these two valleys will give rise to the lowest subbands which are twofold degenerate. The ladder of subbands due to these two valleys is denoted by E_1, E_2, and so on, as in Figure 3.12(b). The other four valleys, numbered 3–6, have their transverse mass normal to the surface and the ladder of subbands, each fourfold degenerate, is denoted by prime symbols: $E_1', E_2' \ldots$ and so on. A simple calculation shows that E_1' is slightly above E_2 (see the "Problems" section of this chapter). In Figure 3.12b, only the subband E_1' has been shown [5]. The subband is shifted upward as shown due to strain effect to be discussed in this chapter.

The above arguments easily follow from Eq. (3.36). For the (100)-oriented surface of Si and for valleys 1 and 2 in Figure 3.11 $w_{33} = m_l^{-1}$ and $w_{11} = w_{22} = m_t^{-1}$, the other

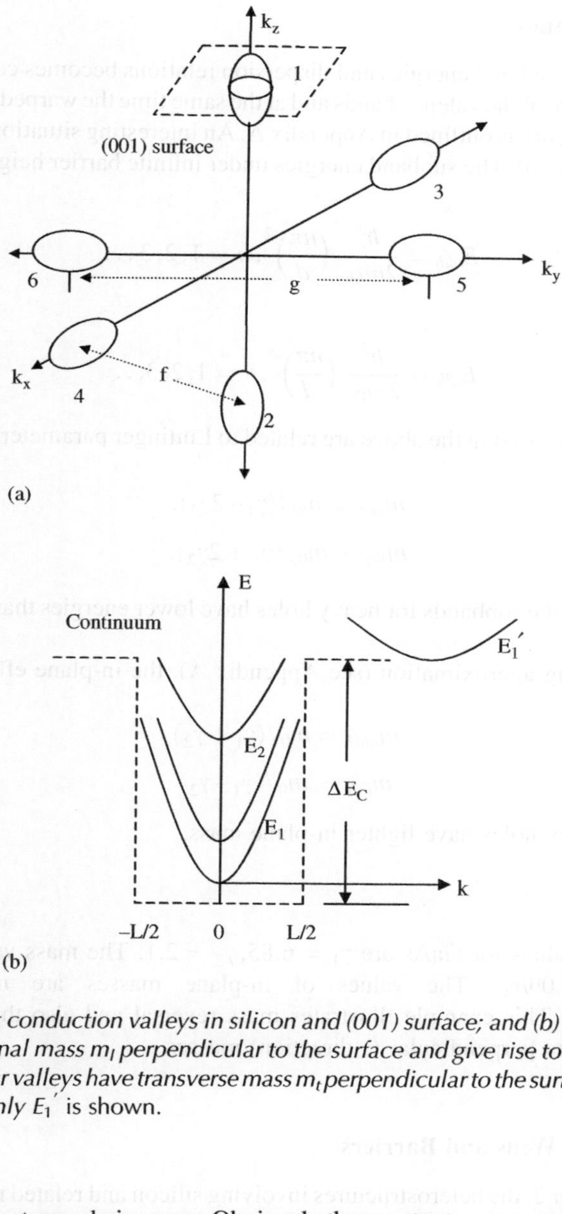

Figure 3.12 *(a) Six conduction valleys in silicon and (001) surface; and (b) the valleys 1 and 2 have their longitudinal mass m_l perpendicular to the surface and give rise to lowest subband E_1 (right). The other four valleys have transverse mass m_t perpendicular to the surface and give rise to primed subband; only E_1'* is shown.

components of the w-tensor being zero. Obviously the constant energy contour in the xy (the layer) plane is a circle corresponding to an isotropic effective mass m_t. For the other four valleys, $w_{33} = m_t^{-1}, w_{11} = m_l^{-1}$, and $w_{22} = m_t^{-1}$ for valleys numbered 3 and 4 and $w_{12} = w_{13} = w_{23} = 0$. The constant energy contours are ellipses. The same argument applies to valleys numbered 5 and 6. For highly symmetric surfaces, therefore, the constant energy contours are obtained by taking a slice of the bulk constant energy ellipsoids along the plane under consideration, as shown in Figure 3.12a.

3.5.2 Hole Subbands

Calculation of hole subband energies and dispersion relations becomes complicated due to the degenerate nature of the valence bands and at the same time the warped nature of the E–$\mathbf{k}$ relationship. The theory is outlined in Appendix A. An interesting situation arises for the in-plane wave vector $\mathbf{k} = 0$. The subband energies under infinite barrier height approximation are given by

$$E_{nhh} = \frac{\hbar^2}{2m_{hhz}}\left(\frac{n\pi}{d}\right)^2, n = 1, 2, 3 \ldots \tag{3.37a}$$

and

$$E_{nlh} = \frac{\hbar^2}{2m_{lhz}}\left(\frac{n\pi}{d}\right)^2, n = 1, 2, 3, \ldots \tag{3.37b}$$

The effective masses used in the above are related to Luttinger parameters by the following expressions:

$$m_{hhz} = m_0/(\gamma_1 - 2\gamma_2), \tag{3.38a}$$

$$m_{lhz} = m_0/(\gamma_1 + 2\gamma_2). \tag{3.38b}$$

Since $m_{hhz} > m_{lhz}$, the subbands for heavy holes have lower energies than the subbands for light holes.

Under decoupling approximation (see Appendix A), the in-plane effective masses are given by

$$m_{hhII} = m_0/(\gamma_1 + \gamma_2) \tag{3.39a}$$

$$m_{lhII} = m_0/(\gamma_1 - \gamma_2) \tag{3.39b}$$

It is seen that heavy holes have lighter in-plane mass.

Example 3.4

The parameter values for GaAs are $\gamma_1 = 6.85, \gamma_2 = 2.1$. The mass values are $m_{hhz} = 0.38m_0, m_{lhz} = 0.09m_0$. The values of in-plane masses are $m_{hhII} = 0.112m_0$, $m_{lhII} = 0.211m_0$. This example illustrates mass reversal and also the changes of in-plane masses from perpendicular (z-direction) masses.

3.5.3 Quantum Wells and Barriers

As noted in Chapter 2, the heterostructures involving silicon and related materials belong to the group of strained heterostructures. In order to develop Si-based heterojunctions, the other semiconductor is invariably the alloy $Si_{1-x}Ge_x$; more recently, however, the ternary alloys $Si_{1-x-y}Ge_xC_y$ and $Si_{1-x-y}Ge_xSn_y$ are also being used for this purpose.

The lattice constant of Si is 5.4311 Å and of Ge is 5.6579 Å. This large lattice mismatch ~4% between the two allows the growth of Ge or the alloy SiGe on Si under restricted conditions. The layers are usually strained, and strain alters the band gap, heterojunctiion band alignment, and band structure in a very complicated way.

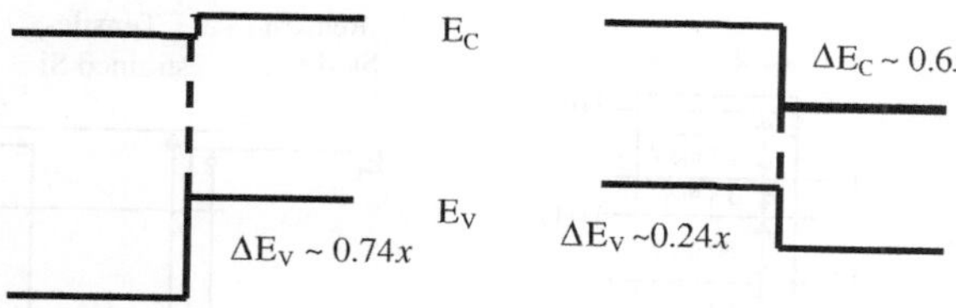

Figure 3.13 *Band alignments in Si–SiGe QWs.*

The simplest pseudomorphic $Si_{1-x}Ge_x$ layer is that grown on a silicon substrate. The alloy layer is then tensile strained, and the bands are aligned in type II configuration. The valence band discontinuity is considerably larger than the conduction band discontinuity, as indicated in Figure 3.13. The QW is produced for holes in the alloy layer, and the small-band discontinuity in the conduction band leads to very poor confinement of the electrons. To achieve enough conduction band discontinuity, a tensile-strained Si or $Si_{1-x}Ge_x$ layer must be grown on a relaxed $Si_{1-y}Ge_y$ substrate with $x < y$. The band alignment is again type II with electrons confined in the Si (or $Si_{1-x}Ge_x$) layer and holes in the $Si_{1-y}Ge_y$ layer.

The presence of both strain and quantum confinement lifts the degeneracy of LH and HH states at $k = 0$.

The confinement energies in the QWs may be calculated by using the simplest infinite barrier approximation. For this purpose, the knowledge of the effective masses perpendicular to QW growth direction is necessary. For Si bands, the situation is complex due to the anisotropic nature of the effective mass tensor. For doped QWs, the subband energies are calculated self-consistently by solving Schrödinger and Poisson equations.

Figure 3.13 gives two examples of QWs with different subband states for both electrons and holes [6]. Figure 3.14a indicates the energy levels for an 8 nm QW formed by a compressive $Si_{0.7}Ge_{0.3}$ layer sandwiched between tensile-strained Si layers. The subband energies and barrier potentials calculated using self-consistent **k.p** theory are shown also.

The situation for a tensile-strained Si layer of 3 nm width sandwiched between relaxed $Si_{0.8}Ge_{0.2}$ layers is shown in Figure 3.14b. In this case, the two valleys having longitudinal mass $m_l = 0.918\, m_0$ perpendicular to the (100) surface give rise to the lowest subband, E_{L1}. The next subband is due to the other four valleys having the transverse mass $m_t = 0.196\, m_0$ perpendicular to the surface and is indicated by E_{T1}. The next subband E_{L2} lies above the QW barrier.

3.6 Effect of Electric Field

The effect of an external electric field in the bulk material is to change the band-to-band and exciton characteristics, which in turn change the absorption coefficient and refractive indices. An external electric field brings about interesting changes in the subband characteristics of quantum structures. These changes give rise to interesting optical phenomena of low-dimensional electron gas, which are exploited to fabricate novel optoelectronic devices. In this section, we shall study the effect of an external field on subband energies and wavefunctions in a QW. The optical processes will be discussed in Chapter 5.

We first assume that the external electric field is applied perpendicular to the QW layer plane and the field is directed along the z-direction. To understand gross features, we rely on

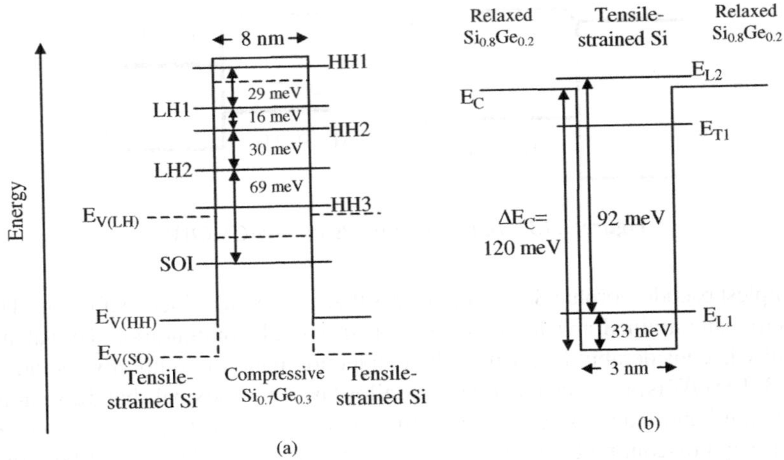

Figure 3.14 *Examples of quantum wells and quantum mechanical barriers for (a) holes and (b) electrons. (a) An 8 nm wide compressively strained $Si_{0.7}Ge_{0.3}$ hole quantum well with tensile-strained Si barriers grown on a relaxed $Si_{0.8}Ge_{0.2}$ virtual substrate. The HH, LH, and SO bands are split due to the strain in both the quantum well and the quantum mechanical barrier with $\Delta E_v(HH) \sim 210$ meV, $\Delta E_v(LH) \sim 101$ meV, and $\Delta E_v(SO) \sim 101$ meV. The first few subband states as calculated by a self-consistent* **k·p** *theory are shown for the HH, LH, and SO bands in the quantum well at $k=0$. (b) A tensile-strained Si layer electron quantum well grown on relaxed $Si_{0.8}Ge_{0.2}$ with the lowest longitudinal (E_L) and transverse (E_T) valley effective mass subband states at $k=0$ Reproduced with permission from [6]. Copyright (2004) Institute of Physics*

analytical expressions obtained for an infinite barrier height [7, 8]. The origin of the coordinate system and of electrostatic potential is taken at the center of a well of width L. The Hamiltonian in the presence of an electric field F is then

$$H = -\frac{\hbar^2}{2m_e}\frac{\partial^2}{\partial z^2} + eFz = H_0 + eFz \tag{3.40}$$

where H_0 is the Hamiltonian without the field and the corresponding eigenvalues are already given in Section 3.2. If the field is weak enough so that the condition $eFL << (\hbar^2/2m_e)(\pi/L)^2$ is valid, then the change in energy may be calculated by using second-order perturbation theory. The change for the nth subband is written as

$$\Delta E_n = \sum_m{}' \frac{|\langle m0|eFz|n0\rangle|^2}{E_{m0}-E_{n0}} = \frac{2e^2F^2m_eL^2}{\hbar^2\pi^2}\sum_m{}' \frac{|\langle m0|z|n0\rangle|^2}{n^2-m^2} \tag{3.41}$$

where

$$E_{n0} = \frac{\hbar^2}{2m_e}\left(\frac{n\pi}{L}\right)^2$$

is the unperturbed eigenvalue (see Eq. (3.10)), and the subscript 0 refers to the zeroth-order eigenfunctions and eigenvalues. Since $|n0\rangle$ and $|m0\rangle$ are sine functions, the matrix element in Eq. (3.30) may easily be calculated to give

$$\langle m0|z|n0\rangle = \frac{8L}{\pi^2}\frac{nm}{n^2-m^2} \tag{3.42}$$

Putting $n=1$ for the lowest subband and summing over all permissible values of m starting from 2, the change in energy may be expressed as

$$\Delta E_1 = -\frac{1}{24\pi^2}\left(\frac{15}{\pi^2}-1\right)\frac{e^2F^2m_eL^4}{\hbar^2} \tag{3.43}$$

It appears therefore that the energy of the lowest subband decreases quadratically with the applied electric field. Since for holes the energy is opposite, the principal effect of the field is to decrease the effective band gap.

Example 3.5

Let a field strength of 10^5 V/cm be applied to a GaAs QW of width $L=10$ nm. The change in subband energy is 1.93 meV, taking $m_e=0.067\,m_0$.

Problems

3.1 Obtain the expression for the mean free path of a 2DEG at 0 K in terms of the sheet carrier density and the mobility, and state the condition for quantization. Calculate the value of the mean free path if $n_s=10^{11}$ cm^{-2} and $\mu=10^6$ V^{-1}s^{-1}.

3.2 Show that the wavefunction in the barrier of a rectangular QW of width L may be expressed as

$$\phi(z) = \left[\frac{1}{\alpha}+\frac{L/2}{\cos^2(kL/2)}+\frac{m_w\alpha}{m_bk^2}\right]^{1/2}\exp\left[\alpha\left(\frac{L}{2}\pm z\right)\right],$$

where $k^2 = 2m_wE/\hbar^2$ and $\alpha^2 = 2m_b(\Delta E-E_1)/\hbar^2$, in which w and b refer to well and barrier, respectively.

3.3 Calculate the values of the Fermi level and the areal electron densities in different subbands in a GaAs QW at 0 K and 300 K, where the sheet carrier density is 10^{12} cm^{-2}.

3.4 Prove that the DOS at the subband energies E_n equals that in the bulk, that is, $\rho_{3D}(E_n)L = \rho_{2D}(L)$.

3.5 Using the expressions for the subband energies for the infinite barriers, prove that $\rho_{2D}\rightarrow\rho_{3D}$, as the well width increases to a very large value.

3.6 The screening constant in a 2D system is expressed as $q_s = (e^2/2\varepsilon)(dN_s/dE_F)$. Obtain an expression for the screening constant at very low temperature.

3.7 The lowest subband energy for a QW of width L is approximately given by

$$E_1 = \frac{\Delta E_c}{2} - \frac{\pi^2\hbar^2}{4m_eL^2}\left[\left(1+\frac{32\Delta E_c^2L^4m_e^2}{\pi^6\hbar^4}\right)^{1/2}-1\right]$$

Calculate E_1 versus the well width in a GaAs$-$Ga$_{0.7}$Al$_{0.3}$As QW, and compare the results with exact values. Assume $\Delta E_c/\Delta E_v = 60/40$.

3.8 Give plots showing the nature of variation of the two sides in Eq. (3.24) and explain how you would obtain the values of subband energies from the graphs.

3.9 Assume that the wavefunction in a finite QW penetrates a distance $\delta = (\hbar^2/2m_B\Delta E_c)^{1/2}$, where m_B is the mass in the barrier. Take $\Delta E_c = 230$ meV and $m_B = 0.093m_0$ for $Ga_{0.7}Al_{0.3}As$. Calculate the first subband energy, assuming the effective width of the QW to be $(d+2\delta)$, and compare the values with those obtained exactly and under infinite barrier approximations.

3.10 The band offset values in a $Si_{1-x-y}Ge_xC_y$/Si QW are $\Delta E_c = -0.09x - 0.5y$ and $\Delta E_v = 0.82x - 2.6y$. Consider a well of $Si_{0.72}Ge_{0.25}C_{0.03}$ having width 10 nm. Calculate the values of lowest conduction and heavy-hole subband energies assuming mass values for unstrained $Si_{0.75}Ge_{0.25}$ alloy.

3.11 Find the principal effective masses in QWs made of Si and Ge when the layer planes are along the (100), (110), and (111) directions.

3.12 Calculate the subband energies for a (100)-oriented Si metal oxide semiconductor field effect transistor (MOSFET), when the sheet carrier density $N_s = 10^{12}$ cm^{-2}. Neglect depletion layer charges, and assume a very low temperature.

3.13 Prove that for a Si QW of infinite barrier height, the subband energies satisfy the condition $E_1 < E_2 < E_1'$ for QW growth along the (100) direction.

3.14 Assume that the potential $V(z) = -eFz$ in Eq. (3.4). Show that the solution for the effective mass Schrödinger equation is the Airy function.

3.15 Use the values of band edges given in Figure 2.15 for Sn concentration = 0.25. Calculate the effective direct band gap when the Ge QW thickness = 10 nm.

3.16 The DOS in 0D is $\rho_{0D} = \sum_{m,n,l} \delta(E - E_{mnl})$. Using m, n, l as continuous variables, and the expressions for subband energies for infinite barriers, show that the expressions for the DOS functions for 1D, 2D, and 3D can successively be obtained.

3.17 Show that the density of electrons in a 1D wire under nondegenerate conditions may be expressed as $n_{1D} = (2m_e k_B T/\pi\hbar^2)^{1/2} \exp[(E_F - E_i)/k_B T]$.

3.18 Show that at low temperature when the Fermi function is box-like, the 1D carrier density is proportional to $(E_F - E_i)^{1/2}$. Obtain the complete expression for n_{1D}.

3.19 From the general definition of DOS function given by Davies [1998], obtain the expressions for DOS in a bulk material.

3.20 Solve the effective mass Schrödinger equation for a quantum box by using separation of variables. Hence, show that the energy states are given by Eq. (3.30). The barrier potentials are infinite.

3.21 The Luttinger parameters for Si are $\gamma_1 = 4.22, \gamma_2 = 0.39$ and the values for Ge are $\gamma_1 = 13.4, \gamma_2 = 4.25$. Calculate the lowest subband energies for HH and LH in the $Si_{0.6}Ge_{0.4}$ layer, assuming that the barrier height is infinite.

3.22 Consider the problem of confinement in a spherical dot of radius R. The potential is a square well of height ΔE_c for $r > R$ and 0 for $r < R$. Consider the radial part of wavefunction and zero angular momentum (see, e.g., Schiff 1968; Basu 2003, 354). Show that there exists a critical radius $R_c^2 > \dfrac{\pi^2\hbar^2}{8m_e\Delta E_c}$ for which an electron possesses a bound state.

References

1. Schiff, L. (1968) *Quantum Mechanics*, McGraw-Hill, New York.
2. Stern, F. and Howard, W.E. (1967) Properties of semiconductor surface inversion layers in the electric quantum limit. *Phys. Rev.*, **163**, 816–835.
3. Stern, F. (1972) Self-consistent results for n-type Si inversion layers. *Phys. Rev. B*, **5**, 4891–4899.
4. Ando, T., Fowler, A.B., and Stern, F. (1982) Electronic properties of two-dimensional systems. *Rev. Mod. Phys.*, **54**, 437–672.
5. Basu, P.K. and Paul, S.K. (1992) Ideal interband absorption spectra in an indirect-gap semiconductor quantum well. *Phys. Rev. B.*, **46**, 13389–13393.
6. Paul, D.J. (2004) Si/SiGe heterostructures: from material and physics to devices and circuits. *Semicond. Sci. Technol.*, **19**, R75–R78.
7. Bastard, G., Mendez, E.E., Chang, L.L., and Esaki, L. (1983) Variational calculation on a quantum well in an electric field. *Phys. Rev. B*, **28**, 3241–3245.
8. Ahn, D. and Chuang, S.L. (1987) Calculation of linear and nonlinear intersubband optical absorptions in a quantum-well model with an applied electric field. *IEEE J. Quantum Electron.*, **QE-23**, 2196–2204. 1987.

Suggested Readings

Balkanski, M. and Wallis, R.F. (2000) *Semiconductor Physics and Applications*, Oxford University Press, Oxford, UK.

Bastard, G. (1988) *Wave Mechanics Applied to Semiconductor Heterostructures*, Les Editions de Physique, Les Ulis, France.

Basu, P.K. (2003) *Theory of Optical Processes in Semiconductors: Bulk and Microstructures*, Clarendon Press, Oxford, UK.

Chuang, S.L. (1995) *Physics of Optoelectronic Devices*, John Wiley & Sons, Inc., New York.

Datta, S. (1989) *Quantum Phenomena*, Addison-Wesley, Reading, MA.

Davies, J.H. (1998) *The Physics of Low-Dimensional Semiconductors: An Introduction*, Cambridge University Press, Cambridge, UK.

Harrison, P. (2000) *Quantum Wells, Wires and Dots: Theoretical and Computational Physics*, John Wiley & Sons, Ltd, Chichester, UK.

Manasreh, O. (2005) *Semiconductor Heterojunctions and Nanostructures*, McGraw-Hill, New York.

Mitin, V., Kochelap, V.A., and Stroscio, M.A. (1999) *Quantum Heterostructures*, John Wiley & Sons, Inc., New York.

Ridley, B.K. (2000) *Quantum Processes in Semiconductors*, 5th edn, Clarendon Press, Oxford, UK.

Roblin, P. and Rohdin, H. (2002) *High-Speed Heterostructure Devices*, Cambridge University Press, Cambridge, UK.

Singh, J. (2003) *Electronic and Optoelectronic Properties of Semiconductor Structures*, Cambridge University Press, Cambridge, New York.

Weisbuch, C. and Vinter, B. (1991) *Quantum Semiconductor Structures*, Academic, San Diego, CA.

4

Optical Processes

4.1 Introduction

This chapter outlines the theory of optical processes in bulk semiconductors. First of all, the relationships between various optical constants, like absorption coefficient, refractive index (RI), and so on, are developed. The three basic processes in light matter interaction, that is, the absorption, spontaneous emission, and stimulated emission, are then introduced. This is followed by a brief description of absorption processes in semiconductors.

The theory of optical processes in semiconductors is developed mostly for direct-gap materials. Though the theory for indirect-gap semiconductors is available, its applications are somewhat limited. Recent developments indicate the possibility of having direct-gap type I band alignment in group IV materials. This necessitates the inclusion of the theory of optical transitions in both direct- and indirect-gap semiconductors.

A semiclassical theory of fundamental band-to-band absorption in semiconductors is developed for both direct and indirect-gap semiconductors. The condition for obtaining gain in semiconductors is then discussed. Other important absorption processes, like intervalence band and free carrier absorption, are then treated. The opposite process of absorption, that is, recombination and luminescence, is then introduced. Emphasis is given to main recombination processes in indirect-gap materials, that is, the trap-assisted and Auger recombination. Finally, the theory of excitonic absorption in direct-gap materials is developed, followed by a brief description of impurity absorption.

4.2 Optical Constants

In this section, the relationship between different optical constants in a conductor or dielectric and their interrelationship will be presented. The starting equations are the

Silicon Photonics: Fundamentals and Devices, First Edition. M. Jamal Deen and P. K. Basu.

following four Maxwell's equations related to the behavior of electromagnetic waves.

$$\nabla \times \mathbf{F} + \frac{\partial \mathbf{B}}{\partial t} = 0 \tag{4.1a}$$

$$\nabla \times \mathbf{H} - \frac{\partial \mathbf{D}}{\partial t} = \mathbf{J} \tag{4.1b}$$

$$\nabla \cdot \mathbf{D} = \rho \tag{4.1c}$$

$$\nabla \cdot \mathbf{B} = 0 \tag{4.1d}$$

where **F** and **H** are the electric and magnetic fields, $\mathbf{D} = \varepsilon\mathbf{F}, \mathbf{B} = \mu\mathbf{H}$, and $\boldsymbol{J}$ and ρ are the current and charge densities, respectively. The electric and magnetic fields are expressed in terms of vector potential **A** and scalar potential ϕ as follows:

$$\mathbf{F} = -\frac{\partial \mathbf{A}}{\partial t} - \nabla\phi \tag{4.2}$$

$$\mathbf{B} = \nabla \times \mathbf{A} \tag{4.3}$$

Taking curl of the second Maxwell equation and using the third one, we may write

$$\frac{1}{\mu_0}\nabla \times \nabla \times \mathbf{A} + \varepsilon\frac{\partial^2 \mathbf{A}}{\partial t^2} + \varepsilon\nabla\frac{\partial \phi}{\partial t} = \mathbf{J} \tag{4.4}$$

We note that $\nabla \times \nabla \times \mathbf{A} = \nabla(\nabla \cdot \mathbf{A}) - \nabla^2\mathbf{A}$. It is usual to assume the Coulomb gage in which $\phi = 0$ and $\nabla \cdot \mathbf{A} = 0$. Therefore,

$$\frac{1}{\mu_0}\nabla^2\mathbf{A} - \varepsilon\frac{\partial^2 \mathbf{A}}{\partial t^2} = \mathbf{J}, \tag{4.5}$$

If $\mathbf{J} = 0$, the solutions for **A** are plane waves expressed as

$$\mathbf{A}(\mathbf{r}, t) = \mathbf{A}_0\{\exp[j(\mathbf{q}.\mathbf{r} - \omega t)] + c.c.\} \tag{4.6}$$

where $c.c.$ stands for *complex conjugate*, $q^2 = \varepsilon\mu_0\omega^2$, and the velocity of light in free space is $c = (\varepsilon_0\mu_0)^{1/2}$. The electric and magnetic fields are given by

$$\mathbf{F} = \frac{\partial \mathbf{A}}{\partial t} = -2\omega\mathbf{A}_0\sin(\mathbf{q}.\mathbf{r} - \omega t) \tag{4.7}$$

$$\mathbf{B} = \nabla \times \mathbf{A} = -2\mathbf{q} \times \mathbf{A}_0\sin(\mathbf{q}.\mathbf{r} - \omega t) \tag{4.8}$$

The Poynting vector **S** representing the optical power is expressed as

$$\mathbf{S} = (\mathbf{F} \times \mathbf{H}) = \frac{4}{\mu_0}vq^2|\mathbf{A}_0|^2\sin^2(\mathbf{q} \cdot \mathbf{r} - \omega t)\hat{\mathbf{q}} \tag{4.9}$$

where $\hat{\mathbf{q}}$ is a unit vector in the direction of **q** and $v = \omega|\mathbf{q}|$ is the velocity of light in the medium having permittivity $\varepsilon = \varepsilon_0\varepsilon_r$, with ε_r being the relative permittivity.

Taking the time-averaged value of **S**, we obtain the expression for energy density as

$$I = \left|\frac{\mathbf{S}}{v}\right| = \frac{2\varepsilon\omega^2|\mathbf{A}_0|^2}{c^2} \tag{4.10}$$

We now consider a conducting material in which the current density $\mathbf{J} = \sigma\mathbf{F}$, where σ is the conductivity. The wave equation for the electric field is obtained as

$$\nabla^2\mathbf{F} = \varepsilon\mu_0\frac{\partial^2\mathbf{F}}{\partial t^2} + \sigma\mu_0\frac{\partial\mathbf{F}}{\partial t} \tag{4.11}$$

The general solution for the field is

$$\mathbf{F} = \mathbf{F}_0\exp\{j(\mathbf{q}\cdot\mathbf{r} - \omega t)\} \tag{4.12}$$

Putting this form in Eq. (4.11) we obtain

$$-q^2 = -\varepsilon\mu_0\omega^2 - j\sigma\mu_0\omega \tag{4.13}$$

Since q is a complex number, we may introduce a complex RI n to account for the changed-phase velocity from the free-space velocity. Thus

$$n = n_r + jn_i = \left(\varepsilon_r + \frac{j\sigma\mu_0}{\omega}\right)^2 \tag{4.14}$$

so that

$$q = \frac{n_r\omega}{c} + jn_i\frac{\omega}{c} \tag{4.15}$$

Here the subscripts r and i of n represent the real and imaginary parts, respectively. The electric field may now be expressed as

$$\mathbf{F} = \mathbf{F}_0\exp\left\{j\omega\left(\frac{n_r z}{c} - t\right)\right\}\exp\left(\frac{-n_i\omega z}{c}\right) \tag{4.16}$$

The field amplitude is damped exponentially as it propagates along the z-direction. The intensity of light decreases exponentially as

$$I(z) = I_0\exp(-\alpha z) \tag{4.17}$$

where α is the absorption coefficient. Since the intensity is proportional to F^2, the absorption coefficient is given by

$$\alpha = \frac{2n_i\omega}{c} \tag{4.18}$$

The relation between wave vector q and angular frequency ω can also be expressed as

$$c^2q^2 = \omega^2\varepsilon_r(\omega). \tag{4.19}$$

The quantity $\varepsilon_r(\omega)$ is called the *dielectric function*. Using Eq. (4.19), we may rewrite Eq. (4.12) in the following form:

$$\mathbf{F}(\mathbf{r},t) = \mathbf{F}_0 \exp\left[j\omega\left(\frac{n(\omega)}{c}\mathbf{q}\cdot\mathbf{r}-t\right)\right]\hat{\mathbf{q}}, \tag{4.20}$$

where the RI $n(\omega)$ is given by

$$n(\omega) = [\varepsilon_r(\omega)]^{1/2} \tag{4.21}$$

When a conductor is considered, the complex conductivity can be related to the permittivity by the following expressions:

$$\bar{\sigma} = \sigma - j\omega\varepsilon_r\varepsilon_0 \quad \text{and} \quad \bar{\varepsilon} = \varepsilon_r + (j/\varepsilon_0\omega)\sigma = (j/\omega\varepsilon_0)\bar{\sigma}. \tag{4.22}$$

The dielectric function $\varepsilon(\omega)$ and the refractive index $n(\omega)$ are complex quantities expressed as follows: $\bar{\varepsilon} = \varepsilon(\omega) = \varepsilon'_r(\omega) + j\varepsilon''_r(\omega)$ and $N(\omega) = n(\omega) + jK(\omega)$.

The real part of RI, $n(\omega)$, and the imaginary part of the RI, $K(\omega)$, called the *extinction coefficient*, are related to the real and imaginary parts of the dielectric function by

$$\varepsilon'_r(\omega) = n^2 - K^2, \tag{4.23a}$$

$$\varepsilon''_r(\omega) = 2nK. \tag{4.23b}$$

If electromagnetic (EM) radiation of intensity I_0 is incident on the surface of a semiconductor, the intensity is reduced along the depth due to absorption in the material according to the relation

$$I(z) = I_0 \exp(-\alpha z), \tag{4.24}$$

where α is called the *absorption coefficient* of the material, which may be related to the above quantities as

$$\alpha(\omega) = \frac{2\omega K(\omega)}{c} = \frac{\omega\varepsilon''_r(\omega)}{cn(\omega)}. \tag{4.25}$$

To complete this section, one should note that the real and imaginary parts of the dielectric function are not independent, but are related to each other by the Kramers–Kronig relation:

$$\begin{aligned} \varepsilon'(\omega) &= \frac{1}{\pi} P \int_{-\infty}^{\infty} \frac{\varepsilon''(\omega')d\omega'}{\omega'-\omega}, \\ \varepsilon''(\omega) &= -\frac{1}{\pi} P \int_{-\infty}^{\infty} \frac{\varepsilon'(\omega')d\omega'}{\omega'-\omega}. \end{aligned} \tag{4.26}$$

where P stands for the Cauchy principal value of the integral that follows.

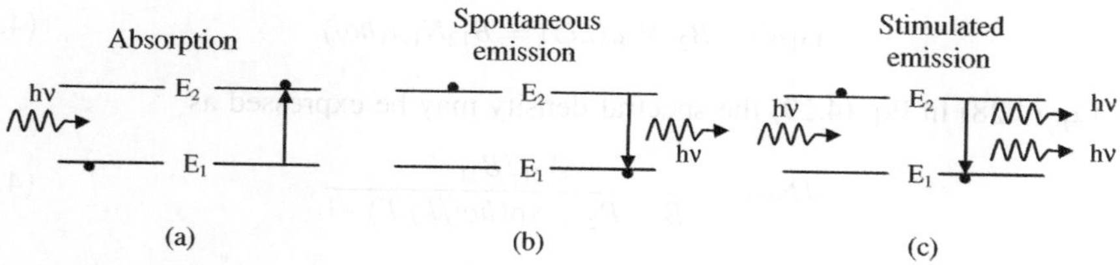

Figure 4.1 *Three basic processes of interaction of radiation with a two-level atomic system: (a) absorption, (b) spontaneous emission, and (c) stimulated emission.*

4.3 Basic Concepts

4.3.1 Absorption and Emission

There are three basic processes through which EM radiation can interact with matter. The processes are illustrated by Figure 4.1 using a hypothetical atom having only two energy levels, E_1 (lower level) and E_2 (upper level). Under normal conditions, all materials absorb light. The absorption process is illustrated by the left part of Figure 4.1, in which a photon of energy $\hbar\omega = E_2 - E_1$ is absorbed, transferring an electron from the lower level to the upper level. The intensity of the incident light gets attenuated due to the absorption. The excited atoms eventually return to their normal ground state, and the excess energy $\hbar\omega$ is emitted in the form of EM radiation. The radiation is directed along a random direction and has no phase relationship among them. The third process is known as *stimulated emission*, which differs from spontaneous emission in that the process is initiated by an incident photon. This process is illustrated by Figure 4.1c. In stimulated emission the frequency, phase, and direction of propagation of the emitted and incident radiation are all identical. All lasers, including semiconductor lasers, emit light through the process of stimulated emission, while the light-emitting diodes (LEDs) emit light by the spontaneous emission process. Stimulated emission gives rise to coherent light, while light emitted by spontaneous emission is incoherent in nature.

4.3.2 Absorption and Emission Rates

We shall first define the absorption and emission rates for the two-level atomic system discussed in Section 4.3.1. The concept will be extended to semiconductors in this chapter.

Let N_1 and N_2 be the atomic densities in the ground and excited states shown in Figure 4.1, and let $\rho(\hbar\omega)$ be the spectral density of radiation energy. The rates of spontaneous emission, stimulated emission, and absorption can be written as

$$R_{\text{spon}} = AN_2, A_{21} = B_{21}N_2\rho(\hbar\omega), \quad \text{and } R_{\text{abs}} = B_{12}N_1\rho(\hbar\omega), \tag{4.27}$$

where A_{21}, B_{12}, and B_{21} are constants known as the *Einstein coefficients*. In thermal equilibrium, atomic densities obey Boltzmann statistics, and accordingly

$$N_2/N_1 = \exp[-(E_2 - E_1)/k_BT] = \exp(-\hbar\omega/k_BT), \tag{4.28}$$

where k_B is the Boltzmann constant and T is the absolute temperature. Under steady state, the upward and downward transition rates should be equal leading to

$$A_{21}N_2 + B_{21}N_2\rho(\hbar\omega) = B_{12}N_1\rho(\hbar\omega). \tag{4.29}$$

Using Eq. (4.28) in Eq. (4.29) the spectral density may be expressed as

$$\rho(\hbar\omega) = \frac{A_{21}/B_{21}}{(B_{12}/B_{21})\exp(\hbar\omega/k_BT)-1}. \tag{4.30}$$

In thermal equilibrium, the radiation spectral density must be identical with that for blackbody radiation given by Planck formula

$$\rho_{BB}(\hbar\omega) = \frac{2\hbar\omega^3}{\pi c^3}\frac{1}{\exp(\hbar\omega/k_BT)-1}. \tag{4.31}$$

Comparing Eqs. (4.30) and (4.31) gives the relations

$$A_{21} = (2\hbar\omega^3/\pi c^3)B_{21} \quad \text{and} \quad B_{12} = B_{21}. \tag{4.32}$$

The ratio of stimulated to spontaneous emission rates is therefore

$$R_{stim}/R_{spon} = [\exp(\hbar\omega/k_BT)-1]^{-1}. \tag{4.33}$$

It appears from Eq. (4.33) that in the visible and infrared region ($\hbar\omega \sim 1\,\text{eV}$), spontaneous emission always dominates over stimulated emission around room temperature ($k_BT \approx 25$ meV). In order that stimulated emission should exceed spontaneous emission, the system must operate away from thermal equilibrium. This is achieved by using an external pump source.

It also appears from Eqs. (4.28) and (4.33) that $R_{\text{stim}} > R_{\text{abs}}$ only when $N_2 > N_1$. This condition is referred to as *population inversion*, which may be realized by external pumping. Population inversion is a necessary condition for laser operation. The reader is referred to standard textbooks for the proof of this statement [1, 2].

4.4 Absorption Processes in Semiconductors

The absorption process in semiconductors can best be understood with the help of quantum mechanical theory. The radiation then is assumed to be a bunch of photons. When incident on a semiconductor, this bunch induces transitions among different energy levels in different bands in the material. A few representative transitions occurring in a semiconductor are illustrated in Figure 4.2 by using the E–$\mathbf{k}$ diagram.

The process indicated by (A) in Figure 4.2 represents direct valence to conduction band transitions (constant $\boldsymbol{k}$ vector) that occurs in a direct band gap semiconductor, like GaAs. Process (B) illustrates indirect valence to conduction band transitions aided by photon–phonon coupling interactions. Process (C) represents intervalence band transitions, for example, transition between heavy-hole (HH) and split-off (SO) bands. Process (D) depicts valence band free-carrier transitions aided by impurities or photon–phonon interactions. Process (E) is for free-carrier transitions in conduction band aided by impurities or photon–phonon interactions. Depending on the photon energy, a single type or a combination of a few types of all the absorption processes indicated in the figure may be important. We shall discuss the origin and characteristics of these processes one by one.

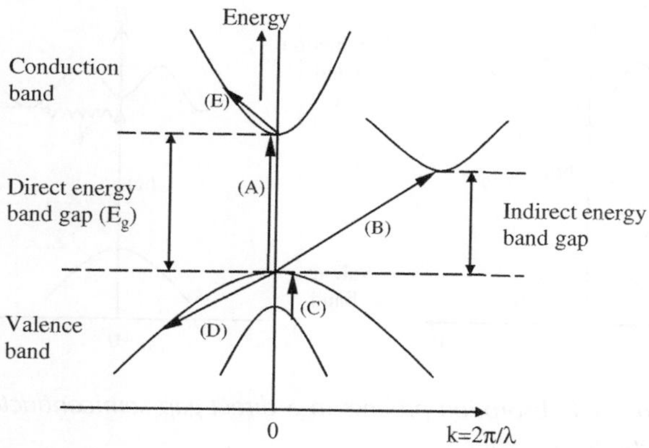

Figure 4.2 *Illustration of the different absorption processes in semiconductors.*

4.5 Fundamental Absorption in Direct Gap

Although Si is an indirect-gap material, the purposes of introducing the section related to direct gap are quite a few. First of all, the theory is used in this chapter to develop the theory for indirect-gap semiconductors. Second, the present theory is modified in Chapter 4 for quantum structures using direct-gap material. Third, indirect- to direct-gap crossover has indeed been observed in quantum structures involving group IV materials. The present theory will be useful in such systems.

We present in this section the basic conservation laws in direct-gap materials and then proceed to obtain the expression for the absorption coefficient for direct-gap semiconductors.

4.5.1 Conservation Laws

When photons of energy greater than the fundamental band gap are incident on a semiconductor sample, electrons from the valence band absorb the photons and move to the conduction band. As a consequence, an empty state or a hole is created in the valence band. The absorption of photons thus leads to the generation of excess electron–hole pairs (EHPs). The absorption coefficient rises rapidly as the photon energy $\hbar\omega \geq E_g$, the band gap energy. This process is shown schematically in Figure 4.3.

Two conditions must be fulfilled for this interband transition. The first one is the energy conservation, that is, the energy of the electron should equal the sum of energies of the hole and photon. This leads to the expression

$$E_c(\mathbf{k}_c) = E_v(\mathbf{k}_v) + \hbar\omega. \tag{4.34}$$

In Eq. (4.34), **k**'s denote the wave vectors for the respective particles.

The absorption threshold occurs when the photon energy equals the band gap energy. We may define a threshold or cutoff wavelength from Eq. (4.34) as follows:

$$\lambda_c = \frac{hc}{E_g} = \frac{1.24}{E_g(\text{eV})}(\mu\text{m}). \tag{4.35}$$

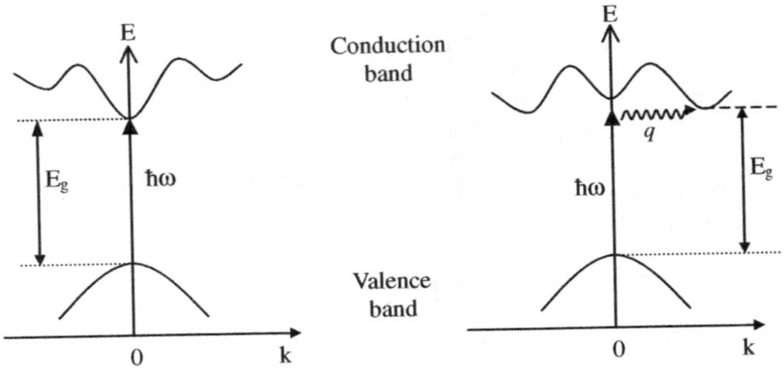

Figure 4.3 *Fundamental absorption process in a direct-gap semiconductor (left) and in an indirect-gap semiconductor (right).*

The last ratio indicates that the value of the cutoff wavelength is in μm, when the energy is in eV.

The second condition to be satisfied is the momentum conservation, stated as follows:

$$\mathbf{k}_c = \mathbf{k}_v + \mathbf{q}. \tag{4.36}$$

However, in most of situations the photon wave vector $\mathbf{q} \ll \mathbf{k}_c$ and $\mathbf{k}_v$. Therefore, we obtain

$$k_c = k_v. \tag{4.37}$$

This **k**–conservation is indicated by a vertical line in the E–**k** diagram in Figures 4.2 and 4.3.

Example 4.3

Let the photon wavelength be 1 μm. The wave vector $q = 2\pi/\lambda = 2\pi \times 10^6\ \mathrm{m}^{-1}$. Consider the electron energy to be 25 meV. The wave vector $k_e = (2m_e E/\hbar^2)^{1/2} = 2.56 \times 10^8\ \mathrm{m}^{-1}$, for $m_e = 0.1\ m_0$. Thus $q \ll k_e$.

4.5.2 Calculation of Absorption Coefficient

The calculation of the absorption coefficient is performed by calculating quantum mechanically the probability of transition from the valence band state to the conduction band state. Here we shall use the semiclassical approach by treating electromagnetic waves classically, but describing the electrons as quantum mechanical (Bloch) waves [3–6]. An alternative way is to consider the radiation as a collection of particles (photons) and to use the photon creation, annihilation, and number operators to describe the absorption, stimulated, and spontaneous emissions [7, 8].

The unperturbed one-electron Hamiltonian in the semiclassical treatment is given by

$$H_0 = p^2/2m_0 + V(\mathbf{r}) \tag{4.38}$$

The electric field and magnetic fields of the EM wave are expressed as

$$\mathbf{E} = \frac{\partial \mathbf{A}}{\partial t} \quad \text{and} \quad \mathbf{B} = \nabla \times \mathbf{A} \tag{4.39}$$

$$H = \frac{1}{2m_0}(\mathbf{p} + e\mathbf{A})^2 + V(r) \tag{4.40}$$

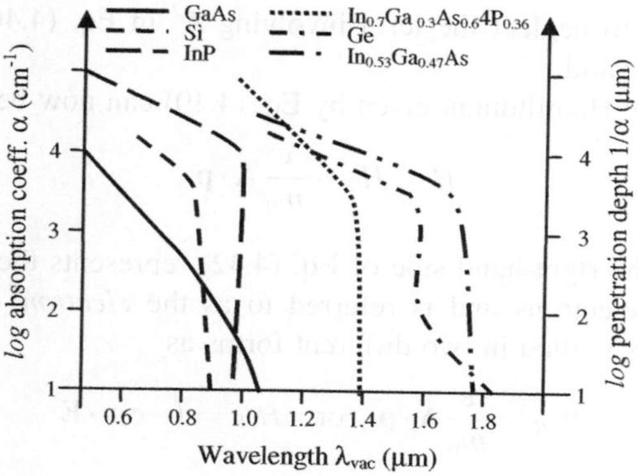

Figure 4.4 Absorption coefficients of a few common semiconductors as a function of wavelength. The penetration depths in μms are shown on the right.

Writing the term $(\mathbf{p} + e\mathbf{A})^2 = (e\mathbf{A}\cdot\mathbf{p}) + (e\mathbf{p}\cdot\mathbf{A})$ and noting that $\mathbf{p} = (\hbar/j)\nabla$, we may write

$$(\mathbf{p}\cdot\mathbf{A})f(r) = \mathbf{A}\cdot\left(\frac{\hbar}{j}\nabla f\right) + \left(\frac{\hbar}{j}\nabla\cdot\mathbf{A}\right)f \tag{4.41}$$

Example 4.4

The ratio of the terms involving A^2 and **A.p** is $\sim 10^{-4}$, as may be seen from Problem 4.8.

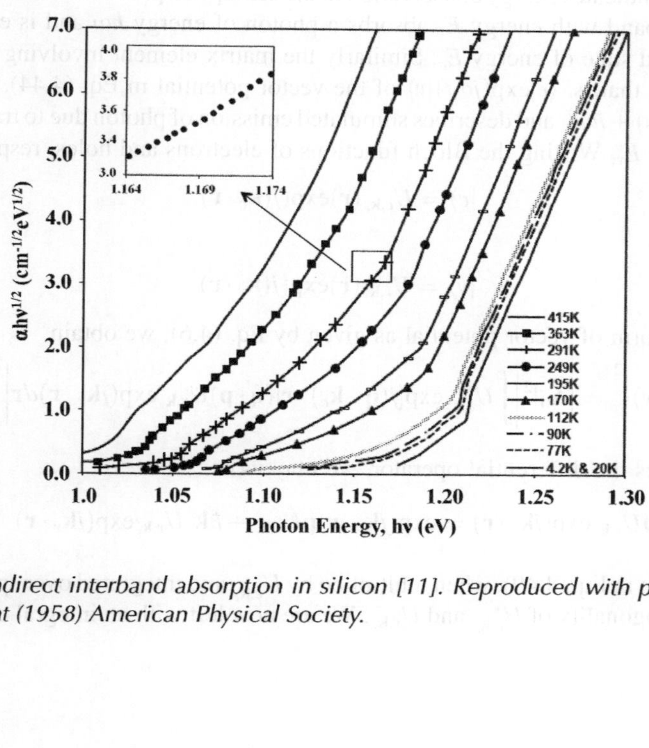

Figure 4.5 Indirect interband absorption in silicon [11]. Reproduced with permission from [11]. Copyright (1958) American Physical Society.

It is safe therefore to neglect the term involving A^2 in Eq. (4.40) so that first-order perturbation theory is valid.

Since $\nabla \cdot \mathbf{A} = 0$, the Hamiltonian given by Eq. (4.40) can now be approximated as

$$H = H_0 + \frac{e}{m_0}\mathbf{A} \cdot \mathbf{p} \tag{4.42}$$

The second term in the right-hand side of Eq. (4.42) represents the interaction between radiation and Bloch electrons and is referred to as the *electron–radiation interaction Hamiltonian* H_{eR}. It is written in two different forms as

$$H_{eR} = \frac{e}{m_0}\mathbf{A} \cdot \mathbf{p} \quad \text{or} \quad H_{eR} = (-e)\mathbf{r} \cdot \mathbf{E} \tag{4.43}$$

The equivalence of the two can be established when the wave vector of the EM field is small, the so-called *electric dipole approximation*. The first form of H_{eR} is more frequently used.

Our next task is to evaluate the rate of transition which involves the following squared matrix element for transition from a valence band state $|v\rangle$ with energy E_v and wave vector $\mathbf{k}_v$ to a conduction band state $|c\rangle$ with energy E_c and wave vector $\mathbf{k}_c$

$$|\langle c|H_{eR}|v\rangle|^2 = (e/m_0)^2|\langle c|\mathbf{A} \cdot \mathbf{p}|v\rangle|^2 \tag{4.44}$$

The matrix element involves integration in space. The time integration involving $\exp(j\omega t)$ and corresponding time dependencies in Bloch functions may be written as

$$\int \exp(jE_c t/\hbar)\exp(-j\omega t)\exp(-jE_v t/\hbar) \propto \delta[E_c(\mathbf{k}_c) - E_v(\mathbf{k}_v) - \hbar\omega] \tag{4.45}$$

The δ-function indicates energy conservation in the absorption process, in which an electron in the valence band with energy E_v absorbs a photon of energy $\hbar\omega$ and is excited to the conduction band state of energy E_c. Similarly the matrix element involving the complex conjugate term, that is, $\langle c|\exp(j\omega t)|v\rangle$ of the vector potential in Eq. (4.44), gives rise to $\delta[E_c(\mathbf{k}_c) - E_v(\mathbf{k}_v) + \hbar\omega]$, and describes stimulated emission of photon due to transition from state E_c to state E_v. Writing the Bloch functions of electrons and holes, respectively, as

$$|c\rangle = U_{c,\mathbf{k}_c}(\mathbf{r})\exp(j(\mathbf{k}_c \cdot \mathbf{r}) \tag{4.46}$$

and

$$|v\rangle = U_{v,\mathbf{k}_v}(\mathbf{r})\exp(j(\mathbf{k}_v \cdot \mathbf{r}) \tag{4.47}$$

and using the form of vector potential as given by Eq. (4.6), we obtain

$$\left|\langle c|\mathbf{A} \cdot \mathbf{p}|\mathbf{v}\rangle\right|^2 = |\mathbf{A}_0|^2 \left|\int U^*_{c,\mathbf{k}_c}\exp[j(\mathbf{q} - \mathbf{k}_c) \cdot \mathbf{r}](\hat{\mathbf{q}} \cdot \mathbf{p})U_{v,\mathbf{k}_v}\exp(j\mathbf{k}_v \cdot \mathbf{r})d\mathbf{r}\right|^2 \tag{4.48}$$

Since $\mathbf{p}$ involves the differential operator, we may write

$$\mathbf{p}U_{v,\mathbf{k}_v}\exp(j\mathbf{k}_v \cdot \mathbf{r}) = \exp(j\mathbf{k}_v \cdot \mathbf{r})\mathbf{p}U_{v,\mathbf{k}_v} + \hbar\mathbf{k}_v U_{v,\mathbf{k}_v}\exp(j\mathbf{k}_v \cdot \mathbf{r}) \tag{4.49}$$

The second term in Eq. (4.49), when multiplied by $U^*_{c,\mathbf{k}_c}$ and integrated over space, vanishes due to the orthogonality of $U^*_{c,\mathbf{k}_c}$ and $U_{v,\mathbf{k}_v}$. We are now left with an integral containing the first term, which can be broken into two parts as follows in view of the periodicity of functions U_c and U_v,

$$\int U^*_{c,\mathbf{k}_c} \exp[j(\mathbf{q}-\mathbf{k}_c+\mathbf{k}_v)\cdot\mathbf{r}]\mathbf{p}U_{v,\mathbf{k}_v} d\mathbf{r} = \left(\sum_l \exp[j(\mathbf{q}-\mathbf{k}_c+\mathbf{k}_v)\cdot\mathbf{R}_l]\right) \int_{unit\ cell} U^*_{c,\mathbf{k}_c} \exp[j(\mathbf{q}-\mathbf{k}_c+\mathbf{k}_v)\cdot\mathbf{r}']\mathbf{p}U_{v,\mathbf{k}_v} d\mathbf{r}' \quad (4.50)$$

In the above, we have written $\mathbf{r} = \mathbf{R}_l + \mathbf{r}'$, where $\mathbf{R}_l$ is a lattice vector and $\mathbf{r}'$ lies within one unit cell. One notes that

$$\sum_l \exp[j(\mathbf{q}-\mathbf{k}_c+\mathbf{k}_v)\cdot\mathbf{R}_l] = \delta(\mathbf{q}-\mathbf{k}_c+\mathbf{k}_v)$$

The delta function ensures wave vector (momentum) conservation in the absorption process. Using the conservation condition in the integral over unit cell in Eq. (4.50) yields the following simplified result:

$$\int_{unit\ cell} U^*_{c,\mathbf{k}_c} \exp[j(\mathbf{q}-\mathbf{k}_c+\mathbf{k}_v)\cdot\mathbf{r}']\mathbf{p}U_{v,\mathbf{k}_v} d\mathbf{r}' = \int_{unit\ cell} U^*_{c,\mathbf{k}_v+\mathbf{q}}\mathbf{p}U_{v,\mathbf{k}_v} d\mathbf{r}' \quad (4.51)$$

This equation is further simplified by assuming $\mathbf{q} \ll \mathbf{k}_v$. Expanding $U_{\mathbf{k}+\mathbf{q}} \approx U_\mathbf{k} + \mathbf{q}\cdot\nabla_\mathbf{k} U_\mathbf{k}$, and keeping only the first term, the matrix element may be expressed as

$$\left|\langle c|\hat{\mathbf{e}}\cdot\mathbf{p}|\mathbf{v}\rangle\right|^2 = \left(\int_{unit\ cell} U^*_{c,\mathbf{k}}(\hat{\mathbf{e}}\cdot\mathbf{p})U_{v,\mathbf{k}} d\mathbf{r}'\right)^2 \quad (4.52)$$

This approximation is known as the *dipole approximation*. In most cases, the momentum matrix element defined above does not depend on $\mathbf{k}$. We therefore write

$$|\langle c|H_{eR}|v\rangle|^2 = (e/m_0)^2|A_0|^2\left|(\hat{\mathbf{e}}\cdot\mathbf{p})_{cv}\right|^2, \quad (4.53)$$

where $\left|(\hat{\mathbf{e}}\cdot\mathbf{p})_{cv}\right|$ denotes the integral over the unit cell. The transition probability R for photon absorption per unit time by using the Fermi golden rule is expressed as

$$R = (2\pi/\hbar)\sum_{\mathbf{k}_c,\mathbf{k}_v} |\langle c|H_{eR}|v\rangle|^2 \delta[E_c(\mathbf{k}_c) - E_v(\mathbf{k}_v) - \hbar\omega] \quad (4.54)$$

Considering the unit volume of the crystal, the above equation gives the rate of transition per unit volume due to absorption process and $R\hbar\omega$ gives the rate of loss of power per unit volume due to absorption of photons. The power loss can also be expressed in terms of the absorption coefficient α. If I denotes the intensity, then the rate of decrease of I may be expressed as

$$-\frac{dI}{dt} = -\left(\frac{dI}{dz}\right)\left(\frac{dz}{dt}\right) = \frac{c}{n}\alpha I \quad (4.55)$$

Thus

$$\alpha = \frac{1}{I}\frac{n}{c}\left(\frac{dI}{dt}\right) = \frac{1}{I}\frac{n}{c}R\hbar\omega \tag{4.56}$$

Both R and I contain $|\mathbf{A}_0|^2$ and therefore α is independent of intensity of light.

Evaluation of the transition rate R involves summation over $\mathbf{k}_c$ and $\mathbf{k}_v$, as given in Eq. (4.54). We write $\mathbf{k}_c = \mathbf{k}_v = \mathbf{k}$ and convert summation into integration by noting $\sum_{\mathbf{k}} \rightarrow \frac{2}{(2\pi)^3}\int 4\pi k^2 dk$. The integral is therefore

$$I_k = \sum_{\mathbf{k}} \delta[E_c(\mathbf{k}) - E_v(\mathbf{k}) - \hbar\omega] = \frac{2}{8\pi^3}\int 4\pi k^2 dk \delta[E_c(\mathbf{k}) - E_v(\mathbf{k}) - \hbar\omega] \tag{4.57}$$

Assuming parabolic E–k relationships, we have

$$E_c(\mathbf{k}) = E_g + \frac{\hbar^2 k^2}{2m_e} \quad \text{and} \quad E_v(\mathbf{k}) = -\frac{\hbar^2 k^2}{2m_h} \tag{4.58}$$

where m_e and m_h are the electron and hole effective masses, respectively, and the argument of the δ-function becomes $E_g + (\hbar^2 k^2/2m_r) - \hbar\omega$, where the reduced mass is given by

$$m_r^{-1} = m_e^{-1} + m_h^{-1}. \tag{4.59}$$

The integration is performed by transforming k to $E(k)$ and then using the δ-function. Using the expressions for R and I, the final expression for absorption coefficient becomes

$$\alpha(\hbar\omega) = \frac{e^2(2m_r)^{3/2}}{2\pi\varepsilon_0 c n m_0^2 \hbar^3 \omega}(\hbar\omega - E_g)^{1/2}\langle |p_{cv}^2| \rangle, \hbar\omega \geq E_g. \tag{4.60}$$

The term $\langle |p_{cv}^2| \rangle$ is the average of the squared momentum matrix element for unpolarized light. The average is obtained by considering the polarization dependence of the term $|(\hat{\mathbf{e}}\cdot\mathbf{p})_{cv}|$.

For transitions involving states near the band edges, both $U_{c,k}$ and $U_{v,k}$ are given by their zone center values. In this approximation,

Conduction band: $U_{c,0} = |s\rangle$

Valence band: heavy-hole states $|3/2, 3/2\rangle = (-1/\sqrt{2})(|p_x\rangle + j|p_y\rangle)|\alpha\rangle$

$$|3/2, -3/2\rangle = \left(1/\sqrt{2}\right)(|p_x\rangle - j|p_y\rangle)|\beta\rangle \tag{4.61}$$

Light-hole (LH) states: $|3/2, 1/2\rangle = (-1/6)(|p_x\rangle + j|p_y\rangle)|\beta\rangle - 2|p_z\rangle|\alpha\rangle$

$$|3/2, -1/2\rangle = (1/6)(|p_x\rangle - j|p_y\rangle)|\alpha\rangle + 2|p_z\rangle|\beta\rangle \tag{4.62}$$

Symmetry allows only the following matrix elements to be nonzero:

$$\langle p_x|p_x|s\rangle = \langle p_y|p_y|s\rangle = \langle p_z|p_z|s\rangle = p_{cv} \tag{4.63}$$

This identifies the following allowed transitions with the corresponding matrix elements:

$$\begin{aligned}\langle HH|p_x|s\rangle &= \langle HH|p_y|s\rangle = (1/\sqrt{2})\langle p_x|p_x|s\rangle \\ \langle LH|p_x|s\rangle &= \langle LH|p_y|s\rangle = (1/\sqrt{6})\langle p_x|p_x|s\rangle \\ \langle LH|p_z|s\rangle &= (2/\sqrt{6})\langle p_x|p_x|s\rangle = (2/\sqrt{6})p_{cv}\end{aligned} \tag{4.64}$$

It is also to be noted that

$$\langle HH|p_z|s\rangle = 0 \tag{4.65}$$

The squared matrix element for light polarized along different directions may now be listed. This dependence on orientation is useful in quantum wells because HH and LH states become no longer degenerate there.

z-polarized light: HH $\rightarrow$ CB: no coupling

$$\text{LH} \rightarrow \text{CB}: |p_{cv}|^2 = (2/3)|\langle p_x|p_x|s\rangle|^2 \tag{4.66a}$$

x-polarized light: HH $\rightarrow$ CB: $|p_{cv}|^2 = (1/2)|\langle p_x|p_x|s\rangle|^2$

$$\text{LH} \rightarrow \text{CB}: |p_{cv}|^2 = (1/6)|\langle p_x|p_x|s\rangle|^2 \tag{4.66b}$$

y-polarized light: HH $\rightarrow$ CB: $|p_{cv}|^2 = (1/2)|\langle p_x|p_x|s\rangle|^2$

$$\text{LH} \rightarrow \text{CB}: |p_{cv}|^2 = (1/6)|\langle p_x|p_x|s\rangle|^2 \tag{4.66c}$$

For finite values of k, there is some mixing of HH and LH states. As may be seen for x–y polarized light, HH–CB coupling is three times stronger than LH–CB coupling.

In the present context, we are considering unpolarized light. Therefore, using Eq. (4.66), we may write

$$\left\langle |p_{cv}|^2 \right\rangle = \frac{2m_0{}^2 E_g(E_g+\Delta)}{3m_e(E_g+2\Delta/3)} \tag{4.67}$$

Example 4.5

For GaAs, $m_e = 0.067\, m_0$, $E_g = 1.424$ eV, and $\Delta = 0.33$ eV; this gives $<|p_{cv}|^2> = 10.8\, m_0 E_g$.

In the literature, a quantity E_P defined as follows is used to describe the momentum matrix element (see Appendix A):

$$E_p = \frac{2}{m_0}\langle p_x|p_x|s\rangle \tag{4.68}$$

The value of E_p for most semiconductors lies in the range of 20–25 eV.

The final expression for the absorption coefficient is therefore

$$\alpha(\hbar\omega) = \frac{e^2(2m_r)^{3/2}}{2\pi\varepsilon_0 c n m_0{}^2\hbar^3\omega}(\hbar\omega - E_g)^{1/2}\left\langle |p_{cv}|^2 \right\rangle. \tag{4.69}$$

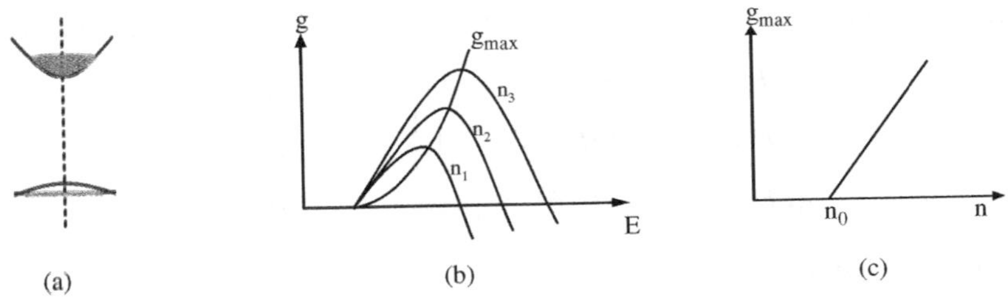

Figure 4.6 *(a) Band state filling under heavy injection; shaded regions represent occupied states. (b) Variation of gain with photon energy with injected electron density as parameter. (c) Plot of maximum gain with injected carrier density; n_0 is the transparency carrier density.*

The absorption coefficients for a few common semiconductors are presented in Figure 4.6 as a function of wavelength [9]. The presence of a cutoff wavelength can easily be identified in the diagram for direct-gap materials like GaAs, InP, and the ternary and quaternary alloys. The rise in absorption below λ_c (above E_g) is also apparent.

The absorption coefficient is sometimes expressed in terms of a dimensionless quantity called the *oscillator strength*, denoted as f_{vc} and defined as

$$f_{vc} = \frac{2P^2}{m_0(E_{\mathbf{k}c} - E_{\mathbf{k}v})}. \tag{4.70}$$

The absorption coefficient may be expressed as follows in terms of the oscillator strength:

$$\alpha(\omega) = \frac{e^2(2\mu)^{3/2}}{4\pi\varepsilon_0 c m_0 n(\omega)\hbar^2} f_{vc}(\hbar\omega - E_g)^{1/2}. \tag{4.71}$$

Example 4.6

Assuming $m_r \approx m_h \approx m_0$, $n = 3.5$, and $f_{vc} \approx 2$, the absorption coefficient in GaAs at photon energy of 1.63 eV($\lambda = 0.76\ \mu m$)) becomes $\alpha = 2.5 \times 10^5\ cm^{-1}$ with $E_g = 1.43$ eV. The experimental value shown in Figure 4.4 is not widely different in spite of the approximations in parameters.

The absorption coefficient expressed by Eq. (4.69) applies to the ideal situation for 0 K, when all states in the valence band are full and all states in the conduction band are empty. The expression becomes modified at finite temperature by including the Fermi functions for both electrons and holes, as will be discussed in Section 4.7.

4.6 Fundamental Absorption in Indirect Gap

The conduction band minima and valence band maxima occur at the center of the Brillouin zone in direct-gap semiconductors. Therefore, the optical absorption process is indicated by a vertical line in Figure 4.3, to indicate that photons provide negligible momentum to the electrons. This happens in most of the III–V and II–VI compound semiconductors and their alloys.

The most important material for microelectronics, Si, is not, however, a direct-gap semiconductor. Therefore, a photon cannot directly lift an electron to the conduction band. In order to conserve crystal momentum or wave vector, it is necessary that some momentum-conserving agency should participate in the absorption process. At present, we shall assume that phonons, that is, quantized lattice vibrations, act as such momentum-conserving agents. The transition is indicated in Figure 4.3. The photon brings an electron in the valence band to an intermediate state, by the process of which there is negligible change of momentum. The electron then has its momentum changed by an amount $\mathbf{q}=\mathbf{k}_0$, by colliding with a phonon, and comes to one of the six degenerate conduction band valleys. The wave vector conservation then takes the form

$$\mathbf{k}_c = \mathbf{k}_v \pm \mathbf{q}, \tag{4.72}$$

where $\mathbf{q}$ is the phonon wave vector. We have used the same symbol $\mathbf{q}$ for the phonon wave vector in this chapter. The energy conservation is now expressed as

$$\hbar\omega = E_c(\mathbf{k}_c) - E_v(\mathbf{k}_v) \pm \hbar\omega_{\mathbf{q}}, \tag{4.73}$$

where $\hbar\omega_{\mathbf{q}}$ is the energy of the phonon. In Eqs. (4.72) and (4.73) the $+$ and $-$ signs refer to phonon absorption and emission processes, respectively.

4.6.1 Theory of Absorption

The analysis of absorption in indirect-gap semiconductors involves the participation of an electron, a photon, and a phonon. The transition is referred to as a *second-order process* in the language of the quantum mechanics [10]. The transition probability is calculated by second-order perturbation theory, and the probability is extremely low in comparison to that in direct-gap semiconductors. The absorption coefficient being directly related to the probability of transition is also very small.

Figure 4.3 shows one of the two types of transitions involving a photon and a phonon. In the process shown, an electron absorbs a photon first and makes a vertical transition to the direct conduction band; this process is completed by a phonon scattering whereby the electron finally comes to an indirect valley. The photon absorption takes the electron to an *intermediate state*, conserving momentum but not the energy because the process is virtual and time–energy uncertainty ensures that there is no need to conserve energy. Energy is necessarily conserved in the overall process. The process is described by second-order perturbation theory, and the transition rate is

$$W_{if} = \frac{2\pi}{\hbar}\int\left|\frac{\sum\langle f|H_{per}|n\rangle\langle n|H_{per}|i\rangle}{E_i - E_n}\right|^2 \delta(E_f - E_i)d\mathbf{k} \tag{4.74}$$

where i, f, and n stand for the initial, final, and intermediate states, respectively; H_{per} is the perturbation which is given by

$$H_{per} = H_{eR} + H_{eP}. \tag{4.75}$$

The electron–phonon interaction is denoted by H_{eP}, and, as already defined, H_{eR} is the electron–photon interaction.

In the second process, an electron in the valence band undergoes first a phonon scattering and reaches a state deep in the valence band; it then makes a vertical transition by absorption of a photon and reaches the indirect conduction band valley. However, the energy denominator $E_i - E_n$ is larger in this process and the contribution from this process to absorption is small and can be neglected. The transition rate is then

$$W_{if}(\mathbf{k}) = \frac{2\pi}{\hbar}\sum_f \sum_n (|S_+|^2 + |S_-|^2)\delta(E_f - E_i) \tag{4.76}$$

The two terms represent phonon absorption and emission processes, which are assumed to be independent. Their expressions are

$$|S_\pm|^2 = \frac{\left|\langle c, \mathbf{k} \pm \mathbf{q}|H_{eP,em}^{eP,ab}|c, \mathbf{k}\rangle\right|^2 \left|\langle c, \mathbf{k}|H_{eR}^{ab}|v, \mathbf{k}\rangle\right|^2}{(E_{g\mathbf{k}} - \hbar\omega)^2} \tag{4.77}$$

The $+(-)$ signs correspond to the superscript (subscript) of the electron–phonon interaction, H_{eP}. The matrix element for absorption of radiation is identical with that for a direct transition used in Section 4.5.2. The phonon-scattering mechanism in the present context is the intervalley phonon scattering connecting the conduction band at zone center and the lowest one (X for Si, L for Ge).

The square of the electron–intervalley phonon interaction matrix element may be written as

$$|\langle c, \mathbf{k} \pm \mathbf{q}|H_{eP}|c, \mathbf{k}\rangle|^2 = M_\mathbf{q}^2 = \frac{\hbar D_{ij}^2}{2\rho V \omega_{ij}} \begin{pmatrix} N_\mathbf{q}(\omega_{ij}) \\ N_\mathbf{q}(\omega_{ij}) + 1 \end{pmatrix} \tag{4.78}$$

where D_{ij} is the deformation potential constant for the intervalley transition from valley i to j, ρ is the mass density, V is the volume of the crystal, ω_{ij} is the angular frequency of the phonon of wave vector $\mathbf{q}$, and the N_q and $(N_q + 1)$ terms are connected with the phonon absorption and emission processes, respectively.

In direct absorption, only one state is coupled to a given initial state because of momentum conservation. In indirect absorption, a given initial state $|v\ \mathbf{k}>$ is connected to a number of final states in the conduction band due to phonon scattering. The transition rate corresponding to a given initial state $|v\ \mathbf{k}>$ is then given by

$$W(\mathbf{k}) = \frac{2\pi}{\hbar} \frac{|M_{eR}|^2}{(E_{g0} - \hbar\omega)^2} \frac{\hbar D_{ij}^2 g_v}{2\rho V \omega_{ij}} \sum_{\mathbf{k}'} (N_\mathbf{q} + 1/2 \mp 1/2)\delta(E_{\mathbf{k}'} - E_\mathbf{k} \mp \hbar\omega_{ij}) \tag{4.79}$$

$$|M_{eR}|^2 = \left|\langle c, \mathbf{k}|H_{eR}^{ab}|v, \mathbf{k}\rangle\right|^2 \tag{4.80}$$

where g_v is the number of equivalent conduction band valleys constituting the final states and $+(-)$ sign refers to the emission (absorption) of phonons.

Eq. (4.79) gives the rate of transition from a given valence band state. To calculate the total rate induced by a given photon energy, we have to take a sum over all the above $W(k)$'s keeping the photon energy constant. This means we have to sum over all initial states ranging from $E_\mathbf{k} = 0$ to $E_\mathbf{k} = E_{\mathbf{k}_{\max}}$, where

$$E_{\mathbf{k}_{\max}} = \hbar\omega - E_g \pm \hbar\omega_{ij} \tag{4.81}$$

where the $+(-)$ sign is valid for phonon absorption (emission). Thus

$$W(\hbar\omega) = \sum_{\mathbf{k}} W(\mathbf{k}) \tag{4.82}$$

gives the total transition rate for absorption of a photon of energy $\hbar\omega$.To convert the summation over integration, we use $\sum_{\mathbf{k}} \to 2V/(2\pi)^3 \int 4\pi k^2 dk$, the factor 2 arising out of two spin states. The final-state wave vector **k'** lies in ellipsoidal constant energy surfaces, which may be converted into equivalent spherical energy surfaces as indicated in Section 2.3. The exercise yields a term $(m_{ed})^{3/2} = (m_l m_t{}^2)^{1/2}$, where m_{ed} is the density-of-states (DOS) effective mass. The valence band is assumed to have an isotropic effective mass m_h. Assuming parabolic E–**k** relationships, one obtains an integral like

$$I = (m_{ed} m_h/\hbar^2)^{3/2} \int_0^{E^{\pm}_{\mathbf{k}\max}} (N_{\mathbf{q}} + 1/2 \mp 1/2) E_{\mathbf{k}}^{1/2} (E^{\pm}_{\mathbf{k}\max} - E_{\mathbf{k}})^{1/2} \, dE_{\mathbf{k}} \tag{4.83}$$

The result of integration is $\pi E^2_{\mathbf{k}\max}$ The final result is

$$W(\hbar\omega) = \frac{|M_{eR}|^2 D^2_{ij} g_v (m_{ed} m_h)^{3/2}}{8\pi^2 (E_{g0} - \hbar\omega)^2 \hbar^6 \rho \omega_{ij}} \begin{bmatrix} N_{\mathbf{q}}(\omega_{ij})(\hbar\omega - E_g + \hbar\omega_{ij})^2 \\ + \{N_{\mathbf{q}}(\omega_{ij}) + 1\}(\hbar\omega - E_g - \hbar\omega_{ij})^2 \end{bmatrix} \tag{4.84}$$

The absorption coefficient may be obtained from the above rate by using the proper group velocity. Note, however, that all the relevant phonons are to be considered while calculating the transition rates as well as the absorption coefficient. In Eq. (4.84), transition rates are to be taken as zero when $\hbar\omega \le (E_g \pm \hbar\omega_{ij})$. If we consider a single phonon, then it is evident from Eq. (4.84) that absorption starts when the photon energy is below the gap by the phonon energy, $\hbar\omega = E_g - \hbar\omega_{ij}$. The phonon is absorbed with a photon then. The plot of $\alpha^{1/2}$ versus $\hbar\omega$ should increase linearly. After the photon energy exceeds the gap energy by the phonon energy, $\hbar\omega = E_g + \hbar\omega_{ij}$, the emission of phonons starts and then there is a change of slope with addition from the phonon emission. Consequently there are breaks in the absorption spectra. It may be concluded from Eq. (4.84) that in each section of the curve, in which the contribution of one phonon is considered, the variation of $\alpha^{1/2}$ with photon energy, $\hbar\omega$, should be a straight line.

4.6.2 Absorption Spectra in Si

A detailed investigation of the absorption in indirect-gap materials like Si and Ge has been made by McFarlane and coworkers [11, 12]. A good review of all the work appears in McLean [13]. Figure 4.5 gives detailed experimental data for the absorption coefficient of Si at different temperatures [11]. For each type of intervalley phonon, there are two contributions from absorption and emission of phonons. Since more than one phonon is allowed by selection rules to participate in the process, the slope of the curve changes as the threshold for optical absorption for a higher energy phonon is reached.

Examination of the plot and comparison with the expected variation in Figure 4.5 reveal that several phonons are involved and the plots have straight-line portions, as discussed in Section 4.6.1. However, there are additional structures in the plot which arise due to the formation of excitons.

A comparison with absorption values for direct-gap materials like GaAs (Example 4.4 and Eq. (4.69)) indicates that whereas the absorption coefficient rises sharply in direct-gap material, the rise is more gradual in Si. Further, the magnitude of absorption coefficient in Si near the band gap is only 10–100 cm^{-1}.

The low values of the absorption in Si makes the absorption length longer, thereby increasing the size of the photodetectors, as we shall see in later chapters of this volume.

The variation can be understood by using the expression for the absorption coefficient obtained by second-order perturbation theory [7]. Instead, we consider the following semi-empirical description of the wavelength dependence of fundamental absorption coefficient given by Bücher *et al.* [14], considering the results obtained by several authors:

$$\alpha(T) = \sum_{\substack{i=1,2\\ j=1,2}} A_j C_i \left\{ \frac{\left[\hbar\omega - E_{gj}(T) + E_{phi}\right]^2}{\exp(E_{phi}/k_B T) - 1} + \frac{\left[\hbar\omega - E_{gj}(T) - E_{phi}\right]^2}{1 - \exp(-E_{phi}/k_B T)} \right\} + \frac{A_d(\hbar\omega - E_{gd})^{3/2}}{\hbar\omega} \tag{4.85}$$

where i and j denote, respectively, the types of phonons and of band gaps. E_{g1} is the difference between the lowest Γ point and the conduction band minima in the Δ point ($\Delta_1 - \Gamma_{25}$) whereas E_{g2} is that for the conduction band minima in the L point ($L_1 - \Gamma_{25}$), E_{gd} is the direct band gap and E_{ph} is the energy of the phonon. In Eq. (4.85), the first term describes the absorption of phonons and the second the emission of phonons. The third term gives the absorption for a direct (forbidden) transition. $E_g(T)$ is the band gap at temperature T related to the band gap at $T = 0$ by Eq. (2.6). The phonons of energy $E_{ph}/k = 212$ K (TA), 670 K (TO), 1030 K, and 1420 K take part in the transitions. However, we have considered only phonons of low energy (TA,TO) as considered by Bücher *et al.* The values of constants are $A_1 = 253\ cm^{-1}\ eV^{-2}$, $A_2 = 3312\ cm^{-1} eV^{-2}$, $A_d = 2.3 \times 10^7\ cm^{-1}\ eV^{-2}$, $E_{g1} = 1.16$ eV, $E_{g2} = 2.25$ eV, and $E_d = 3.2$ eV.

4.6.3 Absorption Spectra in Ge

The absorption spectra in Ge have been studied by a number of workers [13] and the results are more or less similar. However, the magnitude of absorption, the nature of phonons, and the values of phonon energies are different. A noteworthy feature in the absorption in Ge is a sharp rise in the values, due to direct-gap absorption to the conduction band (gap energy ~0.8 eV) which is slightly above the minima at L points.

4.7 Absorption and Gain

In Section 4.5, the transition rate from a valence band state to a conduction band state conserving momentum has been calculated and from the rate the expression for the absorption coefficient has been obtained. The same transition rate can be used to obtain the expression for Einstein's B_{12} coefficient (see Eq. (4.27)), and from it the coefficients A and B_{21}.

Under usual circumstances, a light beam falling on a semiconductor sample is absorbed. However, if a condition for population inversion is created in the semiconductor, then instead of absorption, the electromagnetic radiation is amplified. In this section we shall establish the condition for population inversion in a semiconductor and then obtain the expression for the gain coefficient in terms of the B coefficient. It is straightforward to relate the gain coefficient with the absorption coefficient.

Calculation of the emission and absorption rates in a semiconductor must take into consideration the band picture as well as the Fermi occupational probabilities in the conduction and valence bands. Let us consider an energy level E_2 in the conduction band and an energy level E_1 in the valence band of a direct-gap semiconductor. Spontaneous emission from E_2 to E_1 can occur only when the upper state is occupied and the lower state is empty (i.e., occupied by a hole). The occupational probabilities for the electrons in the conduction and valence bands are expressed by Fermi–Dirac statistics and are given by

$$f_c(E_2) = \{1 + \exp[(E_2 - E_{fc})/k_B T]\}^{-1}, \tag{4.86}$$

and

$$f_v(E_1) = \{1 + \exp[(E_1 - E_{fv})/k_B T]\}^{-1}. \tag{4.87}$$

Here E_{fc} and E_{fv} are the quasi-Fermi levels (under non-equilibrium situations), respectively, for electrons and holes. Now consider small energy intervals dE_2 and dE_1 around levels E_2 and E_1, respectively, in the conduction and valence bands. The number of states in these intervals is, respectively, $S_c(E_2)dE_2$ and $S_v(E_1)dE_1$, where S_c and S_v denote, respectively, the density-of-states in the respective bands. Assuming parabolic DOS functions, one may write for their expressions,

$$S_c(E_2)dE_2 = \frac{(2m_e)^{3/2}}{2\pi^2\hbar^3}(E_2 - E_c)^{1/2}, \tag{4.88a}$$

$$S_v(E_1)dE_1 = \frac{(2m_h)^{3/2}}{2\pi^2\hbar^3}(E_v - E_1)^{1/2}. \tag{4.88b}$$

The rate of spontaneous transition from the band of width dE_2 around E_2 to the cluster of energy levels of width dE_1 around E_1 is dependent on the DOS functions and the probability that the upper states are occupied and the lower states are unoccupied by electrons. Thus, the rate is expressed as

$$r_{sp} dE_1 dE_2 = A\rho(\hbar\omega) S_C(E_2) S_v(E_1) f_c(E_2)[1 - f_v(E_1)\} dE_1 dE_2. \tag{4.89}$$

The total spontaneous emission rate at photon energy $\hbar\omega$ is obtained by summing over all the possible energy levels E_1 and E_2 that are separated by $\hbar\omega$ $(= E_2 - E_1)$. It follows easily that

$$\hbar\omega - E_g = (E_2 - E_c) + (E_v - E_1) = \frac{\hbar^2 k_c^2}{2m_e} + \frac{\hbar^2 k_v^2}{2m_h}. \tag{4.90}$$

If in addition **k**-conservation is taken into account ($k_c = k_v$), we may write the joint DOS function in terms of reduced mass $m_r^{-1} = m_e^{-1} + m_h^{-1}$ as

$$S_{cv} = \frac{(2m_r)^{3/2}}{2\pi^2\hbar^3}(\hbar\omega - E_g)^{1/2}. \tag{4.91}$$

The spontaneous emission rate at $\hbar\omega$ may be written as

$$R_{spon}(\omega) = \int_{E_c}^{\infty} A(E_1, E_2) f_c(E_2)[1 - f_v(E_2 - \hbar\omega)] S_{cv} dE_2. \tag{4.92}$$

The stimulated emission and absorption rates can be obtained in a similar manner and be expressed as

$$R_{stim}(\omega) = \int_{E_c}^{\infty} B(E_1, E_2) f_c(E_2)[1 - f_v(E_2 - \hbar\omega)] S_{cv} \rho(\hbar\omega) dE_2, \tag{4.93}$$

$$R_{abs}(\omega) = \int_{E_c}^{\infty} B(E_1, E_2) f_v(E_2 - \hbar\omega)[1 - f_c(E_2)] S_{cv} \rho(\hbar\omega) dE_2. \tag{4.94}$$

Recall that the rate of absorption has been calculated by using the exp($-j\omega t$) term of the vector potential expressed by Eq. (4.45). It is pointed out that the complex conjugate exp ($j\omega t$) is used to calculate the stimulated emission rate. Both these rates involve an identical factor $B(E_1,E_2)$ in the above expressions.

In order to achieve population inversion condition, $R_{stim} > R_{abs}$, and comparing Eqs. (4.93) and (4.94) it follows easily that $f_c(E_2) > f_v(E_1)$. Using Eqs. (4.86) and (4.87) it is found that this condition is satisfied when

$$E_{fc} - E_{fv} \geq E_2 - E_1 > E_g. \tag{4.95}$$

The condition shown in Eq. (4.95) is known as the Bernard–Duraffourg condition [15]. The separation of the quasi-Fermi levels must therefore exceed the band gap. In equilibrium the Fermi levels align, that is, $E_{fc} = E_{fv}$. In order to separate them a pumping scheme is needed. For semiconductors, the most convenient way to effect this is to use a p–n junction. Application of a forward bias allows injection of both electrons and holes, and the condition of population inversion is established in the junction.

As stated earlier, EM waves grows, rather than decays, when a condition of population inversion is prevalent. One may now express the absorption (or gain coefficient) by using Eqs. (4.16) and (4.17). The difference gives the net rate of stimulated processes. The absorption coefficient may be expressed by dividing the net rate by the velocity of lightwave, *c/n*. Thus,

$$g(\hbar\omega) = \frac{B}{(c/n)} \int_{E_c}^{\infty} [f_c(E) - f_v(E - \hbar\omega)] S_{cv} \rho(\hbar\omega) dE \tag{4.96}$$

where $g(\hbar\omega)$ is the gain coefficient, evaluation of which is usually done numerically due to the presence of two Fermi functions in the integral. It is instructive to have an idea of the gain spectra by using the following approximate expressions for f_c and f_v.

Assuming that the quasi-Fermi levels satisfy the following conditions,

$$E_{Fc} - E_c < 4k_B T \quad \text{and} \quad E_v - E_{Fv} < 4k_B T. \tag{4.97}$$

Then the Fermi functions may be approximated as straight lines such that

$$f_v(E) = \frac{1}{2} + \frac{E_{Fv} - E}{4k_B T} \quad \text{and} \quad f_c(E) = \frac{1}{2} - \frac{E - E_{Fc}}{4k_B T} \tag{4.98}$$

The integration may be performed analytically by considering the parabolic nature of DOS functions. The expression for the gain becomes

$$g(\hbar\omega) = K(\Delta F - \hbar\omega)(\hbar\omega - E_g)^2 \tag{4.99}$$

where all the constants are lumped into the prefactor K, and ΔF is the difference between quasi-Fermi levels. The gain curve therefore covers the range from E_g to ΔF, shows a maximum at $\hbar\omega = (1/3)(E_g + 2\Delta F)$, and then becomes negative (absorption) when $\hbar\omega > \Delta F$. With increasing injection, the values of ΔF increase and the gain curve covers a wider range of photon energy. The qualitative variation of gain spectra is shown in Figure 4.6.

The position of the Fermi levels for a given carrier density is usually calculated numerically. There are, however, approximate formulas which may be used for this purpose (see Problem 4.10).

4.8 Intervalence Band Absorption

As noted already, the valence bands in typical semiconductors including Si comprise LH, HH, and SO bands. At elevated temperatures or in heavily doped p-type materials having Fermi energy below the valence band edge, transitions are possible from the LH band to the HH band or from the SO band to the HH or LH band. An example of such transitions is given in Figure 4.7 as process (C) indicating a transition from the SO to the HH band. The different transitions are further illustrated in Figure 4.7, in which processes A, B, and C denote, respectively, the LH–HH, SO–LH, and SO–HH transitions. The intervalence band transitions are forbidden at $\mathbf{k} = 0$, due to quantum mechanical selection rules. Direct (vertical) transitions become possible at $\mathbf{k} \neq 0$. The momentum matrix elements for these transitions are proportional to k. The absorption bands are broad.

4.9 Free-carrier Absorption

The semiconductors used in devices are doped with impurities, and as a result there exists a substantial amount of free electrons or holes. These carriers may absorb photons and make a transition from a state of wave vector $\mathbf{k}$ to another state of wave vector $\mathbf{k}'$ in the same band. These transitions are shown in Figure 4.2 as process (D) and process (E), respectively, in the

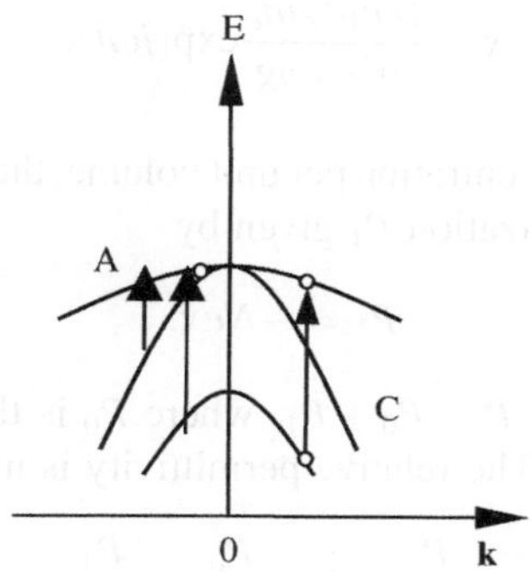

Figure 4.7 Schematic of intervalence band absorption.

valence band and conduction band. These transitions are termed as *free-hole* (D) and *free-electron* (E) *absorption* or in general as *free-carrier absorption*.

A change in wave vector occurs in the absorption process and to conserve the wave vector, one must have $|\mathbf{k}'-\mathbf{k}| = \boldsymbol{\kappa}$. The photon wave vector is related to the frequency of radiation by $\kappa = \omega n(\omega)/c$. In order to satisfy both energy and momentum conservation, the energy of the photon should be prohibitively large, larger than the typical band gap. Therefore the radiative transitions just discussed cannot account for intraband free carrier absorption.

The intraband processes occur with the simultaneous action of a photon and another momentum-conserving agency, that is, scatterers like phonons, impurities, alloy disorder, or other imperfections. Again the process is a second-order transition and the energy of the photon is quite small compared to that needed for interband transitions. The photons introduce negligible change in the **k**-value of the carrier; the main contribution to wave vector change comes from the scattering agencies. Quantum mechanical calculation of the free-carrier absorption has been described in a number of textbooks and monographs [7, 16].

The expressions for the free-carrier absorption coefficient may also be obtained by using classical electromagnetic theory and the Drude model for conduction [7]. There is not much difference between the expressions obtained by classical and quantum methods. Due to its simplicity, the classical method looks attractive. Since the expression for the carrier-induced change in the refractive index (RI) is also obtained from the theory and will be utilized in subsequent discussions in this chapter, the derivation seems useful.

Let us consider the motion of a free electron in the conduction band of a semiconductor under the influence of a sinusoidal electric field. The equation of motion takes the following form:

$$m_e \frac{d^2x}{dt^2} + m_e g \frac{dx}{dt} = -eE_0 \exp(j\omega t), \tag{4.100}$$

where x is the displacement of the electron having effective mass m_e, g is a damping coefficient, and E_o is the amplitude of the impressed electric field varying with angular frequency ω. The first term is the force term; the second represents the damping of electron motion by scattering with lattice vibrations (phonons) and impurities, and so on; and the right-hand side represents the applied force. The steady-state solution of Eq. (4.100) may easily be written as

$$x = \frac{(eE_0)/m_e}{\omega^2 - j\omega g} \exp(j\omega t). \tag{4.101}$$

If N denotes the free-carrier concentration per unit volume, then the displacement of carriers x will produce additional polarization P_1 given by

$$P_1 = -Nex, \tag{4.102}$$

so that the total polarization is $P = P_0 + P_1$, where P_0 is the polarization present in the material without free carriers. The relative permittivity is now given as

$$\varepsilon_r = \frac{\varepsilon}{\varepsilon_0} = 1 + \frac{P}{\varepsilon_0 E} = 1 + \frac{P_0}{\varepsilon_0 E} + \frac{P_1}{\varepsilon_0 E} = n_0^2 + \frac{P_1}{\varepsilon_0 E}, \tag{4.103}$$

where n_0 is the RI without free carriers. Using Eqs. (4.101) and (4.102) in Eq. (4.103), we may obtain

$$\varepsilon_r = n_0^2 - \frac{(Ne^2)/(m_e\varepsilon_0)}{\omega^2 - j\omega g}. \tag{4.104}$$

The real and imaginary parts of the relative permittivity may now be expressed as

$$\varepsilon_{rr} = n_0^2 - \frac{(Ne^2)/(m_e\varepsilon_0)}{\omega^2 + g^2}, \tag{4.105}$$

$$\varepsilon_{ri} = \frac{(Ne^2 g)/(m_e\omega\varepsilon_0)}{\omega^2 + g^2}. \tag{4.106}$$

At steady state $d^2x/dt^2 = 0$, and from Eq. (4.100)

$$m_e g \frac{dx}{dt} = eE. \tag{4.107}$$

Since the drift velocity is related to the mobility μ_e by the relation

$$\frac{dx}{dt} = \mu_e E, \tag{4.108}$$

Eqs. (4.107) and (4.108) yield

$$g = \frac{e}{\mu_e m_e}. \tag{4.109}$$

Assume an electron mobility = 1000 cm^2/V.s for Si; the effective mass is 0.26 m_0, where m_0 is the free electron mass. The value of g is $6.75 \times 10^{12}\,\text{s}^{-1}$, but $\omega \approx 10^{15}\,\text{s}^{-1}$. This allows one to neglect g in the denominators of Eqs. (4.105) and (4.106), which are modified as follows by using Eq. (4.109):

$$\varepsilon_{rr} = n_0^2 - \frac{Ne^2}{m_e\varepsilon_0\omega^2}, \tag{4.110}$$

$$\varepsilon_{ri} = \frac{Ne^3}{m_e^2\varepsilon_0\omega^2\mu_e}. \tag{4.111}$$

The exponential loss coefficient α is related to ε_{ri} by $\alpha = k\varepsilon_{ri}/n$, where n is the RI and k is the light wave vector given by $k = \omega/c$. We may write, therefore,

$$\alpha_{fc} = \frac{Ne^3\lambda_0^2}{4\pi^2 n m_e^2 \mu_e \varepsilon_0 c^3}. \tag{4.112}$$

Eq. (4.110) indicates that there is a change of the RI also due to free carriers. Writing the small change as Δn_e, the corresponding expression may be written as

$$\Delta n_e = -\frac{Ne^2\lambda_0^2}{8\pi^2 c^2 n\varepsilon_0 m_e}. \tag{4.113}$$

Example 4.7

The prefactor for electron concentration in Eq. (4.113) for Si is calculated using $n = 3.5$, $\lambda_0 = 1.55\,\mu m$, and $m_e = 0.26\,m_0$. One may obtain $\Delta n_e = 11.7 \times 10^{-22} N$, when N is in cm^{-3}.

Detailed analysis of the free-carrier absorption coefficient and the related RI change has been done by Soref and Bennet [17] for different wavelengths using the available experimental data. Focussing on communication wavelengths 1.3 µm and 1.55 µm, they expressed the changes in absorption coefficient and RI in terms of the changes in free-carrier densities, ΔN_e and ΔN_h. However, their results showed good agreement with Eqs. (4.112) and (4.113) only for electrons; for holes, the quantities varied as $(\Delta N_h)^{0.8}$. The empirical expressions given by them for 1.3 µm and 1.55 µm are reproduced below. These expressions are now almost universally used by different workers.

At $\lambda = 1.55\,\mu m$:

$$\Delta n = \Delta n_e + \Delta n_h = -[8.8 \times 10^{-22} \Delta N_e + 8.5 \times 10^{-18} (\Delta N_h)^{0.8}], \qquad (4.114)$$

$$\Delta \alpha = \Delta \alpha_e + \Delta \alpha_h = 8.5 \times 10^{-18} \Delta N_e + 6.0 \times 10^{-18} \Delta N_h]. \qquad (4.115)$$

Here Δn and $\Delta \alpha$ are the changes of RI and absorption coefficient introduced by the change in carrier density ΔN, and the subscripts e and h refer, respectively, to electron and hole. Similarly, the expressions for $\lambda = 1.3\,\mu m$ are

$$\Delta n = \Delta n_e + \Delta n_h = -[6.2 \times 10^{-22} \Delta N_e + 8.5 \times 10^{-18} (\Delta N_h)^{0.8}], \qquad (4.116)$$

$$\Delta \alpha = \Delta \alpha_e + \Delta \alpha_h = 6.0 \times 10^{-18} \Delta N_e + 4.0 \times 10^{-18} \Delta N_h]. \qquad (4.117)$$

Example 4.8

Suppose $\Delta N_e = \Delta N_h = 5 \times 10^{17}$ cm^{-3}. The change in RI at 1.3 µm is therefore -1.17×10^{-3}.

4.10 Recombination and Luminescence

The process opposite to absorption is emission. Since in semiconductors the excess electrons and holes recombine, the excess energy is given up in the form of photons. The general name *luminescence* is given to the phenomenon of light emission. In this section, the recombination and luminescence processes will be discussed.

4.10.1 Luminescence Lifetime

Band-to-band absorption leads to the creation of an excess electron–hole pair. These excess carriers have relatively short lifetimes and tend to recombine with the emission of a photon. The recombination processes for direct- and indirect-gap semiconductors are shown in

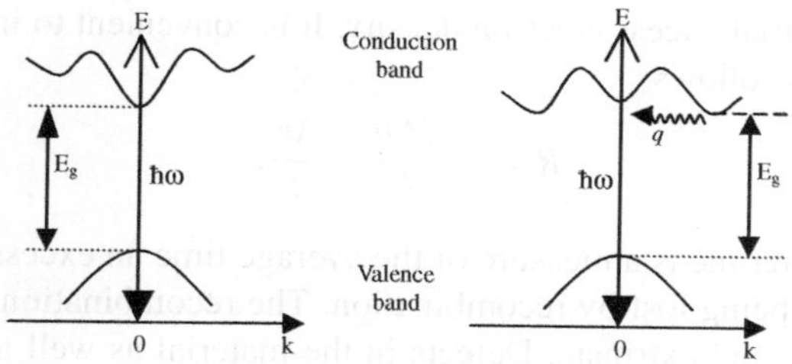

Figure 4.8 Recombination processes in a direct-gap (left) and an indirect-gap (right) semiconductor.

Figure 4.8. Just like absorption, a radiative transition can be direct or indirect depending on the type of semiconductor.

The electrons and holes generated by absorption may lie in a state higher than the band edge in the respective band. Before recombination, the energetic particles thermalize, that is, they come to the lowest lying states in the bands. In direct band gap semiconductors, conduction band and valence band extrema occur at the zone center. The electron can readily make a transition to the valence band. Note that the same momentum conservation condition as in Eq. (4.36) should be satisfied. Since the photon has negligible momentum, the condition given by Eq. (4.37) applies and the transition is a vertical transition as shown in Figure 4.8. The energy of the photon equals the band gap energy.

In the indirect-gap material, the excited carriers thermalize to the lowest lying states in the respective bands. The recombination process now involves a phonon to conserve the momentum. The probability of such transition is extremely low.

The recombination of excess electron–hole pairs gives rise to light emission or luminescence. Depending on the type of generation, the luminescence processes are classified in the following way.

Photoluminescence: this occurs when the excitation is caused by a light beam producing the excess electron–hole population.

Cathodoluminescence: the excitation mechanism for this situation is high-energy electron bombardment.

Electroluminescence: the emission takes place due to the injection of carriers in a biased p–n junction.

In all these cases, excess electrons and holes in excess of the thermal equilibrium values are produced. Let $\Delta n(t)$ be the excess electron density at any instant t. Since excess electrons and holes are created in pairs and they recombine in pairs, $\Delta n(t) = \Delta p(t)$, the excess hole density. The rate equation for excess electron density is

$$\frac{d\Delta n}{dt} = -\frac{\Delta n}{\tau}, \tag{4.118}$$

where τ is the recombination lifetime of the carriers. It is simple to integrate this equation to obtain

$$\Delta n(t) = \Delta n(0)e^{-t/\tau}, \tag{4.119}$$

where $\Delta n(0)$ is the initial excess electron density. It is convenient to introduce a recombination rate defined as follows:

$$R = -\frac{d\Delta n}{dt} = \frac{\Delta n}{\tau}. \tag{4.120}$$

The recombination lifetime is a measure of the average time an excess carrier pair spends in the sample before being lost by recombination. The recombination in a semiconductor may be both intrinsic and extrinsic. Defects in the material as well as the surface of the sample provide recombination channels.

In addition to radiative recombination, nonradiative recombination also occurs, in which the excess energy is given up in the form of phonons or heat waves. These and other defect- and surface-related processes will be discussed in this chapter. We now introduce a quantity named *internal quantum efficiency*, which is an important parameter characterizing the efficiency of light-emitting devices.

Let the radiative recombination rate in a material per unit volume be R_r, while the rate for a nonradiative process is R_{nr}. The total recombination rate for a spontaneous process becomes

$$R_{sr} = R_r + R_{nr}. \tag{4.121}$$

The internal quantum efficiency (IQE) is the ratio of radiative to total recombination rate and is therefore

$$\eta_i = \frac{R_r}{R_r + R_{nr}}. \tag{4.122}$$

Now the lifetime of radiative (τ_r) and nonradiative (τ_{nr}) processes may be defined by Eq. (4.120) and in terms of the lifetimes the IQE may be written as

$$\eta_i = \frac{\tau_r^{-1}}{\tau_r^{-1} + \tau_{nr}^{-1}} = \left(1 + \frac{\tau_r}{\tau_{nr}}\right)^{-1}. \tag{4.123}$$

4.10.2 Carrier Lifetime: Dependence on Carrier Density

We first discuss the dependence of spontaneous recombination lifetime for nondegenerate semiconductors. The spontaneous recombination rate is directly proportional to the product of the electron and hole concentrations as follows:

$$R_{sr} = B_r np, \tag{4.124}$$

where n and p are, respectively, the electron and hole concentrations, and B_r is a recombination coefficient that may be derived quantum mechanically.

The thermal equilibrium electron and hole concentrations, denoted by n_0 and p_0, respectively, are related by $n_0 p_0 = n_i^2$, where n_i is the intrinsic carrier concentration. In the presence of excess carriers, the spontaneous recombination rate becomes

$$R_{sr} = B_r(n_0 + \Delta n)(p_0 + \Delta p). \tag{4.125}$$

Since $\Delta n = \Delta p$ and the spontaneous recombination rate $R_{sr}{}^0 = B_r n_0 p_0$, the recombination rate for the excited carriers R_{sr}^{exc} may be written as

$$R_{sr}^{exc} = R_{sr} - R_{sr}^0 = B_r \Delta n(n_0 + p_0 + \Delta n), \tag{4.126}$$

and therefore the excess carrier lifetime becomes

$$\tau_r = [B_r(n_o + p_0 + \Delta n)]^{-1}. \tag{4.127}$$

Use has been made of Eq. (4.120) to express lifetime in terms of recombination rate.

In the high-injection case, $\Delta n >> n_0$ or p_0, and $\tau_r = [B_r(\Delta n)]^{-1}$, while in the low-injection case, $\Delta n \ll n_0$ or p_0 and $\tau_r = [B_r(n_0 + p_0)]^{-1}$.

4.10.3 Absorption and Recombination

The coefficient B_r depends on the band structure of the material. The value is large in direct-gap material, and hence the recombination lifetime is small. On the other hand, for indirect gap the value of B_r is extremely low and the recombination lifetime is extremely large. The value of the recombination coefficient may be obtained from the absorption data or from microscopic calculation. It is well known that good absorbers are good emitters. Since the absorption coefficient is small in silicon, its emissive characteristics are also very poor. We present here the treatment of van Roosebroeck and Shockley [18] relating the lifetime with the absorption coefficient.

The principle of detailed balance states that under equilibrium, the rate of photoexcitation of carriers across the gap at frequency interval dv should be equal to the rate of generation of photons in the same frequency interval by electron–hole recombination. The rate of photoexcitation is given by

$$R_{pe}^0 = \int \frac{c}{n(v)} \rho(v)\alpha(v)dv, \tag{4.128}$$

where $n(v)$ is the RI of the material, $\rho(v)$ is the photon density, and $\alpha(v)$ is the absorption coefficient. The photon density increases rapidly with wavelength. The main contribution to the integral therefore comes from the vicinity of the absorption edge where the absorption is weak and the dispersion is small. One may therefore assume the RI to be constant and thus write,

$$\rho(v) = \frac{8\pi v^2 n^3}{c^3(e^{hv/k_B T} - 1)}. \tag{4.129}$$

Using the relation between the Planck function for surface emission, $D(v)$ and $\rho(v)$,

$$\rho(v) = 4v^{-2} n^3 D(v), \tag{4.130}$$

replacing $\rho(v)$ in Eq. (4.128) by $D(v)$, and also replacing the integration variable v by wavelength, one obtains

$$R_{pe}^0 = 4n^2 \int_0^\infty D(\lambda)\alpha(\lambda)d\lambda. \tag{4.131}$$

The principle of detailed balance demands that

$$R^0_{sr} = R^0_{pe}. \tag{4.132}$$

Using Eq. (4.124), one may write

$$B_r = \frac{R^0_{pe}}{n_i^2}. \tag{4.133}$$

In Eq. (4.133), valid for the thermal equilibrium condition, the relationship $n_0 p_0 = n_i^2$ has been used. The longest possible lifetime corresponds to the case when $n_0 = n_i$, and it may be proved that

$$\tau_r(\max) = \frac{n_i}{2R^0_{pe}}. \tag{4.134}$$

4.10.4 Microscopic Theory of Recombination

The microscopic theory of recombination is developed by using the perturbation theory of quantum mechanics. The derivation is similar to that followed for absorption. The first-order theory is applicable for direct-gap material, while the second-order theory is used for indirect-gap material. The final expression for the recombination rate contains the *np* product as expected as well as the Boltzmann distribution of the carriers. The reader is referred to standard literature for further details [6–8].

In order to correlate the rate of capture probability coefficient B_r to the optical absorption coefficient α, one can treat the semiconductor as a blackbody radiation source and use the principle of detailed balance under thermal equilibrium conditions. B_r may be determined by setting the rate of radiative recombination equal to the rate of total blackbody radiation absorbed by the semiconductor due to band-to-band recombination, which is given by

$$B_r n_i^2 = \int \frac{n^2\alpha}{(\pi^2 e^2 h^3)} \frac{E^2 dE}{e^{E/k_B T} - 1}, \tag{4.135}$$

where $E = h\nu$, the photon energy. The right-hand side of Eq. (4.135) is obtained from the Planck blackbody radiation formula. From Eq. (4.135) one obtains the rate of capture probability B_r for the direct transition, which reads,

$$B_r = \left(\frac{E_g}{n_i}\right)^2 (2\pi)^{3/2} \left(\frac{he^2}{3m_0^2 c^2}\right) \eta(1 + m_0/\mu) \left(\frac{m_0}{m_e + m_h}\right)^{3/2} (k_B T)^{-3/2} (m_0 c^2)^{-1/2} \tag{4.136}$$

It is important to note from Eq. (4.136) that B_r is inversely proportional to the square of the intrinsic carrier density, which shows an exponential dependence of B_r on temperature. This implies that the band-to-band radiative recombination lifetime is a strong function of temperature.

A similar calculation of the capture probability for the indirect transition involving the absorption and emission of phonons in an indirect band gap semiconductor yields the capture probability coefficient B_i, which is given by

$$B_i = \left(\frac{4\pi h^3}{m_0^3 c^3}\right)(A\delta^2)\left(\frac{m_0^2}{m_e m_h}\right)^{3/2} E_g^2 \coth(\theta/2T), \tag{4.137}$$

where A and δ are adjustable parameters used to fit the measured absorption data. Equation (4.137) shows that B_i depends weakly on temperature. For a direct band gap semiconductor in which recombination is via band-to-band radiative transition, values of B_r can be quite high (i.e., 3×10^{-11} cm^3/sec). On the other hand, for indirect transitions, values of B_i are found 3–4 orders of magnitude smaller than those of B_r for direct transitions (2.0×10^{-15} cm^3/sec for Si at $T = 300$ K).

Example 4.9

The maximum lifetime is $\tau_r(\max) = (2B_r n_i)^{-1}$. Using $n_i = 1.5 \times 10^{10}$ cm^{-3} for Si and $B_i = 2.0 \times 10^{-15}$ cm^3/s, the lifetime becomes 1.67×10^3 s.

4.11 Nonradiative Recombination

The radiative recombination in indirect-gap semiconductors, in particular in Si, is very inefficient. The excess carriers in silicon decay mostly by nonradiative processes involving trap levels or by the Auger recombination process. In the following, these two processes are discussed.

4.11.1 Recombination via Traps

The dominant recombination mechanism in silicon is via traps. The theory of this process was developed by Shockley and Read [19] and modified by Hall [20]. The model is thus referred to as the SRH model. The mechanism involves four electron and hole transitions as shown in Figure 4.9. A trap level E_t first captures an electron (process 1), and then a hole is captured by the trap filled by the electron (process 3): the direction of the arrow indicates the transition of an electron, and the hole moves oppositely. The combined result of electron and hole capture is a recombination of the pair. As shown in the right part of the diagram, the trap is free to capture another EHP after this. The inverse processes are the emission of an electron from the filled trap into the conduction band (process 2) and the emission of a hole from an empty trap into the valence band (process 4).

The rate of electron capture by the traps, R_{nc}, is proportional to the number of electrons and to the number of empty traps, so that

$$R_{nc} = C_n n(1 - f_t)N_t, \tag{4.138}$$

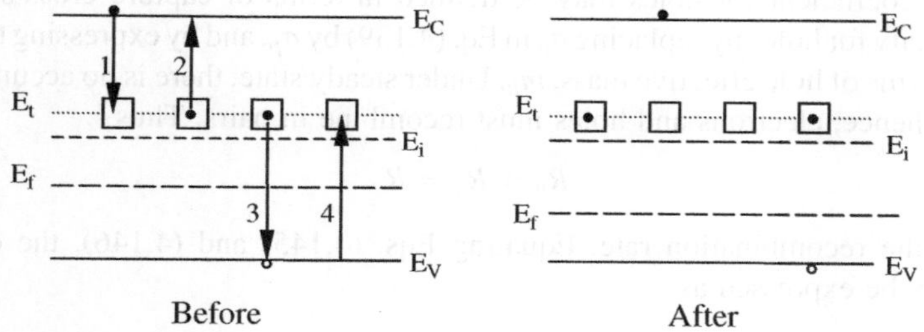

Figure 4.9 Four basic steps in trap-assisted recombination: (1) electron capture, (2) electron emission, (3) hole capture, and (4) hole emission.

where N_t is the trap density, f_t is the occupancy function of the trap level, and C_n is the capture coefficient of the electrons expressed in terms of a capture cross-section, σ_n, for electrons and the electron thermal velocity, $v_{thn} = (3k_BT/m_e)^{1/2}$, as

$$C_n = \sigma_n v_{thn}. \tag{4.139}$$

The rate of electron emission from the traps, R_{ne}, is proportional to the number of filled traps, f_tN_t, and is given by

$$R_{ne} = e_n f_t N_t, \tag{4.140}$$

where e_n is the emission probability of the electron. Under thermal equilibrium $R_{nc} = R_{ne}$, and hence from (4.138) and (4.140),

$$C_n n_0 = e_n f_{t0}/(1-f_{t0}). \tag{4.141}$$

The equilibrium electron concentration, n_0, is expressed as

$$n_0 = N_c \exp[(E_F - E_c)/k_BT], \tag{4.142}$$

and the ratio $f_{t0}/(1 - f_{t0})$, when f_{t0} is the equilibrium occupancy of the trap level, is expressed by using the Fermi–Dirac occupation function as

$$f_{t0}/(1-f_{t0}) = \exp[-(E_t - E_F)/k_BT]. \tag{4.143}$$

Here, E_F, E_c, and E_t, denote, respectively, the Fermi level, the conduction band edge, and the energy of the trap level. Using Eqs. (4.141) and (4.143), we may write

$$e_n = N_c \exp[(E_t - E_c)/k_BT] = n_t C_n. \tag{4.144}$$

The difference between electron capture and electron emission rates is given by

$$R_n = R_{nc} - R_{ne} = C_n N_t[(1-f_t)n - f_t n_t]. \tag{4.145}$$

Proceeding in a similar fashion, the difference between the hole capture and hole emission rates may be written as

$$R_p = R_{pc} - R_{pe} = C_p N_t[f_t p - (1-f_t)p_t]. \tag{4.146}$$

The capture coefficient for holes may be defined in terms of capture cross-section and thermal velocity for holes by replacing σ_n in Eq. (4.139) by σ_p, and by expressing the thermal velocity in terms of hole effective mass, m_h. Under steady state, there is no accumulation of charge and, hence, electrons and holes must recombine in pairs. Thus

$$R_p = R_n = R, \tag{4.147}$$

where R is the recombination rate. Equating Eqs. (4.145) and (4.146), the occupation function may be expressed as

$$f_t = \frac{nC_n + p_tC_p}{C_n(n+n_t) + C_p(p+p_t)}. \tag{4.148}$$

Substituting this in Eq. (4.145), one finds

$$R = \frac{pn - n_i^2}{\tau_{pl}(n+n_t) + \tau_{nl}(p+p_t)}. \tag{4.149}$$

The electron and hole lifetimes are expressed as

$$\tau_{nl} = 1/(v_{thn}\sigma_n N_t), \tag{4.150a}$$

$$\tau_{pl} = 1/(v_{thp}\sigma_p N_t). \tag{4.150b}$$

The recombination rates under special conditions may easily be obtained from Eq. (4.149). Consider p-type materials, in which electrons are minority carriers. Then $n \ll p \approx N_A$: the acceptor density. In this case, $p \gg p_t$ and also $p \gg n_t$. Equation (4.149) now reduces to

$$R = \frac{n - (n_i^2/N_A)}{\tau_{nl}} = \frac{n - n_0}{\tau_{nl}}. \tag{4.151a}$$

Similarly, for n-type materials,

$$R = \frac{p - (n_i^2/N_D)}{\tau_{pl}} = \frac{p - p_0}{\tau_{pl}}. \tag{4.151b}$$

It is noted that the SRH model presented in this section is applicable for describing the nonradiative recombination process via a single deep-level recombination center in the forbidden gap of a semiconductor. Treatment of the nonradiative recombination process via multiple deep-level centers in the forbidden gap of the semiconductor can be found in a classical paper by Sah and Shockley [21].

An example of the variation of hole lifetime in n-type Si is shown in Figure 4.10 as a function of doping density [22]. The lifetime indeed decreases with doping at low

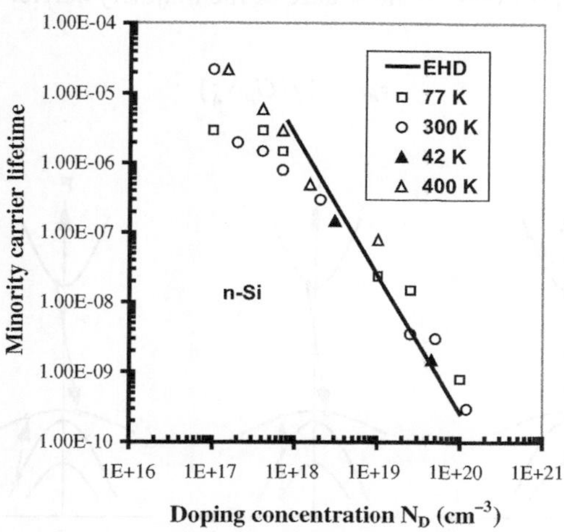

Figure 4.10 *Minority carrier lifetime versus doping density in n-type Si at different temperatures. The lifetime decreases linearly at a low doping level (<10^{17} cm^{-3}) but decreases quadratically (dashed line) with carrier density at a higher doping level due to onset of the Auger effect. Reproduced with permission from [22]. Copyright (1977) American Institute of Physics.*

doping levels (shown by open symbols in Figure 4.10). It is assumed that the trap concentration is proportional to the concentration of dopants, and one expects $\tau \sim 1/N_D$, which is observed.

At higher doping densities, however, τ_{pl} decreases faster than $1/N_D$. The reason for this is the dominance of a different recombination mechanism, known as *Auger recombination*, which is described in Section 4.11.2.

4.11.2 Auger Recombination

The Auger band-to-band recombination is usually the predominant recombination process occurring in the degenerate semiconductors and small–band gap semiconductors such as InSb and HgCdTe materials. The Auger recombination process can also become the predominant recombination mechanism under high-injection conditions. Unlike the non-radiative and radiative recombination processes, which are two-particle processes, the Auger band-to-band recombination is a three-particle process, which involves two electrons and one hole for n-type semiconductors, or one electron and two holes for p-type semiconductors. For an n-type semiconductor, the Auger recombination is accomplished first via electron–electron collisions in the conduction band, and followed by the electron–hole recombination in the valence band. Based on the principle of detailed balance, the rate of recombination is equal to the rate of generation of excess carriers under thermal equilibrium condition, and hence a charge–neutrality condition prevails throughout the semiconductor specimen.

The recombination mechanism is illustrated in Figure 4.11. Such a process is an inverse process of impact ionization, in which an energetic carrier generates an electron–hole pair. As Figure 4.11 suggests, two electrons (in n-type material) and two holes (in p-type material) are involved, and the recombination lifetime associated with Auger recombination is inversely proportional to the square of the majority carrier concentration. Thus, for p-type material,

$$\tau_{nl} = 1/(G_p N_A^2), \tag{4.152}$$

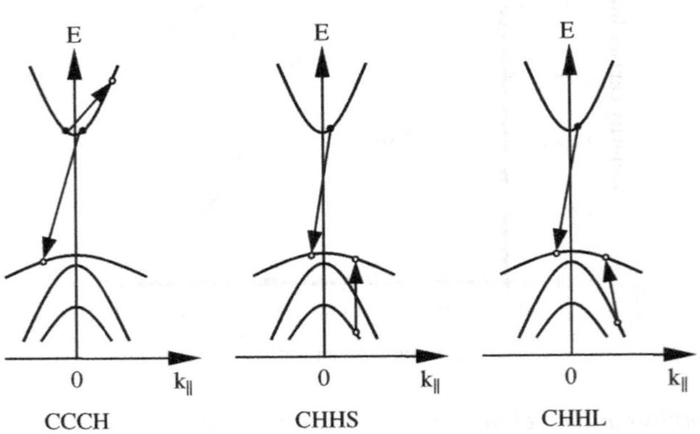

Figure 4.11 *(Left) Auger recombination (CHCC) process in an n-type sample. (Right) CHHS process occurring in a p-type sample.*

and for n-type material,

$$\tau_{pl} = 1/(G_n N_D^2). \tag{4.153}$$

The values of the Auger recombination coefficient for silicon are $G_p = 9.9 \times 10^{-32}$ cm^6/s, and $G_n = 2.28 \times 10^{-31}$ cm^{-6}/s.

Example 4.10

Using the above value of G_p and $N_A = 10^{19}$ cm^{-3}, the value of τ_{nl} is $\sim 10^{-7}$ s. Compare this value with the value in Figure 4.10.

Since a band-to-band Auger process is an intrinsic process, it will be the dominating process in ultrapure materials and the lifetime will be the intrinsic limit to the carrier lifetime in such materials. A detailed study made by Hacker and Hangleiter [23] indicated that the lifetimes predicted by Eqs. (4.152) and (4.153) should contain a Coulomb correlation part which lowers the intrinsic values.

Most of the available results have been included in the study of Kerr and Kuevas [24], and the data have been expressed in parametric forms.

4.11.3 Surface Recombination

The presence of extra defects and trap levels in the surface of semiconductors increases the recombination rates drastically. Surfaces and interfaces typically contain a large number of recombination centers due to abrupt termination of the crystal. A large number of electrically active states are then created at the surface. In addition, a large number of impurities may be present at the surfaces and interfaces, since they are exposed during the device fabrication process. The net recombination rate due to trap-assisted recombination and generation is given by

$$R_s = \frac{pn - n_i^2}{p + n + 2n_i \cosh\left(\dfrac{E_i - E_{st}}{k_B T}\right)} N_{st} v_{th} \sigma_s. \tag{4.154}$$

This expression is almost identical to that for SRH recombination. In this case, however, the two-dimensional density of traps, N_{st}, appears in the expression, as the traps exist only at the surface.

Further simplification of Eq. (4.154) is possible. For example, for electrons in the quasi-neutral p-type region, $p \gg n$ and $p \gg n_{\mathrm{i}}$ so that for $E_{\mathrm{i}} = E_{\mathrm{st}}$, it can be simplified to

$$R_{s,n} = R_{s,nc} - R_{s,ne} = v_s(n - n_0), \tag{4.155}$$

where the surface recombination velocity is expressed as

$$v_s = N_{st} v_{th} \sigma_s. \tag{4.156}$$

4.11.4 Recombination of Complexes

It has been noticed in connection with absorption processes that bound electron–hole pairs or excitons play their role in the low-temperature absorption spectra. Excitonic signature is also observed in the photoluminescence spectra of semiconductors, especially in direct-gap semiconductors. Excitonic effects are also found in the recombination in indirect-gap materials. In particular, the radiative decay of excitons bound to isoelectronic impurities is an important physical process in indirect band-gap semiconductors like GaP and silicon, in which band-to-band radiative transitions are forbidden by the $\boldsymbol{k}$-conservation rule. The introduction of isoelectronic impurities into indirect band-gap semiconductors like GaP and its alloys improves the quantum efficiency of optical emission. Emission due to excitons bound to an isoelectronic center is characterized by a long radiative decay time, which indicates stability against nonradiative Auger processes, and high radiative quantum efficiency. The lifetime can be $> 10^{-5}$ s compared to $< 10^{-7}$ s for isoelectronic bound exciton (IBE) at single donors or acceptors. The binding of both electron and hole is relatively weak, and hence the luminescence from the isoelectronic centers usually has the photon energy close to the electronic gap, but considerably less than for an exciton bound to shallow donors or acceptors.

The BE emission from silicon has attracted considerable interest in the last few years and has now been observed for S- or Be-doped Si.

4.12 Excitonic and Impurity Absorption

In this section, a few of the special cases of absorption, other than the band-to-band transitions discussed in Section 4.11, will be considered.

4.12.1 Excitons

The discussion in Section 4.3 neglects any interaction between an electron in the conduction band and a hole in the valence band. In reality, particularly at low temperatures, the Coulombic interaction between an electron and a hole may modify some physical properties like absorption or recombination. When this interaction is considered, it may be thought that an electron rotates around the heavier hole. The complex is known as an *exciton*, and the electron–hole pair gives rise to hydrogen-like bound states which are described by envelope functions. We shall consider that the envelope function is not confined within a few unit cells, but extends over a larger distance. This type of exciton, the Mott–Wannier excitons, plays an important role in most semiconductors.

The excitons obey the following effective-mass Schrödinger equation:

$$\left[-\frac{\hbar^2}{2m_e}\nabla_e^2-\frac{\hbar^2}{2m_h}\nabla_h^2-\frac{e^2}{4\pi\varepsilon|\mathbf{r}_e-\mathbf{r}_h|}\right]\Psi_{ex}=E\Psi_{ex} \tag{4.157}$$

Here m_e and m_h are the electron and hole masses, respectively; $|\boldsymbol{r}_e-\boldsymbol{r}_h|$ is the separation between an electron and a hole; and Ψ_{ex} is the exciton wavefunction. Using the following transformations,

$$r=r_e-r_h,\mathbf{k}=\frac{m_e\mathbf{k}_e-m_h\mathbf{k}_h}{m_e+m_h},\mathbf{R}=\frac{m_e\mathbf{r}_e+m_h\mathbf{r}_h}{m_e+m_h},\mathbf{K}=\mathbf{k}_e+\mathbf{k}_h, \tag{4.158}$$

where lowercase symbols represent the separation and wave vector in relative coordinates (**r**) and the uppercase symbols are the corresponding quantities in the center-of-mass (CM) coordinates, Eq. (4.157) is written as

$$\left(E_g - \frac{\hbar^2}{2M}\nabla_{\mathbf{R}}^2 - \frac{\hbar^2}{2m_r}\nabla_{\mathbf{r}}^2 - \frac{e^2}{4\pi\varepsilon r}\right)\Psi_{ex}(\mathbf{R},\mathbf{r}) = E\Psi_{ex}(\mathbf{R},\mathbf{r}) \quad (4.159)$$

Here m_r is the reduced mass defined earlier and $M = m_e + m_h$ is the CM mass. Assuming the wavefunction to be a product of two functions describing the CM and relative motions, we easily find that for the CM motion,

$$\psi_{CM} = \exp(j\mathbf{K}\cdot\mathbf{R}). \quad (4.160)$$

The solution for relative motion may be obtained from

$$\left(\frac{\hbar^2}{2m_r}\nabla_{\mathbf{r}}^2 - \frac{e^2}{4\pi\varepsilon|\mathbf{r}|}\right)F(\mathbf{r}) = EF(\mathbf{r}) \quad (4.161)$$

This is the usual hydrogen atom problem. The general form of the exciton wavefunction is

$$\Psi_{n\mathbf{K}_{ex}} = e^{j\mathbf{K}_{ex}\cdot\mathbf{R}}F_n(\mathbf{r})\phi_{\mathbf{c}}(\mathbf{r}_e)\phi_v(\mathbf{r}_h) \quad (4.162)$$

We write $\mathbf{K} = \mathbf{K}_{ex}$, and ϕ is the cell periodic part of Bloch functions at the band edges. The energies of the excitons are then

$$E_{n,\mathbf{K}_{ex}} = E_n + \frac{\hbar^2 {K_{ex}}^2}{2M}. \quad (4.163)$$

The second term represents the kinetic energy due to CM motion of EHPs, and E_n's are the hydrogen atom–like energy levels expressed as

$$E_n = -\frac{m_r e^4}{2(4\pi\varepsilon)^2\hbar^2}\frac{1}{n^2}. \quad (4.164)$$

Example 4.11

Take $m_e = 0.067\, m_0$, $m_h = 0.5\, m_0$, and $\varepsilon = 13\varepsilon_0$ for GaAs. The binding energy of the lowest (1s) exciton is 4.75 meV.

The above theory indicates that the usual hydrogenic model is employed to calculate the binding energy. The expression for the binding energy is similar to Eq. (2.18), with the effective mass replaced by the reduced mass of the EHP given by Eq. (4.59). A photon with energy slightly smaller than the band gap energy can be absorbed with the simultaneous creation of an exciton. The ideal absorption spectra corresponding to two hydogenic levels ($n = 1$ and 2) are shown in Figure 4.12 for a direct-gap semiconductor. The solid curve is for continuum state absorption (large values of n). Note that with increasing photon energy, the spectra show the usual $(\hbar\omega - E_g)^{1/2}$ behavior corresponding to the band-to-band transition.

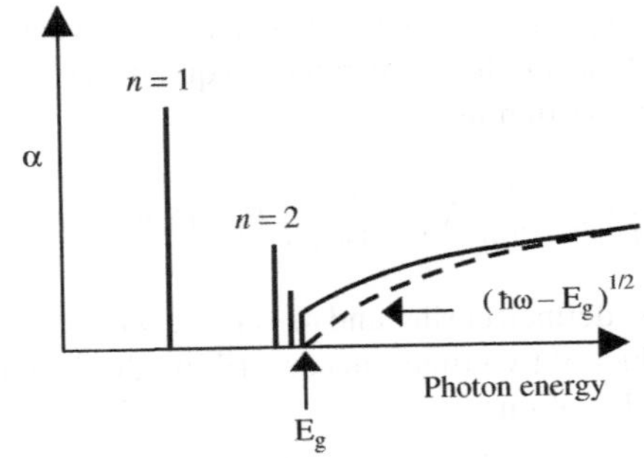

Figure 4.12 *Absorption coefficient in direct-gap semiconductors. The sharp peaks correspond to excitonic absorption for hydrogenic levels 1 and 2. The solid line is the absorption in the continuum states, which merge with the dashed curve representing the expected variation for band-to-band transition without excitonic effect.*

Experimental evidence for an excitonic effect is provided by an absorption peak just below the band gap. See, for example, the absorption spectra for GaAs obtained by Sturge [25]. As the temperature rises, the excitons dissociate into EHPs due to thermal effect.

Excitons are also formed in indirect-gap semiconductors. However, no absorption peak, as observed in GaAs, manifests itself in the absorption spectra. Instead, the excitons change the absorption spectra from the $(\hbar\omega - E_g)^2$ behavior to a simple linear relation. The deviations in the absorption curve due to the excitonic effect are seen at low temperatures in the experimental data for Si shown in Figure 4.5.

4.12.2 Impurity

As noted earlier, impurities give rise to localized states within the band gap of the semiconductors. The energy levels of the donors lie closer to the conduction band edge, while those for acceptors lie near the valence band edge.

Optical transitions can occur between localized states of an impurity atom. These transitions are governed by the same selection rules as applicable for isolated atoms. Often there is splitting of the different energy levels. The transitions usually take place at infrared frequencies.

In addition, optical absorption may take place between a level of an impurity and the continuum states in the band, for example, from a donor impurity level to the conduction band. This gives rise to photo-ionization of the impurity center. A variety of infrared detectors are fabricated by using this phenomenon. The experimental absorption spectra for B-doped Si, for example, show sharp peaks due to transitions between various levels of the impurity. At higher energy, the spectra show a broad tail corresponding to transition from a localized state to the band states.

4.12.3 Bound Excitons

The ground state of the donor or acceptor impurities already contains one bound electronic particle, an electron or a hole, respectively. The excited states for these centers can usually be described in the effective mass approximation based on the hydrogen-like Coulomb potential. Localization of an exciton to these centers creates the bound exciton (BE) excitation. A BE consists of three carriers (two holes and one electron for acceptor BE, and two electrons and one hole for donor BE) bound to a charged impurity. BEs give rise to localized states in the gap, and accordingly the absorption occurs at a photon energy slightly below the fundamental gap. Because in a BE three carriers are localized in the same region of real space, an Auger transition, in which an electron recombines with a hole and the energy is carried by the third carrier, can occur. Auger transitions are believed to limit the lifetimes of BEs in many case.

4.12.4 Isoelectronic Centres

Isoelectronic centers have no particle localized in the ground state, and the analog to the above-mentioned excitations for the donors and acceptors therefore does not exist. Such centers bind electrons (or holes) by a weak interaction (a strain field, or a local change of electronegativity). The cross-section for capture of the first carrier is usually quite small for these centers. Upon localization of the primary particle, the secondary particle, a hole (or an electron), can be captured in the Coulomb field of the first particle. This BE state containing two electronic particles associated with the isoelectronic center usually has very well-defined electronic energy, corresponding to a sharp emission–absorption line in optical spectra (if the phonon coupling in the optical transition is not too strong). These systems have been classified as isoelectronic donors and isoelectronic acceptors on the basis of whether the impurity is attractive for holes or electrons, respectively.

Problems

4.1 Show that the absorption coefficient for a material with finite conductivity $\boldsymbol{\sigma}$ may be related with the absorption coefficient by $\boldsymbol{\alpha} = \sigma/nc\varepsilon_0$, where n is the real part of the RI.

4.2 In the classical model, an electron bound to a nucleus is assumed to be displaced by an amount x from its equilibrium position when an electric field $E(t) = E_0\exp(-j\omega t)$ is applied. Using Newton's law and a damping constant, obtain the expression for x and hence the expressions for polarization and susceptibility.

4.3 Using the expressions for the real and imaginary parts of the susceptibility derived in Problem 4.2, show that the two are related by Kramers–Kronig relationships. Choose a proper contour in the complex ω-plane to arrive at these relationships.

4.4 Prove that $B_{12} = B_{21}$ and $A_{21}/B_{21} = (\hbar\omega^3 n^3)/\pi^2 c^3$.

4.5 Calculate the rate of increase of power *dP/dt* due to combined effect of stimulated emission and absorption. Using the fact that *dP/dt* = *(dP/dz) x velocity*, obtain the expression for α from $P(z) = P_0 exp(-\alpha z)$ and relate this to *dP/dt*. Hence prove that absorption is negative when $N_2 > N_1$, that is, for population inversion.

4.6 Show that the linewidth related to spontaneous emission is $h(A_{21})$.

4.7 Express Einstein's *A* and *B* coefficients in terms of momentum matrix elements.

4.8 The ratio of A^2 and $(A.p)$ terms in Eq. (4.40) is $e|A|/2p = e|F_0|/2p \cong (e/\omega p)$ $(2I/\varepsilon_0 cn)$. Assume intensity $I = 10^4\ \mathrm{Wm}^{-2}$, $f = 10^{14}$ Hz, and use momentum for free electron at 300 K. Hence show that the square term is negligible.

4.9 Establish the equivalence of the two forms of perturbation Hamiltonian as given in Eq. (4.43).

4.10 Using energy and momentum conservations, prove that absorption from the valence band to the degenerate conduction band starts at absolute zero at a photon energy $\hbar\omega = E_g + (E_F - E_c)(1 + m_e^*/m_h^*)$.

4.11 Establish Eq. (4.95): the Bernard–Duraffourg condition.

4.12 Derive Eq. (4.99). Also show that maxima in the gain spectra occur at (1/3(E_g + ΔF).

4.13 Calculate the electron concentration at transparency by using the following formula:

$$\frac{E_{Fc} - E_c}{k_B T} = \ln\left(\frac{n}{N_c}\right) + \sum_i A_i \left(\frac{n}{N_c}\right)^i,$$

where $A_1 = 0.35355$, $A_2 = -4.95009 \times 10^{-3}$, $A_3 = 1.48386 \times 10^{-4}$, $N_c = 2(m_e k_B T/2\pi\hbar^2)^{3/2}$. For holes one has a similar equation with electron mass m_e replaced by $\left(m_{hh}^{3/2} + m_{lh}^{3/2}\right)$. [W.B. Joyce and R.W. Dixon (1977) Appl. Phys. Lett, 31, 354–356 [26]] and the simplified expression for gain $g = (\Delta F - \hbar\omega)^2(\hbar\omega - E_g)$.

4.14 The DOS functions are expressed as $2 \times 10^{54}(E - E_c)^{1/2}$ and $4 \times 10^{55}(E_v - E)^{1/2}$, respectively, for the conduction and valence bands. Calculate the positions of the quasi-Fermi levels at 0 K. Calculate also the value of the gain at 1.43 eV when the band gap is 1.42 eV.

4.15 The maximum gain in a semiconductor is expressed as $g_{max} = \beta(J/d - J_0)$, where J is the current density, J_0 is the transparency current density, and d is the thickness of the active layer. Prove the threshold condition in a double heterojunction laser.

4.16 Obtain the expression for absorption coefficient in an indirect-gap material where alloy disorder scattering provides the momentum for transfer from a direct valley to the indirect one. Use $|M(\mathbf{k}, \mathbf{k}'|^2 = Bx(1-x)$ for the matrix element, where B is a prefactor and x is the alloy composition.

4.17 The second-order perturbation theory used to calculate absorption in an indirect gap may be employed to calculate two-photon absorption (TPA), provided $\hbar(\omega_1 + \omega_2) \geq E_g$. Show that in a direct gap, the TPA absorption may be expressed as $\alpha(\omega_1) = C_{TPA}(\hbar\omega_1 + \hbar\omega_2)^{1/2}$, where C is the proportionality constant and ω's are angular frequencies of the photons participating.

4.18 Following the method for obtaining the expression for absorption in a direct gap, obtain the expression for intervalence band absorption.

4.19 Prove that in an intraband free-carrier absorption process, electrons must absorb or emit phonons in order to conserve momentum.

4.20 Consider Eq. (4.112) for free-carrier absorption. Assuming that the electron mobility is limited by deformation potential acoustic phonon scattering, comment how the free-carrier absorption depends on temperature.

4.21 Consider intraband free-carrier absorption.. Assume that the angle between the initial wave vector **k** and the light wave vector κ is θ. Show that the energy of the photon to satisfy energy and momentum conservation conditions may be expressed as

$$\hbar\omega = 2\left[m^*\left(\frac{c}{n(\omega)}\right)^2 + \hbar k\left(\frac{c}{n(\omega)}\right)\cos\theta\right].$$

Take $m^* = 0.25\, m_0$, $n(\omega) = 3.5$, and $k = 0$. Calculate the value of photon energy needed.

4.22 Prove that the maximum recombination lifetime may be expressed as $\tau_r(\max) = (2B_r n_i)^{-1}$, appropriate for an intrinsic semiconductor, and that the radiative lifetime is reduced if the material is made of n or p type.

4.23 Prove that for a p-type material, the lifetime under low injection is given by $\tau_r = [B_r N_A]^{-1}$, where N_A is the acceptor concentration.

4.24 Assume that the reduced mass in semiconductors is approximately equal to the electron effective mass. Hence show that the exciton binding energy scales roughly with the band gap.

References

1. Bhattacharya, P. (1996) *Semiconductor Optoelectronic Devices*, 2nd edn, Prentice-Hall, Upper Saddle River, NJ.
2. Ghatak, A.K. and Thyagarajan, K. (1989) *Optical Electronics*, Cambridge University Press, Cambridge, UK.
3. Yu, P. and Cardona, M. (1995) *Fundamentals of Semiconductors*, Springer, Berlin.
4. Agrawal, G.P. and Dutta, N.K. (1986) *Long Wavelength Semiconductor Lasers*, Van Nostrand, New York.
5. Yariv, A. (1989) *Quantum Electronics*, 3rd edn, John Wiley & Sons, Inc., New York.
6. Ridley, B.K. (2000) *Quantum Processes in Semiconductors*, 5th edn, Clarendon Press, Oxford, UK.
7. Basu, P.K. (2003) *Theory of Optical Processes in Semiconductors: Bulk and Microstructures*, Clarendon Press, Oxford, UK.
8. Singh, J. (2003) *Electronic and Optoelectronic Properties of Semiconductor Structures*, Cambridge University Press, Cambridge, New York.
9. Shur, M. (1990) *Physics of Semiconductor Devices*, Prentice Hall, Englewood Cliffs, NJ.
10. Loudon, R. (1983) *The Quantum Theory of Light*, 2nd edn, Oxford University Press, Oxford, UK.
11. MacFarlane, G.G., McLean, T.P., Quarrington, J.E., and Roberts, V. (1958) Fine structure in the absorption-edge spectrum of Si. *Phys. Rev.*, **111**, 1245–1254.
12. MacFarlane, G.G., McLean, T.P., Quarrington, J.E., and Roberts, V. (1957) Fine structure in the absorption-edge spectrum of Ge. *Phys. Rev.*, **108**, 1377–1383.
13. McLean, T.P. (1960) *Progress in Semiconductors*, vol. **5**(eds A.F. Gibson, F.A. Kroger, and R.E. Burgess), Heywood, London, pp. 53–102.
14. Bucher, K., Bruns, J., and Wagemann, H.G. (1994) Absorption coefficient of silicon: an assessment of measurements and the simulation of temperature variation. *J. Appl. Phys.*, **75**, 1127–1132.
15. Bernard, M.G.A. and Duraffourg, G. (1961) Laser conditions in semiconductors. *Phys. Stat. Solid.*, **1**, 699–703.
16. Nag, B.R. (1980) *Electron Transport in Compound Semiconductors*, Springer, Berlin.
17. Soref, R.A. and Bennet, B.R. (1987) Electrooptical effects in silicon. *IEEE J. Quant. Electron.*, **QE-23**, 123–129.
18. van Roosebroeck, W. and Shockley, W. (1954) Photon radiative recombination of electrons and holes in Ge. *Phys. Rev.*, **94**, 1558–1560.
19. Shockley, W. and Read, W.T. (1952) Statistics of recombination of electrons and holes. *Phys. Rev.*, **87**, 835–842.
20. Hall, R.N. (1952) Electron-hole recombination in germanium. *Phys. Rev.*, **152**, 387.

21. Sah, C-T. and Shockley, W. (1958) Electron-hole recombination statistics in semiconductors through flaws with many charge conditions. *Phys. Rev.*, **109**, 1103–1115.
22. Dziewior, J. and Schmid, W. (1977) Auger coefficients for highly doped and highly excited silicon. *Appl. Phys. Lett.*, **31**, 346–348.
23. Hacker, R. and Hangleiter, A. (1994) Intrinsic upper limits of the carrier lifetime in silicon. *J. Appl. Phys.*, **75**, 7570–7572.
24. Kerr, M.J. and Cuevas, A. (2002) General parameterization of Auger recombination in crystalline silicon. *J. Appl. Phys.*, **91**, 2473–2480.
25. Sturge, M.D. (1962) Optical absorption in gallium arsenide between 0.6 and 2.75eV. *Phys. Rev.*, **127**, 768.
26. Joyce, W.B. and Dixon, R.W. (1977) Analytical approximations for Fermi energy of an ideal Fermi gas. *Appl. Phys. Lett.*, **31**, 354–356.

5

Optical Processes in Quantum Structures

5.1 Introduction

Optical processes in quantum structures have been intensively studied mostly for direct-gap semiconductors and their alloys. It is now known that photonic devices using quantum wells (QWs), quantum wires (QWRs), and quantum dots (QDs) show improved performance over their bulk counterparts. Optical processes in quantum structures involving Si, Ge, and their alloys are, on the other hand, less studied. Interest in this area took momentum with the announcement of light emission in porous silicon and the conclusion that quantum confinement plays an important role in the operation of the device.

There have been recent results in light emission from silicon-based quantum structures, in interband and intersubband transitions leading to photodetectors, and in strong excitonic absorption and its change with electric field in Ge–SiGe multiple quantum wells (MQWs): the quantum-confined Stark effect has sparked off renewed interest in the study of optical processes in Si-based quantum structures.

The present chapter gives the theory of different optical processes in nanostructures. Though some coverage of the processes in indirect-gap structures will be made, the present trend is to achieve direct-gap transitions in silicon-based structures. Therefore, in most cases, the existing theory for optical processes in direct-gap materials will be presented for understanding recently reported results and theoretical predictions in novel silicon-based structures of laser action and other physical phenomena having potential for device applications.

Various absorption and recombination processes occurring in QWs, QWRs, and QDs may occur in the following three ways:

1. Interband absorption and recombination: transitions between the subbands belonging to different bands
2. Intersubband processes: transition between subbands belonging to the same band, either the conduction or valence band
3. Intrasubband processes: transitions occur between states belonging to the same subband

Silicon Photonics: Fundamentals and Devices, First Edition. M. Jamal Deen and P. K. Basu.

The optical processes may involve electrons and holes as well as excitons. The processes are influenced by the presence of electric fields. As already mentioned, the active layer in which transitions occur may have either a direct or indirect gap.

The present chapter attempts to cover the basic theory involved in all the processes discussed above.

5.2 Optical Processes in QWs

A theory of optical processes in QWs may be developed along the same lines as used for bulk materials [1–5]. There are some modifications due to quantum confinement. We present in the following subsections the theory of absorption due to interband as well as intersubband transitions.

5.2.1 Absorption in Direct-gap QW

We consider that the band alignment is type I, so that both the electron and hole wavefunctions are confined in the well material. Therefore, in calculating the absorption coefficient, the transitions occurring in the well are to be considered. The initial- and final-state wavefunctions, respectively, for the *m*th subband in the valence band and the *n*th subband in the conduction band are given by

$$|1\rangle = |h, m, k_h\rangle = (AL)^{-1/2} U_h(\mathbf{r}, z) \exp(j\mathbf{k}_h \cdot \mathbf{r}) \phi_{hm}(z) \tag{5.1}$$

$$|2\rangle = |c, n, k_e\rangle = (AL)^{-1/2} U_c(\mathbf{r}, z) \exp(j\mathbf{k}_e \cdot \mathbf{r}) \phi_{cn}(z) \tag{5.2}$$

In the above, U's are cell-periodic parts of the Bloch function; ϕ's are envelope functions; **r** and **k** denote, respectively, the two-dimensional (2D) position and wave vectors; A is the area; and L is the width of the QW. The matrix element for transition may be evaluated by assuming that momentum is conserved in the QW layer plane, that is,

$$\mathbf{k_e} = \mathbf{k_h}. \tag{5.3}$$

Since the motion is quantized along the z-direction, there is no selection rule for the conservation of wave vector in that direction. The squared momentum matrix element may therefore be written as

$$p_{12} = (AL)^{-1} \langle \phi_{hm} | \phi_{cn} \rangle \int \exp[\, j(\mathbf{k_e} - \mathbf{k_h}) \cdot \mathbf{r}] \langle U_h | \mathbf{a.p} | \mathbf{U_c} \rangle d^2 r. \tag{5.4}$$

The squared momentum matrix element may therefore be written as

$$|p_{12}|^2 = A_{mn} \left\langle |p_{cv}|^2 \right\rangle C_{mn} \delta_{\mathbf{k_e}, \mathbf{k_h}} \tag{5.5}$$

where

$$C_{mn} = |\langle \phi_{hm} | \phi_{cn} \rangle|^2 = \left| \int \phi_{cn} \phi_{hm} dz \right|^2, \tag{5.6}$$

denotes the overlap between the electron and hole envelope functions. It arises since there is no selection rule for k-conservation along the z-direction. For infinite potential barriers, the envelope functions are sin (cos) functions and the overlap becomes zero, unless $n = m$. In that case, one may write $C_{nm} = \delta_{mn}$. For finite barrier height, the envelope functions leak into the barrier and this selection rule is not valid. Even when $m = n$, the factor C_{mm} is different from its maximum value.

Using parabolic dispersion for the electrons and holes, the argument of the energy-conserving δ-function is written by using 2D **k**-conservation as

$$E_e(\mathbf{k_e}) - E_h(\mathbf{k_h}) - \hbar\omega = (E_g + E_{cn} + E_{hm}) - \frac{\hbar^2 {k_h}^2}{2m_r} - \hbar\omega, \qquad (5.7)$$

where m_r is the reduced mass. Proceeding as in the bulk case, and noting that

$$\sum_{\mathbf{k}} \rightarrow (2A/4\pi^2) \int \delta\left(E_{gmn} - \frac{\hbar^2 k^2}{2m_r} - \hbar\omega\right) 2\pi k dk [f_h(k) - f_e(k)],$$

the integration may easily be performed due to the presence of the δ-function. The final expression for the absorption coefficient is

$$\alpha(\hbar\omega) = \frac{e^2 \left\langle |p_{cv}|^2 \right\rangle}{\varepsilon_0 {m_0}^2 c\hbar n \hbar\omega L} m_r C_{mn} A_{mn} (f_h - f_e) H(\hbar\omega - E_{gmn}) = \alpha_0 (f_h - f_e) H(\hbar\omega - E_{gmn}) \quad (5.8)$$

where $H(x)$ is the Heaviside step function and $E_{gmn} = (E_g + E_{cn} + E_{hm})$ is the effective band gap in the QW, and f's denote the occupancy of the subbands in terms of quasi-Fermi levels.

In both the cases of bulk and QW, the absorption coefficients are proportional to the joint density-of-states. The absorption in QW is, however, step-like as indicated by the presence of step function. Figure 5.1 shows the transitions allowed by selection rule $m = n$ and the nature of variation of the absorption coefficient as given by Eq. (5.8)

Example 5.1

Consider a GaAs QW with L = 10 nm having infinite barriers. Using standard parameters $m_{ez} = 0.067m_0$, $m_{hhz} = 0.45m_0$, $m_{ell} = 0.067m_0$, $mm_{hhll} = 0.112m_0$, $|p_{cv}|^2 = 2.7\, m_0 E_g C_{mn} = 1$, $A_{11} = 1.5$, $n = 3.6$, the value of absorption becomes $4.13 \times 10^5\ \mathrm{m}^{-1}$. The effective band gap is 1.49 eV. It is assumed that $f_e = 1$ and $f_h = 0$.

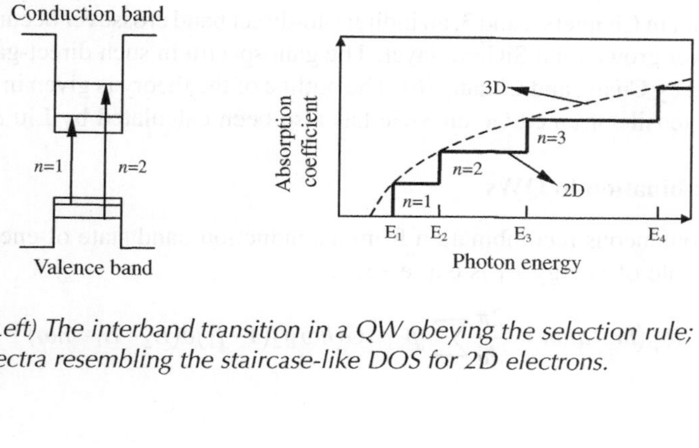

Figure 5.1 (Left) The interband transition in a QW obeying the selection rule; and (right) the absorption spectra resembling the staircase-like DOS for 2D electrons.

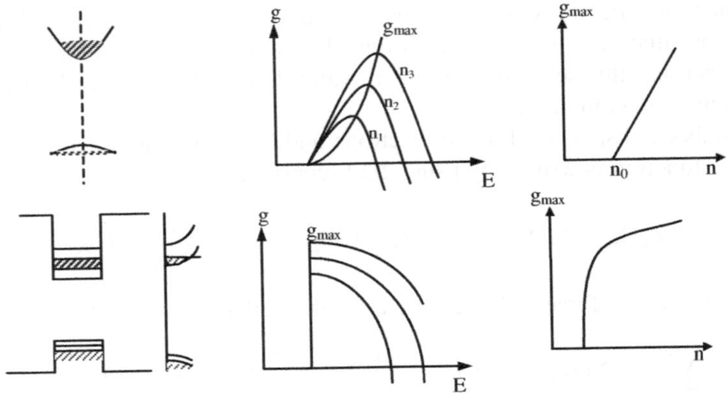

Figure 5.2 Qualitative variation of gain spectra and of g_{max} in a QW (bottom). The variation for bulk materials is included for comparison (top).

5.2.2 Gain in QW

It appears from Eq. (5.8) that when $f_e = f_h$, the absorption coefficient is zero. Assuming the quasi-neutrality condition, that is, injected electron and hole densities are equal, $n = p$, the condition $f_e = f_h$ is satisfied at the transparency carrier density n_0. Above this value $f_e > f_h$, $F_e - F_h > \hbar\omega$ is satisfied and amplification of electromagnetic waves occurs. Examining the forms for f_e and f_h, one may conclude that the maximum gain occurs at $\hbar\omega = E_{gmn}$ and the expression for maximum gain is

$$g_{\max} = \alpha_0[f_e(E_{cn}) - f_h(E_{mh})], \tag{5.9}$$

where α_0 is given in Eq. (5.8).

The qualitative variation of gain may be understood by examining the approximate expression $g(\hbar\omega) = K(\Delta F - \hbar\omega)^2 H(\hbar\omega - E_{mn})$, obtained by using the linear approximation for f's as given by Eq. (4.98) (see "Problems" section of this chapter).

Figure 5.2 gives the plot of gain as a function of photon energy in a QW for different values of injected carrier density. The variation of maximum gain with carrier density is shown also. For the sake of comparison, variations of similar quantities for bulk materials are included. The basic differences can be easily identified from the plots.

As pointed out in Chapters 2 and 3, an indirect-to-direct band crossover occurs in a tensile-strained Ge layer grown on a SiGeSn layer. The gain spectra in such direct-gap QWs have been calculated by Chang and Chuang [6]. The outline of the theory is given in Appendix A.

The gain in tensile-strained Ge on SiGe has also been calculated by Liu *et al.* [7].

5.2.3 Recombination in QWs

The rate of spontaneous recombination from a conduction band state of energy E_2 to the valence band state of energy E_1 is expressed as

$$R_{sp}(\hbar\omega)d\hbar\omega = \frac{2\pi}{\hbar}\sum |p_{21}|^2 G(\hbar\omega) f_2(1-f_1)\delta(E_2 - E_1 - \hbar\omega) \tag{5.10}$$

Using the expression for the optical density-of-states function, the wavefunctions for electrons in the nth conduction subband, and that for holes in the mth valence subband, and expressing f_1 and f_2 in terms of quasi-Fermi levels, the spectrally dependent spontaneous emission rate becomes

$$R_{sp}(\hbar\omega) = \frac{e^2 n m_r \hbar\omega}{\pi^2 {m_0}^2 \varepsilon_0 c^3 \hbar^4 L} \sum_{m,n} \left\langle |p_{cv}|^2 \right\rangle C_{mn}\, f_e(E_{cn} - F_e)[1 - f_h(E_{vm} - F_h)], \quad (5.11)$$

where **k**-conservation has been assumed to be obeyed. The emitted radiation is unpolarized, and therefore the term $A_{mn} = 1$.

In order to calculate the total recombination rate, the above expression should be integrated over all values of photon energy. Therefore,

$$R_{sp} = B \int f_e[1 - f_h]dE, \quad (5.12)$$

where all the terms that are independent of photon energy in Eq. (5.11) have been taken out of the integral and lumped into the symbol B. The probabilities are expressed in terms of electron density n, such that

$$f_e = \frac{n\hbar^2\pi}{m_e k_B T} \frac{1}{\exp[(E - F_e)/k_B T] + 1} \quad (5.13)$$

A similar equation can be written for f_h in terms of hole density p, hole effective mass m_h, and quasi-Fermi level F_h. Substituting f_e and f_h in Eq. (5.12), the total recombination rate may be calculated. A useful approximation is to assume that the distributions f_e and f_h are nondegenerate. In that situation,

$$f_e[1 - f_h] \to \exp\left[-(\hbar\omega - E_{gmn})/k_B T\right] \exp\left[-(E_{gmn} - \Delta F)/k_B T\right], \quad \Delta F = F_e - F_h$$

Inserting this in the expression for $R_{sp}(\hbar\omega)$ and noting that the prefactor is a slowly varying function of $\hbar\omega$, the integral is of the form $\int_0^\infty \exp(-x)dx = 1$. It is also noted that $\exp\left[-(E_{gmn} - \Delta F)/k_B T\right]$ is proportional to product np. Therefore the total spontaneous emission rate may be expressed as

$$R_{sp} = \frac{e^2 n_r \hbar\omega m_r}{\pi^2 {m_0}^2 \varepsilon_0 L c^3 \hbar^4} \left\langle |p_{cv}|^2 \right\rangle C_{mn} \frac{(\pi\hbar^2)^2}{m_e m_h k_B T} np = B_r np \quad (5.14)$$

The lifetime of excess carriers in a QW is defined as

$$\tau = (B_r n_0)^{-1} \quad (5.15)$$

where n_0 is the electron density under equilibrium.

5.2.4 Polarization-Dependent Momentum Matrix Element

The anisotropy factor A_{mn} appearing in the expression for gain and absorption is absent in the bulk material. In QW, however, its value depends on the orientation of the electric field vector of the emitted radiation with respect to the direction of growth of the QW.

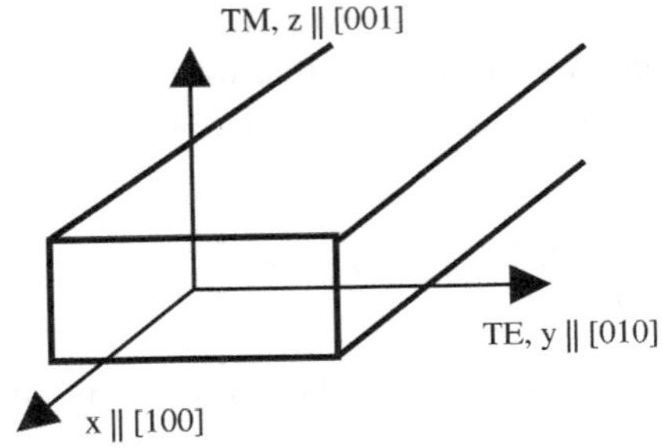

Figure 5.3 Geometry of a QW layer plane. The electric field vectors for TE and TM polarized lights are shown.

The matrix elements for various polarization directions in bulk semiconductors have been expressed in Eqs. (4.55) and (4.56).

The QW is assumed to grow along the (001) direction, which we denote as the z-direction in Figure 5.3. Emission is along the x-direction in the QW layer plane. The light is transverse electric (TE) polarized if the electric field vector is along the $y = [010]$ direction, while it is transverse magnetic (TM) if the light field vector is along the z-direction. We may therefore rewrite the above results as follows:

$$\begin{aligned} \text{TE polarization:}\quad & \text{HH} \rightarrow \text{CB} : |p_{if}|^2 = (1/2)p_{cv}^2;\ A_{11} = 3/2; \\ & \text{LH} \rightarrow \text{CB} : |p_{if}|^2 = (1/6)p_{cv}^2; \\ \text{TM polarization:}\quad & \text{HH} \rightarrow \text{CB} : |p_{if}|^2 = 0; \\ & \text{LH} \rightarrow \text{CB} : |p_{if}|^2 = (2/3)p_{cv}^2 \end{aligned} \tag{5.16a}$$

For finite values of $\boldsymbol{k}$, the values of A_{mn} are altered. If we define a quantity $\cos\theta = E_{cn}(\mathbf{k})/E_{cn}$, the anisotropy factors may be expressed as follows:

$$\begin{array}{lll} & \text{CB} \rightarrow \text{HH} & \text{CB} \rightarrow \text{LH} \\ \text{TE} & (3/4)(1+\cos^2\theta) & (1/4)(5-3\cos^2\theta) \\ \text{TM} & (3/2)\sin^2\theta & (1/4)(4-3\sin^2\theta) \end{array} \tag{5.16b}$$

The derivation of these results is given by Asada *et al.* [8] and Yamanishi and Suemune [9].

5.2.5 Absorption in the Indirect Gap

Optical processes in QWs formed by indirect-gap materials like Si, Ge, and their alloys are of current interest. The most studied system is a QW made of Si and $Si_{1-x}Ge_x$. In this section, we will investigate the nature of absorption in such systems [10].

We shall assume, as in the case of bulk Si, that the momentum needed for transfer of electrons from the intermediate state to the Δ-like conduction band valleys is provided by intervalley phonons. The transition rate is expressed as

$$W_2(\mathbf{k}) = \frac{2\pi}{\hbar}\frac{M^2}{(E_{g0}-\hbar\omega)^2}\frac{\hbar D_{ij}^2 g_v}{2\rho V\omega_{ij}}\sum_{\mathbf{k}'}(\mathbf{N_Q}+1/2\mp 1/2)\delta(E_{\mathbf{k}'}-E_{\mathbf{k}}\mp\hbar\omega_{ij}) \tag{5.17}$$

where the rate is for bulk materials and the symbols are already defined. We now seek a modification of the expression for the case of QW.

Let us assume that an electron of 2D wave vector **k** in the lth subband of the heavy-hole band absorbs a photon of energy $\hbar\omega$ and reaches a virtual intermediate state in the conduction band at the zone center. The electron finally reaches a state of wave vector $\mathbf{k}'$ in the mth subband of the conduction band through the emission or absorption of an intervalley phonon of energy $\hbar\omega_{ij}$. The wavefunctions for these initial, intermediate, and final states may be written, respectively, as

$$|v\mathbf{k}\rangle = U_v \exp(j\mathbf{k}\cdot\mathbf{r})\phi_{hl}(z) \tag{5.18a}$$

$$|\Gamma\mathbf{k}\rangle = U_\Gamma \exp(j\mathbf{k}\cdot\mathbf{r})\phi_\Gamma(z) \tag{5.18b}$$

$$|c\mathbf{k}'\rangle = U_c \exp(j|\mathbf{k}'-\mathbf{k_0}|\cdot\mathbf{r})\exp(-jk_{0z}z)\phi_{cm}(z) \tag{5.18c}$$

The intermediate state is in the zone center conduction band and is well above the hetero-barrier. The electrons in that band, however, see a change in effective mass at the hetero-interfaces and are therefore reflected. We are not interested in the proper form of $\phi_\Gamma(z)$ at present. The indirect valleys are located at points ($\mathbf{k}_0$, k_{0z}) in **k**-space.

With the above definitions of the wavefunctions, one may write the squared matrix element for photonic processes as

$$M_\lambda^2 = \left(\frac{eA_0}{m_0}\right)^2\left\langle|p_{\Gamma v}|^2\right\rangle A_{l\Gamma}C_{l\Gamma} \tag{5.19}$$

where

$$C_{l\Gamma} = \left|\int \phi_\Gamma(z)\phi_{hl}(z)dz\right|^2, \quad p_{\Gamma v} = \langle v\mathbf{k}|\mathbf{a}\cdot\mathbf{p}|\Gamma\mathbf{k}\rangle. \tag{5.20}$$

and where A is the polarization-dependent part and $\left\langle|p_{\Gamma v}|^2\right\rangle$ is the average squared matrix element. In obtaining Eq. (5.20), the arguments given in Section 5.2.2 are followed.

The matrix element for a transition from the intermediate state to the final state via emission and absorption of a phonon is given by

$$M_{ij}^2 = |\langle c\mathbf{k}'|H_{ep}|\Gamma\mathbf{k}\rangle|^2 = \sum_{\mathbf{k}',q_z}\left(\frac{\hbar D_{ij}^2}{2\rho V\omega_{ij}}\right)(N_\mathbf{Q}+1/2\mp 1/2)\delta_{\mathbf{k}+\mathbf{q},\mathbf{k}'-\mathbf{k_0}}|I(q_z)|^2. \tag{5.21}$$

The interaction potential is assumed to be independent of the phonon wave vector $\mathbf{Q} = (\mathbf{q}, q_z)$, and $-$ ($+$) refers to phonon absorption (emission) processes. The Kronecker delta signifies that the momentum is conserved in the plane of the QW. On the other hand, there is no selection rule for the z-direction and hence the term $I(z)$, known as the *overlap integral* for electron–phonon interaction and as expressed below, appears:

$$I(q_z) = \int_0^L \phi_{cm}(z)\exp(jq_z z)\phi_\Gamma(z)dz. \tag{5.22}$$

Since we are not interested in the actual form for different scattering processes, we write the sum over q_z in the following form:

$$\sum_{q_z} |I(q_z)|^2 = \frac{L}{2\pi} \int dq_z |I(q_z|^2 = \frac{L}{2\pi} F \tag{5.23}$$

In Eq. (5.16b), $W_2(\mathbf{k})$ is the probability of transition from a fixed state $|v\mathbf{k}\rangle$ to all the final states $|c\mathbf{k}'\rangle$. To calculate the total probability W at energy $\hbar\omega$, one needs to sum over all the initial states $\mathbf{k}$, ensuring energy conservation. Thus the expression for the transition probability becomes

$$\begin{aligned} W(\hbar\omega) = & \frac{2\pi}{\hbar}\left(\frac{eA_0}{m_0}\right)^2 \frac{A_{l\Gamma} C_{l\Gamma} \left\langle |p_{\Gamma v}|^2 \right\rangle}{(E_{g0} - \hbar\omega)^2} \left(\frac{\hbar D_{ij}{}^2}{2\rho V \omega_{ij}}\right) \\ & \times \sum_{\mathbf{k},\mathbf{k}',q_z} |I(q_z)|^2 (N_\mathbf{Q} + 1/2 \mp 1/2) \delta(E_{\mathbf{k}'} - E_\mathbf{k} \mp \hbar\omega_{ij}) \end{aligned} \tag{5.24}$$

The summation over q_z is converted into an integral. The summations over $\mathbf{k}$ and $\mathbf{k}'$ may be converted into the following integral:

$$J = \int \delta(E_{\mathbf{k}'} - E_\mathbf{k} \mp \hbar\omega_{ij}) k' dk' d\theta' k dk d\theta \tag{5.25}$$

Using parabolic dispersion relations, the k$'$ integration is easily performed by using the δ-function, yielding a constant. The integral over k is converted to an integration over $E_k = \hbar^2 k^2 / 2m_h$ with limits 0 and $(\hbar\omega - E_g + \hbar\omega_{ij})$. Collecting all the constant factors, taking a factor 2 for spin degeneracy and a term g_v for valley degeneracy, one obtains

$$\alpha(\hbar\omega) = W(\hbar\omega) n_r / c = \frac{B_{bbph}}{\hbar\omega} \frac{1}{L} (N_\mathbf{Q} + 1/1 \mp 1/2)(\hbar\omega - E_g \pm \hbar\omega_{ij}) H(\hbar\omega - E_g \pm \hbar\omega_{ij}), \tag{5.26}$$

where

$$B_{bbph} = \frac{e^2 A_{l\Gamma} C_{l\Gamma} \left\langle |p_{\Gamma v}|^2 \right\rangle D_{ij}^2 m_e m_h g_v F(\Gamma, c)}{8\pi^2 m_0{}^2 \varepsilon_0 n \rho c \omega_{ij} \hbar^2 (E_{g0} - \hbar\omega)^2}. \tag{5.27}$$

In Eqs. (5.26) and (5.27), $H(x)$ is a step function, and E_g is the effective band gap plus the subband energies for an electron and a hole. The absorption coefficient in the above denotes the intensity reduction of light as it propagates through the QW layer plane.

Equation (5.27) for the absorption coefficient holds good at absolute zero. There will be modifications if broadening of the energy levels and finite temperature effects are included. Also, the expressions should be changed if the anisotropy of the band structures is included. These refinements are necessary only when the experimental data need thorough scrutiny. The only noteworthy feature is that the absorption coefficient varies as $(\hbar\omega - E')$ instead of $(\hbar\omega - E')^2$, as in bulk materials. The simple linear dependence also alters when excitonic effects are included, as will be pointed out in Section 5.4.

The absorption spectra, as obtained qualitatively from Eq. (5.26), are shown in Figure 5.4.

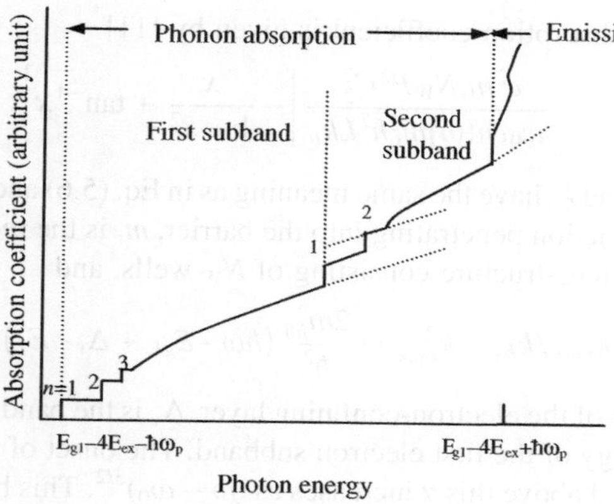

Figure 5.4 *Ideal absorption spectra in an indirect-gap QW.*

5.2.6 Absorption in Type II QWs

The spatial separation of electron and hole wavefunctions is illustrated schematically in Figure 5.5. It is evident that the overlap of the envelope functions given by Eq. (5.6) is due only to the exponential tails of these two functions. The contribution is zero if the barrier potential is infinite. The overlap also increases with increase in the subband index n or m.

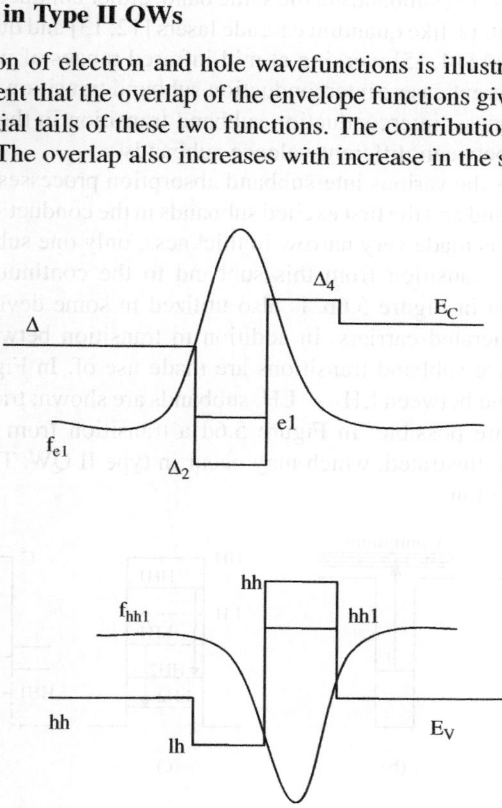

Figure 5.5 *A spatially indirect transition in a type II QW. The overlap between electron and hole envelope functions occurs due to the exponential tails.*

The expression for the absorption coefficient is given by [11]

$$\alpha(\omega) = \frac{e^2 m_r N_W P^2 C_2^2}{\pi \varepsilon_0 c n(\omega) m_0^2 \hbar^2 L k_b} \left[-\frac{x}{1+x^2} + \tan^{-1} x \right]. \quad (5.28)$$

where the constants C_2 and k_b have the same meaning as in Eq. (5.6) and are associated with the electron envelope function penetrating into the barrier, m_r is the reduced mass, L is the total thickness of the heterostructure consisting of N_W wells, and

$$x = k_{\max}/k_b, \quad k_{\max}^2 = \frac{2m_{hh}^*}{\hbar^2} \left(\hbar\omega - E_{gE} + \Delta_v - E_1 \right). \quad (5.29)$$

Here E_{gE} is the band gap of the electron-confining layer, Δ_v is the band offset in the valence band, and E_1 is the energy of the first electron subband. The onset of absorption occurs at $\hbar\omega_0 = E_{gE} - \Delta_v + E_1$, and above this α increases as $(\omega - \omega_0)^{3/2}$. This behavior is similar to that observed in bulk indirect-gap semiconductors.

5.3 Intersubband Transitions

Transition between different subbands in the same band, either conduction or valence band, is involved in light emitters, like quantum cascade lasers [12, 13] and quantum well infrared photodetectors (QWIPs) [14, 15] working at mid-infrared ranges of wavelength. Here we shall discuss three different cases: interconduction subband transition in a material having an isotropic effective mass, interconduction subband transition in the case of anisotropic mass, and transitions between different-valence subbands.

Figure 5.6 illustrates the various intersubband absorption processes. In Figure 5.6a, the transition between ground and the first excited subbands in the conduction band of the QW is shown. When the QW is made very narrow in thickness, only one subband appears in the conduction band. The transition from this subband to the continuum states above the heterobarrier, as shown in Figure 5.6b, is also utilized in some devices that offer better collection of photogenerated carriers. In addition to transition between interconduction band states, intervalence subband transitions are made use of. In Figure 5.6c transitions between HH → HH and between LH → LH, subbands are shown; transitions like HH → LH and HH → SO are possible. In Figure 5.6d a transition from a HH subband to a conduction subband is illustrated, which may occur in type II QW. This is an example of spatially indirect transition.

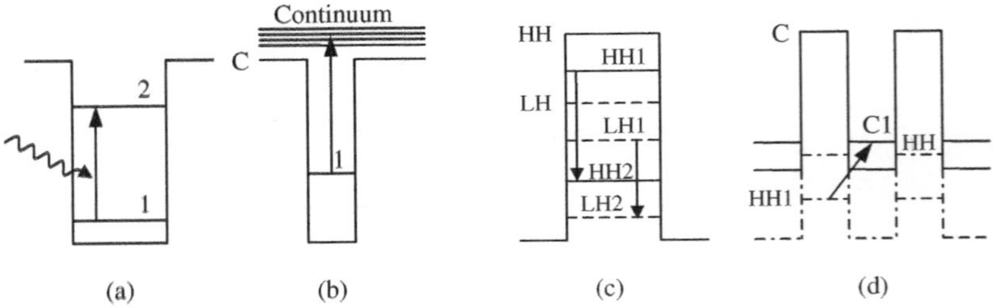

Figure 5.6 *Various intersubband transitions: (a) from subband 1 to subband 2 in the conduction band; (b) from subband 1 to continuum states; (c) between hole subbands; and (d) an indirect transition in real space from a hole subband to electron subband in type II QWs.*

Example 5.2

Consider the values of subband energies given in Example 3.3. The difference is 169 meV, and the wavelength of photon involved in the transition is 7.34 μm.

5.3.1 Conduction Subbands: Isotropic Mass

Assume that the QW is grown along the z-direction. We consider two subbands, i and $f, f > i$ in the conduction band, for which the wavefunctions may be written as

$$\psi_i(\mathbf{k}, z) = (AL)^{-1/2} U_{ci}(\mathbf{r}) \exp(j\mathbf{k}\cdot\mathbf{r})\phi_i(z), \tag{5.30}$$

$$\psi_f(\mathbf{k}, z) = (AL)^{-1/2} U_{cf}(\mathbf{r}) \exp(j\mathbf{k}\cdot\mathbf{r})\phi_f(z). \tag{5.31}$$

As defined, A is the area, L is the well width, $\mathbf{k}$, and $\mathbf{r}$ are 2D vectors, and ϕ's are envelope functions. We assume that the central cell function U remains unchanged for both the subbands. Since ϕ's vary slowly in comparison to U_c, the matrix element may be written as

$$\langle\psi_f|V_p|\psi_i\rangle = \langle U_{cf}|V_p|U_{ci}\rangle_{cell}\langle\phi_f|\phi_i\rangle + \langle U_{cf}|U_{ci}\rangle_{cell}\langle\phi_f|V_p|\phi_i\rangle, \tag{5.32}$$

where the subscript *cell* means that the scalar product is evaluated over a unit cell and V_p stands for the interaction potential.

For transitions within the same band, the first term is zero since ϕ_f and ϕ_i are the eigenfunctions of the same effective mass Hamiltonian and hence are orthogonal to each other.

The cell periodic functions U's are expressed in **k.p** theory as

$$U_{cnk} = U_{n0} + \frac{\hbar}{m_0}\sum_{m0}\frac{\mathbf{k}\cdot\langle \mathbf{m}0|\mathbf{p}|n0\rangle U_{m0}}{E_{no} - E_{m0}}. \tag{5.33}$$

These eigenfunctions are used to calculate the scalar product in the second term in Eq. (5.32). Noting that the effective mass in **k.p** theory is

$$\frac{1}{m_e} = \delta_{ij} + \frac{2}{m_0}\sum_{m0}\frac{\langle m0|\mathbf{p}|n0\rangle\langle n0|\mathbf{p}|m0\rangle}{E_{n0} - E_{m0}}, \tag{5.34}$$

the matrix element as given by Eq. (5.32) is now expressed as

$$\langle\psi_f|V_p|\psi_i\rangle = -\frac{eA_0\langle\phi_f|\mathbf{a}\cdot\mathbf{p}|\phi_i\rangle}{m_e}. \tag{5.35}$$

We have used the standard form of V_p. Notice that the effective mass m_e appears in the expression now, instead of free-electron mass m_O appearing in the matrix element for interband transitions. The above matrix element involving momentum may be converted into an expression involving a position operator and thus

$$\langle\psi_f|V_p|\psi_i\rangle = -\frac{eA_0(E_i - E_f)}{j\hbar}\mathbf{a}\cdot\langle\phi_f|\mathbf{r}|\phi_i\rangle. \tag{5.36}$$

Since the envelope functions ϕ_m and ϕ_n are functions of z only and orthogonal to each other, it follows that only the matrix element $\langle\phi_f|z|\phi_i\rangle$ survives and the other matrix elements $\langle\phi_f|x|\phi_i\rangle$ and $\langle\phi_f|y|\phi_i\rangle$ are zero.

Using the expression for A_0 and following the standard steps, the absorption coefficient may be expressed as

$$\alpha_{mn}(\hbar\omega) = \frac{\omega e^2}{Ln_r\varepsilon_0 c}\sum_n (f_{ei}-f_{ef})^2\langle z\rangle^2\delta(E_f-E_i-\hbar\omega); \quad \langle z\rangle = \langle\phi_f|z|\phi_i\rangle. \tag{5.37}$$

In Eq. (5.37), f_{em} and f_{en} denote the Fermi occupational probabilities for electrons in the initial and final subbands, respectively, and the summation is over all the final states satisfying momentum and energy conservation. The summation is converted into integration over the 2D wave vector **k**, and making use of the δ-function the integration is easily performed to yield

$$\alpha(\hbar\omega) = \sum_{i,f}\frac{\omega e^2 m_e k_B T}{Ln_r\varepsilon_0 c\hbar^2}\langle z\rangle^2 \ln\frac{1+\exp[(E_F-E_i)k_BT]}{1+\exp[(E_F-E_f)k_BT]}\frac{(\Gamma/2)}{(\hbar\omega-E_{fi})^2+(\Gamma/2)^2}. \tag{5.38}$$

In this expression E_F is the Fermi level, $E_{fi}=E_f-E_i$; and Γ is the linewidth for transition. Now the ratio related to the logarithmic term is proportional to the difference in population in the two subbands (N_i-N_f). We rewrite the above expression as

$$\alpha(\hbar\omega) = \frac{\omega e^2}{n_r\varepsilon_0 cL}\langle z\rangle^2(N_i-N_f)L(\hbar\omega, E_{fi}), \tag{5.39}$$

where $L(\hbar\omega, E_{fi})$ stands for the Lorentzian lineshape given above. The peak of the absorption occurs at E_{fi}.

The oscillator strength for transition from the ground subband to the first excited subband is expressed as $f_{osc}=(2m_0E_{21}/\hbar^2)\langle z\rangle^2$. In a QW with infinite barrier, $\langle z\rangle = 16L/9\pi^2$. The value of f_{osc} is $f_{osc}=(m_0/m_e)(256/27\pi^2)=14.34$, when $m_e=0.067\,m_0$ as in GaAs.

Example 5.3

The value of the peak absorption coefficient will be calculated for GaAs QWs using Eq. (5.39). The actual width is 6.5 nm; however, the same values of subband energies are obtained if $L=10.13$ nm and barrier height is infinite. For an electron density of $1.6\times10^{17}\,\text{cm}^{-3}$, $E_1=54.7$ meV, $E_2=218.8$ meV, and $E_F=6.49$ meV. Taking half width $=5$ meV and expressing $\langle z\rangle^2$ in terms of f_{osc}, the peak absorption coefficient becomes $2.9\times10^5\,\text{m}^{-1}$.

5.3.2 Anisotropic Mass

The theory in Section 5.3.1 needs some modification for Si, Ge, or SiGe alloys, since the effective mass is not isotropic, but depends on the chosen direction. As noted, the interaction potential between an electron and the radiation field is

$$V_p = (eA_0/m_0)\mathbf{a}\cdot\mathbf{p}. \tag{5.40}$$

The matrix element for transition from a Bloch state $\psi_{\mathbf{k}}$ to another state $\psi_{\mathbf{k}'}$ may be written as [1]

$$M(\mathbf{k},\mathbf{k}') = (jeA_0/\hbar)\mathbf{a}\cdot\int \psi_{\mathbf{k}'}(\nabla_{\mathbf{k}}E_{\mathbf{k}})\psi_{\mathbf{k}}d^3r, \tag{5.41}$$

where the wavefunctions may be written as in Eqs. (5.30) and (5.31) in Section 5.3.1. In a multivalley semiconductor, the *E–k* relationship is given by

$$E_{\mathbf{k}} = \sum_{i,j}\frac{\hbar^2 k_i k_j}{2m_{ij}}, \quad i,j = x,y,z. \tag{5.42}$$

Then the matrix element defined above reduces to [16]

$$M(\mathbf{k},\mathbf{k}') = (je\hbar\mathbf{A_0})\delta_{\mathbf{k},\mathbf{k}'}\begin{bmatrix}\langle k_x\rangle\left\{\dfrac{\mathbf{a}\cdot\mathbf{i_x}}{m_{xx}}+\dfrac{\mathbf{a}\cdot\mathbf{i_y}}{m_{xy}}+\dfrac{\mathbf{a}\cdot\mathbf{i}_z}{m_{xz}}\right\}+\langle k_y\rangle\left\{\dfrac{\mathbf{a}\cdot\mathbf{i_x}}{m_{yx}}+\dfrac{\mathbf{a}\cdot\mathbf{i_y}}{m_{yy}}+\dfrac{\mathbf{a}\cdot\mathbf{i}_z}{m_{yz}}\right\}+\\ \langle k_z\rangle\left\{\dfrac{\mathbf{a}\cdot\mathbf{i_x}}{m_{zx}}+\dfrac{\mathbf{a}\cdot\mathbf{i_y}}{m_{zy}}+\dfrac{\mathbf{a}\cdot\mathbf{i}_z}{m_{zz}}\right\}\end{bmatrix}. \tag{5.43}$$

In Eq. (5.43), $\mathbf{i}_x$ and so on are the unit vectors along the respective directions, $\langle k_x\rangle = \langle\phi_f|k_x|\phi_i\rangle$, and $\langle k_y\rangle$ and $\langle k_z\rangle$ are similarly defined. We note, however, that $\langle k_y\rangle = \langle k_x\rangle = 0$, since ϕ_f and ϕ_i are orthogonal. The matrix element thus reduces to

$$M(\mathbf{k},\mathbf{k}') = -je\hbar A_0\langle k_z\rangle\delta_{\mathbf{k},\mathbf{k}'}\left\{\frac{\mathbf{a}\cdot\mathbf{i_x}}{m_{zx}}+\frac{\mathbf{a}\cdot\mathbf{i_y}}{m_{zy}}+\frac{\mathbf{a}\cdot\mathbf{i}_z}{m_{zz}}\right\}. \tag{5.44}$$

Since in direct-gap materials, the nondiagonal elements in the effective mass tensor are zero, the above matrix element contains only the term involving m_{zz}. Therefore, replacing m_{zz} by m_e the isotropic mass, the above matrix element reduces to the expression for the case with isotropic effective mass. A noteworthy feature of Eq. (5.44) is that unlike the earlier case of isotropic effective mass, the *xy* polarized electric field of the radiation can cause the motion of electrons along the *z*-direction through the terms involving m_{xz} and m_{yz} and thus induce intersubband transitions.

Writing $\langle k_z\rangle$ in terms of $\langle z\rangle$ and following the prescribed method, the expression for the absorption coefficient takes the form

$$\alpha(\hbar\omega) = \frac{\omega e^2\langle z\rangle^2}{n_r\varepsilon_0 cL}\frac{m_{zz}}{m_0}\left\{\frac{\mathbf{a}\cdot\mathbf{i_x}}{m_{zx}}+\frac{\mathbf{a}\cdot\mathbf{i_y}}{m_{zy}}+\frac{\mathbf{a}\cdot\mathbf{i}_z}{m_{zz}}\right\}(N_i - N_f)L(\hbar\omega, E_{fi}). \tag{5.45}$$

An estimate of the absorption coefficient for different QW growth directions and for different polarization directions has been made by Yang *et al.* [16]. It is mentioned that the value of α for a GaAs QW at 10 μm may be 4800 cm^{-1}, whereas for Si or Ge QWs the peak values may be 20 000 cm^{-1} for [100] growth and *x*-polarization and even 81000 cm^{-1} for [111] growth and *z*-polarization. The experimental values obtained so far are indeed quite large ($\sim 4\times10^4$ cm^{-1}) [17], at wavelengths near 5 μm. The transition energies depend on the strain generated due to lattice mismatch between the active layer and the substrate.

5.3.3 Intervalence Band Absorption

Intervalence band absorption is an important loss mechanism particularly when the band gap is small. The photon emitted due to transition from the conduction to the heavy-hole band is reabsorbed by an electron in the split-off band, to arrive at the heavy-hole band. In other words a hole is transferred from the heavy-hole to the split-off subband. It is straightforward to show that the absorption will occur at a wave vector **k**, and the corresponding kinetic energy for the heavy hole $\hbar^2 k^2/2m_h$ is expressed as

$$E_{IVBA} = m_s(\hbar\omega - \Delta_n)/(m_h - m_s), \quad \Delta_n = \Delta + E_{sn} - E_{hn}, \tag{5.46}$$

where Δ is the spin–orbit splitting energy in the bulk and E_{sn} and E_{hn} are the nth subband energies for split-off and heavy-hole bands, respectively.

The theory of intervalence band absorption may be developed in the same way as in intersubband absorption. Assuming that the matrix element is independent of k, the coefficient is expressed as

$$\alpha(\hbar\omega) = \frac{B}{L}\frac{m_s m_h}{m_s - m_h}\left\langle |p_{sh}|^2 \right\rangle (f_s - f_h), \tag{5.47}$$

where B is a constant of proportionality, p_{sh} is the momentum matrix element, and the f's are occupational probabilities. Usually $f_s \approx 1$ and $(1-f_h)$ may be approximated by the Boltzmann distribution. Therefore,

$$\alpha_{sh}(QW) = \frac{e^2\omega}{cn_r\varepsilon_0 L\hbar^2}\frac{m_s m_h}{m_h - m_s}p\frac{\sum_n \langle z_n\rangle^2 \exp\{(-E_{hn} - E(\mathbf{k}))k_B T\}}{\sum_n \exp(-E_{hn}/k_B T)} = K_0 p. \tag{5.48}$$

Each term in the summation is a product of the dipole matrix element and the Boltzmann factor, and p is the total hole density.

The intersubband absorption coefficient is a useful parameter in the design of QWIPs using QWs with Si–SiGe as well as with compound semiconductor materials [12–15].

5.4 Excitonic Processes in QWs

The study of excitonic processes in indirect band gap semiconductors has been mainly of academic interest so far. The same is true for QWs formed by Si and its alloys. Recently there have been reports of strong excitonic absorption due to transition from the valence band to the zone center conduction band in Ge–SiGe MQWs. The following theory of excitonic processes in direct-gap materials is presented in that context, giving in most cases the bare outline.

5.4.1 Excitons in 2D: Preliminary Concepts

In bulk semiconductors, mutual Coulomb interaction between electrons and holes becomes prominent at low temperatures. The excitons dissociate at higher temperatures owing to their small binding energy. In a QW or other low-dimensional systems, the situation is altogether different. Take, for example, the values of binding energy and Bohr radius in bulk GaAs which are, respectively, 4.2 meV and 15 nm. As shown in Figure 5.7,

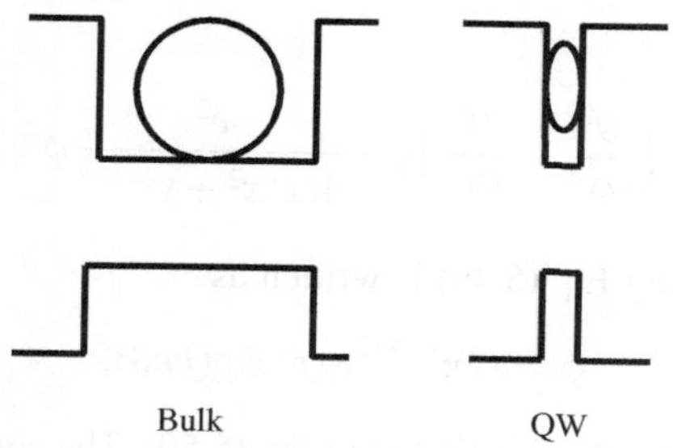

Figure 5.7 *Exciton orbits in bulk and QWs.*

the excitonic orbit is spherical in bulk GaAs. However, if the pair is created in a QW of width less than 15 nm, the Bohr radius, the orbit becomes squeezed as shown in Figure 5.7. The confinement of electrons and holes reduces the physical separation between the particles along the z-direction, as a result of which the Coulombic interaction and hence the binding energy increase. Shinada and Sugano [18] (1966) first showed that the binding energy for pure 2D systems increases fourfold. The measured optical absorption spectra shown in Figure 5.8 indicate that sharp excitonic peaks occur for different subbands (denoted by the symbol n). In addition, as the light and heavy holes get separated in QWs, there are two excitonic peaks. These peaks are barely resolved in the spectra for a 14 nm wide sample.

5.4.2 Excitons in Purely 2D Systems

The hydrogenic levels in a purely 2D system were calculated by Shinada and Sugano [18]. Although a real system differs in many ways from the ideal system they considered, we prefer to give a brief sketch of their work. The equation describing the relative motion in a

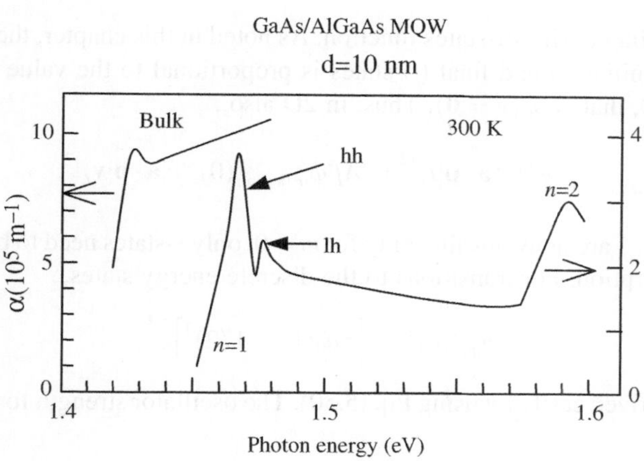

Figure 5.8 *Excitonic absorption spectra in bulk GaAs and GaAs–AlGaAs MQWs. Reproduced with permission from [19]. Copyright (2003) Elsevier Ltd.*

purely 2D system is

$$-\frac{\hbar^2}{2m_r}\left(\frac{\partial^2}{\partial x^2}+\frac{\partial^2}{\partial y^2}\right)\phi-\frac{e^2}{4\pi\varepsilon(x^2+y^2)^{1/2}}\phi=E\phi. \tag{5.49}$$

The wavefunction satisfying Eq. (5.49) is written as

$$\phi=(2\pi)^{-1/2}R(\mathbf{r})\exp(jm\vartheta). \tag{5.50}$$

The radial function $R(\mathbf{r})$ may be substituted in Eq. (5.50). The complete treatment will not be given here. It is found that there are different solutions giving rise to bound-state and discrete (quasi-continuum) and continuum-state eigenvalues. The bound-state energies are expressed as

$$E_n=-\frac{R_x}{(n+1/2)^2},\quad n=0,1,2\ldots\quad R_x=\frac{m_re^4}{32\pi^2\hbar^2\varepsilon^2}, \tag{5.51}$$

where R_x is the effective Rydberg constant. The normalized ground state wavefunction is written as

$$R_{0,0}(r)=(4/a_0)\exp(-2r/a_0), \tag{5.52}$$

The effective Bohr radius is given by $a_0=4\pi\varepsilon\hbar^2/m_re^2$. From Eq. (5.51), the binding energy of the lowest excitonic state is $4R_x$ ($n=0$). For discrete states, the oscillator strength is

$$f_{osc}=\frac{2}{Am_0\hbar\omega}|\langle f|e^{j\mathbf{k}_\mathbf{p}\cdot\mathbf{r}}||i\rangle^2, \tag{5.53}$$

while for continuous and quasi-continuous states, the absorption coefficient is given by

$$\alpha(\hbar\omega)=\frac{\pi\varepsilon}{m_0cn_r\omega a_0}|\langle f|e^{j\mathbf{k}_\mathbf{p}\cdot\mathbf{r}}||i\rangle^2D(\hbar\omega). \tag{5.54}$$

D denotes the joint density-of-states function. As noted in this chapter, the matrix element connecting the initial (i) and final (f) states is proportional to the value of the envelope function at $\mathbf{r}=0$, that is, $\phi(\mathbf{r}=\mathbf{0})$. Thus, in 2D also,

$$\left|\langle f|e^{j\mathbf{k}_\mathbf{p}\cdot\mathbf{r}}\mathbf{a}\cdot\mathbf{p}|i\rangle\right|^2=A\beta\phi_{n(\mathrm{or})k,m}(0)|\langle c|\mathbf{a}\cdot\mathbf{p}|\mathbf{v}\rangle|^2. \tag{5.55}$$

Since ϕ_{nm} and ϕ_{km} are nonvanishing only for $m=0$, only *s*-states need to be considered for calculating absorption. For transitions to the discrete energy states,

$$|\phi_{n0}(0)|^2=\left[\pi a_0{}^2(n+1/2)^3\right]^{-1}, \tag{5.56}$$

which can be derived easily by using Eq. (5.50). The oscillator strength for transition to the nth state is

$$f_0=\frac{2}{m_0\pi a_0{}^2\hbar\omega}(n+1/2)^{-3}\left|\left\langle|p_{cv}|^2\right\rangle\right|, \tag{5.57}$$

taking into account the spin degeneracy factor 2. When n is large, the states may be assumed to be distributed quasi-continuously and the density-of-states function may be defined as

$$D(E) = 2\left|\left(A\frac{\partial E}{\partial n}\right)^{-1}\right| = \frac{(n+1/2)^3}{AR_x}, \tag{5.58}$$

with a factor 2 for spin degeneracy included. Using (5.54) and (5.58), the absorption coefficient is expressed as

$$\alpha_{qc}(\hbar\omega) = \frac{8\pi\varepsilon}{m_0 c n_r a_0 \omega}\left\langle |p_{cv}|^2 \right\rangle. \tag{5.59}$$

The quantum number n does not appear in the expression. The unbound states form a continuum of energy, and for these states one obtains from Eq. (5.50),

$$|\phi_{\mathbf{k}0}(0)|^2 = \frac{\exp(\pi\gamma)}{A\cosh(\pi\gamma)}, \tag{5.60}$$

and the joint density-of-states function is a step function given by

$$D(E) = \frac{m_r}{\pi\hbar^2}, \quad (E > 0). \tag{5.61}$$

By using Eqs. (5.54) and (5.61), the absorption coefficient may be expressed as

$$\alpha_{cont}(\hbar\omega) = \frac{4\pi\varepsilon}{m_0^2 c n_r a_0 \omega}\frac{\exp(\pi\gamma)}{\cosh(\pi\gamma)}\left\langle |p_{cv}|^2 \right\rangle. \tag{5.62}$$

This may be compared with the expression for the absorption coefficient for a purely 2D system for band-to-band transitions, which is readily obtained as

$$\alpha_{bb}(\hbar\omega) = \frac{4\pi\varepsilon}{m_0{}^2 c n_r a_0 \omega}\left\langle |p_{cv}|^2 \right\rangle = \alpha_{cont}(\hbar\omega)[\cosh(\pi\gamma)/\exp(\pi\gamma)]. \tag{5.63}$$

In the limit $\gamma = (R_x/E)^{1/2} \to 0$, the factor $[\cosh(\pi\gamma)/\exp(\pi\gamma)] = 1$.

Thus in the limit $R_x \to 0$ or $E >> R_x$, one obtains $\lim_{\gamma\to 0}(\alpha_{cont} - \alpha_{bb}) = 0$, which should be expected from physical consideration. This conclusion was also drawn for bulk materials, where the absorption spectra for the continuum states merge with band-to-band spectra obeying the $(\hbar\omega - E_g)^{1/2}$ law at high values of photon energies.

Example 5.4

The binding energy for 2D excitons in GaAs is $4 \times 4.2 = 16$ meV. The values in real QWs are ~11 meV.

5.4.3 Excitonic Absorption in Direct-gap QWs

The simplified picture for purely 2D systems given in Section 5.4.2 is modified in a real QW. First, the binding energy does not increase fourfold. However, the increase is high enough so that excitons survice at room temperature. The excitonic absorption spectra for a GaAs–AlGaAs MQW are shown in Figure 5.8, and sharp excitonic absorption peaks are clearly visible [19]. In the absorption spectra for bulk GaAs, included in Figure 5.8, the excitonic peak is barely visible at room temperature. The spectra for QW for the first subband ($n = 1$) show two peaks for HH and LH absorption. However, the peaks are not separated for the $n = 2$ level.

The theory of absorption developed for a purely 2D system is modified for a real QW since the excitonic envelope functions are modified due to finite width of the QW and finite barrier height, and in addition the binding energy of excitons should be calculated as a function of separation of electrons and holes in both the xy plane and z-direction. We refer the interested reader to the existing literature.

5.4.4 Excitonic Processes in Indirect-gap QWs

The calculation of absorption coefficient for transitions to discrete excitonic levels has been made by Basu and Paul [10]. In this subsection, we quote the final results for absorption coefficient for discrete excitonic states:

$$\alpha(\hbar\omega) = \frac{B_{bxph}}{\hbar\omega}\frac{1}{L}\sum_{n=1}^{\infty}(2n-1)^{-3}(N_Q + 1/2 \mp 1/2)H(y), \tag{5.64}$$

$$B_{bxph} = (16M\hbar^2/m_c m_v a_0^2)B_{bbph}, \tag{5.65}$$

where $H(y)$ is a step function and B_{bbph} has been defined in Eq. (5.27).

For transition into continuum states, the approximate Sommerfeld factor as given by Eq. (5.60) is used. The final expression becomes

$$\alpha(\hbar\omega) = \frac{B_{cxph}}{\hbar\omega}\frac{1}{L}(N_Q + 1/2 \mp 1/2)I_{\pm}(\hbar\omega), \tag{5.66}$$

$$B_{cxph} = (M\hbar^2/m_c m_v)B_{bbph}, \tag{5.67a}$$

$$I_{\pm} = \int_0^{\sqrt{\Delta}} \{1 + \exp(-2\pi/a_0 k)\}^{-1} 2k dk, \Delta = \frac{2m_r}{\hbar^2}(\hbar\omega - E_{gmn} \pm \hbar\omega_p) \tag{5.67b}$$

When the photon energy and consequently the upper limit of integration are large, the value of the integral approaches $(\hbar\omega - E_{gmn} \pm \hbar\omega_p)$, the form expected for band-to-band transition.

The expected qualitative variation of absorption coefficient is shown in Figure 5.4. At the threshold there is a sharp rise of absorption due to transition to the 1s excitonic state attached to the first subband. There are a few more steps for the higher bound states. The absorption then rises due to the involvement of continuum states following a $(\hbar\omega - E_{gmn} \pm \hbar\omega_p)$ variation. New steps come in when second-subband and phonon emission processes are involved. The nature of variation is qualitatively different from the same in bulk material like Si or Ge.

The predicted variation as given in Figure 5.4 has been observed by Ait-Ouali *et al.* [20]. However, even when the experimental results for Si–SiGe QWs as shown in Figure 5.9 [21] are examined, there is no qualitative agreement. The experimental results rather follow the bulk variation $(\hbar\omega - E_{gmn} \pm \hbar\omega_p)^2$. In the theory, however, only a single phonon was considered and the subbands are assumed to be wide apart. Calculations taking into account

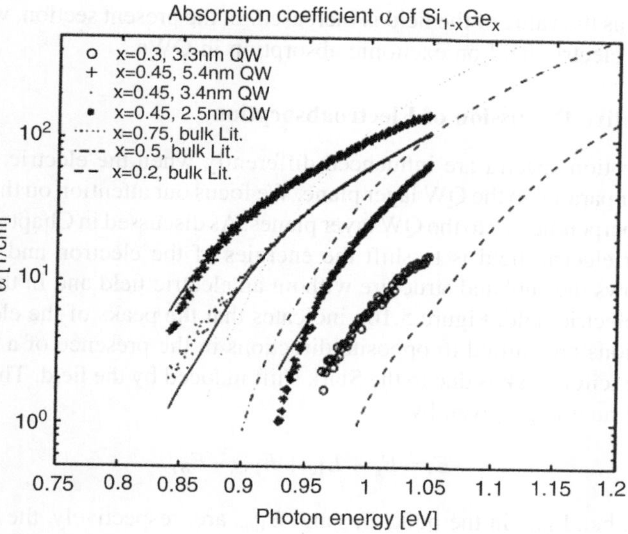

Figure 5.9 *Si1–xGex absorption coefficient of the measured samples in comparison with bulk literature data. The solid lines are least-square fits of parameters in Eq. (5.2) to the measured data (after Palfinger et al.* [21] *Physica,* © Elsevier).

actual situations including the broadening of levels and finite temperature effects are therefore needed to resolve the anomaly.

5.4.5 Photoluminescence in QWs

Luminescence is the opposite process of absorption, and in it the particle (an electron or an exciton) makes a transition from the excited state to the ground state. The process is characterized by the same momentum matrix element as in the absorption process.

Luminescence in QWs obeys the same polarization selection rule and same selection rule for subband indices as followed in the absorption processes. For example, the intersubband emission is maximum when $m - n =$ odd integer. For light emitted in the layer plane with E‖ z, transitions from the conduction subband to LH subband are allowed, but those to the HH subband are forbidden.

The luminescence in pure QWs shows sharp excitonic peaks, and usually the LH and HH excitons are separated even in the room temperature spectra. The first peak occurs at an energy $\hbar\omega = E_g + E_{e1} + E_{hh1} - E_x(QW)$. The peak is slightly broadened due to the collision of excitons with phonons, disorder, and interface roughness.

Interface roughness or the presence of defects may localize the excitons or give rise to excitons bound to impurities.

5.5 Effect of Electric Fields

As mentioned in Chapter 4, the effect of an electric field applied perpendicular to the QW layer plane is to shift the subband energy levels. As a result, the effective band gap decreases more and more, as the value of the field is increased. In the present section, we shall discuss the effect of an electric field on excitonic absorption in QWs.

5.5.1 Qualitative Discussion of Electroabsorption

Excitonic absorption spectra are influenced differently when the electric field is applied perpendicular or parallel to the QW layer plane. We focus our attention on the case when the field is applied perpendicular to the QW layer planes. As discussed in Chapter 3, the effect of a perpendicular electric field is to shift the energies of the electron and hole subbands. Figure 5.10 shows the subband structure without an electric field and in the presence of a perpendicular electric field. Figure 5.10b indicates that the peaks of the electron and hole envelope functions are shifted to opposite directions in the presence of a field.

The shift in exciton peak is due to the Stark shift induced by the field. The exciton peaks should occur at an energy given by

$$E = E_g + E_{1e} + E_{1,hh} - E_B, \tag{5.68}$$

where E_g is the band gap in the bulk; $E_{1,e}$ and $E_{1,hh}$ are, respectively, the field-dependent subband energies for the lowest electron and lowest heavy-hole subbands (see Eq. (3.43)); and E_B is the binding energy of the exciton, which is also dependent on the field. The reason that the exciton binding energy should change may be understood from Figure 5.10. Since the envelope functions for electrons and holes move in opposite directions, there is a reduction of Coulombic attraction in the presence of the field and hence a decrease in the binding energy. The change is, however, small from the value without field.

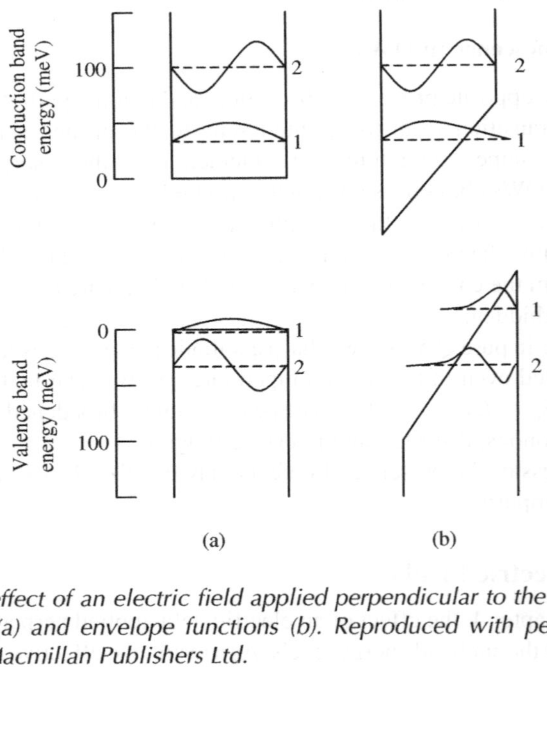

Figure 5.10 The effect of an electric field applied perpendicular to the QW layer plane on subband energies (a) and envelope functions (b). Reproduced with permission from [24]. Copyright (2005) Macmillan Publishers Ltd.

Example 5.5

The changes in subband energies E_{1e} and $E_{1,hh}$ for a field of $10^5\,\mathrm{V\,cm^{-1}}$ are approximately 3 and 12 meV from Eq. (3.63). The binding energy $E_B \approx 10\,\mathrm{meV}$ without field and is slightly lower with field, hence the shift is mainly due to changes in subband energies.

The effect of a perpendicular electric field on the change in peak exciton absorption position is called the quantum confined Stark effect (QCSE) because of the prominence of the Stark shift. The reason that exciton peaks survive at a very high field is that the walls of the wells prevent the electrons and holes from being torn apart, although the electric field tends to separate them from each other. When the field is applied parallel to the QW plane, there are no such walls to confine the electron and the hole, and as a result field ionization becomes easy. As is evident from the nature of wavefunctions shown in Figure 5.10, the overlap between the electron and the hole decreases with increasing field, decreasing thereby the absorption oscillator strength. The peak absorption value decreases with field. Typical absorption spectra for GaAs–AlGaAs systems are given in Schmitt-Rink *et al.* [19] as well as in several books [1–5].

Different methods have been used by several authors to model the absorption in both the presence and absence of the electric field [22].

5.5.2 Electroabsorption and Electrorefraction in SiGe QWs

The QCSE was first observed in GaAs–GaAlAs QWs and subsequently in other QW systems, mainly using direct band gap semiconductors. The effect has been exploited to fabricate high-speed modulators, called electroabsorption modulators (EAMs), that are integral parts of today's fiber-optic communications systems.

Attempts were made to have similar devices made with pseudomorphic SiGe–Si QWs without much success, however. The reason for this is the small conduction band offset of SiGe–Si heterostructures in type I configuration. The electron wavefunction is very weakly localized in the conduction band well even under flat-band (zero-field) conditions. Thus, even a small transverse electric field reduces the exciton binding energy, leading to a blue shift of the transition energy. This blue shift more than balances the field-induced red shift due to the Stark effect, inhibiting the observation of QCSE in SiGe–Si modulators. SiGe–Si type-II QWs have large exciton-binding energy and may produce good modulators. However, such devices are yet to be realized.

Qasaimeh *et al.* [23] noted, however, that the valence band discontinuity between $Si_{1-x}Ge_x$ and Si is large ($\sim 0.74x\,\mathrm{eV}$), and thus the hole wavefunctions are almost localized in the QW. On the other hand, the conduction band discontinuity is only 20 meV. The electron wavefunction is localized in the shallow QW under zero electric field (flat-band condition). Under these conditions, the overlap integral between the electron and the hole wavefunction is large. As the bands are bent with applied bias, the electron wavefunction moves out of the well region. This reduces the overlap integral between the electron and the hole. As a result, the absorption coefficient decreases to a small value with a small change in the electric field. The absolute values of the absorption coefficient are small due to the indirect nature of the transition, but the relative changes are large.

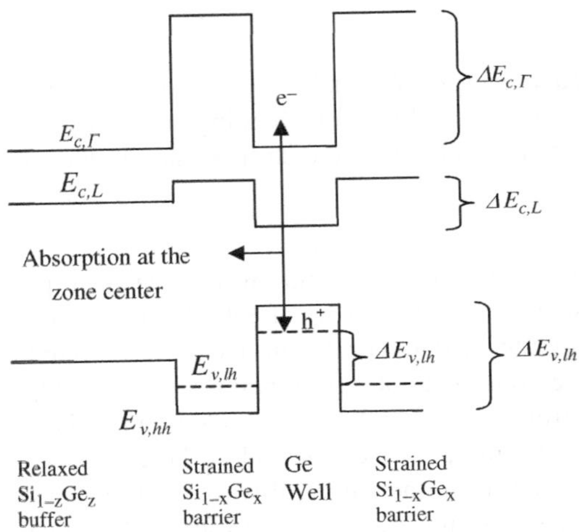

Figure 5.11 *Positions of conduction band and valence band extrema in a strain-symmetric Ge–SiGe MQW. Reproduced with permission from [24]. Copyright (2005) Macmillan Publishers Ltd.*

A strong QCSE has been reported in Si-based systems by Kuo et al [24]; the active material is, however, the Ge QW. Though the lowest band gap in Ge is indirect, it has a direct band gap at 0.8 eV at room temperature, which is ~0.14 eV above the indirect L-point conduction band. The band structure associated with this direct band gap is qualitatively identical to that in QWs formed by III–V materials that exhibit strong QCSE. It should be noted that indirect absorption is also present at the same and lower photon energies. However, its contribution is expected to be and indeed is much weaker, allowing the direct optical absorption to dominate.

The authors grew strain-balanced Ge–SiGe MQWs on a relaxed Ge-rich SiGe buffer on silicon, giving type I alignment at the Γ point. The advantage of strain-balanced structures is that thick structures can be grown, as the average silicon concentration in the Ge–SiGe MQW layers equals that of the buffer layer.

Figure 5.11 shows the resulting band gap alignment. The band discontinuities between well and barrier calculated by Kuo *et al.* are 101 meV for HH, 47 meV for LH, and 400 meV for electrons at Γ-point. The 400 meV band discontinuity at Γ-point provides strong electron quantum confinement in the conduction band. There are, of course, lower energy conduction band minima (L valleys) in the SiGe barriers. However strong tunneling can exist between states in the well and barrier with similar unit cell symmetries. Since the Γ point in Ge wells is higher than the L valleys in the SiGe barriers, the Γ point electrons generated in the Ge wells by optical absorption will be rapidly scattered out into the L valleys, from which they can be swept out by electric fields.

The Ge–GeSi MQW is inserted as the intrinsic (i) layer in a p–i–n structure. The photocurrent spectra give a measure of the absorption coefficient of the intrinsic (i) layer. The corresponding effective absorption coefficient spectra, shown in Figure 5.12, exhibit clear quantum confinement. In Figure 5.12, the spectra for different applied bias are vertically shifted by 2000 cm^{-1}, the purpose of which is explained in this chapter. The exciton peaks are assigned to electron-to-heavy-hole (e–hh; 0.88 eV at 0 V) and electron-to-light-hole

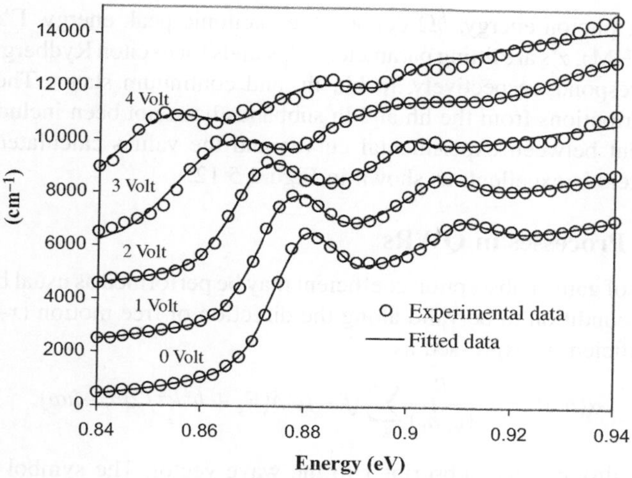

Figure 5.12 *Excitonic electroabsorption in Si–SiGe MQWs. Each of the curves for 1–4 volts has been shifted by 2000 cm^{-1} from the curve for an immediate lower value of bias. Reproduced with permission from [24]. Copyright (2005) Macmillan Publishers Ltd.*

(e–lh; 0.91 eV at 0 V) transitions. The absorption edge at zero bias is 80 meV higher than that of bulk unstrained Ge. This shift closely agrees with the calculated sum from quantum well confinement (56 meV) and strain-induced shifts (36 meV). As the quantum confinement energy originates primarily from the electron, the clarity of this quantum shift shows that the (quantum mechanical) confinement at the Γ point is strong. The photocurrent (or absorption) spectra at a higher temperature have been reported by the same group in another publication [25]. The peak corresponds to a telecommunications wavelength of 1550 nm. The change in absorption peak with field has been modeled recently [26].

The authors calculated the binding energy of the excitons by using a simple variational wavefunction. The change in binding energy with electric field is, however, very small. The change in absorption peak energy may therefore be attributed to the shift of subband energies due to a perpendicular electric field.

In a recent work, the subband energies in the MQW have been calculated by employing a tight binding approximation [27]. The experimental data are also modeled in that work, and satisfactory agreement between the calculated and experimentally obtained data is obtained.

The absorption spectra have been empirically modeled [28] by using the following expression that takes into account two discrete excitonic transitions (hh–e and lh–e) as well as absorption by the 2D continuum states:

$$\alpha(\hbar\omega) = \alpha_h \exp\left[\frac{(\hbar\omega\ \hbar\Omega_h)^2}{2(\hbar\Gamma_h)^2}\right] + \alpha_l \exp\left[\frac{(\hbar\omega\ \hbar\Omega_l)^2}{2(\hbar\Gamma_l)^2}\right] + \frac{\alpha_c}{1+\exp\left(\dfrac{\hbar\Omega_c\ \hbar\omega}{\hbar\Gamma_c}\right)} \times \frac{2}{1+\exp\left\{2\pi\left[|\hbar\Omega_c\ \hbar\omega|/R_y\right]^{1/2}\right\}} \tag{5.69}$$

where $\hbar\omega$ is the photon energy, $\hbar\Omega$ denotes the excitonic peak energy, Γ's represent the linewidths (HWHM), α's are fitting parameters, R_y stands for exciton Rydberg, and subscripts *h*, *l*, and *c* correspond, respectively, to hh, lh, and continuum states. The two different continuum contributions from the hh and lh subbands have not been included separately.

The agreement between experimental curves and the values calculated with suitable chosen parameters is excellent, as shown in Figure 5.12.

5.6 Optical Processes in QWRs

The calculation of gain or absorption coefficient may be performed as usual by assuming the **k**-conservation condition to be valid along the direction of free motion (*x*-direction). The absorption coefficient is expressed as

$$\alpha(\hbar\omega) = -\frac{B}{(c/n_r)}\sum_k (f_e - f_h)\delta(E_g + \hbar^2k^2/2m_r - \hbar\omega), \tag{5.70}$$

where we have dropped the subscript *x* to the wave vector. The symbol E_g denotes the effective gap which is the bulk band gap plus the electron and hole subband energies. The summation over *k* may be converted into an integral and, assuming $f_e - f_h = 1$, the integration may be carried out to yield [29]

$$\alpha(\hbar\omega) = B_1(\hbar\omega - E_g)^{-1/2}, \tag{5.71}$$

where

$$B_1 = \frac{e^2 A_{1D} C_{1D}\left\langle |p_{cv}|^2\right\rangle (2m_r)^{1/2}}{2m_0^2\varepsilon_0 n_r \hbar\omega c A}. \tag{5.72}$$

The coefficients A_{1D} and C_{1D} have similar meanings to A_{mn} and C_{mn} in QWs, and *A* is the cross-sectional area of the wire. The polarization dependence of the matrix element is discussed in [1].

Eq. (5.71) shows that the absorption coefficient is proportional to the joint density-of-states function. The absorption coefficient shows a singularity at photon energy equal to the effective band gap energy and falls with increasing photon energy. The sharp feature is smoothed out by including the effect of broadening. The expression for the gain is written by replacing the δ-function in Eq. (5.70) by a Lorentzian with an intraband relaxation time τ_{in} as

$$g(\hbar\omega) = B_1\int (f_e - f_h)\frac{(\hbar/\tau_{in})dE}{(E - \hbar\omega)^2 + (\hbar/\tau_{in})^2}. \tag{5.73}$$

Example 5.6

For the purpose of estimating the order-of-magnitude values of **α**, we derive an approximate expression as follows. The δ-function in Eq. (5.70) is replaced by a Lorentzian with width γ. Taking $f_e - f_h = 1$, the integration over *k* may be made analytically and the result becomes [29]

$$\alpha(\hbar\omega) = B_1\left\{\frac{\Delta + (\Delta^2 + \gamma^2)^{1/2}}{\Delta^2 + \gamma^2}\right\}^{1/2}, \quad \Delta = \hbar\omega - E_g$$

Let the cross-section of the GaAs wire be $L_y = L_z = 10\,nm$, $m_e = 0.067\,m_0$, and $m_h = 0.112\,m_0$, giving $m_r = 0.042\,m_0$. Also let $E_g = 1.6\,eV$ and $\gamma = 6.6\,meV$ ($\tau_{in} = 0.1\,ps$). Then $\alpha = 6.4 \times 10^5\,m^{-1}$.

The sharply peaked DOS in QWR ensures that carriers are confined to a very narrow spectral region. It may easily be concluded that the transparency carrier density is lower in QWR than in a QW laser. Based on this argument, the peak gain and differential gain in QWR lasers should be higher. However, in the actual situation, inhomogeneities, particularly due to size fluctuations, introduce larger broadening of the gain spectrum.

On theoretical grounds, a lower threshold current, higher modulation bandwidth, and weaker temperature dependence of the threshold current are expected from QWR lasers.

There had been a lot of activities around porous Si after the announcement of visible light emission from such structures. It was revealed that porous Si structures are in cylindrical wire-like form, in dot form, or as an admixture of both.

The excitonic absorption in QWRs made of indirect-gap materials such as silicon has been studied [29]. The absorption spectra corresponding to phonon scattering show a peak which is enhanced from the values for QWs.

5.7 Optical Processes in QDS

We consider a quantum box of cubic shape with each side length a. The absorption coefficient for this box is expressed as [30]

$$\alpha = \frac{A}{a^3}\sum_{n^2} g(n^2)\delta\left[\hbar\omega - E_g - (\pi^2\hbar^2 n^2/2m_r a^2)\right], \quad A = \frac{2\pi e^2\left\langle |p_{cv}|^2 \right\rangle}{m_0{}^2 n_r \varepsilon_0 c\omega} \tag{5.74}$$

where $g(n^2)$ is the degeneracy of the energy level determined by n^2. Only transitions satisfying $\Delta n = 0$ are allowed. The above equation indicates that the interband absorption will be a series of discrete lines representing the reduced density-of-states function of a zero-dimensional (0D) system. The discrete lines will occur at photon energies

$$\hbar\omega = E_g + (\pi^2\hbar^2 n^2/2m_r a^2). \tag{5.75}$$

The discrete lines are broadened due to several reasons. As for example, we consider a family of dots whose sides show a fluctuation described by the following Gaussian distribution:

$$P(a) = \frac{1}{D\sqrt{2\pi}}\exp\left\{-(a - a_0)^2/2D^2\right\}, \tag{5.76}$$

where a_0 is the average side length, and the standard deviation D is given by $D^2 = \langle (a - a_0)^2 \rangle$. Using Eqs. (5.74) and (5.76), the absorption coefficient for a non-uniform dot size distribution is

$$\alpha = \frac{A}{D}\frac{1}{\sqrt{2\pi}}\sum g(n^2)\int_0^\infty a^{-3}\exp\left\{-(a - a_0)^2/2D^2\right\}\delta\left[\hbar\omega - E_g - (\pi^2\hbar^2 n^2/2m_r a^2)\right]da. \tag{5.77}$$

Let us define the following reduced variables:

$$x^2 = \frac{\hbar\omega - E_g}{\pi^2\hbar^2/2m_r a_0{}^2}, \quad \text{and} \quad \sigma = D/a_0. \tag{5.78}$$

The absorption coefficient may then be expressed as

$$\alpha = \frac{\beta}{a_0}\sum_{n^2}\frac{g(n^2)}{\sigma n^2}\exp\left[-\{(n/x)-1\}^2/2\sigma^2\right], \quad \beta = (1/\sqrt{2\pi})(Am_r/\pi^2\hbar^2). \tag{5.79}$$

Eq. (5.79) represents a set of absorption peaks, with peak positions at $n/x = 1$. The peak positions are not affected by dot size distribution. The linewidth is given by

$$\Gamma = 4\sigma n^2(\pi^2\hbar^2/2m_r a_0{}^2) = 4\sigma(\hbar\omega - E_g). \tag{5.80}$$

The line broadening is also governed by a phonon-scattering process.

The momentum matrix element is polarization sensitive and depends on the shape of the box. In the special case of cubic boxes, it equals the value for bulk material. The matrix element reaches its maximum when the electric field vector is parallel to the longest side. The calculated gain spectra [31] are shown in Figure 5.13. Also included in Figure 5.13 are the gain for bulk, QWs, and QWRs using InGaAs as the active material.

The physical origin of the peaked absorption and gain of enhanced value lies in the concentration of carriers near the subband edge in a one-dimensional setting (1D) and within a very small energy interval in a 0D setting. The energy levels look like very sharp atomic levels, and the resonant transitions between two levels are extremely sharp. It is to be noted that the integrated gain for fully inverted conduction and valence bands is constant

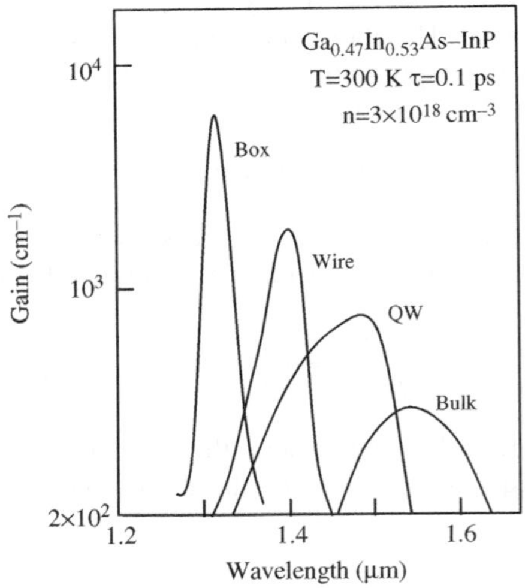

Figure 5.13 *Calculated gain spectra for bulk, QWs, QWRs, and QBs at 300 K for $Ga_{0.47}In_{0.53}As$–InP system. The injected electron density is 3×10^{18} cm^{-3}, and the intraband relaxation time is 0.1 ps (after Asada et al. [31] © IEEE).*

independent of dimensionality. However, the concentration of carriers in a narrower range of energy increases the gain peak enormously in 1D and 0D systems.

The first consequence of enhanced gain in 1D and 0D is a drastic reduction of threshold current density in QWR and QD lasers. A practical limitation in achieving ultra-low threshold current density is the interface roughness and dot size distribution, which tend to reduce the peak value of gain by increasing the line width of gain spectra. Nevertheless, by careful fabrication, the threshold current in QD lasers has been reduced to a fraction of microamperes. Several other advantages of QD laser over QW lasers have been pointed out both theoretically and experimentally. These are increased modulation bandwidth, and better temperature independence of threshold current density.

Problems

5.1 Work out the steps leading to Eq. (5.8).

5.2 By using the approximate expressions for the Fermi functions, show that the gain in QW may be expressed as $g(\hbar\omega) = K(\Delta F - \hbar\omega)^2 H(\hbar\omega - E_{mn})$, where H is a step function and E_{mn} is the gap corresponding to mth conduction and nth valence subbands. Sketch the gain curves for different injected carrier densities.

5.3 Calculate the overlap integral for a parabolic QW in which the envelope functions are given by $\phi_n(z) = (\alpha/\sqrt{\pi}2^n n!)^{1/2}\exp(-\alpha^2/2)zH_n[\alpha(z)]$.

5.4 Calculate the absorption coefficient in a GaAs QW with $L = 10$ nm and infinite barrier height at an energy E_{g11}. Take $C_{11} = 1$, $A_{11} = 1.5$, $n = 3.6$, $f_h = 1$, and $f_e = 0$.

5.5 Obtain the value of recombination lifetime in a GaAs QW from Eqs. (5.14) and (5.15). Use the parameter values given in Example 5.1.

5.6 Compare the gain spectra and the variation of g_{max} versus the injected carrier density for bulk material and QWs, by using the approximate expressions for $g(\hbar\omega)$ obtained for linear approximation of Fermi functions.

5.7 Calculate the values of gain in a direct gap of Ge as a function of injected carrier density. The parameter values may be taken from [7]. Compare your results with the values in fig. 2a of [7].

5.8 The difference between Γ and L valleys in a tensile-strained Ge layer is 115 meV. The Ge layer is doped heavily so that the Fermi level touches the Γ valley. Find out the required donor density using parameter values for unstrained Ge. Compare your value with that given in [7].

5.9 Obtain the expression for $\langle z \rangle$ for a transition from subband m to subband n for a QW with infinite barrier. Discuss how the matrix element changes with difference $(n–m)$. Show that $\langle z \rangle = 16L/9\pi^2$ for transition from the lowest subband to the first excited subband.

5.10 Using the expression for $\langle z \rangle$ derived in Problem 5.9, show that $f_{osc} = (m_0/m_e)(256/27\pi^2)$.

5.11 Design a $Si–Si_{0.6}Ge_{0.4}$ QW intersubband detector for 10 μm radiation. Assume infinite barriers. Also calculate the well width for finite barriers. Take the band offset values given in Chapter 2. Consider both conduction and heavy-hole subbands.

5.12 Calculate the intersubband absorption coefficient in Si and Ge QWs for light polarized along the (100), (110), and (111) directions. Take $\Gamma = 10$ meV.

5.13 Prove that the recombination lifetime for a 2 $\rightarrow$ 1 intersubband transition is given by $\tau_r^{-1} = 2\pi n e^2 f_{12}/m_0\varepsilon_0\lambda^2 c$. Calculate the value assuming $f_{12} = 15$, $n = 3.6$, and $\lambda = 20$ μm.

5.14 Obtain an expression for the intervalence band absorption coefficient assuming that the matrix element is independent of **k**.

5.15 Calculate the binding energy for hh and lh excitons in Ge QWs using the parameter values given in [26].

5.16 The excitonic absorption and recombination linewidths are determined by interface roughness and other scattering processes. Consider that the well width L has a mean fluctuation denoted by δL. Obtain the expression for linewidth assuming infinite barrier height.

5.17 The linewidth of excitons in GaAs QW is expressed as $\gamma = \gamma_{in} + \gamma_0 T + \gamma_{LO}/[\exp(\hbar\omega_{LO}/k_B T) - 1]$, where the first term stands for the inhomogeneous part. Assume $\gamma_{\text{in}} = 4.5$ meV, $\gamma_0 = 20$ meV and energy of longitudinal optical (LO) phonon $\hbar\omega_{LO} = 36$ meV. Calculate the lifetime of excitons at 300 K.

References

1. Basu, P.K. (2003) *Theory of Optical Processes in Semiconductors: Bulk and Microstructures*, Clarendon Press, Oxford, UK.
2. Singh, J. (2003) *Electronic and Optoelectronic Properties of Semiconductor Structures*, Cambridge University Press, Cambridge, New York.
3. Harrison, P. (2000) *Quantum Wells, Wires and Dots: Theoretical and Computational Physics*, John Wiley & Sons, Ltd., Chichester, UK.
4. Mitin, V., Stroscio, M.A., and Kochelap, V.A. (1999) *Quantum Heterostructures*, John Wiley & Sons, Inc., New York.
5. Manasreh, O. (2005) *Semiconductor Heterojunctions and Nanostructures*, McGraw-Hill, New York.
6. Chang, S-W. and Chuang, S.L. (2007) Theory of optical gain of Ge–Si_x Ge_y Sn_{1-x-y} quantum-well lasers. *IEEE J. Quantum. Electron.*, **43**, 249–256.
7. Liu, J., Sun, X., Pan, D. *et al.* (2007) Tensile-strained n-type Ge as a gain medium for monolithic laser integration on Si. *Opt. Exp.*, **15**, 11272.
8. Asada, M., Kameyama, A., and Suematsu, Y. (1984) Gain and intervalence band absorption in quantum well lasers. *IEEE J. Quant. Electron.*, **QE-20**, 745–753.
9. Yamanishi, M. and Suemune, I. (1984) Comment on polarization dependent momentum matrix element in quantum well lasers. *Japan J. Appl. Phys.*, **23**, L35–L36.
10. Basu, P.K. and Paul, S.K. (1992) Ideal interband absorption spectra in indirect gap quantum wells. *Phys. Rev. B*, **46**, 13389–13393.
11. Bastard, G. (1988) *Wave Mechanics Applied to Semiconductor Heterostructures*, Les Editions de Physique, Les Ulis, France.
12. Lynch, S.A. *et al.* (2006) Toward Si based lasers for THz sources. *IEEE J. Sel. Top. Quant. Electron.*, **12**, 1570–1578. (2006).
13. Han, G., Yu, J., and Liu, Y. (2008) n-Type Ge–SiGe quantum cascade structure utilizing quantum wells for electrons in the L and Γ valleys. *IEEE Photonic. Tech. L.*, **20**, 419–422.
14. Loehr, J.P. and Manasreh, M.O. (1993) in *Semiconductor Quantum Wells and Superlattices for Long Wavelength Infrared Detectors*(ed. M.O. Manasreh), Artech, Boston, p. 19, Chapter 2.
15. Paiella, R. (2006) *Intersubband Transitions in Quantum Structures*, McGraw-Hill Professional, New York.

16. Yang, C.L., Pan, D.S., and Somano, R. (1989) Advantages of indirect semiconductor quantum well system for infrared detection. *J. Appl. Phys.*, **65**, 3253–3258.
17. Wang, K.L. and Karunasiri, R.P.G. (1993) in *Semiconductor Quantum Wells and Superlattices for Long Wavelength Infrared Detectors*(ed. M.O. Manasreh), Artech, Boston, p. 139, Chapter 5.
18. Shinada, M. and Sugano, S. (1966) Interband optical transitions in extremely anisotropic semiconductors. I. bound and unbound excitons. *J. Phys. Soc. Japan*, **21**, 1936–1946.
19. Schmitt-Rink, S., Chemla, D.S., and Miller, D.A.B. (1989) Linear and nonlinear optical properties of semiconductor quantum wells. *Adv. Phys.*, **38**, 89–188.
20. Ait-Ouali, A. and Jandl, S. (1994) Two dimensional indirect excitons in layer type trichalcogenide ZrS_3. *Phys. Rev. B*, **49**, 1813–1817.
21. Palfinger, G., Bitnar, B., Sigg, H. *et al.* (2003) Absorption measurement of strained SiGe nanostructures deposited by UHV-CVD. *Physica E*, **16**, 481–488.
22. Bandyopadhyay, A. and Basu, P.K. (1993) Modeling of excitonic electrorefraction in InGaAsP multiple quantum wells. *IEEE J. Quantum. Electron.*, **QE29**, 2724–2730.
23. Qasaimeh, O., Bhattacharya, P., and Croke, E.T. (1998) SiGe–Si quantum-well electroabsorption modulators. *IEEE Photon. Tech. Lett.*, **10**, 807–809.
24. Kuo, Y.H., Lee, Y.K., Ge, Y. *et al.* (2005) Strong quantum confined stark effect in Ge quantum well structures on Si. *Nature*, **437**, 1334–1336.
25. Kuo, Y.H., Lee, Y.K., Ge, Y. *et al.* (2006) Quantum confined stark effect in Ge/SiGe quantum wells on Si for optical modulators. *IEEE J. Sel. Top. Quant. Electron.*, **12**, 1503–1513.
26. Schaevitz, R.K., Roth, J.E., Ren, S. *et al.* (2008) Material properties of Si-Ge/Ge quantum wells. *IEEE J. Sel. Top. Quant. Electron.*, **14**, 1082–1089.
27. Virgilio, M. and Grosso, G. (2008) Quantum confined stark effect in Ge/SiGe quantum wells: a tight binding calculation. *Phys. Rev. B*, **77**(165315), 1–7.
28. Sen, G., Mukhopadhyay, B., and Basu, P.K. (2008) Modeling electroabsorption and electro-refraction in Ge/SiGe multiple quantum wells for application as modulators. Proc. Intl. Conf Fiber Optics and Photonics (Photonics 2008), New Delhi, Dec. 15–17.
29. Ray, P. and Basu, P.K. (1993) Peaked nature of excitonic absorption in quantum well wires of indirect gap semiconductors. *Phys. Rev. B*, **48**, 11420–11422.
30. Wu, W.Y., Schulman, J.N., Hsu, T.Y., and Efron, U. (1987) Effect of size nonuniformity on the absorption spectrum of a semiconductor quantum dot system. *Appl. Phys. Lett.*, **51**, 710–712.
31. Asada, M., Miyamoto, Y., and Suematsu, Y. (1986) Gain and the threshold of three-dimensional quantum box lasers. *IEEE J. Quantum. Electron.*, **QE-22**, 1915–1921.

6

Light Emitters in Si

6.1 Introduction

The main obstacle in the widespread use of silicon and related materials lies in the poor light emission properties of these indirect-gap materials. Active research on the optoelectronic properties of silicon and its alloy SiGe was undertaken soon after the announcement of transistors. At first, the motivation behind the work was to know the basic properties. However, with the emergence of silicon as the sole material for microelectronics, work on the realization of an efficient light emitter was intensified. The goal is to develop truly monolithic optoelectronic integrated circuits exhibiting the magic performance of silicon-based microelectronics. The main applications of silicon-based light-emitting devices, that is, light-emitting diodes (LEDs), lasers, and optical amplifiers, are in electronic and optical systems and displays. Optical devices in the wavelength range from 0.4 to 1.6 µm are needed for full-color displays and for sources and amplifiers for standard telecommunications wavelength bands around 1.3 and 1.55 µm. Specific applications for such sources are as transmitters for optical fiber communication, optical interconnects within and between computer chips and boards, optical controllers for phased-array microwave antennas, information display screens, printing elements for xerography, and writing and readout in optical compact disc systems.

Unfortunately, however, the progress in achieving the coveted goal is far from a satisfactory level, in spite of the large volume of work performed in this area. Whereas semiconductor lasers and LEDs grown with III–V compounds and alloys have shown quantum efficiencies >80%, the highest value of this quantity for Si-based light sources, mainly LEDs, barely exceeds 10% in the laboratory scale. High-volume production of LEDs is still an unattained task. No report of laser action using silicon-based materials has appeared in the literature, although many novel suggestions have been put forward. Very recently, there are reports of achieving laser action by stimulated Raman scattering in silicon microstructures, which will form the subject of discussion in Chapter 9.

Silicon Photonics: Fundamentals and Devices, First Edition. M. Jamal Deen and P. K. Basu.

It appears from this discussion that the topic *light emission from silicon and related materials* is still an immature subject. In view of its importance, however, it is worthwhile to discuss the basic problems involved, to examine the suggestions put forward to increase the quantum efficiency of the light emitters, and to mention the specialties in structures and modification of optical properties in them, and the results reported by various authors using these novel structures. In the present chapter, such a presentation will be given.

The various methods proposed so far to improve emission are band structure engineering, quantum confinement, nanocrystals, quantum dots (QDs), and impurity doping, in particular rare-earth Er doping. All these methods will be described in separate sections, in each of which the basic principle will be described first, followed by a description of the structure and device, and finally the results obtained by various authors. In addition to band-to-band processes, transitions between subbands, in both the conduction band and valence band, have been studied with a view to achieving light emission in mid-infrared wavelengths and at THz frequencies. The study has successfully led to the development of quantum cascade lasers (QCLs) in quantum wells made of compound semiconductors. Similar work in Si has also been initiated and pursued. This chapter will include a discussion of the basic principle, the structures, and the results obtained.

The subject is still growing, and useful work is being reported continuously. It is not possible to include all such work; however, some representative papers are included to give the reader an idea of the activities going on.

6.2 Basic Theory of Light Emission

Light emission in a semiconductor occurs due to the creation and recombination of excess electron–hole pairs (EHPs). The pairs are created by photon absorption, by the injection of carriers in a forward-biased p–n junction, or by bombardment of the material by energetic particles. The recombination process is illustrated by considering a direct-gap semiconductor, the energy (E) – wave vector ($\mathbf{k}$) (E–$\mathbf{k}$) diagram of which is shown in Figure 4.3a. An excess electron in the conduction band jumps to the valence band to occupy an empty state there, and thereby the excess energy is liberated in the form of a photon. Two conditions must be satisfied in this recombination process. The first, the energy conservation, requires that

$$E_e(\mathbf{k}) = E_h(\mathbf{k}) + \hbar\omega, \tag{6.1}$$

where E_e and E_h denote, respectively, the electron and hole energies and $\hbar\omega$ is the energy of the emitted photon. The second condition to be fulfilled is the conservation of (crystal) momentum, which is expressed as

$$\mathbf{k}_e = \mathbf{k}_h \pm \mathbf{k}_{\text{photon}}. \tag{6.2}$$

It has been shown in Chapter 4 that the wave vector of the photon, $\mathbf{k}_{\text{photon}}$, is small compared with the wave vectors for both electrons and holes, $\mathbf{k}_e$ and $\mathbf{k}_h$, leading to the condition for momentum conservation

$$\mathbf{k}_e = \mathbf{k}_h. \tag{6.3}$$

The excess carriers recombine with a characteristic excess carrier lifetime, which is approximately given by

$$\tau_{rr} = \frac{1}{B_r N_D}, \tag{6.4}$$

where N_D is the donor density if the material is n-type, and B_r is a constant that may be calculated quantum mechanically. The value of recombination lifetime τ_{rr} is of the order of nanoseconds.

In addition to the recombination process described in this section, excess EHPs can also recombine via traps, impurities, and other defects. These processes are generally non-radiative, that is, photons are not generated by these processes. A similar nonradiative recombination lifetime τ_{nr} may be assigned to these processes. It is natural that the light-emitting devices must convert the EHPs injected to as many photons as possible. One may define internal quantum efficiency (IQE), η_{int}, as follows:

$$\eta_{\text{int}} = \frac{\text{no of photons generated}}{\text{no of EHPs injected}} = \frac{\tau_{nr}}{\tau_{rr} + \tau_{nr}}. \tag{6.5}$$

In direct-gap semiconductors, the nonradiative processes can be minimized by careful design, so that the probability of nonradiative events to occur is extremely low and the corresponding lifetime is quite large. This makes $\tau_{nr} >> \tau_{rr}$, and from Eq. (6.5) the ideal internal quantum efficiency may be 100%.

Another useful parameter related to devices is the power conversion efficiency or plug-in efficiency, which is the ratio of the optical power emitted by, say, an LED, to the electrical driving power. The desired value is 100%.

Consider now the situation in an indirect band semiconductor like Si, the *E*–**k** diagram for which is shown in Figure 4.8b. Here the conduction band minimum does not occur at the center of the Brillouin zone; rather, it is located approximately at a distance equal to $\mathbf{k}_0 = (0.85\,\mathbf{k}_{\text{max}})$ along the Δ axis; $\mathbf{k}_{\text{max}} = 2\pi/a$, where a is the lattice constant. Now an excess electron in the conduction band cannot directly recombine with an excess hole in the valence band having a **k** vector near the zone center. Conservation of momentum is fulfilled if the excess momentum $\mathbf{k}_0$ needed for the recombination to occur is provided by a momentum conserving agency-like phonons. The lifetime for the overall process may be calculated quantum mechanically by using second-order perturbation theory. and the value for τ_{rr} turns out to be on the order of milliseconds. Recombination of excess EHPs in Si and other indirect-gap semiconductors occurs predominantly via recombination or trap centers, as mentioned in Chapter 2. The lifetime of such nonradiative processes is of the order of microseconds, and hence the IQE for light emission in Si is extremely low.

Example 6.1

Consider $\tau_{rr} = 1$ msec and $\tau_{nr} = 1\ \mu$sec. The value of $\text{IQE} = 10^{-3}$ from Eq. (6.5). However, if $\tau_{rr} = 1$ nsec, as in a direct-gap semiconductor, the value of $\text{IQE} = 1$.

The discussion in this section along with Example 6.1 clearly point out the fundamental limitations of using silicon as a light source. Different proposals have been given in the literature to overcome this limitation. These proposals, along with the basic principles and results obtained so far, will be the subject of discussions in the next few sections.

6.3 Early Efforts: Zone Folding

Most textbooks on solid-state physics illustrate the formation of energy bands by employing a simple Kronig–Penney model, considering a one-dimensional (1D) chain of atoms separated by an interatomic distance d. The E–$\mathbf{k}$ relationship obtained from this model is periodic with gaps occurring at values of $\mathbf{k} = \pi/a$. This variation is depicted in the reciprocal space within the first Brillouin zone bounded by $\pm\pi/a$ in Figure 3.11 with reference to a superlattice.

The idea of zone folding in a 1D superlattice whose periodicity is of the order of a lattice constant was advanced long ago. The concept was revisited by People and Jackson [1] as well as by People [2], after the successful growth of a high-quality atomic layer $(Si_mGe_n)_p$ superlattice, in which m and n are the number of monolayers of Si and Ge in each period and p is the number of periods. The larger period d in the superlattice gives rise to a minizone having boundaries at $\pm\pi/d$. It may be proved, assuming that the minimum in conduction band is at 0.8 $\mathbf{k}_{max}$ as in Figure 6.1, that if $d \approx 5a$, where a is the lattice constant, then the E–$\mathbf{k}$ diagram is folded back in the new reduced minizone and the gap is converted into a direct one along the superlattice axis. The folding process and the resulting miniband structure are illustrated in Figure 6.1. As the direct gap occurs along one dimension only, the transition probability is still insignificant compared to compound semiconductor materials.

Experimental verification for the modified band structure as indicated in this section had to wait for the successful growth of a strain-symmetrized Si_mGe_n superlattice on a

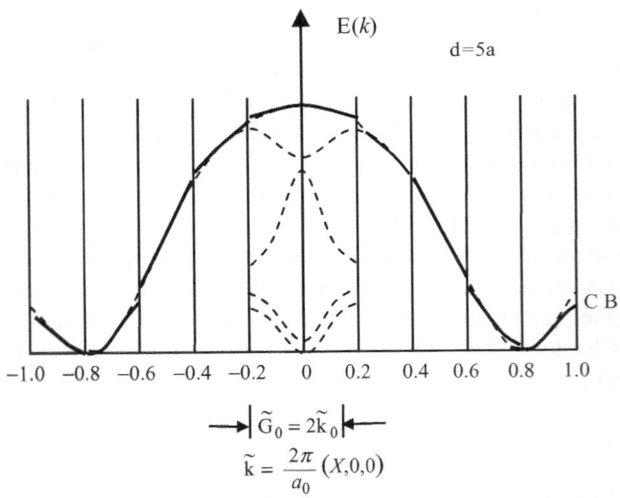

Figure 6.1 *E–k diagram of CB electrons along Δ (solid) in the Brillouin zone. Dashed curve: Si-induced zone folding of Δ_1 point into zone center for d = 5a. A direct gap is opened between VB minima at k = 0 and the folded CB at k = 0. G_0 is the primitive reciprocal lattice vector for the superlattice. Reproduced with permission from [2]. Copyright (1986) IEEE.*

strain-relaxed thick $Si_{1-x}Ge_x$ buffer layer on Si substrate. Enhanced photoluminescence (PL) was indeed observed with improved structures [3]. Unfortunately, the enhanced emission occurs at low temperatures only; with a rise in temperature, PL and electroluminescence (EL) are essentially quenched due to excitonic dissociation. Room temperature EL structures have been reported by using wavy superlattices. However, the IQE for pin structures at room temperature is only $\approx 10^{-5}$ [4]. Therefore, unless drastic improvements in materials quality take place, the prospect of using zone-folded structures as light emitters does not seem too encouraging.

6.4 Band Structure Engineering Using Alloys

As mentioned already, a momentum-conserving agent is needed in order to complete the recombination of excess EHPs in an indirect band gap semiconductor. It has been stated that quantized lattice vibrations, that is, phonons, may provide the required momentum in materials like silicon. Defects or disorder may also provide the momentum needed for a recombination process to occur. An example of defect is the alloy disorder in an alloy semiconductor. For example, in $Si_{1-x}Ge_x$, the atomic potentials are different at the Si and Ge sites. If one assigns an average potential V_{alloy} to a virtual crystal, then deviations amounting to $|V_{alloy} - V_{Si/Ge}| = \Delta V_{Si/Ge}$ will be found respectively at the Si and Ge sites. It is usual to assume this fluctuation as a spherical square well potential, the Fourier transform of which may be written as

$$\Delta V_q = \sum_q \Delta V \exp\left[-\mathbf{q}.|\mathbf{r}-\mathbf{r_i}|\right]. \tag{6.6}$$

If $\mathbf{q} = \mathbf{k_0}$, the required momentum may be obtained from this alloy disorder potential. Following the theory of absorption in an indirect gap as outlined in Chapter 4, the expression for absorption coefficient due to alloy disorder becomes

$$\alpha(\hbar\omega) = Ax(1-x)(\hbar\omega - E_g)^m, \tag{6.7}$$

where $m = {}^1\!/_2$ for excitonic absorption but $m = 2$ for band-to-band transition. The corresponding equation for a recombination rate may be obtained, but it contains the same factor $x(1-x)$. It is evident that absorption and recombination are maximum when the alloy composition $x = 0.5$. Such a contribution due to alloy disorder has been observed by Weber and Alonso [5] for $Si_{1-x}Ge_x$, with maximum contribution coming when $x = 0.5$.

Engineering of the band structure of silicon can be accomplished by using a heterojunction of Si and its alloy $Si_{1-x}Ge_x$ or $Si_{1-x-y}Ge_xC_y$. As noted already, due to a large lattice mismatch between Si and its alloys, the grown layers are invariably strained, giving rise to band gap alteration with composition. A noteworthy feature is that the alloy band gap corresponds to emission wavelengths of 1.3 and 1.55 μm, as shown in Figure 2.12.

A disadvantage of the SiGe and Si heterostructures is the small value of critical layer thickness, particularly for larger values of x. The growth of pseudomorphic layers on a substrate or thick buffer layer has been discussed in Chapter 2. For example, Si may be grown on a thick $Si_{1-z}Ge_z$ buffer layer grown on Si substrate. In that case, the top Si layer is strained. As already noted, depending on the nature of the film, virtual (buffer) layer, or substrate, the band alignments may be of type I or type II. Also, a strain-symmetrized

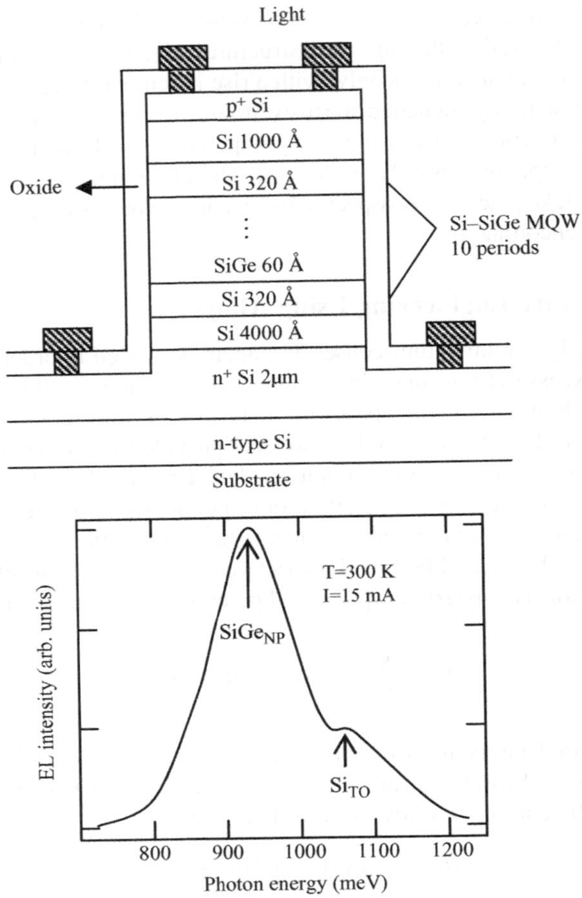

Figure 6.2 *(a) The pin structure used to observe EL due to NP transition in SiGe. The I layer consists of 10 periods of 32 nm Si and 6 nm SiGe layers. (b) Room temperature PL spectra with peak at 1.34 μm. The peak at higher energy occurs due to TA-assisted transition in Si. Reproduced with permission from [6]. Copyright (1969) American Institute of Physics.*

superlattice may alleviate the problem of low value of critical layer thickness. In all situations, however, the band gaps of Si and its alloy remain indirect.

Despite the indirect nature of the band gap, many workers studied the EL and PL of Si and SiGe heterostructures exhibiting type I band alignment. A disadvantage of type I alignment is that the confining potential for electrons is very small.

EL and PL from a SiGe single layer and superlattice exhibited enhancement compared to bulk silicon. The recombination mechanism is either excitonic or band to band, depending on the quality of the interface and thickness of the alloy layer. In earlier studies, the intensity of NP emission persisted up to 80 K or so, but the EL intensity from pin diodes was quenched at a higher temperature. With improved material quality, room temperature EL has been observed in pin structures, the intrinsic layer of which comprises a MQW formed by Si and SiGe. The layer structure and the EL spectra obtained [6] are shown in Figure 6.2.

The problem with the structure is low IQE $\sim 2 \times 10^{-4}$. Also, for type I alignment, the potential barrier is not high and in most cases transitions occur in the neighboring silicon layer. Growth on a buffer layer may give rise to type II alignment, with electrons staying in the conduction band of Si and holes in the valence band of the alloy. The transition probability in these structures, which is indirect in real space, decreases due to a reduced overlap of electron and hole wavefunctions.

A larger IQE $\sim$1% at room temperature was obtained by Presting *et al.* [7] by using a Ge–Si–Ge double barrier structure. The alignment is type II; however, the thin barrier layers allow penetration of the wavefunction into adjacent layers and sufficient spatial overlap between electron and hole envelope functions. The recombination is believed to be associated with excitons bound to the interface states. The enhancement of overlap between electrons and holes may be understood from a simple theory [8].

The effects of introducing C in SiGe alloys have been studied in recent years on the lattice constant, strain, and band gap [9]. Considering that the lattice constant $a(x,y)$ of the alloy $Si_{1-x-y}Ge_xC_y$ linearly varies with the composition of the constituent atoms, one may write,

$$a(x,y) = a_{Si} + x(a_{Ge} - a_{Si}) + y(a_C - a_{Si}) \tag{6.8}$$

Since the lattice constant of C is less than that of Si, the last term is negative and the lattice constant of the ternary may perfectly match that of Si, giving rise to strain-free growth of the alloy. It is estimated that a Ge:C ratio of $\sim$9:1 will ensure perfect strain-free growth. The band gap of the alloy decreases as 26 meV/%C. Better conduction band offset is also predicted, and the alignment may be type I. However, the alloy is still an indirect-gap material, and the solubility of C does not exceed 3%. Weak PL has so far been observed from this material. There is still controversy over the role of C: whether it remains in interstitial sites acting as a recombination center, whether perfect strain-free growth is possible, and whether it changes the band gap and band alignment in heterojunctions. It is to be expected that the addition of substitutional C will increase the alloy-scattering contribution and hence the NP luminescence intensity.

6.5 Quantum Confinement

A vast body of literature has accumulated on the studies of light emission in quantum nanostructures. The impetus for such studies came immediately after the announcement of strong light emission from porous silicon observed by Canham [10].

Porous silicon belongs to the general class of quantum-confined nanostructures. In addition, other forms of nanostructures like quantum wires and quantum dots prepared by standard epitaxial process and self-assembly, are examined to obtain efficient light emission. Si nanocrystals embedded in glass and other matrixes also form an important class of materials showing prospects for efficient luminescence.

The basic idea behind the use of nanometer-sized luminescent material is increased overlap between electron and hole envelope functions leading to enhanced absorption and recombination. In Sections 6.5.1, 6.5.2, 6.5.3, and 6.5.4, we shall first explain the phenomena of increased overlap and then consider different structures and systems and their light emission characteristics.

6.5.1 Quasi-direct No-phonon Transition

The spreads in momentum and position of any quantum particle are related by Heissenberg's uncertainty relationship, that is,

$$\Delta r \cdot \Delta p \sim \hbar \tag{6.9}$$

This means that a particle having a precisely defined position ($\Delta r = 0$) will have an infinite spread in its momentum. In a nanostructure, the wavefunction of an electron is confined within a nanometer dimension. Hence, the above argument suggests that the envelope function will have a substantial amount of Fourier component with large $\boldsymbol{k}$. Provided the component with wave vector $\mathbf{k}_0$ is appreciable, the electron wavefunction will spread near $\mathbf{k} = 0$ and the overlap between electron and hole in $\mathbf{k}$-space will increase, giving rise to augmented absorption and recombination.

To illustrate this increased overlap due to quantum confinement, let us consider a 1D potential well of infinite height for both electrons and holes. The envelope functions are written by the well-known expressions:

$$\phi_{e,h} = \sqrt{2/L}\sin(\pi z/L), \tag{6.10}$$

where L is the width of the quantum structure along the z-direction. The squared overlap function is written as

$$|\langle\phi_e|\phi_h\rangle|^2 = \int_0^L \phi_e\phi_h \exp[j|k-k_0|z]dz = \left(\frac{2}{L}\right)^6\left[\frac{1}{1-(k_0L/\pi)^2}\cdot\frac{\sin(k_0L/2)}{(k_0L/2)}\right]^2. \tag{6.11}$$

The variation of the squared overlap function, or the contributing factor to the oscillator strength obtained from Eq. (6.11), is shown as a function of the length L of the nanostructure in Figure 6.3a using $\mathbf{k}_0 = (1.7\ \pi/a)$. It is found to be oscillatory; however, when the size is reduced below 1 nm, the function increases rapidly, supporting the above arguments.

An improved calculation has been presented by Takagahara and Takeda [11], who considered the excitonic enhancement of the oscillator strength in Si and Ge QDs. The calculated excitonic decay rates are shown in Figure 6.3b as a function of dot radius. It appears from the plots that for radii below 1 nm, the recombination lifetime rapidly decreases to a few tens of nanoseconds, thereby ensuring dominance of the NP process. Similar arguments may also be found in the work of Hybertsen and coworkers [12, 13].

The above theoretical models are supported by experimental results obtained by Kovalev *et al.* [14]. The authors used naturally oxidized Si nanocrystals (NCs), as well as NCs surrounded by SiO_2. The ratio of NP to phonon-assisted recombination rates versus confinement energy is plotted in Figure 6.4.

As may be found from Figure 6.4, for smaller confinement energy, that is, larger nanocrystal width, the phonon-assisted process dominates. However, when the confinement energy exceeds 0.3 eV, the NP process takes over the TA phonon-assisted process, validating

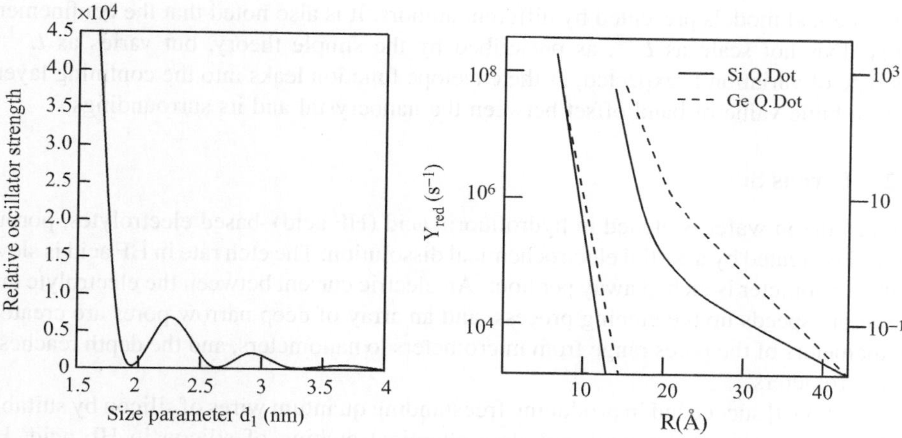

Figure 6.3 *(a) The contribution of confinement to the oscillator strength as obtained from Eq. (6.11). (b) Radiative decay rates of excitons in Si (solid line) and Ge QDs (dashed lines) as a function of dot radius. Reproduced with permission from [11]. Copyright (1992) American Physical Society.*

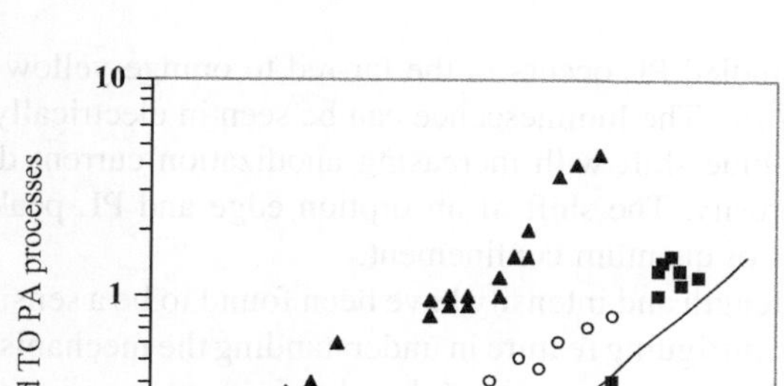

Figure 6.4 *The ratio of NP to TA phonon-assisted (solid triangle) and TO phonon-assisted process (solid squares) versus confinement energy for naturally oxidized Si nanocrystals. Open circles represent the NP:TO ratio for Si NCs embedded in a silica matrix. The dotted and dashed lines represent theoretical fit. Assuming confinement energy varying as L^{-m}, the dotted ($m = 2$) and solid lines ($m = 1.3$) represent the theoretical fits. Reproduced with permission from [14]. Copyright (1998) American Physical Society.*

the theoretical models presented by different authors. It is also noted that the confinement energy does not scale as L^{-2}, as prescribed by the simple theory, but varies as $L^{-1.3}$. This type of variation is expected, as the envelope function leaks into the confining layers due to a finite value of band offset between the nanocrystal and its surroundings.

6.5.2 Porous Si

When a silicon wafer is etched in hydrofluoric acid (HF acid)–based electrolytes, porous material is created by a partial electrochemical dissolution. The etch rate in HF acid is slow; about a nanometer is etched away per hour. An electric current between the electrolyte and the sample speeds up the etching process, and an array of deep narrow pores are created. The diameters of the pores range from micrometers to nanometers, and the depth reaches a few micrometers.

Canham [10] succeeded in producing freestanding quantum wires of silicon by suitably combining the electrochemical and slow chemical etching of silicon in HF acid. He observed intense light emission from carefully prepared samples. Visible PL from green to red was soon reported by a number of workers. The measurement of optical absorption by Lehmann and Gösele [15] showed a blueshift of the edge from that in the bulk. All these observations led to the conclusion that porous silicon is a quantum wire and the absorption and luminescence at higher energy are due to quantum confinement effect. Interested readers are referred to a few excellent reviews and books dealing with porous Si (por-Si), for a detailed description of the various aspects of por-Si and a list of useful references [16].

The most widely studied PL occurs in the far-red to orange-yellow wavelength range, usually termed as *red PL*. The luminescence can be seen in electrically excited materials. The spectra shows a blue shift with increasing anodization current density, or, in other words, increasing porosity. The shift of absorption edge and PL peak provided the first evidence of the effect of quantum confinement.

The PL peak wavelength and intensity have been found to be a sensitive function of the surface chemistry. An intriguing feature in understanding the mechanism of PL in por-Si is the large Stokes shift between absorption (photoluminescence emission, PLE) and the PL peak. The expected Stokes shift is ~0.1 meV; on the other hand, the observed shift is ~0.1 eV for a diameter = 1.5 nm. The optical absorption gaps agree well with the calculated values, but the luminescence energies are practically independent of the size, while the predicted behavior is $d^{-1.47}$. To explain this behavior, the existence of deep luminescent centers, such as surface states, must be postulated. Allan *et al.* [17] worked out a theory for that by considering self-trapped excitons. These excitons are stabilized due to large increase in the gap due to confinement. The transitions involved are shown in Figure 6.5.

In the model, absorption of a photon brings an exciton from the ground state (G) to an excited state (E). It then relaxes to the self-trapped excitonic (STE) state. The transition to ground state gives the characteristic PL of energy less than the absorbed photon energy.

Oxidation of the por-Si surface produces blue PL. Blue PL has also been observed in oxide-free por-Si by decreasing the crystallite size, in accordance with the quantum confinement mechanism. The PL wavelength can be tuned from blue (~3eV) to the band gap of crystalline Si (1.1 eV) by a judicious choice of the porosity of the unoxidized por-Si.

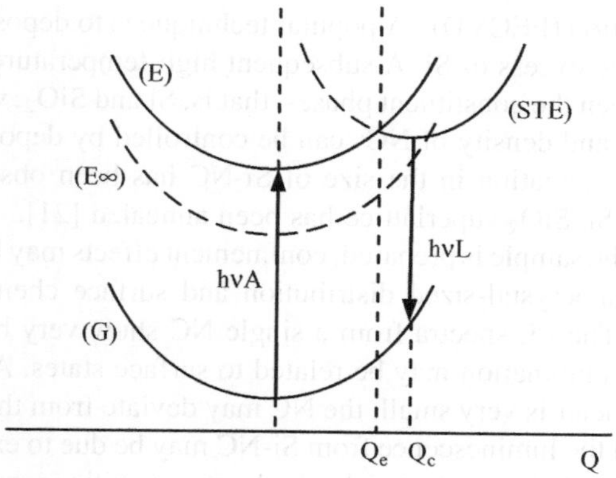

Figure 6.5 Schematic configuration coordinate diagram. The energies of the ground state (G), the normal excitonic state (E), and the self-trapped exciton state (STE) are shown. The curve (E_∞) corresponds to a very large crystallite with no blueshift, showing that the STE state might not exist for large crystallites. Reproduced with permission from [17]. Copyright (1996) American Physical Society.

It appears, therefore, that the PL in por-Si is very sensitive to the chemistry of preparation and treatment. The fabrication technique is not fully compatible with the standard CMOS technology, however. Nevertheless, it is easy to produce por-Si, the external quantum efficiency is quite high (1–10%), and the wavelength is tunable over a wide range. These are the incentives to produce practical room temperature LEDs.

Fauchet and coworkers [18] reported fabrication of por-Si LEDs. The power efficiency of the devices was ~0.01% for the early LEDs. After further device optimization, power efficiency in the ~0.1% range was reported and stability exceeding 100 hours was achieved. Integration of a por-Si LED with a transistor was reported in 1996.

Although the external quantum efficiency (EQE) of LEDs was as high as 1%, the power conversion efficiency was still low (~0.1%) due to high drive voltages for the LEDs (~10 V). Other problems with these LEDs, that is, poor operating lifetime and an aging effect, have been overcome by using proper passivation techniques. By anodically oxidizing por-Si, Gelloz and Koshida [19] improved the stability further and, using ITO for contact, they scaled down the operating voltage below 5 V. The reported EQE was >1%, and power conversion efficiency was ~0.37%. The authors used surface passivation techniques based on Si–C and Si–C–O bonds to produce highly stable LEDs. The efficiency and other properties are yet to be reported.

Such LEDs have been integrated with Si microphotonic circuits to produce an addressable display. The modulation bandwidth of the LEDs is quite low ~1 MHz. By inserting the LED in a por-Si microcavity, some improvement in the performance has been achieved [20].

6.5.3 Silicon Nanocrystals

Silicon nanocrystals (Si-NC) are grown by a variety of techniques: by (1) decomposition of silane, (2) laser ablation, (3) ion implantation, (4) sputtering, and (5) plasma-enhanced chemical vapor deposition (PECVD). A popular technique is to deposit substoichiometric silica films, with a large excess of Si. A subsequent high-temperature annealing causes a phase separation between the constituent phases, that is, Si and SiO_2, with the formation of small Si-NC. The size and density of NCs can be controlled by deposition and annealing parameters. Very little deviation in the size of Si-NC has been observed in a proposed technique, in which a Si–SiO_2 superlattice has been annealed [21].

Depending on how the sample is prepared, confinement effects may be seen or may not be seen, as in por-Si. Nanocrystal-sized distribution and surface chemistry determine the nature of PL spectra. The PL spectra from a single NC show very broad characteristics, suggesting that the recombination may be related to surface states. Another possibility is that when the NC diameter is very small, the NC may deviate from the diamond structure. Calculations show that the luminescence from Si-NC may be due to excitons trapped at the surface of the NC, which is passivated by hydrogen or silicon oxide. The model for absorption and recombination has already been described with reference to Figure 6.5. Allan *et al.* [17] proposed that localized radiative levels may be formed due to a Si–O (silanone) double bond at the Si-NC surface. The red PL quantum efficiency and lifetime are similar to that found for por-Si. Kovalev *et al.*, on the other hand, concluded that the emission is due to a quasi-direct NP transition characteristic of NCs having smaller diameter. As already stated, the NP process has been found by them to dominate when the confinement energy exceeds 0.7 eV (see Figure 6.4).

A detailed theoretical and experimental investigation has been reported by Daldosso *et al.* [22] on the nature of the silicon NC surrounded by silica matrix. They noted a transition from a crystalline to amorphous state from Si-NC to the interface and then a strained silica layer. They concluded that the amorphous layer plays a significant role in determining the emission characteristics of the NC.

LEDs have been made using annealed silicon-rich SiO_2 [23]. The structure is metal oxide semiconductor (MOS)-like in that the silica layer contains the NCs. The bias applied to the gate controls the EL properties of the NCs. Here the design depends on the distance between two NCs to allow charge transport. Passivation of Si by oxide increases the drive current. Again too many dots in the layer increase the leakage current. A suitable design for efficient operation has been investigated [24].

The effect of embedding the Si-NCs in a micro-cavity was studied by Amans *et al.* [25]. The NCs were placed on top of a half-cavity, that is, a mirror; they were also embedded in a full cavity which is a Fabry–Perot resonator. Both the cavities are made of dielectric-distributed Bragg reflectors. Efficient spectral and angular narrowing is observed for the full cavity. In addition, the vertically emitted light showed an intensity enhanced by a factor of about 20. The result indicates that enhanced light emission is possible by using the cavity effect. Results for EL were not reported, however.

In another report by Valenta *et al.* [26], EL from individual Si-NCs was measured. The earlier papers reported the emission characteristics of an ensemble of NCs. The authors observed a spectral narrowing of emission, and a high degree of linear polarization. They concluded that the yield of radiative emission was as high as 19%. The findings are believed to open a route to all silicon light emitters.

SiGe dots are also investigated for light emission. Work has started using Ge dots grown in silica, or Ge dots grown in a Si matrix [27]. The nature of luminescence is being studied at present.

6.5.4 Quantum Wells, Wires, and Dots

Both por-Si and Si-NCs exhibit inhomogeneities in their structures, and the PL and EL spectra are broadened thereby. On the other hand, uniformity in the grown structure may be ensured in QWs, QWRs, and QDs. Modern epitaxy is used to grow QWs; QWRs can grow on patterned substrate, or by side etching the QW structure. QDs can be fabricated by side etching or by self-assembly.

Thin QWs of Si may be grown by using wide band gap barriers like, SiO_2, CaF_2, or Al_2O_3. Si–SiO_2 superlattices grown using molecular beam epitaxy (MBE) showed interesting PL across the visible range for silicon-layer thicknesses $1 < d < 3$ nm. Figure 6.6 shows the PL peak energy versus Si thickness [28] in a Si–SiO_2 SL. Since SiO_2 has a large barrier, the band gap for infinite barrier, obtained by effective mass theory, is expressed as

$$E_G(\mathrm{QW}) = E_G(\mathrm{Bulk}) + \frac{\pi^2\hbar^2}{2d^2}\left(\frac{1}{m_e} + \frac{1}{m_h}\right). \tag{6.12}$$

The fitted peak emission energy is $E\,(\mathrm{eV}) = 1.60 + 0.72\,d^{-2}$. The bulk band gap 1.6 eV agrees well with the values for amorphous Si, and the prefactor 0.72 is also explained by using effective masses for electrons and holes in bulk amorphous Si. The agreement is clear evidence of quantum confinement as well as of the direct nature of the gap, which is further supported by X-ray analysis.

Further refined calculation reveals that in these superlattices, the Si layer indeed behaves as a pseudo direct-gap material. The bright PL observed motivated researchers to fabricate emitters in the range from 500 to 800 nm. Several prototype LEDs have been constructed, and all report visible EL. There is no strong evidence that the emission originates from the

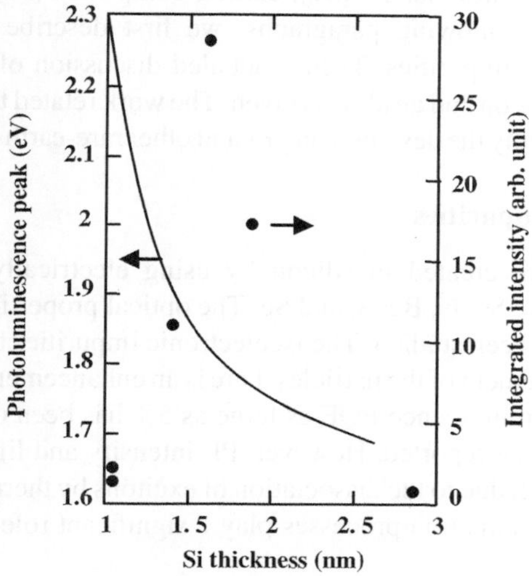

Figure 6.6 The PL peak energy versus silicon thickness in a Si and SiO_2 superlattice (open circles). The solid line is the energy gap calculated by effective mass theory. Solid dots indicate the integrated intensity at room temperature. Reproduced with permission from [28]. Copyright (1996) American Physical Society.

confined states, however. As an example, the work of Gaburro *et al.* [29] is reported here. The authors fabricated Si LEDs, in which the multilayers are embodied by the dielectric in a MOS capacitor. The PL is due to e–h recombination, but the EL was due to hot-electron relaxation in the substrate. The measured EQE of the EL was only $\sim 5 \times 10^{-5}$.

Several workers studied light emission from Si and SiGe QWRs to examine if quantum confinement leads to improved performance and then to fabricate LEDs. The diameters of the wires need to be substantially less than 10 nm to obtain any improvement due to confinement. As an example, we quote the work by Nassiopoulos *et al.* [30]. They fabricated silicon nanopillars by using deep-UV lithography, highly anisotropic RIE based on fluorine chemistry, and high-temperature thermal oxidation and etching for further thinning below 10 nm diameter. They used a contact layer on top and also on the bottom Si substrate. The devices showed rectifying behavior and EL were observed for forward voltages exceeding 10–12 V. The emission with a peak around 650 nm could be seen with the naked eye.

Several theoretical calculations indicated the direct nature of the band gap in silicon nanopillars. A number of workers studied PL and EL in Si and SiGe QDs recently and have found encouraging light emission. It is believed that these QDs are the best candidates for observing laser action. The detailed discussion about emission in QDs will therefore be undertaken in a separate subsection.

6.6 Impurities in Silicon

The role of isoelectronic impurities, like N, Be, and so on, in GaP is well known. These impurities bind excitons in the indirect-gap semiconductors, and the dissociation of bound excitons gives rise to the characteristic light emission from GaP or its alloy $Ga_{1-x}As_xP$. This approach has been undertaken to fabricate LEDs with silicon, by doping the materials with rare-earth impurities, in particular Er [31], carbon complexes [32], and sulphur–oxygen complexes [33]. In the following paragraphs, we first describe briefly the results of introducing isoelectronic impurities. Then a detailed discussion of the energy levels and transition in Er-doped silicon and an alloy is given. The work related to Er-doped Si and SiGe is then described, and finally the devices using Er and other rare-earth elements are presented.

6.6.1 Isoelectronic Impurities

Isoelectronic centers are created in silicon by using electrically neutral dopants like isovalent elements C, Ge, Sn, N, Be, S, and Se. The optical properties of all these elements in Si and $Si_{1-x}Ge_x$ have been studied. The isoelectronic impurities bind free excitons in Si, and due to spatial confinement of the particles there is an enhancement of the recombination. In all these studies, a luminescence EQE as large as 5% has been observed and a lifetime greater than 1 ms has been reported. However, PL intensity and lifetime decrease rapidly with a rise in temperature, due to the dissociation of excitons by thermal energy. In addition, other nonradiative recombination processes play a significant role.

6.6.2 Rare-earth Luminescence

Much work has been done on the study of optical processes in rare earth (RE)–doped semiconductors and insulators. Among all the RE elements, Er has received most attention. It is well known that Er introduced into the core of optical fibers leads to amplification of lightwaves, and Er-doped fiber amplifiers (EDFAs) are essential components of present-day dense-wavelength division multiplexing (DWDM) communication.

6.6.2.1 *Energy Levels*

The most important characteristic of luminescence in RE elements is that the transitions occur between the internal energy levels of the RE ions, and hence the host material does not play any significant role in the optical processes.

The RE elements occupy the 3+ valence electronic states in a host material. The observed transitions from the RE^{3+} ions are internal transitions of the 4*f* states. In most illustrations, Er will be taken as the representative of the RE element. The sharp energy levels of a free Er^{3+} ion split into a manifold of sublevels due to crystal field splitting. The energy levels relevant for the present discussion, for both free ions and ions in crystal field, are shown in Figure 6.7. The transition between $^4I_{15/2}$ and $^4I_{13/2}$ manifolds is indicated in Figure 6.7. The downward transition shown in Figure 6.7 leads to emission at 1.54 μm, independent of the host material.

6.6.2.2 *Transition Rates: Simple Model*

It is instructive to get the idea of spontaneous and stimulated emission and the corresponding requirements for LEDs and lasers from a simple analysis presented by Xie *et al.* [34].

We first consider spontaneous emission. Er^{3+} in Si can be loosely treated as a three-level system, in which the ground state, level 1, is the $^4I_{15/2}$ manifold. The first excited state, level 2, is the lowest sublevel in the $^4I_{13/2}$ manifold. The higher excited level, level 3, may be any sublevel above level 2. The populations in the levels are denoted by N_1, N_2, and N_3. A pump excites the system to level 3 first, from which state it quickly decays to level 2. The transition from level 2 to level 1 is radiative with a rate equation given by

$$\frac{dN_2}{dt} = R - \frac{N_2}{\tau}, \text{ and } \frac{1}{\tau} = \sum_i \frac{1}{\tau_i}. \tag{6.13}$$

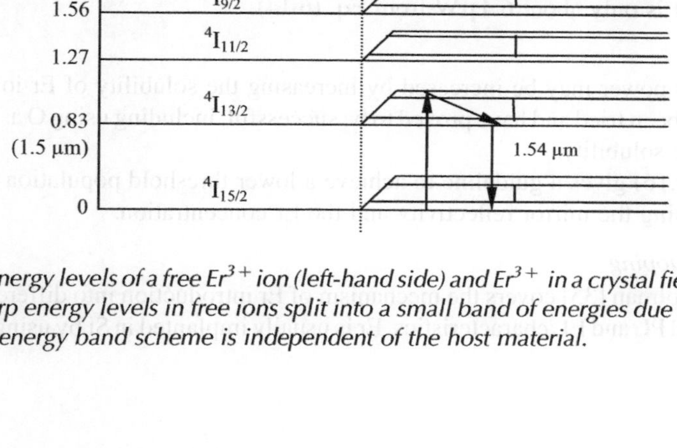

Figure 6.7 *Energy levels of a free Er^{3+} ion (left-hand side) and Er^{3+} in a crystal field (right-hand side). The sharp energy levels in free ions split into a small band of energies due to crystal field splitting. The energy band scheme is independent of the host material.*

In Eq. (6.13), R denotes the pumping rate, τ is the overall lifetime in level 2, and τ_i are lifetimes due to different transition paths from level 2 to level 1. The upper limit to the pumping rate and output power is obtained by replacing τ by the spontaneous emission lifetime τ_{sp} in Eq. (6.13). Assuming that the pump rate R is large enough to transfer all the Er ions into level 2, so that $N_2 = N_{Er}$, one obtains the following expression for the steady-state output power:

$$P = (N_{Er}/\tau_{sp})(hc/\lambda)V., \tag{6.14}$$

where λ is the free space wavelength of the emitted photon and V is the optically active volume of the LED.

On the other hand, the gain coefficient related to stimulated emission in a system in which population inversion occurs may be expressed as

$$g = (N_2 - N_1)(\lambda^4/8\pi n^2 \tau_{sp} c \Delta\lambda) = \sigma(N_2 - N_1), \tag{6.15}$$

where n is the RI of the material, $\Delta\lambda$ is the full width at half maximum (FWHM) value of the gain spectrum, and σ is the interaction cross-section. As usual, when $N_2 > N_1$, there is positive gain. Laser action is possible when the gain exceeds the mirror losses and material losses, which in this case is predominantly due to free-carrier absorption. The threshold population inversion N_{th} may be written as

$$N_{th} = (N_2 - N_1)_{th} = (1/\sigma)[\alpha - (1/2L)\ln(R_1 R_2)], \tag{6.16}$$

where α is the material loss, L is the length of the cavity, and R's are the reflectivities of the two mirrors forming the laser cavity.

Although the transitions involved and the pumping mechanism in actual situations are more complicated, the analysis given here presents some idea about the requirements for a good device.

It appears from Eq. (6.14) that the output power of a LED depends on the Er ion concentration. The limit to this is governed by the solid solubility of Er in Si, which is less than $1 \times 10^{18}\ \text{cm}^{-3}$.

Example 6.2

The spontaneous emission lifetime is about 1 ms for most of the semiconductors. With a diode geometry of 400 μm diameter and 0.4 μm active layer thicknesses, the maximum power output is only about 6.4 μW from Eq. (6.14).

It appears that power may be increased by increasing the solubility of Er ions. Different methods have been tried and have proved to be successful, including using O as a co-dopant, to increase the solubility.

Equation (6.16) gives a guideline to achieve a lower threshold population inversion by suitably choosing the mirror reflectivity and the Er concentration.

6.6.2.3 Er Doping

A review by Polman [35] covers the mechanism of Er introduction into different materials, co-doping, and PL and EL characteristics. Er is usually implanted in Si by using a high dose.

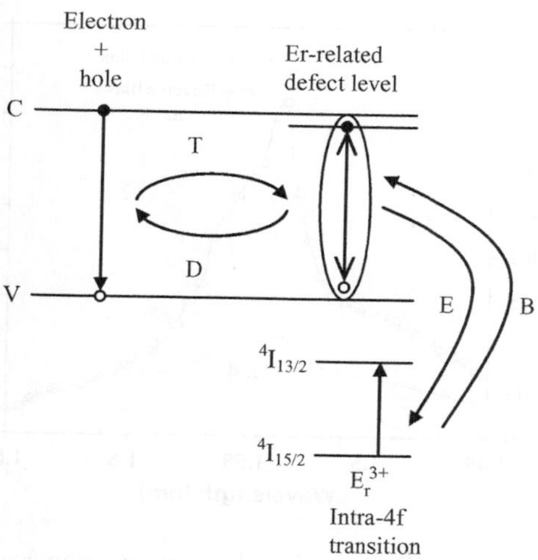

Figure 6.8 *Excitation of Er ions in Si. Reproduced with permission from [35]. Copyright (1969) American Institute of Physics.*

The solid solubility under normal situations does not exceed 10^{16} cm^{-3}. Co-doping with O increases the concentration of Er. However, the maximum Er concentration that can be optically activated is only 3×10^{17} cm^{-3}. Incorporation of O co-doping increases the Er concentration and the quantum efficiency.

It is found that by using multiple-energy Er and O co-implant, uniform Er and O concentrations of 1×10^{19} and 1×10^{20} cm^{-3}, respectively, can be achieved.

The excitation of Er ions in Si can be explained with the help of Figure 6.8. It is assumed that the Er–O complex provides a defect level in the gap of Si. An electron may then be trapped (T in Figure 6.8), and subsequently a hole may be bound to it. The energy released in recombination of this pair is transferred by the Auger process to the Er ion, which is then promoted to the excited level. Finally, the Er ion may decay to the ground level by emitting 1.54 μm photons.

Er-doped p–n junctions are fabricated by using Er and O co-implantation. The EL spectra of the diodes are shown in Figure 6.9. The forward-bias EL is attributed to the excitation of Er ions by e–h recombination, as discussed in this subsection. However, the reverse-bias EL is due to excitation of Er ions by hot electrons released by impact ionization under a reverse bias of −5 V. The IQE of the diodes is only 10^{-4}.

6.6.2.4 Er Ions in Nanocrystals

It appears that the performance of Er-doped bulk Si LEDs cannot be improved further due to a few fundamental reasons. The maximum energy of hot electrons in exciting the Er ion is limited. There is also substantial energy loss due to Auger recombination of free carriers in Si. Lastly, there is an energy back transfer process, as denoted by B in Figure 6.8. The density of Er ions in the material also should be kept below a certain limit to avoid clustering. A partial solution to all these problems has been made by using silicon dots embedded in SiO_2 in place of bulk Si. The QDs of Si act as mere sensitizers. The larger band gap of QDs

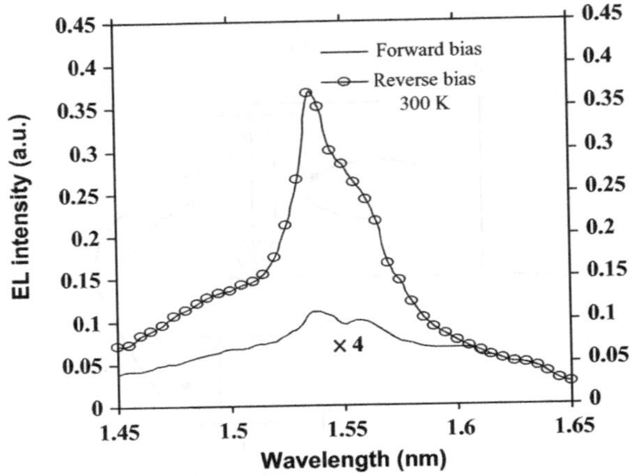

Figure 6.9 EL intensity of Er:Si LED. Reproduced with permission from [35]. Copyright (1969) American Institute of Physics.

minimizes the energy back transfer to Si from Er. As there is no free carrier in SiO_2, energy loss by Auger recombination is ruled out.

A detailed comparison of room temperature PL of Er in Si-NC, in SiO_2, and of Si-NC without Er has been made by Pacifici *et al.* [36]. Figure 6.10 shows the comparison. The density of Si-NC is $\sim 1 \times 10^{19}$ cm^{-3}, while the Er concentration is 6.5×10^{20} cm^{-3}. The PL intensity with Er in silica is about two orders higher with respect to the sample with Er in SiO_2. Each Si-NC absorbs a photon and promptly transfers its energy to a nearby Er ion,

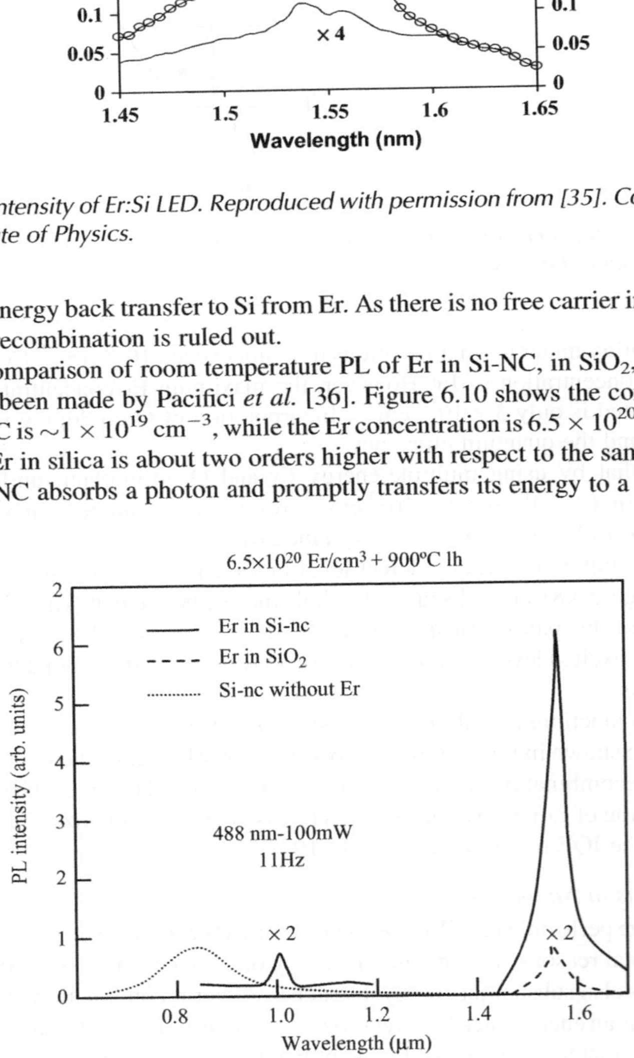

Figure 6.10 Room temperature PL spectra of Si-NC, and Si-NC with Er and Er in silica. Reproduced with permission from [36]. Copyright (2003) Elsevier Ltd.

which then returns to its ground state by emitting 1.54 μm wavelength radiation. The measured effective excitation cross-section for Si-NC mediated excitation is about three orders of magnitude higher than that of Er in oxide only. The emission in 1.54 μm is stronger than the emission at 0.85 μm by a Si-NC. A model for energy transfer to the Er ions has been discussed, which is not included here.

The PL spectra are further intensified by enclosing the active material in a microcavity. The active material containing the Si-NC is $\lambda/2n$ thick and is sandwiched between two distributed Bragg mirrors formed by four $Si-SiO_2$ pairs on the bottom and three such pairs at the top. Each Si and SiO_2 layer is λ/4n thick. The PL spectrum is extremely sharp with $\Delta\lambda = 5.5$ nm, as compared to ~40 nm in Er:Si-NC without the microcavity. The combination thus gives higher PL by about three orders of magnitude.

The authors fabricated MOS structures in which the Er ions and Si-NCs are incorporated in the oxide. Room temperature EL exhibited enhancement of about two orders of magnitude. The results are quite promising for the fabrication of high-efficiency LEDs with Er-doped Si-NC operating at room temperature.

Encouraging results are reported by Castagna *et al.* [37] about Er-doped Si-NC LEDS. In their MOS structures, Er was implanted in the thin gate oxide. The EQE was 10%, comparable to that of standard LEDs made with III–V compounds. However, the excitation of Er ions is caused by hot electrons, and the reliability of the devices suffers due to oxide wear-out. Much more stable LEDs have been fabricated by using Si-rich oxide films as the gate dielectric. Although reliability is increased, the EQE now reduces to 1%. In these devices, the energy transfers from Si-NCs to the Er ions. By using Tb ($\lambda = 540$ nm) and Yb ($\lambda = 980$ nm) in the oxide, EQEs of 10% and 0.1% respectively were achieved by the same authors. Some recent results have been reported by dal Negro *et al.* [38]. Light emission has been found to be enhanced in photonic crystal nanocavities [39].

6.7 Stimulated Emission: Prospect

Although some encouraging results on enhanced PL from Si-based NCs or Er-doped systems have been obtained and also good EQE has been reported from LEDs, laser action from Si-based structures have not yet been observed. Some recent work reports appreciable values of optical gain by stimulated emission. It is expected that these findings will encourage researchers to design suitable structures for realizing laser action. Many workers believe that Si lasers will soon become a reality. It is to be noted that Si-based lasers based on stimulated Raman scattering were unveiled by Intel in 2004, the working principle of which will be described in Chapter 9. In this section, we report recent findings on positive gain coefficients in some of the structures discussed in this chapter.

6.7.1 Si Nanocrystals

Pavesi *et al.* [40] reported for the first time optical gain in Si-NCs. Since the first observation, a number of reports have been published by the group, and all the work has been summarized in Dal Negro *et al.* [41]; see also Pavesi [42] and Fauchet [43]. In this subsection, we shall essentially reproduce their findings.

Dal Negro *et al.* used Si-NC samples produced by high-temperature annealing of substoichiometric silicon oxide (SiO_x) thin films grown by the PECVD process. The

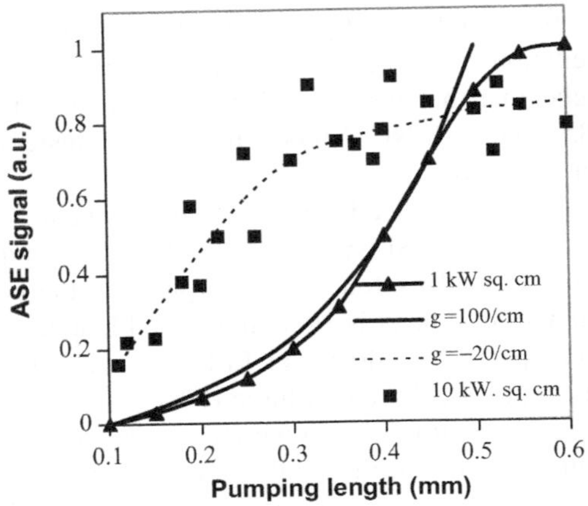

Figure 6.11 Amplified spontaneous emission (ASE) signal versus pumping length for two pumping powers. Reproduced with permission from [42]. Copyright (2005) Elsevier Ltd.

amplified spontaneous emission has been measured by a variable stripe length (VSL) geometry, and by using a pump–probe technique. In the VSL method, the length of the pumped region is varied and the ASE signal coming from the edge of the waveguide whose core is rich in Si-NC is measured. Data in Figure 6.11 show that the ASE intensity grows sublinearly with pump power, when the pump power is less than the threshold. However, when the pump power exceeds the threshold, the ASE signal increases more than exponentially. In time-resolved measurement, the ASE decay line shape shows fast decay corresponding to stimulated emission.

Figure 6.12 shows the wavelength dependence of the luminescence, absorption, and gain spectra in a sample with 4 nm Si-NC. The gain spectrum is on the high-energy side of the emission band, and absorption is negligible in the region of gain and luminescence. All these facts can be explained by a four-level model of the gain as depicted in Figure 6.13.

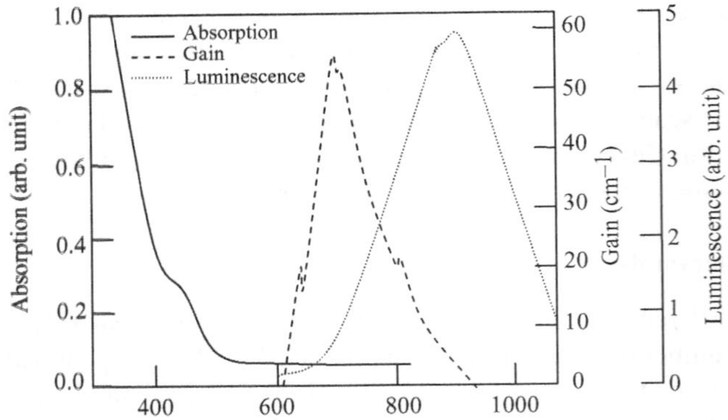

Figure 6.12 Luminescence, absorption, and gain spectra for a Si-NC rich waveguide at room temperature. Reproduced with permission from [42]. Copyright (2005) Elsevier Ltd.

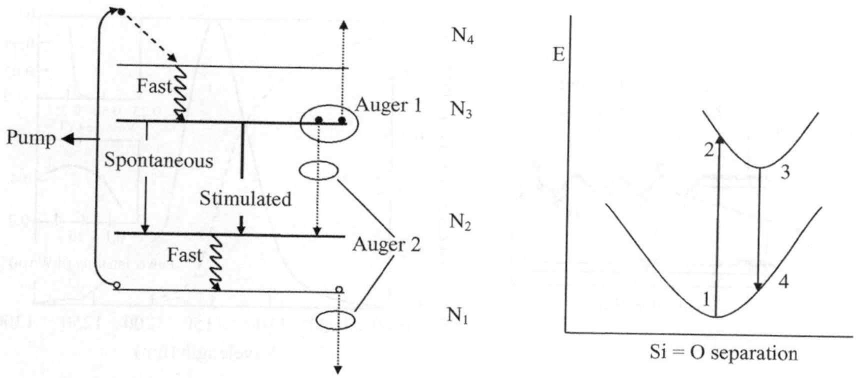

Figure 6.13 *(Left) Model to explain recombination dynamics under gain conditions. The right-hand side gives the energy level diagram for a Si-NC in an oxygen-rich matrix. Radiative states exist inside the NC band gap due to interface O atoms. The excited NC states occur at different lattice coordinates with respect to the ground state. Reproduced with permission from [44]. Copyright (2002) American Physical Society.*

The levels shown are associated with different Si-NC populations or to radiative states associated with Si–O double bonds. Optical excitation causes a large lattice relaxation as in the silanone molecule [44]. Calculations show that Auger recombination is a strongly competitive process.

Although gains in the range of 10–100 cm^{-1} have been measured, no observation of lasing in cavity geometry or of net gain in a pump–probe waveguide configuration has been reported.

Evidence for optical gain is provided further by the group [45] from Si-NC planar waveguides grown via phase separation and thermal recrystallization of SiO–SiO_2 superlattices. An efficient optical confinement was obtained by using multilayered mirrors and using NCs of proper size. The large gain values measured were attributed to small size distributions of Si-NCs.

The design of a nanocavity for a Si-NC based laser [39] has been reported and the narrowing of linewidth to a few nanometers has been observed. The major obstacle to the realization of a silicon laser is the need to pump it electrically, not optically. As observed in discussions throughout this chapter, electrical pumping is in most cases less efficient than optical pumping.

Since the EL efficiency shown by Er:Si-NC reaches quite a high value, there are good prospects for use of this material system in realizing laser action.

6.7.2 Bulk Si

The common belief that bulk Si cannot be a light-emitting material has been put to question in a work by Green *et al.* [46], who noticed that some solar cells possess an extremely long recombination lifetime (~ a few milliseconds). Since the recombination lifetime equals the radiative lifetime, the IQE must be ~1 (see Eq. (6.5)). The cell behaves as an LED when forward biased. The structure of the device and its room temperature emission spectrum are shown in Figure 6.14.

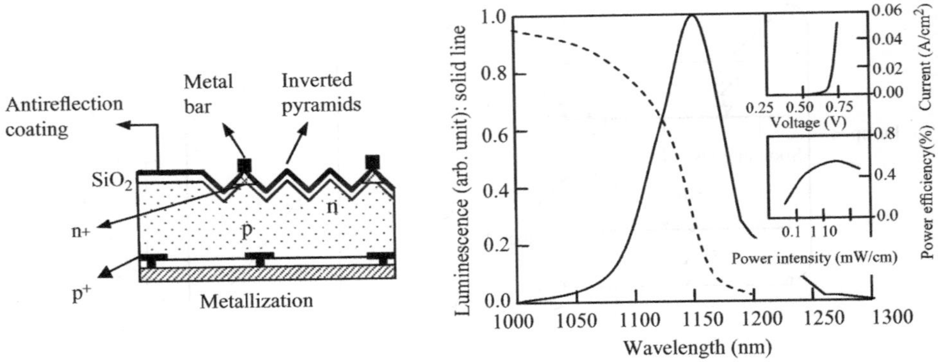

Figure 6.14 *(Left) Schematic of the LED geometry. (Right) Luminescence spectra (peaked), and absorption spectrum; insets show (1) the power efficiency versus injected power density, and (2) the I–V characteristics at room temperature. Reproduced with permission from [46]. Copyright (2001) McMillan Ltd.*

As shown, the LED structure is textured so that most of the internally generated light falls on the external surface of the cell with an incident angle lower than the critical angle for total internal reflection. Thus the light-extracting efficiency is increased from a small percentage, characteristic of a flat surface, to about 100% in the texturized surface. The heavily doped contact regions are confined in very thin small lines, to reduce free-carrier absorption. Also the nonradiative recombination rates are reduced by using high-quality intrinsic Si substrates. All these factors led to the highest reported power conversion efficiency of ~1% for Si LEDs.

The difficulties related to using this approach for the fabrication of integrated emitters are as listed. The high-purity silicon substrate proves to be costly. Texturizing of the surface is not compatible with standard complementary MOS (CMOS) technology. There may be problems of integrating the bulk device with an optical cavity. The strong contribution of free-carrier absorption is not estimated. Finally, since the recombination lifetime is a few milliseconds, the modulation speed of the LED or laser may be limited to a few kilohertz only.

The same group also presented a theoretical analysis that modifies the earlier theory about the nonfeasibility of optical gain in an indirect-gap material [47]. Optical gain, according to these authors, is theoretically possible in the sub-bandgap region where phonon-assisted processes can help achieve gain. These theoretical arguments have been partially confirmed by a recent study [48], the results of which are shown in Figure 6.15.

As mentioned earlier, a short nonradiative lifetime is detrimental to achieving efficient emission. The authors therefore tried to avoid carrier diffusion by spatially localizing free carriers in a small region where nonradiative recombination centers can be saturated. Ion implantation was used to induce dislocation loops at the junction of the p–n diode. The dislocation loops cause local strain fields, which increase the energy gap locally, thereby producing a potential barrier for carrier diffusion. LEDs based on this idea have been realized [49].

The power current and EL characteristics of the diode shown in Figure 6.15 were obtained by Chen *et al.* in the diodes where carrier localization was realized by spin-on doping of

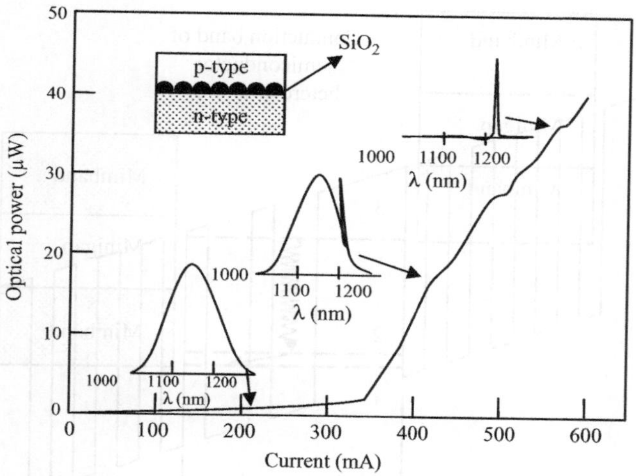

Figure 6.15 Optical power versus injected current for an LED containing silica nanoclusters in the junction region (shown as inset). The EL spectra at different injection currents (indicated by arrows) are also shown. Reproduced with permission from [48]. Copyright (2004) American Institute of Physics.

small silica nanoparticles at the p–n junction. The I–V curve of the diode shows a clear rectifying nature with a clear threshold in the light–current curve. The spectrum is broad at low current, but a very sharp peak occurs at higher current showing the signature of stimulated band-to-band emission. The origin of stimulated emission is ascribed to a two-phonon indirect transition, as theoretically predicted.

These results are very encouraging since the device is an electrically operated p–n junction. However, the gain achieved in this system must be adequate to overcome the large free-carrier absorption.

6.8 Intersubband Emission

While momentum needs to be conserved in band-to-band transitions in indirect-gap semiconductors, transitions between two subbands in a QW may be direct without involvement of a phonon or any other momentum-conserving agent. For example, in the conduction band of Si or SiGe, **k**-conservation is satisfied for E_1 and E_2 subbands belonging to valleys located along the same directions (see Figure 3.12). For transition from unprimed subbands to prime subbands, participation of a phonon is necessary. The situation is simpler for transitions between two subbands in the valence band, for example from HH1 to HH2.

Intersubband absorption in Si and SiGe QW systems has been studied in detail in order to design quantum well infrared photodetectors (QWIPs). Although intersubband emission was investigated by several workers in both quantized Si inversion layers and QWs, the purpose was mainly to understand basic optoelectronic processes. The lasing action involving intersubband transitions was conceived quite early. However, the practical realization of the concept was achieved only in 1994 with the announcement

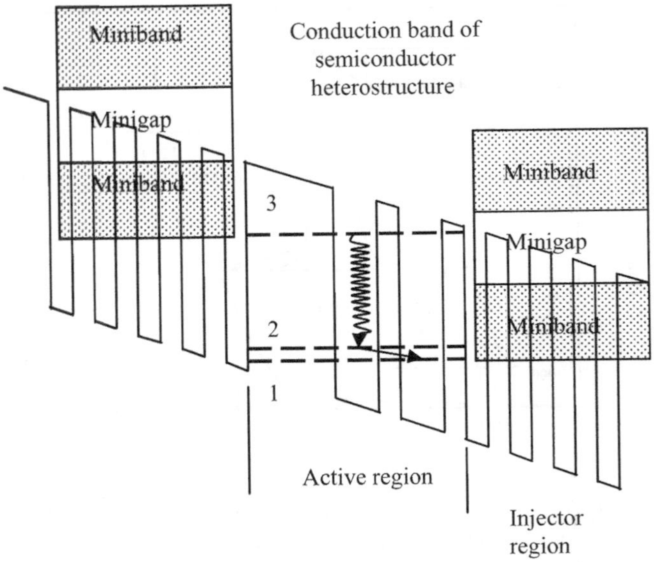

Figure 6.16 *Basic QCL structure using conduction subbands.*

of quantum cascade lasers (QCLs) involving III–V compounds by Faist *et al.* [50]. A lot of work has since then been reported for QWs made of III–V and II–VI compounds. In a QCL structure there exists an injector, a QW in which transition occurs, and then a collector to collect the carrier. The pattern is repeated so that the number of emitted photons is multiplied in the cascade structure by utilizing the same carrier over many periods. Population inversion is achieved through careful control of the lifetimes of the upper and lower states. The band diagram associated with a QCL structure is shown in Figure 6.16. The superlattice in the left part of the figure contains a miniband which is aligned with energy level 3 in the QW in which the transition occurs. The electrons are injected to level 3, from which it makes a radiative transition to level 2 in the active QW. The electron then quickly transits to level 1, and by resonant tunneling it goes to the miniband in the next superlattice. The process then repeats itself. The advantage is that the same electron can be utilized to create another photon in the next active QW. Although the process is illustrated using conduction subbands, it may readily be applied to hole subbands also.

The possibility of achieving QCL action in interminiband transitions in the valence band was addressed in [51], and in the conduction band in [52]. Recent theoretical works [53, 54] involve transitions in the conduction subbands in Ge QWs.

Until now, experimental work has been confined to p-type materials. Stimulated emission has been observed in the mid-infrared and THz frequency ranges. In addition, THz emission has also been observed for band impurity transitions. The following discussion is divided into two subsections discussing: (1) emission in mid-infrared involving transition between two heavy-hole subbands, and (2) THz emission involving LH-to-HH subband transitions as well as band impurity transitions.

6.8.1 Emission at Mid-infrared

To achieve a conduction band discontinuity, the Si–Ge superlattice should be grown on a relaxed Si–Ge buffer layer. For pseudomorphic growth on a Si substrate, most of the band offset occurs at the valence band. Dehlinger *et al.* [55] were the first to observe QCL action involving the valence band of a Si–SiGe multilayered structure.

The basic block consists of an active quantum well and injector and collector wells. The main well, a $Si_{0.68}Ge_{0.32}$ quantum well 40 Å wide (w1), contains two quasi-bound heavy-hole states (HH1 and HH2) separated by 130 meV, as well as a light-hole state at 60 meV below HH1. Situated next to w1 are four coupled SiGe quantum wells (w2 – w5), forming a "mini"-superlattice. The width and Ge content of the four coupled wells are designed in such a way that, for an electric field of 50 kV/cm, a series of states in the minisuperlattice have a near-overlapping energy with the HH1 state. This allows for an easy tunneling out from this state and into the injector states in the following period. In contrast, there is a minigap for both heavy-hole and light-hole states around the energy of the HH2 state. This minigap prevents the transport out from this state, increasing the probability for a radiative transition between the HH1 and HH2 states. In the usual pattern for a cascade structure, this block is repeated 12 times, although a Si layer (1000 Å wide) is inserted after each fourth repeat to limit the total strain.

Several other groups also reported EL from structures fabricated differently. The EL spectra persist at low temperature; the peak broadens with a rise in temperature above 90 K and finally disappears at ∼150 K. The quantum efficiency (QE) is $\sim 10^{-5}$.

Several attempts have been made to extend the operating temperature and QE of the structures. A strain symmetric structure has been grown and studied by growing Ge-rich $Si_{0.2}Ge_{0.8}$ and Si QWs on a $Si_{0.5}Ge_{0.5}$ buffer layer. The structure leads to higher valence band offset and a larger number of periods [56].

Although the PL results for SiGe and Si QCLs are comparable to the corresponding structures grown with III–V compounds, there are inherent limitations in using SiGe systems. Since the thickness limit is governed by critical layer thickness, the number of periods is restricted. As the gain per single period is low, a large number of periods are needed to achieve sizable gain to overcome loss and to give rise to laser action. In fact, no stimulated emission in the structures has yet been reported. The proper waveguide structure for the superlattice is now being developed.

The emission wavelength of the QCL does not match the communication wavelengths. Some authors propose to use these lasers, if and when developed, in free-space optical interconnects and pollution monitoring, thermal imaging. and so on.

6.8.2 Terahertz Emission

6.8.2.1 Quantum Cascade Structures

Semiconductor coherent sources working at far-infrared wavelengths (∼30–300 μm) or roughly at 1–10 THz frequencies are not available at present. For lower frequencies, electronic devices like high-electron mobility transistors (HEMTs), heterojunction bipolar transistors (HBTs), or resonant tunneling devices are used, while at optical frequencies usual lasers meet the requirements. The mid-infrared range of the electromagnetic spectrum is taken care of by QCLs. Good sources and detectors at the THz range may find numerous applications including chemical and biological spectroscopy, medical imaging, gas sensing, wireless telecommunication, and so on.

A THz laser using a III–V compound was reported first by Kohler *et al.* [57]. The work was a major step toward a compact and practical electrically pumped terahertz emitter. Quantum cascade lasers offer high power and control of layer thickness, allowing the emission wavelength to be designed. The reader is referred to [58] for advances in the subject. It may be mentioned that room temperature operation has been reported only recently [59].

Usually in one period of silicon–germanium quantum cascade lasers, the carriers enter the upper energy level, emit a photon upon falling to the lower level, and then move rapidly through the continuum to be reinjected into the upper level of the next period. The transition involves HH and LH subbands. A terahertz quantum cascade laser can have more than 100 such periods. Population inversion is achieved by designing the upper level to have a longer lifetime than the lower one, which is rapidly depopulated by the continuum.

In silicon–germanium structures, time-resolved experiments have shown constant lifetimes up to ~150 K [60]. Silicon also has a higher thermal conductivity than III–V materials. A silicon-based quantum cascade laser therefore promises to be a good candidate for a room temperature THz source. Because of material considerations, all silicon–germanium quantum cascade structures investigated to date have been based on transitions in the valence band. Unfortunately, the valence band is made up of many interacting subbands, and the carriers are holes (as opposed to electrons in the conduction band) with a very high effective mass. These and other factors make the design of successful silicon–germanium quantum cascade structures more challenging than is the case for III–V materials. The progress in Si-based THz sources based on QCL has been reviewed in [61].

6.8.2.2 *Impurity Transitions*

Transitions between impurity states in doped silicon give rise to terahertz emission. Pavlov and coworkers have demonstrated lasing at wavelengths of 50–60 μm (5–6 THz) from bulk silicon doped with phosphorus, bismuth, and antimony at temperatures of up to 30 K [62–65].

Figure 6.17 illustrates the transitions involved in a terahertz phosphorus-doped silicon (Si:P) laser. A sharp emission is seen at ~5.5 THz. Such terahertz silicon lasers may eventually be used on telescopes such as the Stratospheric Observatory for Infrared Astronomy. As shown in the left inset, an optical pumping is used and the pump photons excite carriers from the 1s ground state to the conduction band (CB). The carriers fall to the $2p_0$ state by nonradiative transitions. A photon is then emitted due to an optical transition between the $2p_0$ and 1s states (depicted by an arrow). Finally, the carriers return to the 1s state. It may be recalled from Chapter 2 that $1s(A_1)$, $1s(T_2)$, and $2p_0$ are the atomic-like states of the phosphorus dopant atoms. The ionization energies for these states are given in Table 2.1. Using the values given, the emission due to the $2p_0 \rightarrow 1s$ transition occurs at 5.42 THz (see Example 6.3).

Example 6.3

Consider the ionization energies of phosphorus donor in Si as given in Table 2.1. The energies are $1s\,(T_2) = 33.9$ meV and for $2p_0 = 11.5$ meV. The energy difference 22.4 meV corresponds to a wavelength of 0.0553×10^{-3} m and a frequency of 5.42 THz.

In these systems, the emitted photon energy depends on the energy spacing of the impurity states and hence on the dopant species used. The operation of any laser depends on achieving

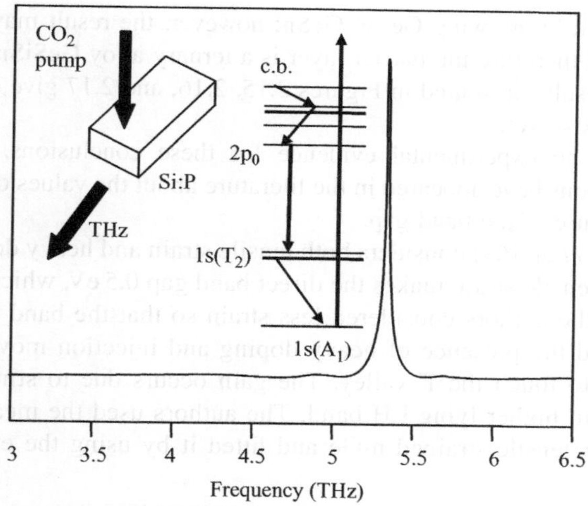

Figure 6.17 *Transitions between the impurity states for P in Si, responsible for emission at ~5.5 THz.*

population inversion, where there are more carriers in an upper energy level than in a lower one. In doped bulk silicon, this can occur as a result of the relatively long lifetime of the excited state. Hübers *et al.* have reported lasing at similar wavelengths from silicon monocrystals doped with arsenic [65].

These silicon-based terahertz lasers are being developed for applications in astronomy and atmospheric spectroscopy. However, they require a separate optical pump laser and operate only at low temperatures. An ideal laser would be electrically pumped, directly emit terahertz radiation, and operate at room temperature. Impurity emission from doped silicon can also be initiated by electrical pumping. Lv *et al.* [66] demonstrated impurity-related electroluminescence from 20 to 50 μm (6 to 14 THz) from Ga-doped silicon under pulsed currents. The peak power was 12 μW, the highest temperature was 30 K, and the transition involved acceptor states. The subject has been reviewed in [67]. Results are encouraging and the devices are easy to fabricate, but the emission degrades above 20 K. Furthermore, if the devices are to lase, population inversion is required, and no evidence of this was seen for these structures.

Optically pumped silicon impurity lasers in the terahertz range have been around for some years, but a compact, electrically pumped terahertz laser operating at room temperature remains elusive. The quantum cascade approach is arguably the most promising; here, silicon–germanium structures may offer key advantages over III–V materials for high-temperature operation. However, serious obstacles must be overcome before a working silicon quantum cascade laser can be produced.

6.9 Tensile-Strained Ge Layers

It was pointed out in Section 2.8 that the band gaps for the Γ valley and L valley in Ge differ by only about 140 meV. The application of tensile strain to Ge alters these two valleys at different rates, and at about 2% tensile strain Ge becomes a direct-gap material. The tensile strain can be applied by growing Ge on GeSn; however, the result may not be a type I heterostructure, and therefore the barrier layer is a ternary alloy GeSiSn.

The calculated results presented in Figures 2.15, 2.16, and 2.17 give clear evidence of indirect-to-direct crossover.

So far, there is no experimental evidence for these conclusions. However, some theoretical calculations have appeared in the literature about the values of gain coefficient using the direct nature of the band gap.

The work by Liu *et al.* [68] considers both tensile strain and heavy doping in the n-Ge layer. The needed tensile strain makes the direct band gap 0.5 eV, which does not emit a 1550 nm. Instead, the authors considered less strain so that the band gap difference is about 115 meV and the presence of heavy doping and injection moves the electronic quasi-Fermi level to touch the Γ valley. The gain occurs due to stimulated emission from the Γ valley to higher lying LH band. The authors used the measured direct-gap absorption data for tensile-strained n-Ge and fitted it by using the expression

$$\alpha_\Gamma(\hbar\omega) = A\left(\sqrt{\hbar\omega - E_g^\Gamma(lh)} + \sqrt{\hbar\omega - E_g^\Gamma(hh)}\right)/\hbar\omega, \tag{6.17}$$

where A is the fitting parameter and other symbols are self-explanatory.

The gain is calculated from the absorption spectra by using

$$g_\Gamma(\hbar\omega) = \alpha_\Gamma(\hbar\omega)(f_c - f_v) \tag{6.18}$$

where f's denote the Fermi occupation probabilities.

The authors found that with a doping level of $\sim 7.6 \times 10^{19}$ cm^{-3} and with an injection level of $\sim 8.0 \times 10^{18}$ cm^{-3}, a gain of 930 cm^{-1} may be obtained. The reader may study the work to have values of loss, threshold current level, and so on.

In their work. Chang and Chuang [69] considered Ge and $Si_xGe_ySn_{1\text{-}x\text{-}y}$ MQWs. The theory is based on actual envelope functions obtained by solving a Hamiltonian matrix (see Appendix A), the polarization-dependent momentum matrix element, quasi-Fermi levels calculated for different injection levels, and a broadening parameter. They could calculate the gain and threshold current considering leakage into L valleys which are in close proximity. This puts a severe limitation to the laser action, as the authors concluded.

There is no experimental report so far on laser action in tensile-strained Ge. The only experimental report on a room temperature, electrically pumped Ge laser is reported in [70]. The authors used a metal–insulator–semiconductor structure in which Ge is the semiconductor and SiO_2 is the insulator. The injection into Ge is due to tunneling. Laser action in the structure has indeed been observed in the form of intense emission, a narrowing spectral line, a highly directional beam, and so on. Peak power was ~2.0 μW over a 1600–2200 nm range. Although spontaneous emission from the LED was stable, stimulated emission from the laser spiked sharply at intervals of several seconds even though the drive current was steady. It has been suspected that carrier trapping and detrapping cycles contribute to the spiking because they're on the same time scale.

It has been suggested that the work should be continued using GeSn and other materials [71].

Problems

6.1 Follow the methods outlined in solid-state physics textbooks to obtain the E–$\mathbf{k}$ diagram in the reduced space. Repeat the process for the E–$\mathbf{k}$ diagram shown in Figure 6.1, and verify that the superlattice period should be five times the lattice constant.

6.2 Modify the theory of phonon-assisted absorption as presented in Chapter 4 for the alloy disorder potential given by Eq. (6.6). Show that the absorption coefficient due to alloy disorder scattering is expressed by Eq. (6.7).

6.3 Calculate the overlap function as given by Eq. (6.11).

6.4 Using a coupled QW structure with Si and $Si_{0.4}Ge_{0.6}$ for which the widths of each layer are 2 nm, calculate the overlap integral. Use proper values of band offsets. Show that the overlap integral increases from the value in type II alignment.

6.5 Obtain the expression for power output as given by Eq. (6.14).

6.6 Obtain the expression for gain in a two-level system as given by Eq. (6.15).

6.7 Obtain the complete expression for the fitting parameter A in Eq. (6.17).

6.8 Use the absorption curves for unstrained and strained Ge given in [68]. Obtain the values of A given in Eq. (6.17) for both cases, and compare the values.

References

1. People, R. and Jackson, S.A. (1987) Indirect, quasi direct and direct optical transitions in the pseudomorphic (44)-monolayer Si-Ge strained layer superlattices on Si (001). *Phys. Rev. B*, **36**, 1310–1313.
2. People, R. (1986) Physics and applications of Ge_xSi_{1-x}/Si strained layer heterostructures. *IEEE J. Quantum. Electron.*, **QE-22**, 1696–1710.
3. Menczigar, U., Abstreiter, G., Olajos, J. *et al.* (1993) Enhanced band gap luminescence in strain-symmetrized $(Si)_m/(Ge)_n$ superlattices. *Phys. Rev. B*, **47**, 4099–4102.
4. Vescan, L. and Stoica, T. (1998) Room-temperature SiGe light emitting diodes. *J. Lumin.*, **80**, 485–489.
5. Weber, J. and Alonso, M.J. (1989) Near band gap photoluminescence of Si-Ge alloys. *Phys. Rev. B*, **40**, 5683–5693.
6. Mi, Q., Xiao, X., Sturm, J.C. *et al.* (1992) Room temperature 1.3μm electroluminescence from strained $Si_{1-x}Ge_x$/Si quantum wells. *Appl. Phys. Lett.*, **60**, 3177–3179.
7. Presting, H., Zinke, T., Splett, A. *et al.* (1996) Room temperature electroluminescence from Si/Ge/$Si_{1-x}Ge_x$ quantum well diodes grown by molecular-beam epitaxy. *Appl. Phys. Lett.*, **69**, 2376–2378.
8. Ghosh, S. and Basu, P.K. (2002) Light emission potential of Si-based nanostructures from calculated absorption coefficient. *Microwave Opt. Technol. Lett.*, **30**, 352–355.
9. Pantelides, S.T. and Zollner, S. (eds) (2002) *Silicon–Germanium Carbon Alloys: Growth, Properties and Applications*, vol. **15**, in Series Optoelectronic Properties of Semiconductors and Super-Lattices, Series editor M.O. Manasreh, Taylor & Francis Books, Inc., New York.
10. Canham, L.T. (1990) Silicon quantum wire array fabrication by electrochemical and chemical dissolution of wafers. *Appl. Phys. Lett.*, **57**, 1046–1048.
11. Takagahara, T. and Takeda, K. (1992) Theory of quantum confinement effect on excitons in quantum dots of indirect-gap materials. *Phys. Rev. B.*, **46**, 15578–15580.
12. Hybertsen, M.S. (1994) Absorption and emission of light in nanoscale silicon structures. *Phys. Rev. Lett.*, **72**, 1514–1517.

13. Xie, Y.H., Hybertsen, M.S., Wilson, W.R. *et al.* (1994) Absorption and luminescence studies of free standing porous silicon films. *Phys. Rev. B*, **49**, 5386–5401.
14. Kovalev, D., Heckler, H., Ben-Chorin, M. *et al.* (1998) Breakdown of k-conservation rules in Si nanocrystals. *Phys. Rev. Lett.*, **81**, 2803–2806.
15. Lehmann, V. and Gösele, U. (1991) Porous Si formation: a quantum wire effect. *Appl. Phys. Lett.*, **58**, 856–858.
16. Cullis, A.G., Canham, L.T., and Calcott, P.D.J. (1997) The structural and luminescence properties of porous silicon. *J. Appl. Phys.*, **82**, 909–965 andCanham, L. (ed.) (1997) *Properties of Porous Silicon*, INSPEC, Salford, UK.
17. Allan, G., Delerue, C., and Lannoo, M. (1996) Nature of luminescent surface states of semiconductor nanocrystallites. *Phys. Rev. Lett.*, **76**, 2961–2964.
18. Fauchet, P.M. (2004) Monolithic silicon light sources, in *Si Photonics* (eds L. Pavesiand D.J. Lockwood), Springer, Berlin, pp. 177–199.
19. Gelloz, B. and Koshida, N. (2000) Electroluminescence with high and stable quantum efficiency and low threshold voltage from anodically oxidized thin porous silicon diode. *J. Appl. Phys.*, **88**, 4319–4324.
20. Pavesi, L., Guardini, R., and Mazzolini, C. (1996) Porous silicon resonant cavity light emitting diodes. *Solid State. Commun.*, **97**, 1051–1053.
21. Lenchyshyn, L.C., Thewalt, M.L.W., Houghton, D.C. *et al.* (1993) Photoluminescence mechanism in thin $Si_{1-x}Ge_x$ quantum wells. *Phys. Rev. B*, **47**, 16655–16658.
22. Daldosso, N., Luppi, M., Ossicini, S. *et al.* (2003) Role of interface region on the optoelectronic properties of silicon nanocrystals embedded in SiO_2. *Phys. Rev. B*, **68**, 085327 (1–8).
23. Irrera, A., Pacifici, D., Miritello, M. *et al.* (2003) Electroluminescence properties of light emitting devices based on silicon nanocrystals. *Physica E*, **16**, 395–399.
24. De la Torre, J., Souifi, A., Poncel, A. *et al.* (2003) Optical properties of Si nanocrystal LEDs. *Physica E*, **16**, 326–330.
25. Amans, D., Callard, S., Gagnaire, A. *et al.* (2004) Spectral and spatial narrowing of the emission of silicon nanocrystals. *J. Appl. Phys.*, **95**, 5010–5013.
26. Valenta, J., Lalic, N., and Linros, J. (2004) Electroluminescence of single silicon nanocrystallites. *Appl. Phys. Lett.*, **84**, 1459–1461.
27. Beyer, A., Muller, E., Sigg, H. *et al.* (2000) Size control of carbon-induced Ge quantum dots. *Appl. Phys. Lett.*, **77**, 3218–3220.
28. Lockwood, D.J., Lu, Z.H., and Baribeau, J-M. (1996) Quantum confined luminescence in Si/SiO2 superlattices. *Phys. Rev. Lett.*, **76**, 539–541.
29. Gaburro, Z., Pucker, G., Bellutti, P., and Pavesi, L. (2000) Electroluminescence in MOS structures with Si/SiO_2 nanometric multilayers. *Solid St. Commun.*, **114**, 33–37.
30. Nassiopoulos, A.G., Grigoropoulos, S., and Papadimitrion, D. (1996) Electroluminescence device based on silicon nanopillars. *Appl. Phys. Lett.*, **69**, 2267–2269.
31. Ennen, H., Pomrenke, G., Axmann, A. *et al.* (1985) 1.54 μm electroluminescence of erbium doped silicon grown by molecular beam epitaxy. *Appl. Phys. Lett.*, **46**, 381–383.
32. Canham, L.T., Barraclough, K.G., and Robbins, D.J. (1987) 1.3 μm light emitting diode for silicon electron irradiated at its damage threshold. *Appl. Phys. Lett.*, **51**, 1509–1511.
33. Bradfield, P.L., Brown, T.G., and Hall, D.G. (1989) Electroluminescence from sulfur impurities in a p-n junction formed in epitaxial silicon. *Appl. Phys. Lett.*, **55**, 100–102.
34. Xie, Y.H., Fitzgerald, E.A., and Mii, Y.J. (1991) Evaluation of erbium-doped silicon for optoelectronic applications. *J. Appl. Phys.*, **70**, 3223–3228.
35. Polman, A. (1997) Erbium implanted thin film photonic materials. *J. Appl. Phys.*, **82**, 1–39.
36. Pacifici, D., Irrera, A., Franzo, G. *et al.* (2003) Erbium doped silicon nanocrystals: optical properties and electroluminescent devices. *Physica E*, **16**, 331–340.

37. Castagna, M.E., Coffa, S., Monaco, M. *et al.* (2003) Si based materials and devices for light emission in silicon. *Physica E*, **16**, 547–553.
38. Iacona, F., Irrera, A., Franco, G. *et al.* (2006) Si-based light emitting devices: properties and applications of crystalline, amorphous and Er doped nanoclusters. *IEEE J. Sel. Top. Quantum. Electron.*, **12**, 1596–1606.
39. Makarova, M., Sih, V., Waga, J. *et al.* (2008) Enhanced light emission in the photonic crystal nanocavities with Er doped Si nanocrystals. *Appl. Phys. Lett.*, **92**, 161107 (1–3).
40. Pavesi, L., Dal Negro, L., Mazzoleni, C. *et al.* (2000) Optical gain in silicon nanocrystals. *Nature*, **408**, 440–444.
41. Dal Negro, L., Cazzanelli, M., Daldosso, N. *et al.* (2003) Stimulated emission in plasma enhanced chemical vapour deposited silicon nanocrystals. *Physica E*, **16**, 297–308.
42. Pavesi, L. (2005) Routes toward Si based lasers. *Mater. Today*, **8**, 18–25.
43. Fauchet, P.M. (2005) Light emission from Si quantum dots. *Mater. Today*, **8**, 26–33.
44. Filonov, A.B., Ossicini, S., Bassani, F., and d'Avitaya, F.A. (2002) Effect of oxygen on the optical properties of small silicon pyramidal clusters. *Phys. Rev. B*, **65**, 195317 (1–9).
45. Cazzanelli, M., Navarro-Urios, D., Riboli, F. *et al.* (2004) Optical gain in monodispersed silicon nanocrystals. *J. Appl. Phys.*, **96**, 3164–3171.
46. Green, M.A., Zhao, J., Wang, A. *et al.* (2001) Efficient silicon light-emitting diodes. *Nature*, **412**, 805–808.
47. Trupke, T. and Green, M.A. (2003) Optical gain in materials with indirect transitions. *J. Appl. Phys.*, **93**, 9058–9061 and Very efficient light emission from bulk crystalline silicon. *Appl. Phys. Lett.*, **82**, 2996–2998.
48. Chen, M.J., Yen, J.L., Li, J.Y. *et al.* (2004) Stimulated emission in nanostructured silicon p-n junction diode using current injection. *Appl. Phys. Lett.*, **84**, 2163–2165.
49. Homewood, K.P. and Lourenco, M.A. (2005) Light from Si via dislocation loops. *Mater. Today*, **8**, 34–39.
50. Faist, J., Capasso, F., Sivco, D.L. *et al.* (1994) Quantum cascade lasers. *Science*, **264**, 553–556.
51. Friedman, L., Soref, R.A., and Sun, G. (1998) Silicon based interminiband infrared lasers. *J. Appl. Phys.*, **83**, 3480–3485.
52. Ghosh, S., Mukhopadhyay, B., and Basu, P.K. (2002) Calculated gain and threshold current density for interconduction-subband transition in Si triple quantum well structures. *Microwave Opt. Tech. Lett.*, **35**, 470–475.
53. Driscoll, K. and Paiella, R. (2007) Design of n-type Si based quantum cascade lasers for terahertz light sources. *J. Appl. Phys.*, **102**, 093103 (1–8).
54. Han, G., Yu, J., and Liu, Y. (2008) n-Type Ge–SiGe quantum cascade structure utilizing quantum wells for electrons in the L and Γ valleys. *IEEE Photonic. Tech. L.*, **20**, 419–422.
55. Dehlinger, G., Diehl, L., Gennser, U. *et al.* (2000) Intersubband electroluminescence from silicon-based quantum cascade structures. *Science*, **290**, 2277–2280.
56. Diehl, L., Mentese, S., Müller, E. *et al.* (2002) Electroluminescence from strain-compensated Si Ge/Si quantum cascade structures based on a bound-to-continuum transition. *Appl. Phys. Lett.*, **81**, 4700–4702.
57. Kohler, R., Tredicucci, A., Beltram, F. *et al.* (2002) Terahertz semiconductor-heterostructure laser. *Nature*, **417**, 156–159.
58. Paiella, R. (2006) *Intersubband Transitions in Quantum Structures*, McGraw-Hill Professional, New York.
59. Belkin, M.A., Capasso, F., Xie, F. *et al.* (2008) Room temperature terahertz quantum cascade laser sources based on intracavity difference-frequency generation. *Appl. Phys. Lett.*, **92**, 201101 (1–3).
60. Murzyn, P., Pidgeon, C.R., Wells, J-P.R. *et al.* (2002) Picosecond intersubband dynamics in p-Si/SiGe quantum-well emitter structures. *Appl. Phys. Lett.*, **80**, 1456–1458.

61. Lynch, S.A. *et al.* (2006) Toward Si based lasers for THz sources. *IEEE J. Sel. Top. Quantum. Electron.*, **12**, 1570–1578.
62. Pavlov, S.G., Zhukavin, R.K., Orlova, E.E. *et al.* (2000) Stimulated emission from donor transitions in silicon. *Phys. Rev. Lett.*, **84**, 5220–5223.
63. Pavlov, S.G., Hübers, H-W., Rümmeli, M.H. *et al.* (2002) Far-infrared stimulated emission from optically excited bismuth donors in silicon. *Appl. Phys. Lett.*, **80**, 4717–4719.
64. Pavlov, S.G., Hubers, H.W., Reimann, H. *et al.* (2002) Terahertz optically pumped Si:Sb laser. *J. Appl. Phys.*, **92**, 5632–5634.
65. Hübers, H-W., Pavlov, S.G., Reimaqnn, H. *et al.* (2004) Stimulated terahertz emission from arsenic donors in silicon. *Appl. Phys. Lett.*, **84**, 3600–3602.
66. Lv, P-C., Troeger, R.T., Kim, S., Ray, S.K. *et al.* (2004) Terahertz emission from electrically pumped Ga doped silicon device. *Appl. Phys. Lett.*, **85**, 3660–3662.
67. Hubers, H.W., Pavlov, S.G., and Shastin, V.N. (2005) Terahertz lasers based on germanium and silicon. *Semicond. Sci. Tech.*, **20**, S211–S221.
68. Liu, J., Sun, X., Pan, D. *et al.* (2007) Tensile-strained n-type Ge as a gain medium for monolithic laser integration on Si. *Opt. Express*, **15**, 11272–11277.
69. Chang, S-W. and Chuang, S.L. (2007) Theory of optical gain of Ge–Si_x Ge_y Sn_{1-x-y} quantum-well lasers. *IEEE J. Quantum. Electron.*, **43**, 249–256.
70. Cheng, T-H., Kuo, P-S., Lee, C.T. *et al.* (2007) Electrically pumped Ge laser at room temperature. in Proceedings of the IEEE International Electron Devices Meeting (IEDM '07), Washington, DC, December 2007, pp. 659–662.
71. Soref, R.A. (2008) The advances and challenges of silicon photonics. *Adv. Opt. Technol.*, art. ID 472305.

7

Si Light Modulators

7.1 Introduction

Light sources like LEDs and lasers emit radiation continuously with a constant power in which one needs to control the output power of the sources. The most noteworthy example is the modulation of light in communications systems. The signals to be transmitted from one end of the fiber to the remote end are usually audio and video, which are basically analog signals. In modern telephony, the analog signals are converted into digital signals which are then used to modulate the light signal from the source. In most situations, on-off keying is employed, that is, the light signal is transmitted or blocked according to the bit pattern derived from the original analog signals. The total bit rate of the multiplexed signal is at least 10 Gb/s, whereas 40 Gb/s bit rate systems are not uncommon. In addition to digital modulation, analog modulation of light signals is also used in some niche applications. The prime requirement for the modulators in communication is the high speed of operation, preferably in the GHz range.

Another important area regarding control of light: either on-off switching or continuous change is in optical interconnects (ICs) between chips or within a single chip. The interconnection is the standard wiring using copper as the interconnecting metal. As the device dimension shrinks to produce higher and higher speeds for the ICs, the delay time arising from resistor–capacitator (RC) paths and the noise, both due to electrical interconnects, rise to a high level, limiting the speed. At the same time, the ohmic loss of the interconnects must be reduced to keep the overall power consumption of the chip below a desired value. Both the speed and low power consumption requirements are believed to be solved by designing optical interconnects. Use of photons instead of electrons allows high bandwidth and high interconnection density, while maintaining the low power consumption and high computational capability of Si microelectronic devices.

The LEDs and lasers in use in present-day fiber-optic communication systems can be directly modulated with satisfactory performance for bit rates up to 15–20 Gb/s. For higher

Silicon Photonics: Fundamentals and Devices, First Edition. M. Jamal Deen and P. K. Basu.

speed systems, external modulators are employed. Such modulators may be fabricated by using either $LiNbO_3$ as the substrate or heterojunctions and MQWs grown on semiconducting InP.

One of the major challenges in the area of Si photonics is to develop, in addition to an efficient light source, a good electrically driven modulator working at very high speed and consuming little power. A major reason that makes the task difficult is the absence of a linear EO effect. There have been numerous attempts, over the last two decades or so, to realize a suitable Si-based modulator with all the desired properties. The present chapter will outline the physical effects, the problems related to material silicon, the ingenious ways to overcome the problems, and a description of the work done and results obtained in the past and present.

In the present chapter, the physical effects that lead to modulation of light will be discussed first, and then the importance of these effects in silicon will be examined. The material properties affecting the modulation speed and bandwidth and power consumption will be addressed next. After identifying the preferred method of modulating light using Si-based structures, actual structures will then be introduced and the performance achieved so far will be described. Finally, some very recent results using novel concepts and structures, which seem encouraging, will be described.

7.2 Physical Effects

In an intensity modulator, the light intensity is controlled or modulated in accordance with the change of a controlling or modulating signal, usually in the form of an electrical wave. In optical communication, the intensity of the carrier wave is changed by the modulating signal derived from speech, music, a video signal, or computer data. The phase modulator changes the phase of the carrier wave, in conformity with the amplitude of the modulating signal impressed onto the material. The modulating signal basically changes either the absorption coefficient or the refractive index (RI) of the material. When an electrical signal causes such changes, the change in absorption is termed as *electroabsorption*, while the change in RI is called *electrorefraction*.

Other methods exist for introducing a change in the RI of a semiconductor. An acoustic wave impressed upon certain materials may change their RI through the acousto-optic effect. In order that the effect produces perceptible change in the RI, the acousto-optic coefficient of the material must be very high. Another viable method of producing a RI change, particularly suitable for Si and its alloys, is to heat the material. The RI change occurs due to a thermo-optic effect. An optical field of high intensity may induce a change in RI of a material by nonlinear effects; sometimes the effect is called an *opto-optic effect*.

In the present section, we shall discuss the methods of changing the absorption and RI in semiconductors in general, mentioning the physical phenomena, and then present the relationship between the change and the impressed signal. The present section applies to semiconductors in general. We shall discuss mainly the electric field-induced effects. Carrier-induced changes in absorption and refraction are also included in the discussion, as in most situations the carriers are injected or depleted by an external electric field. The thermo-optic effect will be described very briefly here, relegating the detailed discussion to Section 7.3 dealing exclusively with silicon. The acousto-optic effect is not important in silicon and therefore will not be discussed at all. The nonlinear effects will also not be considered.

7.2.1 Electroabsorption and Electrorefraction

It was mentioned in Section 4.2 that the complex RI of a material can be written as $n_r + jn_i$, where n_r is the real part and n_i is the extinction coefficient related to the absorption coefficient α by the relation $\alpha = 2n_i\omega/c$. (see Eq. (4.18)). It is known that n_r and n_i are interrelated by Kramers–Kronig relations, and the same relation holds good between changes Δn_r and Δn_i.

Eq (4.26) expresses the relationship between the real and imaginary parts of the dielectric constant or relative permittivity. A similar relationship between Δn and $\Delta\alpha$ has been specified in textbooks [1–4] as follows:

$$\Delta n(\omega) = \frac{c}{\pi} P \int \frac{\Delta\alpha(\omega')d\omega'}{\omega'^2 - \omega^2}, \tag{7.1}$$

where P denotes the principal value of the integral. The change in absorption may be accomplished in a variety of ways, which will be discussed in this section. A convenient way is to apply an electric field F. The change in absorption (the electroabsorption $\Delta\alpha$) and the change in RI (the electrorefraction Δn) are then defined as follows:

$$\Delta\alpha(\omega, F) = \alpha(\omega, F) - \alpha(\omega, 0), \tag{7.2a}$$

$$\Delta n(\omega, F) = n(\omega, F) - n(\omega, 0). \tag{7.2b}$$

7.2.2 Electro-Optic Effect

Application of an electric field, direct current (DC) or radiofrequency (RF), to a material induces a change in the RI of the material. The effect is known as the electro-optic (EO) effect. The presence of such a field distorts the electron motion in the atoms or molecules of a substance, or distorts the crystal structure, resulting in a change of the optical properties. Application of an electric field may convert an isotropic crystal like GaAs to a birefringent material, or introduce a rotation of the optic axes in a naturally birefringent crystal like a potassium dihydrogen phosphate (KDP) crystal. The EO effects are classified as first- and second-order effects.

The RI in the presence of an electric field may be expressed as

$$n'(F) = n_0 + a_1 F + a_2 F^2 + \ldots, \tag{7.3}$$

where n_0 is the RI of the material without the electric field and the coefficients a_1 and a_2 are called the linear EO effect and second-order EO effect coefficients, respectively. The effects of higher order coefficients are very small and are not considered in Eq. (7.3). The change in n proportional to the electric field F is called the *Pockels effect*, while the change proportional to F^2 is called the *Kerr effect*. The two effects are described by the following two equations:

$$\Delta n = a_1 F : \text{ Pockels effect} \tag{7.4}$$

$$\Delta n = a_2 F^2 = (\lambda K) F^2 : \text{ Kerr effect.} \tag{7.5}$$

The symbol K is called the *Kerr coefficient*. All materials exhibit the Kerr effect. On the other hand, the Pockels effect is exhibited by only certain crystalline solids. Eq. (7.4) indicates that if the direction of field **F** is reversed, then the change in RI should be negative. The material therefore responds differently to **F** and −**F**. Some kind of asymmetry must therefore be present in the crystal to distinguish between +**F** and −**F**. If the crystal has a center of symmetry, then reversing the direction of the field has an identical effect and the coefficient a_1 must be zero. Silicon belongs to this type of crystal. On the other hand, noncentrosymmetric crystals exhibit the Pockels effect. The GaAs crystals belong to this class and are therefore characterized by a finite linear EO (Pockels) coefficient.

In general, the linear EO coefficient is not a scalar quantity, but is a tensor. While calculating the change of RI, due consideration is to be given for the particular crystal direction along which the field is applied as well as the propagation direction and polarization of light.

The Pockels effect is utilized to construct phase modulators in crystals exhibiting a large EO coefficient. Well-known examples are $LiNbO_3$ and GaAs. The principle of operation of an EO phase and intensity modulator using bulk crystals is explained in many textbooks of optoelectronics [1–4].

Consider a KDP crystal in which an EM wave propagates along its optic axis, the z-axis. When an electric field is applied along the principal axis, say the y-axis, the principal axes are rotated by 45°, so that the new axes are x' and y'. The new RIs are given by

$$n_{x'} = n_0 + \frac{1}{2} n_0^3 r_{22} F \quad \text{and} \quad n_{y'} = n_0 - \frac{1}{2} n_0^3 r_{22} F,$$

where r_{22} is the appropriate EO coefficient.

A light ray polarized along y may be decomposed into two components along the two new optic axes. As these two components travel with different velocities along the crystal, the phase difference between the two components makes the lightwave elliptically polarized as it emerges out of the crystal. The axes of the ellipse depend on the phase difference and hence on the magnitude of the electric field. An analyzer with its axis along the horizontal direction will allow different amounts of elliptically polarized light to pass through it, depending on the orientation of the ellipse with respect to its own axis. This is the principle of intensity modulation described in detail in optoelectronics textbooks [1–4].

When a strong electric field is applied to an otherwise optically isotropic material, the change in RI will be due to the Kerr effect, a second-order effect. Let the electric field be applied along the z-direction. The applied field distorts the electron motion in the constituent atoms and molecules, and the motion of valence electrons in covalent bonds, in such a way that it becomes more difficult for the electric field in the lightwave to displace electrons along the direction of the applied field. Thus a lightwave polarized along the z-direction will see a smaller RI $n_e < n_0$, but a wave-polarized orthogonal to the z-axis will experience the original RI n_0. The applied field therefore induces birefringence with an optic axis parallel to the z-axis. The principle of operation of a Kerr modulator is the same as the Pockels effect modulator.

As mentioned, the Kerr effect is exhibited by all materials; however, the magnitude of the Kerr coefficient is in general small compared to that of the linear EO coefficient.

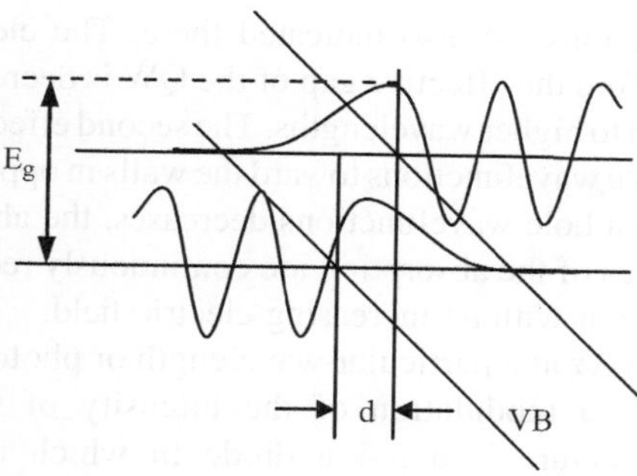

Figure 7.1 *Tilting of the band edges due to applied electric field in the depletion layer. The overlap between electrons and holes increases (photon-induced tunneling), enhancing absorption at lower photon energy.*

7.2.3 Franz–Keldysh Effect

In the presence of a strong electric field, the absorption edge of a semiconductor shifts to longer wavelengths for a direct-gap semiconductor like GaAs. The electric field is created in the depletion layer of a p–n junction by applying a strong reverse bias.

The change in absorption edge due to increasing field may be explained by considering the band diagram shown in Figure 7.1. The reverse bias in the depletion layer tilts the bands. The wavefunctions of electrons and holes penetrate into the gap as indicated in Figure 7.1, and there is increased overlap between the wavefunctions at a lower energy, $h\nu < E_g$. The overlap increases with higher field which reduces the spatial separation of electrons and holes (decreasing d).

The value of the absorption coefficient can be increased from a few cm^{-1} to more than $1000\ \mathrm{cm}^{-1}$ for a wavelength within a few nm from the absorption edge.

Keeping the carrier wavelength fixed and near the absorption edge, changes in the reverse bias introduce a change in absorption and hence the transmitted intensity of the light beam.

The light beam may thus be intensity modulated due to electroabsorption. There is an associated change in the RI also; however, usually the electroabsorption mechanism dominates.

The Franz–Keldysh effect has also been found in indirect-gap semiconductors like Si and Ge. However due to the indirect nature of the band gap, the photon-assisted tunneling involves phonons also [5, 6].

7.2.4 Quantum-Confined Stark Effect

Although the Franz–Keldysh effect may be present in QWs, the dominant mechanism of electroabsorption, as well as electrorefraction, in a MQW structure is known as the quantum-confined Stark effect (QCSE) [7, 8]. As discussed in Chapter 5, the optical absorption in QWs is due to an enhanced excitonic effect.

When an electric field is applied along the direction of growth of the QW, the band edges are tilted as shown in the right-hand side of Figure 5.10. The position of the subbands and the nature of the envelope functions are also indicated there. The electric field lowers the subband energies, and therefore the effective gap of the QW is decreased. This means that the absorption edge is shifted to higher wavelengths. The second effect produced by the field is to push the electron and hole wavefunctions toward the walls in opposite directions. As the overlap between electron and hole wavefunctions decreases, the absorption coefficient is also reduced. The peak values of the absorption are continuously reduced, and the exciton peaks increasingly broaden out with an increasing electric field.

The change in absorption $\Delta\alpha$ at a particular wavelength or photon energy with change in electric field introduces a modulation of the intensity of the transmitted light. Usually, the modulator structure is a p–i–n diode in which the intrinsic (i) layer embodies the MQWs. The applied reverse bias to the pin diode alters the electric field in the MQW region and also introduces electroabsorption by QCSE. The modulators are known as electroabsorption (EA) modulators, and the devices grown on InP are increasingly being used as external modulators at 1.55 μm in high-bit-rate optical communications systems [1].

7.2.5 Carrier-Induced Effects

Introduction and extraction of free carriers change the value of the absorption coefficient of a semiconductor. The free electrons or holes absorb a photon, and then absorb or emit a phonon to make a transition to a higher lying state in the same band. The free-carrier absorption process has been discussed in Section 4.9 and the classical model for the calculation of an absorption coefficient and RI change has been introduced.

The injection and removal of free carriers from a region of semiconductor structure, in most cases the depletion region of a p–n junction, are accomplished by applying an electric field. The carrier induced effects will be discussed in detail in Section 7.3, which is devoted specifically to silicon.

7.2.6 Thermo-Optic Effect

A change in temperature affects the optical properties of all materials. An increase in temperature changes the lattice constant as well as the band gap, and consequently the absorption and refraction are changed. The change of RI with temperature is of prime interest in the present context. The variation of RI with temperature at a constant pressure is called the thermo-optic (TO) coefficient, dn/dT, where n and T are, respectively, the RI and temperature. Normally the value of TO coefficient is very small, of the order of $10^{-3}\ K^{-1}$ to $10^{-6}\ K^{-1}$. In spite of its small value, it affects the tunability of devices like distributed feedback lasers. It can also be exploited for the design of thermally activated devices, like switches, modulators, tunable lasers, demultiplexers, filters, and so on.

All materials, including silica and silicon, exhibit the thermo-optic effect. The change in RI due to heating has been utilized in several passive devices like arrayed waveguide gratings (AWGs). A RI change by TO effect is also utilized in fabricating modulators. The effect will be discussed in more detail in relation to silicon, and the values of TO coefficients in silicon and other materials will be presented in Section 7.3.

7.3 Electrorefraction in Silicon

Having discussed in general terms some of the optical processes in semiconductors and how they affect the absorption and refraction in these materials, we now focus our attention to material silicon. We consider first the effects produced by an electric field (discussed in Section 7.2) and examine the importance of each of them in changing the RI of silicon. In this connection, change in absorption, or electroabsorption, in silicon will also be examined. The thermo-optic effect will be discussed in Section 7.4.

The EO effects in silicon were first examined critically by Soref and Bennett [9], and the following discussion is based mostly on their findings.

7.3.1 Electro-Optic Effects

Since silicon is a centro-symmetric crystal, the linear EO coefficient of it is zero. Its optical absorption spectrum is modified by the Franz–Keldysh (FK) effect. The electroabsorption spectra at the indirect edge as measured by different authors are utilized by Soref and Bennett, who performed a numerical Kramers–Kronig analysis of the spectra to obtain the values of electrorefraction at different wavelengths.

Soref and Bennett gave a plot of the RI change Δn as a function of wavelength ranging from 1.00 to 1.60 μm. Starting at 1.3 μm, as the wavelength decreases, the value of Δn rises and attains a maximum at 1.07 μm, a wavelength slightly below the gap wavelength. It then decreases, crosses zero at 1.05 μm, and becomes negative. The value of $\Delta n = 1.3 \times 10^{-5}$ is obtained at 1.07 μm for an external field of 100 kV/cm.

Although the Pockels effect is absent in silicon, the contribution of the quadratic effect, the Kerr effect, in silicon has been estimated by using an anharmonic oscillator model and the following expression:

$$\Delta n = -3e^2(n^2-1)\frac{F^2}{2nM^2\omega_0^4 x^2}, \tag{7.6}$$

where e is the electronic charge, n is the unperturbed RI, M is the effective mass, ω_0 is the resonant frequency of the oscillator, and x denotes the average oscillator displacement. The values of Δn obtained from this model are given in Figure 7.2. As may be noticed, the value of Δn is 10^{-4} at a field of 10^6 V/cm. The change is quite small in comparison to carrier-induced change, which is discussed in Section 7.3.2.

7.3.2 Carrier Effect

The optical properties of silicon are strongly influenced by the injection of charge carriers into an undoped sample or by the removal of free carriers from a doped sample. The change in RI is more or less unaffected whether the carriers come from impurities or from injection.

The carrier effects are of three categories. First is the traditional free-carrier absorption. The second is the Burstein–Moss effect which occurs due to band filling at high doping levels, due to which the absorption edge shifts to higher photon energies. The third is the Coulomb interaction of carriers with impurities, due to which the spectrum is red-shifted. The experimental results indicate a red shift. This indicates a stronger Coulomb effect, overriding the smaller blue shift due to band filling.

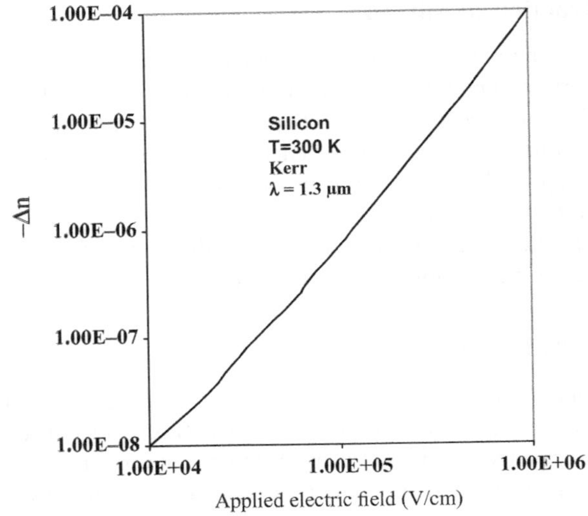

Figure 7.2 *The Kerr effect in crystalline silicon as evaluated by the anharmonic oscillator model. Reproduced with permission from [9]. Copyright (1987) IEEE.*

In the work of Soref and Bennett, the experimental absorption data obtained in the doping range $1 \times 10^{18} - 1 \times 10^{20}\,\mathrm{cm}^{-3}$ and the photon energy range 0.6–1.5 eV, as well as the data obtained from 0.025 eV to 1.1 eV, were used. A composite curve is obtained by using data presented by various authors for undoped and doped samples over different ranges of photon energies, and a suitable extrapolation of the experimental data is made to obtain a smooth variation.

The variation of the absorption coefficient and the RI with carrier density can be described by a Drude model developed in Section 4.9. We reproduce here the empirical expressions given by Soref and Bennet for the changes in absorption and RI, $\Delta\alpha$ and Δn, respectively, for $\lambda = 1.55$ and $1.3\,\mu\mathrm{m}$.

At $\lambda = 1.55\,\mu\mathrm{m}$:

$$\Delta n = \Delta n_e + \Delta n_h = -[8.8 \times 10^{-22}\Delta N_e + 8.5 \times 10^{-18}(\Delta N_h)^{0.8}], \tag{7.7}$$

$$\Delta\alpha = \Delta\alpha_e + \Delta\alpha_h = 8.5 \times 10^{-18}\Delta N_e + 6.0 \times 10^{-18}\Delta N_h. \tag{7.8}$$

Similarly, the expressions for $\lambda = 1.3\,\mu\mathrm{m}$ are

$$\Delta n = \Delta n_e + \Delta n_h = -[6.2 \times 10^{-22}\Delta N_e + 8.5 \times 10^{-18}(\Delta N_h)^{0.8}], \tag{7.9}$$

$$\Delta\alpha = \Delta\alpha_e + \Delta\alpha_h = 6.0 \times 10^{-18}\Delta N_e + 4.0 \times 10^{-18}\Delta N_h. \tag{7.10}$$

These expressions are now almost universally used by different workers. Example 4.8 gives a value of nearly 10^{-3} for the RI change. The maximum change in RI due to the Kerr effect shown in Figure 7.2 is 10^{-4}, whereas the change due to the FK effect is $\sim 10^{-5}$.

7.3.3 Quantum Confined Stark Effect

Electroabsorption modulators based on QCSE have been fabricated using III–V compounds and their alloys. These modulators show highly successful operation.

QCSE in SiGe and Si heterojunctions and QWs has been discussed in Section 5.5. It has been mentioned there that the effect is quite small in type I band alignment. Type II QWs give large exciton-binding energy, and although the absorption is small, the relative change in absorption is large. The situation has drastically changed with the discovery by Kuo *et al.* [10] of strong QCSE in strain symmetric Ge and SiGe MQWs. The excitonic enhancement of absorption is due to a transition to the direct conduction band (Γ point) in Ge, which is slightly above the indirect conduction band valleys at the L point. The absorption is comparable to the values observed in MQWs formed by III–V compounds and alloys.

7.4 Thermo-Optic Effects in Si

For proper design of the devices, it is necessary to have knowledge of the thermal behavior of silicon, especially the effect of temperature on absorption and refraction near the fundamental absorption edge.

Cocorullo *et al.* [11] undertook a detailed measurement of the TO coefficient of a number of silicon samples with different crystallographic orientations and thicknesses at 1.55 μm. The values of *dn*/*dT* have been obtained from room temperature to 550 K.

The authors also provided the following polynomial interpolation of the experimental data:

$$\frac{dn}{dT} = 9.48 \times 10^{-5} + 3.47 \times 10^{-7} \times T - 1.49 \times 10^{-10} \times T^2 (\mathrm{K}^{-1}). \tag{7.11}$$

Example 7.1

At 300 K, the value of *dn*/*dT* is $1.86 \times 10^{-4}\ \mathrm{K}^{-1}$ from Eq. (7.11).

A theoretical paper on physical modeling of the TO effect has been published [12]. The essential points in the theory are now described.

The optical properties of matter are usually determined by considering coupling between various types of oscillators and the electromagnetic radiation field. The amplitude of these oscillations depends on (1) the frequency of the incident electromagnetic field, (2) the oscillator eigenfrequencies, (3) the different coupling strengths *f* between the electromagnetic field and the oscillators, and (4) their damping functions. In a simplified model to analyze the RI of a material in its transparency region, only the real part of the dielectric function may be considered, obtaining

$$\varepsilon = n^2 = 1 + E_p^2 \sum_{\mathbf{k}} \frac{f_{cv}(\mathbf{k})}{E_{cv}^2(\mathbf{k}) - E^2}, \tag{7.12}$$

where $E_p = \sqrt{4\pi N \hbar^2 e^2 / m^*}$ is the electronic plasma energy; *N* is the number of oscillators per unit volume; *E* is the photon energy; and E_{cv} (**k**) and f_{cv} (**k**) are, respectively, the transition energy and the interband oscillator strength between the valence and conduction band. The summation is restricted to the allowed **k**'s per unit volume, with **k** being the reciprocal lattice vector. In a single oscillator model,

$$n^2 = 1 + \frac{E_p^2}{(E_g^2 - E^2)}, \tag{7.13}$$

Considering that the plasma energy is inversely proportional to the volume, one obtains

$$\frac{dE_p^2}{dT} = \frac{4\pi\hbar^2 e^2}{m^*}\frac{dN}{dT} = \frac{4\pi\hbar^2 e^2}{m^*}(-3k_{ex}N) = -3k_{ex}E_p^2. \tag{7.14}$$

where k_{ex} is the thermal expansion coefficient of the material. The TO coefficient may now be written as

$$\frac{dn}{dT} = \frac{n^2-1}{2n}\left[-3k_{ex} - \frac{2}{E_g}\frac{dE_g}{dT}\frac{1}{1-(E/E_g)^2}\right], \tag{7.15}$$

where both n and k_{ex} depend on temperature. Using appropriate fitting parameters the single-oscillator model, as well as the double-oscillator model introduced in the same work, was found to explain the variation of TO coefficients obtained experimentally.

7.5 Modulators: Some Useful Characteristics

Before discussing the construction and performance of the actual Si-based modulators, it may be useful to have some introductory discussion on the basic figures of merit of a modulator or switch [13].

In many cases the same device can function as either a modulator or a switch, depending on the strength of interaction between the optical wave and the controlling electrical signal. The device is a modulator if the controlling signal changes one of the properties of the medium to effect a change in the light signal emerging out of the device. A switch, on the other hand, changes the spatial position of the light, or else turns it off and on. Both the devices possess the same figures of merit, some of which are discussed throughout this section.

7.5.1 Modulation Depth

The presence of a controlling signal $s(t)$ causes the optical output of the device, I_0, to vary between maximum and minimum values. The modulation depth is defined as

$$\eta_m = \frac{I_{0,\max} - I_{0,\min}}{I_{0,\max}}. \tag{7.16}$$

For a switch, this modulation depth should approach unity. For analog modulation, a modulation depth much less than unity may sometimes prove satisfactory.

7.5.2 Modulation Bandwidth

For analog modulation, the bandwidth, that is, the range of frequencies over which the device can be operated, is an important parameter. By convention, the bandwidth of a modulator is the difference between upper- and lower-half power frequencies, at which the modulation depth falls to 50% of the maximum value. In case of a switch, the frequency response is replaced by the bit rate.

An alternate way to describe the speed of a switch is the switching time. If a step function control signal is applied to the device, and the amplitude of the control signal is sufficient to change the transmittance of the light from its minimum to maximum value, then the switching time is the time required for the transmittance to respond to the control signal step.

Minimizing switching time is an important requirement for large-scale arrays of switches to route optical waves over desired paths. Similarly, a large modulation bandwidth is a desirable feature when many information channels are to be multiplexed onto the same optical beam.

7.5.3 Insertion Loss

When the modulating signal decreases the intensity, the insertion loss is defined as

$$L_i = 10 \log (I_t/I_0), \tag{7.17}$$

where I_t is the optical intensity that could be transmitted by the waveguide if the modulator was absent, and I_0 is the intensity when the modulator is inserted but no signal is applied. On the other hand, when the applied signal increases the transmitted intensity, the insertion loss is given by

$$L_i = 10 \log (I_t/I_m), \tag{7.18}$$

where I_m is the transmitted intensity when the maximum signal is applied. The loss is actually a loss in optical power.

7.5.4 Power Consumption

Electrical power is needed to drive a modulator or a switch. In the case of modulators, the required power increases with modulation frequency. Hence a useful figure of merit is the drive power per unit bandwidth, usually expressed in milliwatts per megahertz.

The power requirements of optical switches operating at high bit rates for time division multiplexed (TDM) systems are evaluated in the same manner as discussed. For lower bit rates, however, a more important quantity is the power needed to hold the switch at a given state. An ideal switch should possess no holding power; power is consumed only when there is a change in state.

7.5.5 Isolation

The degree of isolation between input and output of a switch is an important design consideration. The isolation between two ports is expressed as

$$\text{Isolation in dB} = 10 \log (I_2/I_1), \tag{7.19}$$

where I_1 is the optical intensity at the driving port and I_2 is the intensity at the driven port when the switch is in the off state with respect to ports 1 and 2. Thus, a switch with signal leakage or cross-talk of 1% with respect to two ports would have −20 dB isolation.

7.6 Modulation Bandwidth under Injection

Since the linear and quadratic EO effects in silicon are not significant, in most of the studies of modulation the injection of carriers is accepted as the most effective means of introducing RI change in the material. The structures actually used will be discussed in detail in Section 7.7. For the present context, it is sufficient to note that injection or carrier removal from a region of silicon material is accomplished by using a biased p–n or, rather, a p–i–n diode. Application of a forward bias injects electrons and holes into the depletion layer in the p–n diode or into the intrinsic (*i*) layer in the p–i–n diode. The modulation speed of the device depends on the recombination lifetime of the electron–hole pairs as may be understood by the following analysis [1].

The carrier lifetime τ is defined such that it represents the total recombination time of charged carriers in the absence of stimulated recombination. It is defined by the relation

$$R_r + R_{nr} = n/\tau, \tag{7.20}$$

where n is the carrier density, and R_r and R_{nr} denote, respectively, the radiative and nonradiative recombination rates.

To find out the modulation response, one needs to set up a rate equation for carriers. Since electrons and holes are injected in pairs and also recombine in pairs, it is enough to consider one type of carrier. The rate equation should include all the mechanisms by which carriers enter into and disappear from the active region. Since stimulated emission is not present, the rate equation takes the following simple form:

$$\frac{dn}{dt} = \frac{I}{eV} - \frac{n}{\tau}, \tag{7.21}$$

where I is the current and V is the volume of the active region. Now consider that the current is composed of a DC bias current I_0 and a sinusoidally modulating component of amplitude I_m having angular frequency ω_m, so that the total current is

$$I(t) = I_0 + I_m \exp(j\omega_m t). \tag{7.22}$$

Since Eq. (7.21) is linear, its general solution may be written as

$$n(t) = n_0 + n_m \exp(j\omega_m t), \tag{7.23}$$

where $n_0 = \tau I_0/eV$, as may easily be obtained from Eq. (7.21) putting $dn/dt = 0$ under steady state. Using Eq. (7.23) in Eq. (7.21), differentiating and equating the time-varying parts on both sides give the following:

$$n_m(\omega_m) = \frac{\tau I_m}{eV(1 + j\omega_m \tau)}. \tag{7.24}$$

The modulated power P_m is linearly related to the carrier density n_m. A transfer function $H(\omega)$ is defined as

$$H(\omega_m) = \frac{n_m(\omega_m)}{n_m(0)} = \frac{1}{1 + j\omega_m \tau}. \tag{7.25}$$

The *modulation bandwidth* is defined as the modulation frequency at which $|H(\omega_m)|$ is reduced by a factor of 2 or the power is reduced by 3 dB. The bandwidth is then expressed as

$$f_{3-dB} = (2\pi\tau)^{-1}. \tag{7.26}$$

The modulation bandwidth is therefore inversely proportional to the total recombination lifetime of the injected carriers. The smaller its value is, the higher is the bandwidth.

Example 7.2

Consider that the recombination lifetime of electron–hole pairs in silicon is 1 μsec. Then the modulation bandwidth is 160 kHz. A value of $\tau = 1$ ns makes $f_{3-dB} \sim 0.16$ GHz.

This example suggests that the inefficient recombination process in an indirect gap limits the speed of Si-based modulators. The speed discussed here is extremely poor in comparison to the modulation bandwidth of 40 GHz or more in III–V compound–based devices. Attempts to improve the bandwidth will now be described.

7.7 Optical Structures

Si-based modulators can be classified in different ways. The optical structures may differ from device to device. In fact, starting from 1987, many different optical structures have been employed to date. The electrical structures have also evolved since the first practical realization of the modulator. The classification may be made in terms of the method of introducing the RI variation. Mainly two different methods are employed: EO and TO methods. In the EO method also, different ways exist. Finally, the modulators may differ in terms of materials. As noted already, silicon-based devices used silica on Si or Si–SiGe heterostructures initially; in recent years, the emphasis is on using silicon-on-insulator (SOI) structures. The electrical structures will be considered in Section 7.8. In this section, we shall consider the basic optical structures used to construct the modulators.

Broadly there are two categories. First is the Mach–Zehnder interferometer (MZI) configuration. Second is the resonant structure which may be either a Fabry–Perot or a ring resonator. Again the mirrors in a Fabry–Perot structure may be formed in different ways; presently the mirrors are Bragg reflectors. In all the configurations the basic idea is to introduce a change in phase of the lightwave by changing the RI. Either the EO effect or TO effect is introduced to bring about the RI changes

7.7.1 MZI

The basic operating principle of the Mach–Zehnder interferometer (MZI) will be described in Chapter 11. Here we introduce the essential points. As shown in Figure 7.3, the input lightwave coupled into MZI is split into two equal parts which propagate along the two arms. Both of the arms, or one of the two arms, contain an active element, the RI of which is modulated by the modulating signal employing either the EO or TO effect. The two parts of the input light travel with different propagation velocities, and therefore a phase difference is introduced between them. When the two components combine at the output terminal, the relative phase difference in the two arms causes an intensity modulation of the output beam. If E_1 and E_2 are the electric field in the two arms and if ϕ is the phase difference between

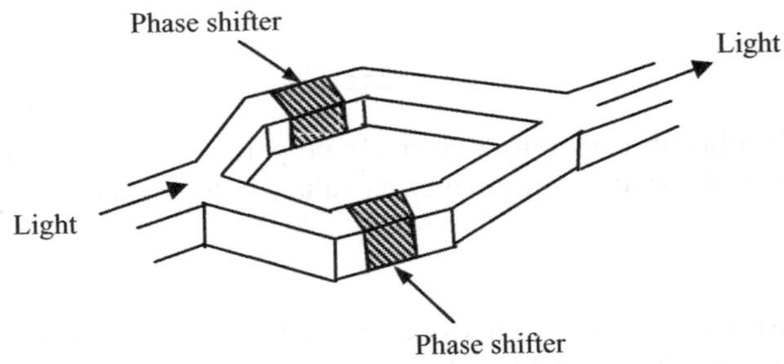

Figure 7.3 The schematic structure of a Mach–Zehnder interferometer used for intensity modulation of light.

them, then the output light intensity will be proportional to the square of the resultant electric field E, expressed as

$$I_0 = |E|^2 = |E_1|^2 + |E_2|^2 + 2E_1E_2\cos^2\phi. \quad (7.27)$$

The intensity varies between $(E_1 + E_2)^2$ and $(E_1 - E_2)^2$ when phase changes vary from 0 to π. When $E_1 = E_2$, the intensity changes from maximum to zero and the modulation depth is 100%. If the modulating voltage is large, then the output intensity may be switched from the maximum to zero and the device then acts as a perfect switch, with an infinite on-off ratio.

As mentioned, the phase change is accomplished by the EO or TO effect. In the former situation, a p–i–n diode in which carriers are injected into the intrinsic (i) region forms one of the arms. Alternately, a metal oxide semiconductor (MOS) capacitor or a modified version of a field effect transistor is inserted in one of the arms.

7.7.2 Fabry–Perot Resonator

A Fabry–Perot resonator is an essential component of all lasers in which the two plane parallel mirrors forming the resonator enclose a medium in which population inversion occurs. The structure also acts as a filter for a multiwavelength input and is employed for demultiplexing wavelength-division multiplexing (WDM) signals. Its tuning characteristics will be discussed in Chapter 11.

The transmission of a resonator is given approximately as

$$T = \frac{T_{\max}}{1 + \left(\frac{4nL}{\pi\Delta\lambda}\right)^2 \sin^2\left(\frac{2\pi nL}{\lambda}\right)} \quad (7.28)$$

where λ is the wavelength of light, and $\Delta\lambda$ is the full width at half maximum (FWHM) of the transmission at a resonance determined by the quality of the resonator and its geometrical configuration.

Because of its sharp tuning characteristics, FP resonators can also be used as intensity modulators of light. As noted earlier, the tuned frequency or wavelength depends on the separation between the mirrors and the RI of the medium inside the FP cavity. If the RI is given slight alteration, then the tuning frequency will be shifted. This is illustrated in

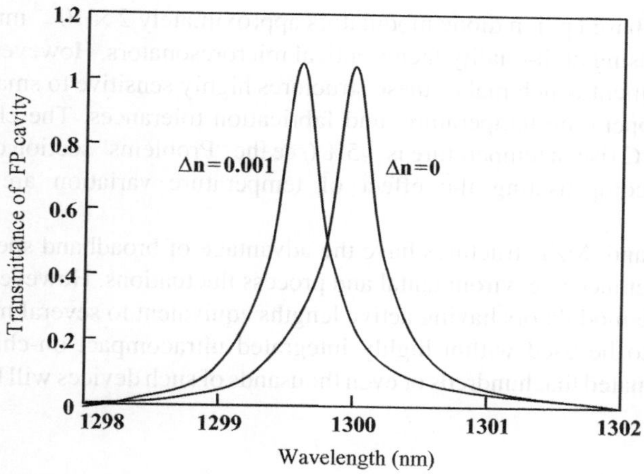

Figure 7.4 *Detuning of the resonance of a FP resonator by a small RI change. The transmittance at 1300 nm is reduced to a mere 20% of the maximum.*

Figure 7.4, in which the transmittance of the cavity for two different RIs is shown. The curve for $\Delta n = 0$ corresponds to the transmittance under normal conditions, and the tuned frequency occurs at 1300 nm, when the transmittance is maximum. If a slight change Δn of the RI of the medium is produced, the tuned wavelength shifts to a lower value causing a drastic reduction to about 20% of the transmittance at 1300 nm. The transmitted intensity is thereby modulated.

The resonator structure is realized in several ways. The traditional method is to employ a pair of plane parallel mirrors. For reasons to be discussed presently, the plane parallel mirrors are replaced by Bragg gratings or ring resonators or photonic band gap mirrors.

In some of the reported studies, neither of the above structures is employed. The input light is allowed to propagate along the length of the guide in which a change in RI is produced by the injection of carriers. The presence of free carriers alters the absorption, and the structures are simply referred to as the free-carrier absorption modulators (FCAMs).

7.7.3 MZI versus a Resonator

A special feature of the resonator is that the light circulates within the cavity to satisfy the resonance condition. Thus the optical path length increases without an increase in the physical length of the device. In other words, the physical length of the MZI is much larger than that of a resonator. The lightwave circulates in the resonator over the photon lifetime given by $\tau_{ph} = \lambda L/\Delta\lambda c$ to satisfy the resonance condition and then exits with the maximum transmittivity given by Eq. (7.28).

Example 7.3

Let $L = 1\,\mu m$, $\lambda/\Delta\lambda = 10^2$. Then the photon lifetime is 0.33 ps. This ensures fast operation of the modulator.

Charge injection–based p–i–n diode modulators approximately 2×10^{-4} mm^2 in size have been realized by using high-quality factor optical microresonators. However, it is the same resonant enhancement which makes these structures highly sensitive to small variations in bias conditions, operating temperature, and fabrication tolerances. The change in transmittivity with a 1°C rise in temperature is ~5% (see the "Problems" section of this chapter).

Methods for compensating the effect of temperature variation are discussed by Lipson [14].

On the other hand, MZI structures have the advantage of broadband spectral operation and improved tolerance to environmental and process fluctuations. However, the large size occupied by these modulators having active lengths equivalent to several millimeters may not allow them to be used within highly integrated ultracompact on-chip interconnect systems. It is estimated that hundreds or even thousands of such devices will be required on a single die.

7.8 Electrical Structures

The most widely used structure is a p–i–n diode, which is basically a two-terminal device. The structure gave way to three terminal structures which are either p–i–n devices or a bipolar mode field effect transistors (BMFETs). The most recent configuration giving encouraging results is a metal–oxide–silicon (MOS) capacitor. In the following, we discuss the features of the structures giving the schematic diagrams. The discussion is mostly based on the review by Reed and Png [15].

7.8.1 p–i–n Structures

The waveguide intensity modulator using free-carrier absorption was first fabricated by Treyz *et al.* [16] to work in the 1.3–1.55 μm range. The schematic diagram of the structure used is shown in Figure 7.5.

The intrinsic region is a p^- region having a doping concentration $< 5 \times 10^{15}\,\mathrm{cm}^{-3}$, while the p^+ region has a doping concentration $> 5 \times 10^{19}\,\mathrm{cm}^{-3}$. By applying a forward bias to the device, free carriers are injected into the p-type Si-guiding layer. The characteristic of the modulator is given in Table 7.1. The switching time corresponds to change from 90% to 10%.

The waveguide possesses a high loss due to relatively weak confinement in the vertical direction and hence absorption of the guided mode in the doped silicon substrate. The

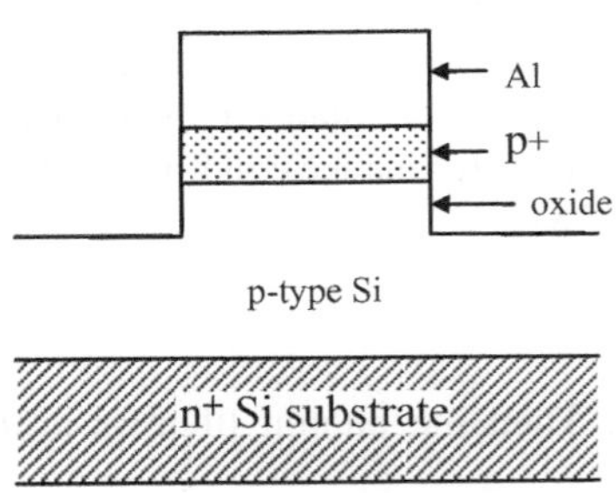

Figure 7.5 *Schematic diagram of the p–i–n structure used by Treyz et al.* (EDL (1991) [16]).

Table 7.1 *All-silicon electro-optic modulators reported in the literature*

Year	Author	Electrical	Optical	M (%)	J (kA/cm^2)	Power (mW)	t_s (ns)	Length (μm)	D/P
1991	Treyz *et al.*	p$^+$–p(i)–n	FCAM	75	3.0	—	50	500	D
1991	Treyz *et al.*	p–i–n	MZ	65	1.6	—	<50	500	D
1991	Xiao *et al.*	p–i–n	FP	10	6.0	—	25	v.d.	D
1994	Liu *et al.*	p–i–n	*y*-switch	>90	9.0	—	200	800	D
1994	Liu *et al.*	p–i–n	TIR switch	>90	12.5	—	100	200	D
1995	Tang *et al.*	p–i–n	FCAM		0.175		20 MHz		D
1995	Zhao *et al.*	p–i–n	MZ	98	—	—	200	816	D
1995	Liu *et al.*	p–i–n	FP	80	—	—	—	20.9	P
1996	Zhao *et al.*	p–i–n	ZGDC	97.2	1.027	123.7	210	1103	D
1997	Cutolo *et al.*	p–i–n	Bragg reflector	50	—	4	24.7	3200	P
1997	Cutolo	BMFET	FCAM	20	2.3	126	6	1000	P
2000	Irace *et al.*	BMFET	*y*-junction	92	—	~350	16	5000	D
2003	Sciuto *et al.*	BMFET	FCAM	75	—	160	—	400	D
2003	Barrios *et al.*	p–i–n	FP	80	0.116	0.025	21	20	P
2003	Barrios *et al.*	p–i–n	FP	80	0.61	0.014	1.3	10	P
2003	Irace *et al.*	p–i–n	BG				1.4 GHz	3000	D
2004	Barrios *et al.*	p–i–n	FP/BR	53		20 mW	—	20	D
2004	Liu *et al.*, *Nature* 427,615,04	MOS	MZ				1.4 GHz	2.500	D
2005	Liao *et al.*, *Opt Exp.* 13,3129 2005	MOS	MZ				10 GHz	3450	D
	Liao *et al.*	MOS	MZ				40		D
2006	Xu *et al.* *Opt Exp.* 2007	Pin	Ring	100%			12.5 gGb/s	~ few μm ϕ	D
2007	Liu	MOS	MZ			613			D
2007	Green *et al.*	p–i–n	MZ			287 μW	10 Gb/s	200 μm	D
2007	Green *et al.*		Ring						D
2008	Liu *et al.*	p–n	MZ	15 dB			30 Gb/s	1 mm	D

BMFET: bipolar mode field-effect transistor; FCAM: free carrier absorption modulator; FP: Fabry–Perot; MOS: Metal Oxide Semiconductor; MZ: Mach Zehnder; TIR: total internal reflection; ZGDC: zero gap directional coupler; M: amplitude modulation depth; J: current density; t_s: switching time; v.d.: vertical device; D: demonstrated; and P: proposed.

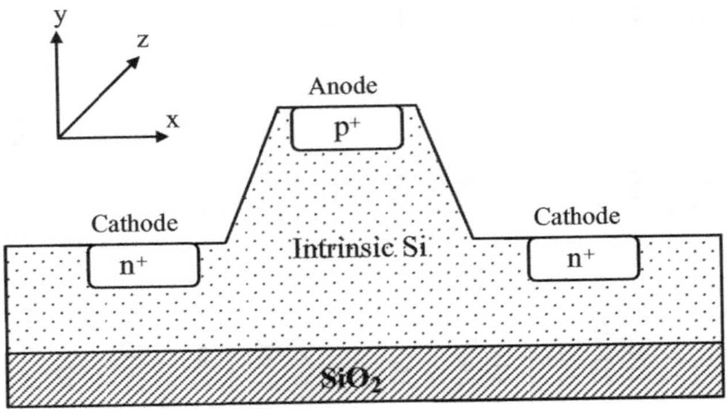

Figure 7.6 *Three-terminal phase modulator with angled rib walls (Tang and Reed, EL (1995) [17]).*

confinement is improved in the SOI structure, but vertical structures are difficult to fabricate as substrate contacts could not be fabricated.

Tang and Reed [17] showed that it was possible to obtain a 30% increase in the concentration of injected carriers onto the waveguiding region of a phase modulator by changing the sidewall angle of the rib from vertical to 54.7°, as indicated in Figure 7.6. The structure is a three-terminal p–i–n structure.

7.8.2 Three-Terminal Structures

By the mid-1990s, Si modulator work was being carried out by using SOI material. A silicon modulator was proposed by Cutolo *et al.* [18] in which the three-terminal device was integrated in a SOI waveguide. The buried oxide was removed from the active region of the modulator, and the substrate acted as the drain. The three-terminal device was shown by the authors to be more efficient than the two-terminal pin modulator. The plasma is injected into the optical channel by applying a forward bias between the source and drain, and the free carriers in the active region are controlled by means of an applied voltage to the control terminal. Hence the carriers can be displaced from the optical mode rather than be depleted from the active region. The total charge remains fixed, but switching is achieved by drifting the carriers. As free-carrier drift is faster than injection, a shorter switching time is expected.

The device described above is called the BMFET, which has been used by Sciuto *et al.* [19] to fabricate a modulator.

7.8.3 Smaller Structures

It is evident that reduction of device dimensions leads to improved device performance. The reductions in length, rib width, and height all contribute to higher modulation bandwidth and lower drive current. Irace *et al.* [20] modeled a 1.4 GHz operating bandwidth of a two-terminal Bragg reflector rib waveguide. The device has a waveguide height and rib width of 1 μm, and an interaction length of 3000 μm. The reduction of Si overlayer thickness and rib width increases the bandwidth from MHz to GHz.

It is obvious that longer devices disallow high integration levels and increase the cost. Barrios *et al.* [21] modeled a modulator based on a Fabry–Perot microcavity with Bragg reflectors. The microcavity confines the optical field in a smaller region. In addition, the transmission of the device near resonance is highly sensitive to small changes in RI. As a result, a lower concentration of injected carriers may switch the device to a nonresonant position, thereby reducing the current drive. The predicted DC power for this 20 μm long device is only 25 μW. Other parameters are given in Table 7.1.

In a separate publication, the same authors [22] have proposed another modulator with only 250 nm in height. The device consists of two uniformly doped contacts, p^+ and n^+ as shown in Figure 11.22. It was predicted that to achieve a RI change of -10^{-3}, a DC power of 1.53 μW/μm is needed. The estimated switching time is 1.3 ns. The DC power is the lowest predicted to date.

The authors have reported a 20 μm long tunable optical resonator integrated on a SOI waveguide. The microresonator consists of a planar Fabry–Perot microcavity formed by $Si–SiO_2$ Bragg reflectors having finesse as large as 11.2. The device is electrically driven. The measured modulation depth is 53%, and power consumption is only 20 mW.

7.8.4 MOS Capacitors

Researchers at Intel Corporation [23] have demonstrated a Si-based optical modulator with a bandwidth exceeding 1 GHz for the first time. This is believed to be a breakthrough in the area of Si photonics. The schematic of the reported device is shown in Figure 7.7 and bears close resemblance to a complementary MOS (CMOS) transistor. The structure consists of *n*-type crystalline silicon and the upper rib is *p*-type poly-Si. The *n*-type and p-type regions are separated by a thin insulating oxide layer. A positive voltage applied to *p*-type poly-Si creates an accumulation of charges at the oxide interface. This in turn changes the RI and induces a phase shift in the optical wave propagating through the device.

The voltage-induced charge density ΔN_e or ΔN_h is related to the drive voltage V_D by

$$\Delta N_e = \Delta N_h = \frac{\varepsilon_0 \varepsilon_r}{e t_{ox} t}[V_D - V_{FB}], \qquad (7.29)$$

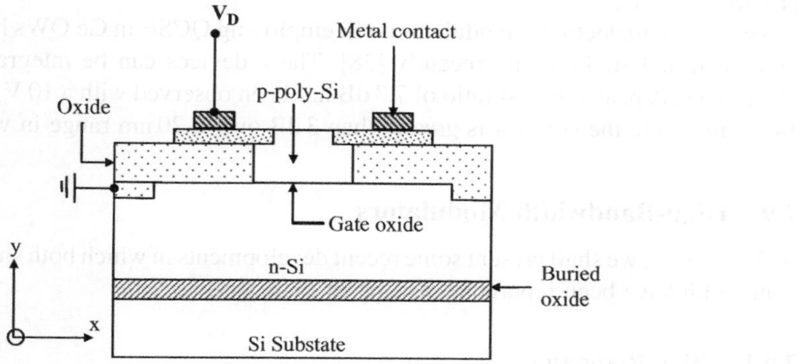

Figure 7.7 Structure of Intel CMOS modulator. Reproduced with permission from [23]. Copyright (2004) Macmillan.

where t_{ox} is the gate oxide thickness, t is the effective charge layer thickness, and V_{FB} is the flat band voltage of the MOS capacitor. The change in RI is still given by Eqs. (7.7) and (7.9) using only one type of carrier. The RI change is due to free carriers, no doubt; however, since the charge transport in MOS structure is governed by the majority carriers, the device bandwidth is not limited by the relatively slow carrier recombination processes in p–i–n diode-based modulators.

The bandwidth of a 2.5 mm long phase modulator using an asymmetric MZI was found to exceed 1 GHz. The on-chip loss of the device is very high (~6.7 dB) and the device was also highly polarization dependent.

A simulation study published later by the same group of workers [24] modified the device configuration in three ways: (1) the poly-Si layer is replaced by c-Si, which is expected to reduce the loss by 5 dB; (2) shrinking the waveguide size and reducing the oxide thickness reduce capacitance; and (3) using a graded doping profile in the vertical direction with higher doping near the gate oxide leads to a balance between speed and optical loss. The predicted bandwidth for a TE-polarized phase modulator is 10 GHz and an on-chip optical loss <2 dB.

The structure has been fabricated accordingly and a silicon modulator with an intrinsic bandwidth of 10 GHz and data transmission from 6 to 10 Gb/s has been demonstrated. The extinction ratio is 3.8 dB, and on-chip loss is ~10 dB [25]. Further improvement in the bandwidth is discussed in Section 7.9.

7.8.5 MQW Structures

Attempts have been made to fabricate modulators by using QCSE in Si and SiGe MQWs [26]. However, the results are not very promising due to the small value of the EO coefficient in indirect-gap material and the small spatial overlap. A 100 μm device showed an extinction ratio of only 1.43 dB and a high insertion loss of 28.5 dB.

A theoretical study [27] has been made of the RI change due to the electric field in Si and SiGe MQWs. The increasing field depletes holes successively from different QWs thereby changing the RI. The effect is not QCSE, nor is electron–hole recombination involved. The predicted escape time of holes from the 10 nm thick QW due to thermionic emission is 10 ps, and therefore a high modulation bandwidth is expected. No experimental support has been provided to date.

The first optoelectronic modulator on Si employing QCSE in Ge QWs has actually been fabricated and studied only recently [28]. These devices can be integrated with silicon electronics. A peak contrast ratio of 7.3 dB has been observed with a 10 V voltage swing at 1457 nm, while the contrast is greater than 3 dB over a 20 nm range in wavelength.

7.9 High-Bandwidth Modulators

In this section, we shall present some recent developments in which both small size and high bandwidth have been reported.

7.9.1 Ring Resonator

A compact modulator using a ring resonator has been reported by Xu *et al.* [29], the bandwidth of which exceeds 12.1 Gb/s, and by Li *et al.* [30]. The basic structure by Xu *et al.*

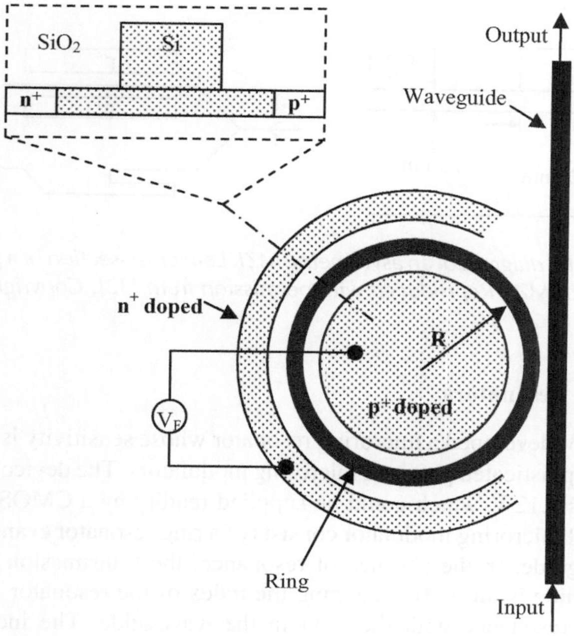

Figure 7.8 *Structure of a ring resonator. The inset shows the pin structure to inject carriers. Reproduced from [29]. Copyright (2007) Optical Society of America.*

is shown in Figure 7.8. The inset gives a schematic cross-section of a ridge waveguide with an integrated lateral p–i–n diode. The applied bias injects carriers into the ring, thereby changing its RI. The high Q of the resonator $\sim 4 \times 10^4$; the authors could achieve modulation depth of >15 dB and a speed of 12.1 Gb/s. The diameter of the ring was 12 μm.

7.9.2 MZ Modulators at 10 Gb/s and Above

As mentioned earlier, the speed of the modulator is determined by how fast the carriers decay after the modulating voltage is turned off. A reverse bias applied to the pin diode grown around the waveguide sweeps the carriers out of the waveguide very quickly, so as not to affect the speed, which is ultimately governed by the *RC* time constant.

The technique has been applied by Intel workers to achieve a high speed of 30 Gb/s [31]. However, the size is of the order of millimeters in length. An alternative is to use microring resonators, but the resonance is susceptible to changes in temperature and other environmental variables.

William M. J. Green *et al.* [32] recently reported a tiny MZ modulator, whose silicon rib waveguides are 550 nm wide and 220 nm high, and with a cross-sectional area of $\sim 0.12\,\mu m^2$. The schematic diagram of the structure is shown in Figure 7.9. The product of half-wave voltage and length for a 200 μm long device is 0.18 V.mm. The index change is 4×10^{-3} corresponding to a charge density of $1.5 \times 10^{18}\,cm^{-3}$. A modulation rate higher than the achieved modulation rate of 10 Gb/s is limited due to the ~16 GHz *RC* time constant.

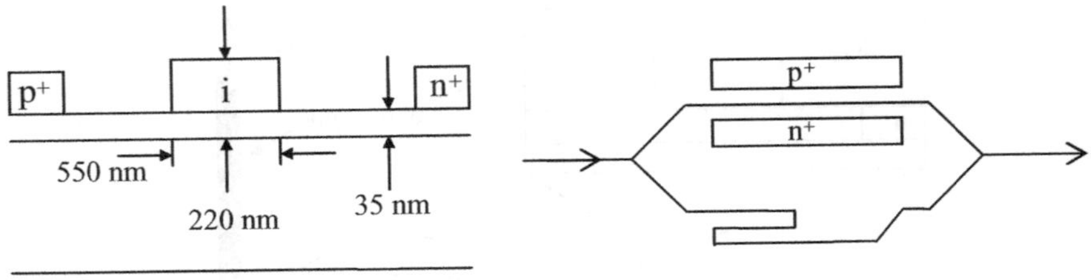

Figure 7.9 Schematic diagram of an asymmetric MZI. Left: cross-section of a pin structure, and (right) the top view of MZI. Reproduced with permission from [32]. Copyright (2007) Optical Society of America.

7.9.3 Microring Resonators

The same group [33] developed a microring resonator whose sensitivity is 25 times greater than that of most sophisticated previous microring modulators. The device can be driven by a signal as small as 0.12 V, which could be supplied readily by a CMOS circuit.

The conventional microring modulator consists of a ring resonator evanescently coupled to a straight waveguide. In the absence of resonance, the transmission of light through the straight waveguide is unity. By changing the index of the resonator with injection, it can be tuned into resonance with the light in the waveguide. The increased coupling reduces the transmission through the guide significantly. Since evanescent coupling between the ring and waveguide is sensitive to fabrication error, an alternative device was developed. In this design, the ring is fabricated to be in resonance with the guide and the coupling between the two is modulated with a MZI. The voltage requirement is only one tenth of that in the first type.

Further reduction in drive voltage is accomplished by introducing a second microring as in Figure 7.10. The second microring imposes a strongly nonlinear phase shift on light

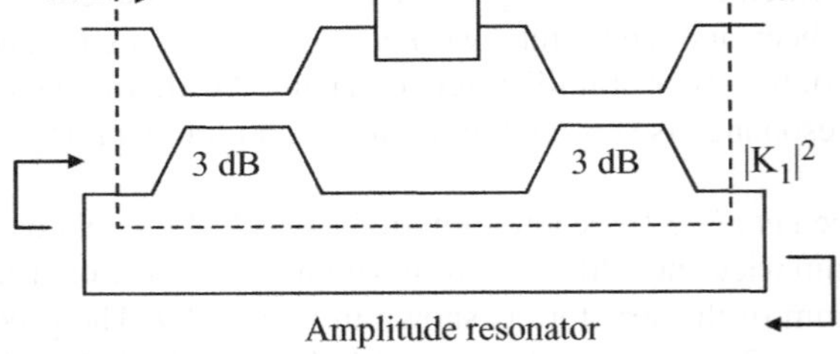

Figure 7.10 Miniature microring resonator modulator. Reproduced with permission from [33]. Copyright (2007) Optical Society of America.

traveling in the upper arm of the MZI, and the drive voltage needed for on-off modulation is reduced 25 times from the voltage for structure in (b). Although free carriers were created by illumination, using a pin structure seems straightforward.

7.9.4 Reverse Biased p–n Diode

Liu *et al.* [31, 34] reported a MZI modulator that contains a 1 mm long p–n junction phase shifter in each arm. They utilized carrier depletion effect in a reverse-biased p–n junction, rather carrier injection using a p–i–n structure, to realize a modulator with a 3 dB optical bandwidth of 30 GHz and a data transmission rate of 40 Gb/s. Recent results for both MOS capacitor and p–n junction–based modulators are discussed in [35].

7.10 Performance of EO Modulators

The earlier two sections attempted to describe the representative optical and electrical structures employed for the fabrication of modulator structures. The change in RI is accomplished via the EO effect: in most situations by free-carrier plasma dispersion effect, in which an electric field injects, extracts, or induces carriers of single or both the polarities.

In this section, a summary of the performance of various types of modulators theoretically studied or experimentally reported has been presented. Table 7.1 gives most of the performance characteristics. This kind of summary is also presented by Lipson [36]. The present table includes all the data reported in that work and in addition collects recent data appearing in the literature. The entries are by no means exhaustive; rather, they are representative values.

Problems

7.1 Consider a bulk effect modulator using KDP as in Section 7.2.2. Show that the ratio of output intensity (I_o) and input intensity (I_i) is

$$I_o/I_i = \sin^2\left[\frac{\pi}{2}\frac{V}{V_\pi}\right], \; V_\pi = \lambda/2n_0^3 r$$

where r is the linear EO coefficient.

7.2 Calculate V_π for a KDP crystal, given $\lambda = 1\ \mu\text{m}$, $n_0 = 1.51$, and $r = 10.6 \times 10^{-12}$ m/V.

7.3 Draw a diagram illustrating the principle of intensity modulation of light using an EO crystal as described in Section 7.2.2. Indicate the state of polarization at each stage. Draw the transmission curve I_o/I_i as a function of applied voltage V. Illustrate how a sinusoidal electrical signal can produce a sinusoidal variation of intensity of light.

7.4 Assume that the modulator in Problem 7.3 is biased by a fixed retardation $\pi/2$. If a sinusoidal voltage $V_m \sin \omega_m t$ is applied to the crystal around this "bias," show that $I_0/I_i = \frac{1}{2}(1 + A \sin \omega_m t)$, where A is related to V_m. How in practice do you introduce a constant phase retardation ("bias") between x' and y' components?

7.5 Consider that the phase difference between two arms is introduced by applying a voltage to one arm which shows the linear–electro-optic (LEO) effect. Derive the expression for the voltage needed to switch off the light at the output.

7.6 Consider that voltage is applied to electrodes over both the arms and the electric fields are $+V/d$ and $-V/d$, where d is the separation between two electrodes. Obtain the expression for V_π.

7.7 Assume that two arms of MZI consist of MOS capacitors. The change in RI is given by Eqs. (7.7) and (7.28). Calculate V_π for $V_d = 1$ V, when $V_{FB} = 0.2$ V, $t_{ox} = 100$ nm, $t = 50$ nm, and $\varepsilon_r = 3.7$.

7.8 Obtain tuning characteristics of a Fabry–Perot resonator for $\Delta n = 0$, 0.001 and 0.005. The operating wavelength is 1550 nm.

7.9 The induced electrons in a MOS capacitor are swept by the transverse electric field. Let the channel length be 0.1. Calculate the time needed if the electrons attain saturation drift velocity.

7.10 Consider the effect of temperature on the transmission of a resonator. Use Eqs. (7.11) and (7.26) to find the ratio T/T_{max} per degree rise in temperature at T = 300 K.

7.11 Calculate the change in the refractive index in a reverse-biased p–n junction modulator when the bias is increased from 2.0 to 2.5 V. The doping densities are $N_D = 3 \times 10^{18}$ cm^{-3} and $N_A = 1.5 \times 10^{17}$ cm^{-3}.

7.12 Assume that the injected current density in a Si p–n junction modulator is 1.6×10^3 A/cm^2. Calculate the electron density injected into the p-layer of thickness 7.7 μm, where the electron lifetime is 50 ns. From this, calculate the values of change in absorption and RI.

7.13 Consider a p–n junction phase modulator and assume that change in RI is due to LEO and electrorefractive effects only. Show that the incremental change in phase may be expressed as

$$\Delta\phi/\Delta V = (\pi L/\lambda W)[-\Gamma(r + 2(V_{bi} - V)s/W],$$

where V is the applied bias, V_{bi} is the built-in voltage, W and L are respectively the width and length of the depletion layer, r and s are respectively the LEO and Kerr coefficients, and Γ is the overlap between the electric and optical fields.

7.14 A MZI contains in its two arms Ge and Si MQWs. The voltages applied to one arm and the corresponding changes in RI due to QCSE are as follows: 0.5 V (0.0404), 1.0 V (0.0612), 1.5 V (0.0876), and 2 V (0.0472), where the RI changes are given in parentheses. Calculate the lengths of the MQWs needed to produce a 180° phase difference between the two arms. The operating wavelength is 1550 nm.

References

1. Agrawal, G.P. (2002) *Fiber Optic Communication Systems*, John Wiley & Sons, Inc., New York.
2. Yariv, A. (1998) *Quantum Electronics*, 5th edn, John Wiley& Sons, Inc., New York.
3. Bhattacharya, P. (1994) *Semiconductor Optoelectronic Devices*, Prentice Hall International, Englewood Cliffs, NJ.
4. Saleh, B.E.A. and Teich, M.C. (1991) *Fundamentals of Photonics*, John Wiley& Sons, Inc., New York.
5. Seraphin, B.O. and Hess, R.B. (1965) Franz Keldysh effect above the fundamental edge in Ge. *Phys. Rev. Lett.*, **14**, 138–140.
6. Seraphin, B.O. (1965) Optical field effect in silicon. *Phys. Rev.*, **140**, A1716–A1725.

7. Miller, D.A.B., Chemla, D.S., Damen, T.C. *et al.* (1985) Electric field dependence of optical absorption near the band gap of quantum well structures. *Phys. Rev. B*, **32**, 1043–1060.
8. Schmitt-Rink, S., Chemla, D.S., and Miller, D.A.B. (1989) Linear and nonlinear optical properties of semiconductor quantum wells. *Adv. Phys.*, **38**, 89–188.
9. Soref, R.A. and Bennett, B.R. (1987) Electrooptical effects in silicon. *IEEE J. Quantum Electron.*, **QE-23**, 123–129.
10. Kuo, Y., Lee, Y., Ge, Y. *et al.* (2005) Strong quantum-confined Stark effect in germanium quantum well structures on silicon. *Nature*, **437**, 1334–1336.
11. Cocorullo, G., Della Corte, F.G., and Rendina, I. (1999) Temperature dependence of thermo-optic coefficient in crystalline silicon between room temperature and 550K at the wavelength of 1523nm. *Appl. Phys. Lett.*, **74**, 3338–3340.
12. Della Corte, F.G., Montefusco, M.E., Moretti, L. *et al.* (2000) Temperature dependent analysis of the thermo-optic coefficient in silicon by single and double oscillator model. *J. Appl. Phys.*, **88**, 7115–7119.
13. Buckman, A.B. (1992) *Guided-Wave Photonics*, Saunders College Publishing, Orlando, FL.
14. Lipson, M. (2006) Compact electro-optic modulators on a silicon chip. *IEEE J. Sel. Top. Quant. Electron.*, **12**, 1520–1526.
15. Reed, G.T. and Jason Png, C.E. (2005) Silicon optical modulators. *Mater. Today*, **8**, 40–50.
16. Treyz, G.V., May, P.G., and Halbout, J.-M. (1991) Silicon optical modulators at 1.3 pm based on free-carrier absorption. *IEEE Electron. Device Lett.*, **12**, 276–278.
17. Tang, C.K. and Reed, G.T. (1995) Highly efficient optical phase modulator in SOI waveguides. *IEEE Electronics Lett.*, **31**, 451–452.
18. Cutolo, A., Iodice, M., Spirito, P., and Zeni, L. (1997) Silicon electro-optic modulator based on a three terminal device integrated in a low-loss single-mode SOI waveguide. *J. Lightwave Technol.*, **15**, 505–518.
19. Sciuto., A., Libertino, S., Assandria, A. *et al.* (2003) Design, fabrication, and testing of an integrated Si-based light modulator. *J. Lightwave Technol.*, **21**, 228–235.
20. Irace, A., Breglio, G., and Cutolo, A. (2003) All-silicon optoelectronic modulator with 1GHz switching capability. *IEEE Eletron. Lett.*, **39**, 232–233.
21. Barrios., C.A., de Almeida, V.R., and Lipson, M. (2003) Low-power-consumption short-length and high-modulation-depth silicon electrooptic modulator. *J. Lightwave Technol.*, **21**, 1089–1098.
22. Barrios, C.A., Almeida, V.R., Panepucci, R., and Lipson, M. (2003) Electrooptic modulation of silicon-on-insulator submicrometer-size waveguide devices. *J. Lightwave Technol.*, **21**, 2332–2339.
23. Liu, A.S., Jones, R., Liao, L. *et al.* (2004) A highspeed silicon optical modulator based on a metal-oxide semiconductor capacitor. *Nature*, **427**, 615–618.
24. Liu, A., Samara-Rubio, D., Liao, L., and Paniccia, M. (2005) Scaling the modulation bandwidth and phase efficiency of a silicon optical modulator. *IEEE J. Select. Top. Quantum. Electron.*, **11**, 367–372.
25. Liao, L., Samara-Rubio, D., Morse, M. *et al.* (2005) High speed silicon Mach Zehnder modulator. *Opt. Express*, **13**, 3129–3135.
26. Qasaimeh, O., Bhattacharya, P., and Croke, E.T. (1998) SiGe-Si quantum well electroabsorption modulators. *IEEE Photon. Technol. Lett.*, **10**, 807–809.
27. Marris, D., Cordat, A., Pascal, D. *et al.* (2003) Design of a SiGe-Si quantum well optical modulator. *IEEE J. Sel. Top. Quantum Electron.*, **9**(3), 747–754.
28. Roth, J.E., Fidaner, O., Schaevitz, R.K. *et al.* (2007) Optical modulator on silicon employing germanium quantum wells. *Opt. Express*, **15**, 5851–5859.
29. Xu, Q., Manipatruni, S., Schmidt, B. *et al.* (2007) 12.5 Gbit/s carrier-injection-based silicon micro-ring silicon modulator. *Opt. Express*, **15**, 403–406.

30. Li, C., Zhou, L., and Poon, A.W. (2007) Silicon microring carrier-injection-based modulators/switches with tunable extinction ratios and OR-logic switching by using waveguide cross-coupling. *Opt. Express*, **15**, 5069–5076.
31. Liu, A., Liao, L., Rubin, D. *et al.* (2007) High-speed optical modulation based on carrier depletion in a silicon waveguide. *Opt. Express*, **15**, 660–668.
32. Green, W.M.J., Rooks, M.J., Sekaric, L., and Vlasov, Y.A. (2007) Ultra-compact, low RF power, 10 Gb/s silicon Mach-Zehnder modulator. *Opt Exp.*, **15**, 17106–17113.
33. Green, W.M.J., Rooks, M.J., Sekaric, L., and Vlasov, Y.A. (2007) Optical modulators using anti-crossing between paired amplitude and phase resonators. *Opt. Exp.*, **15**, 17264–17272.
34. Liu, A., Liao, L., Rubin, D. *et al.* (2008) Recent development in a high-speed silicon optical modulator based on reverse-biased pn diode in a silicon waveguide. *Semicond. Sc. & Technol.*, **23**, 064011(1-7).
35. Basak, J., Liao, L., Liu, A. *et al.* (2008) Developments in gigascale silicon optical modulators using free carrier dispersion mechanisms. *Adv. Opt. Technologies*, Art. ID: 678948 (10 pgs).
36. Lipson, M. (2005) Switching light on a silicon chip. *Opt. Mater.*, **27**, 731–737.

8
Silicon Photodetectors

8.1 Introduction

A (silicon) photodetector is a solid-state transducer that is used for converting light energy into electrical energy [1, 2]. This conversion of the transduced incident light into an electronic representation allows for a diversity of applications because processing, storing, or transmitting electronic images [3] now becomes relatively easy and fast with modern semiconductor integrated circuits and electronic systems. As an example, compared to the earlier photographic methods, solid-state imagers allow for much faster viewing of the images and their easier manipulations since they are in digital format [4–11]. As historical information, we note that some of the earliest digital images of the Apollo and Explorer space missions opened new horizons for digital-imaging technologies. This was further solidi ed by later deep-space pictures from the Hubble space telescope. In fact, in the past four decades, digital imaging has opened up fantastic new horizons in space explorations and in science.

Another major application of photodetectors is in optical ber communications systems – the high-speed communications technology that currently brings information to us over long distances [1, 2]. The basic components of an optical ber communications systems are the transmitter, the transmission medium, and the photoreceiver, and these are schematically shown in Figure 8.1. In the transmitter, there is also an optical encoder, through which the information signal in electronic format is converted to an optical signal by modulating the optical emission from the optical transmitter. This optical transmitter is usually a light-emitting diode (LED) or a laser diode together with related modulating and driving circuits. Once the signal is in the optical domain, it is then transmitted through an optical ber with optical ampli ers and repeaters depending on its length. Finally, at the photoreceiver, the optical signal is converted back into an electrical signal. Here, the photodetector is a key component that is used to convert an optical signal into an electrical signal that is subsequently ampli ed by the preampli er before further processing in the electronic domain.

Silicon Photonics: Fundamentals and Devices, First Edition. M. Jamal Deen and P. K. Basu.

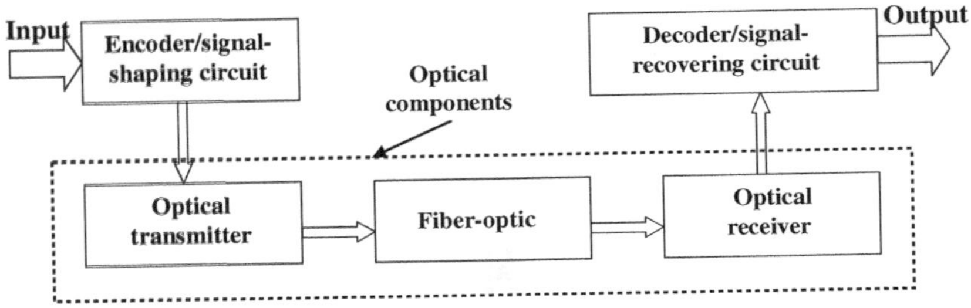

Figure 8.1 *Schematic representation of an optical fiber communication system.*

The photoreceiver contains a photodetector and preampli er with relevant biasing and demodulating circuits. A schematic of the photoreceiver is shown in Figure 8.2. In a ber-optic communication system, when considering signal attenuation along the link, a key contributor to excellent system performance is determined at the photodetector. For example, improvement of the photodetector characteristics and performance means that fewer repeater stations are required.

Photodetectors manufactured in silicon technology have been used in a variety of applications ranging from low-cost optical receivers, scienti c imaging systems (using charge-coupled devices (CCDs) or high-performance complementary metal oxide semiconductor (CMOS) imagers), bar code scanners, and standard CMOS imagers for commercial and industrial use, to medical imaging systems and imaging systems for chemical and biological applications. CCDs are typically used in applications such as electronics inspection due to their excellent read noise performance, wide dynamic range, high light sensitivity, high signal-to-noise ratio, and high shutter rejection ratio. However, CMOS-based photodection systems are rapidly gaining ground, and in 2010 iSuppli (a market research rm) reported that CMOS-based image sensors have more than 90% of the world market share. Silicon-based photodetectors are also used in area-scan sensors for machine vision systems, line-scan and time-delay-integration devices, and cell phone cameras.

Silicon-based photodetectors can be of different structures, for example, simple *p–n junction diodes*, *p–i–n diodes*, *Schottky barrier diodes*, and *phototransistors*, or fabricated in customized technologies for *CCDs* or *photodetectors with added structures such as resonant cavities* to increase quantum ef ciency without compromising high-speed performance. Many of these types of photodetectors are described in Section 8.4. Also, depending on the type of application and wavelength range of interest, a speci c photodetector design is chosen to meet the required speci cations.

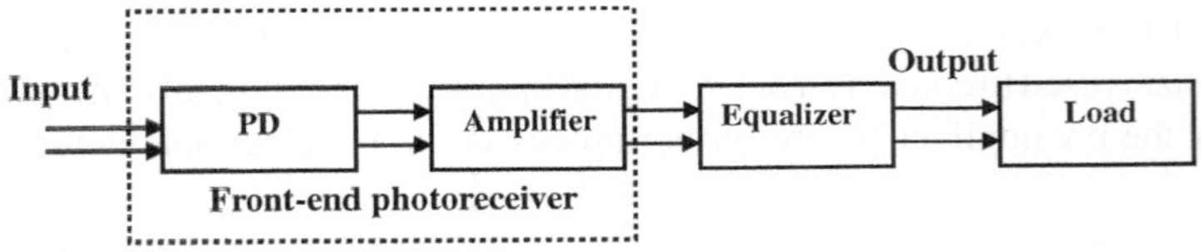

Figure 8.2 *Schematic representation of the photoreceiver of an optical communication system.*

One important application that has been pursued for silicon-based photodetectors is for high-bandwidth, short-distance applications such as in optical interconnections for interchip communications in computer systems, and ber-to-the-home receivers in the 800–900 nm wavelength range. These two applications, in particular, are driven by the low cost of silicon detectors and the feasibility of integrating the detector with postprocessing electronics to create low-cost, monolithic optical receivers.

8.2 Optical Detection

The photodetectors that are used in imaging technologies or in optical communication systems utilize the internal photoemission effect based on Einstein's theory of photoelectric effect and Planck's quantum theory of radiation. In a semiconductor photodetector, photons from the incident optical signal are absorbed as they travel through the semiconductor material. If the photon energy is equal to or larger than the band gap energy (E_G) of the optical absorption (or active) region of the photodetector, then electron–hole (e–h) pairs are created, with the photoelectrons in the conduction band and holes in the valence band. Also, these photoexcited carriers – electrons and holes – are free to move within the semiconductor material, and under an externally applied electric eld, a photocurrent in the external circuit ows.

In this book, we are primarily concerned with silicon or silicon-based photonics. Silicon is an indirect band gap semiconductor with a small absorption coef cient for light with a wavelength above 0.5 μm. However, it has a high ratio of the ionization coef cients of the holes (β_i) and electrons (α_i), so it is a good choice for avalanche photodetectors, especially for local area networks or for short-wavelength applications. However, it is not suitable for long-haul applications because the minimum attenuation in current silica-based bers is around 1.5 μm. But if silicon is combined with germanium, then it is possible to extend the applications of silicon–germanium photodetectors to longer wavelengths. This will be discussed in more detail in this chapter.

Based on what was discussed in this section, the semiconductor is transparent (non-absorbing) for incident light with a wavelength longer than

$$\lambda_{CO} = \frac{hc}{E_G}, \tag{8.1}$$

where h is Planck's constant (6.62×10^{-34} Js), c is the speed of light (3×10^8m/s), λ is the wavelength (m), and E_G is the energy band gap in joules (divide by q, the electronic charge to convert to electron volts, or eV). For silicon with a room temperature energy gap of $\sim$1.12 eV, this means that only light shorter than 1.1 μm (termed the *cutoff wavelength*, or λ_{CO}) will be absorbed or can be detected. Also, silicon is an indirect band gap semiconductor, so a photoexcited electron transition into the conduction band requires an extra source of momentum to satisfy energy and momentum conservation. The additional source of momentum is typically supplied by the quantized lattice vibrations – phonons. Also, for wavelengths near 1.1 μm, since a phonon is required for momentum conservation, then the probability of light absorption decreases signi cantly near λ_{CO}.

Example 8.1

The band gap energy of silicon and germanium are 1.12 eV and 0.66eV, respectively. Calculate the long-wavelength cutoff of these two important elemental semiconductors.

The long-wavelength cutoff is given by

$$\lambda_{co}(Silicon) = \frac{hc}{E_G} = \frac{6.6262 \times 10^{-34} Js \times 3 \times 10^8 \text{ m/s}}{1.12 \text{ eV} \times 1.6 \times 10^{-19} \text{ J/eV}}$$

$$= 1.11 \times 10^{-6} \text{ m} = 1.11 \text{ μm}$$

and

$$\lambda_{co}(Germanium) = \frac{hc}{E_G} = \frac{6.6262 \times 10^{-34} \text{ Js} \times 3 \times 10^8 \text{ m/s}}{0.66 \text{ eV} \times 1.6 \times 10^{-19} \text{ J/eV}} = 1.88 \times 10^{-6} \text{ m} = 1.88 \text{ μm}.$$

Figure 8.3 shows the variation of the absorption coef cient α (in cm^{-1}) with wavelength (in μm, the bottom x-axis) or photon energy (in eV, the top x-axis) for both elemental and compound semiconductors used to make commercial, widely used photodetectors. The cutoff wavelengths of these semiconductors are shown to be between 0.8 and ~1.8 μm. Also, a few observations can be made about the variation of α with λ. In direct band gap

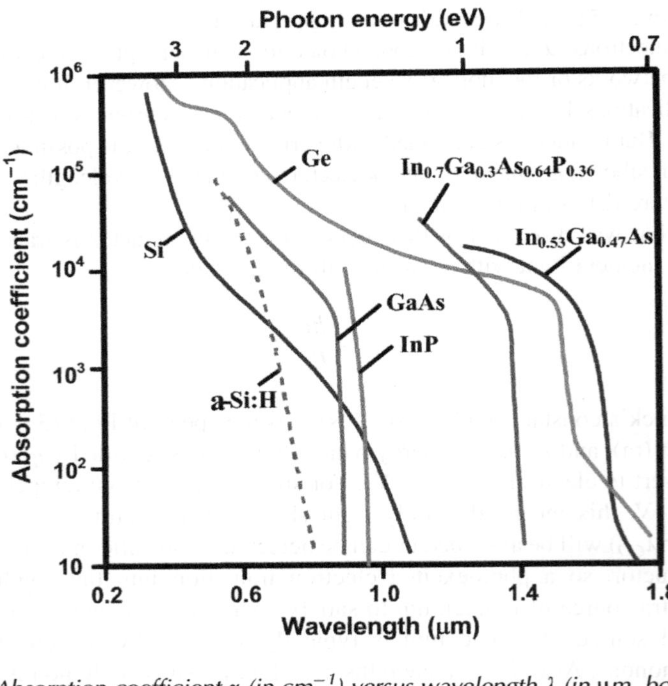

Figure 8.3 *Absorption coefficient α (in cm^{-1}) versus wavelength λ (in μm, bottom* x-*axis)* or photon energy (in eV, top x-axis) for several semiconductors used to make photodetectors. Note the different variation of α with λ for direct versus indirect band gap semiconductors.

semiconductors (such as GaAs or InP), α increases sharply as the wavelengths decrease. This is in contrast to the weaker increase of α with decreasing λ for indirect band gap semiconductors. Also, note that the inverse of the absorption coef cient is termed the *penetration depth* of incident radiation into the semiconductor. For silicon or silicon–germanium photodetectors, depending on the material composition of the absorption layer, the photoresponse can be tuned from infrared (typically 1.55 µm) to ultraviolet (around 0.4 µm) light.

For practical applications of photodetectors, typically the main performance parameters of interest are sensitivity to light at speci ed wavelengths, speed of response, and noise. However, for some applications, other parameters such as gain or cost may also be important. In Section 8.3, we discuss the most important characteristics of photodetectors.

As mentioned before, the wavelength response range of silicon photodetectors can be increased by adding germanium. In this case, the indirect band gap of $Si_{1-x}Ge_x$, $E_G(x)$, is given by [15]

$$E_G(x) = E_{GSi-300K} + \alpha_T \left[\frac{300^2}{300+\beta_T} - \frac{T^2}{T+\beta_T} \right] - 0.740 \cdot x, \tag{8.2}$$

where $E_{GSi-300K}$ is the Si band gap at 300 K (taken in this case as 1.09 eV), and $\alpha_T$and β_T are tting parameters given in Table 8.1. By changing the Si band gap by adding Ge, the absorption coef cient as a function of wavelength will change and longer wavelengths in the infrared domain can be absorbed and detected.

Also, the dependence of the absorption coef cient on the band gap of the alloy, based on a one-phonon McFarlane model, is given by [15]

$$\alpha = A \cdot \left[\frac{\left(h\nu - E_G - \overline{E_{pn}}\right)^2}{1-\exp\left(-\overline{E_{pn}}/kT\right)} + \frac{\left(h\nu - E_G + \overline{E_{pn}}\right)^2}{\exp\left(-\overline{E_{pn}}/kT\right)-1} \right] \tag{8.3}$$

where A is constant, that is, the average weighting factor of the absorption coef cients of each phonon; and E_{pn} is the average energy of all participating phonons (transverse acoustical (TA), longitudinal acoustical (LA), transverse optical (TO), or longitudinal optical (LO)) and is taken as 50 meV (see Table 8.2). In Eq. (8.3), the threshold for a phonon-assisted absorption process is $\left(E_G - \overline{E_{pn}}\right)$, and the term $\left(h\nu - E_G + \overline{E_{pn}}\right)$ describes an absorption process that is induced below the band gap due to a phonon absorption. Note that the typical phonon energies are in the 40–60 meV range [15]. The tting parameters for Eq. (8.3) are given in Table 8.2 for two combinations of $Si_{1-x}Ge_x$, with $x = 025$ and $x = 0.4$.

Figure 8.4 shows the absorption coef cient for pure Si and $Si_{1-x}Ge_x$ alloy versus photon energy. Photons of lower energy can be absorbed by increasing the germanium fraction in the alloy.

Table 8.1 *Coefficients for α_T and β_T for the energy gap expression (8.2)*

	Silicon	Germanium
α_T (meV/K)	0.473	0.477
β_T (K)	636	235

Table 8.2 *Fitting parameters for Eq. (8.3) from [15]*

	Silicon	$x=0.25$	$x=0.4$
A (cm^{-1}))	1	1.309	2.069
$\overline{E_{pn}}$ (meV)	40.8	37.8	32.8
Corrected $\overline{E_{pn}}$ (meV)	50	46	41

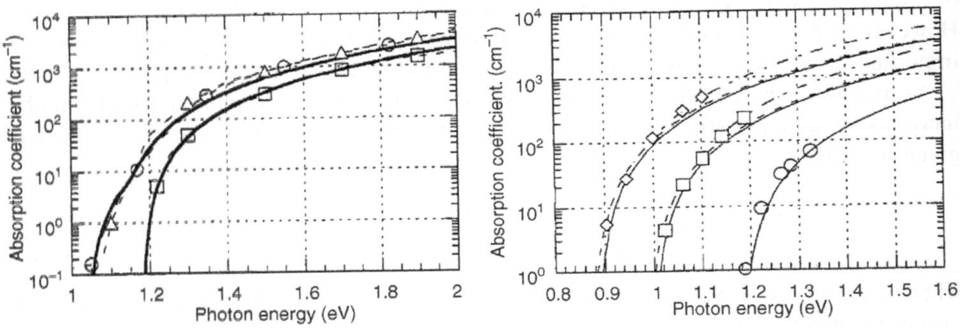

Figure 8.4 *(Left) The absorption coefficient of Si at 300 K (data as triangles from [16], data as circles from [17], and 77 K (squares) from [16]). (Right) The absorption coefficient of SiGe alloy at 90 K for Ge fractions* $x=0$ (circles), $x=0.25$ (squares), and $x=0.4$ (diamond). The data are from [18]. Both figures were put together and fitted using Eq. (8.3) in [15].

8.3 Important Characteristics of Photodetectors

Depending on its application, speci c performance measures of the photodetector are emphasized. For example, for optical communication receivers, some of the key performance characteristics are quantum ef ciency, speed, dynamic range, and minimum detectable power. In this section, we discuss some of the important characteristics of (silicon) photodetectors.

8.3.1 Quantum Efficiency

Quantum ef ciency (QE) (sometimes called *internal QE*) is the ratio of the number of generated electron–hole pairs per incident photon that are collected at the contacts. QE is an important measure of high-sensitivity photodetectors since it should produce the maximum electrical signal per unit optical power.

External quantum ef ciency (EQE or η), the more practical measure, compares the collected photocurrent to the number of incident photons in a real photodetector. It is less than the internal quantum ef ciency because of effects such as surface re ections decreasing the amount of light on the absorption region and surface and nonradiative recombinations of carriers (holes and electrons) before they are collected. The external quantum ef ciency is a function of wavelength (λ) of the incident light because both the absorption coef cient and penetration depth of the light vary with λ.

In a photodetector, its quantum ef ciency η is given by

$$\eta = \frac{\text{electron generation rate}}{\text{incident photon rate}} = \frac{I_{PC}/q}{P_I/(hv)} = \frac{I_{PC}}{P_I} \cdot \frac{hc}{q\lambda} = \frac{1.23948}{\lambda\,(\text{in}\ \mu\text{m})} \cdot \frac{I_{PC}}{P_I}. \tag{8.4}$$

where I_{PC} is primary photocurrent (without internal gain), P_I is the incident optical power, (hc/λ) is the photon energy, h is Planck's constant (6.62×10^{-34} Js), c is the speed of light (3×10^8 m/s), λ is the wavelength (in micrometers for the right-most expression in Eq. (8.4)), and q is the electronic charge (1.602×10^{-19} C).

Example 8.2

Calculate the incident ux of photons Φ for a light source illuminating a photodetector of area $(100\,\mu\text{m})^2$ with $1\,\mu$W of light. Assume that the light is ltered, its wavelength is 600 nm, and all the incident photons are converted into electron–hole pairs.

The light intensity is the light power per unity area,

$$1\,\mu\text{W}/(100\,\mu\text{m})^2 = 10^{-2}\text{W}/\text{cm}^2.$$

The photon energy is given by

$$hv = hc/\lambda = (6.62 \times 10^{-34}\ \text{Js} \times 3 \times 10^{10}\ \text{cm/s})/600 \times 10^{-7}\ \text{cm} = 3.31 \times 10^{-19}\ \text{J}$$

$$\text{The photon ux}\ \Phi = \frac{\text{Light Intensity}}{\text{Photon Energy}} = \frac{10^{-2}\ \text{Wcm}^{-2}}{3.31 \times 10^{-19}\ \text{J}} = 3.02 \times 10^{16}\ \text{Photons}/(\text{cm}^2 \cdot \text{s})$$

At short wavelengths, the QE of silicon photodiodes decreases because the photogenerated charges recombine with surface charges. At the other end, at longer wavelengths, the QE again decreases because of the decrease in the absorption coef cient close to the indirect band gap energy.

According to the Lambert–Beer law, the rate of photogenerated carriers G within silicon has a spatial distribution given by

$$G(x) = \alpha(\lambda) \cdot (1 - R(\lambda)) \cdot \frac{\lambda}{hc} P_I \cdot e^{-\alpha(\lambda)x}. \tag{8.5}$$

where α is the absorption coef cient that is a strong function of the incident wavelength λ, P_I is the incident optical power density (incident power per unit area), and R is the re ectance of the semiconductor–antire ection coating or air surface. Very often, an antire ection coating is used to increase the amount of light absorbed into the active or absorption region of the photodetector.

The external quantum ef ciency depends on three factors – the fraction of photogenerated carriers that reach the outer circuit without recombination (ζ), the fraction of light absorbed and converted to electron–hole pairs $(1 - e^{-\alpha d})$ where d is the thickness of the absorption region, and the amount of light transmitted into the semiconductor that depends on the re ectance (R) from the semiconductor surface. Therefore, we can write η as

$$\eta = \zeta(1 - e^{-\alpha d})(1 - R). \tag{8.6}$$

Example 8.3

Calculate the incident power on a simple pn photodiode with diameter 50 μm illuminated with incident light of intensity 10 μW/cm^2. The wavelength is $\lambda = 0.6\,\mu\text{m}$. Calculate also the number of photons arriving at the surface of the photodiode.

The incident power on the photodiode is

$$P_I = I \times Area = \left(10 \times 10^{-6}\,\text{W} \cdot \text{cm}^{-2}\right) \cdot \pi \left(\frac{50 \times 10^{-4}\,\text{cm}}{2}\right)^2 = 1.96 \times 10^{-9}\,\text{W}$$

The incident photon rate is given by

$$\text{Incident photon rate} = P_I/h\nu = \frac{1.96 \times 10^{-10}\,\text{W} \times 0.6 \times 10^{-6}\,\text{m}}{6.6262 \times 10^{-34}\,\text{Js} \times 3 \times 10^8\,\text{m/s}} = 5.9 \times 10^8\,\text{photons/s}$$

Example 8.4

On a photodiode, when 5×10^{11} photons/s at a wavelength of 0.85 μm are incident on its active area, it was found that a current of 32 nA ows at its terminals. Determine the quantum ef ciency of this photodiode.

The quantum ef ciency is expressed as $\eta = \dfrac{\text{electron generation rate}}{\text{incident photon rate}} = \dfrac{I_{PC}/q}{P_I/h\nu}$.

$$\text{Therefore } \eta = \frac{32 \times 10^{-9} A/1.6 \times 10^{-19} C}{5 \times 10^{11}\,\text{photons/s}} = 0.40.$$

8.3.2 Responsivity

Responsivity or sensitivity ($\Re$) is an alternative gure of merit that may be used instead of quantum ef ciency. It is de ned as the ratio of the primary photocurrent I_{PC} (without internal gain) to the incident optical power P_I, and $\Re$ is related to η as follows:

$$\Re = \frac{I_{PC}}{P_I} = \eta \frac{1}{\left(hc/_{\lambda}\right)} q = \eta(\lambda) \frac{\lambda\,(\text{in}\,\mu\text{m})}{1.24}. \tag{8.7}$$

Note that since the quantum ef ciency is a function of wavelength (through its dependence of the absorption coef cient), then $\Re$ will also be a function of wavelength.

Example 8.5

On a photodiode, when 5×10^{11} photons at a wavelength of 0.85 μm are incident on its active area, it was found that a current of 32 nA ows at its terminals. Determine the responsivity of this photodiode.

From Example 8.4, we found that $\eta = 0.40$.

$$\text{The responsivity is } \Re = \frac{I_{PC}}{P_I} = \eta \frac{1}{\left(hc/_{\lambda}\right)} q = \frac{0.4 \times 1.6 \times 10^{-19}\,\text{C} \times 0.85 \times 10^{-6}\,\text{m}}{6.6262 \times 10^{-34}\,\text{Js} \times 3 \times 10^8\,\text{m/s}}$$

$$= 0.27\,A/W$$

Example 8.6

Calculate the responsivity of a simple pn photodiode assuming that the re ectance $R=0$, that is, all the incident light is transmitted into the photodetector. The quantum ef ciency $\eta=0.8$ for incident light of wavelength $\lambda=0.6\,\mu\text{m}$. If the wavelength increases to $0.7\,\mu\text{m}$, and for the same assumptions, what is the new responsivity?

The responsivity is given by

$$\Re = \frac{I_{PC}}{P_I} = \eta \frac{1}{\left(hc/_{\lambda}\right)} q$$

$$= 0.8 \frac{0.6 \times 10^{-6} m}{6.6262 \times 10^{-34}\,\text{Js} \times 3 \times 10^{8}\,\text{m/s}} \times 1.6 \times 10^{-19}\,\text{C} = 0.386\,A/W$$

For an increase in the wavelength to $0.7\,\mu\text{m}$ and under the same conditions, the responsivity will increase to 0.451. This is because the responsivity is proportional to wavelength for the same quantum ef ciency.

In photodetectors, the sensitivity can be increased by increasing the absorption region thickness and minimizing re ections from its surface. Very often, the absorption region is made $2/\alpha$ thick and an antire ection dielectric coating is used to minimize surface re ections. The antire ection coating is made of dielectric whose refractive index n_{AR} is the geometric mean of the refractive indices of air n_{air} and the semiconductor n_{sc}, that is,

$$n_{AR} = \sqrt{n_{air} n_{sc}}. \tag{8.8}$$

The optimal thickness of the antire ection coating is $d_{AR} = \lambda/4n_{AR}$.

Example 8.7

The refractive index of silicon is 3.49 and that of air is 1. What antire ection (AR) coating can be used for a photodetector operating at 600 nm?

The dielectric to be used as an antire ection coating should have a refractive index.

$$n_{AR} = \sqrt{1 \times 3.49} = 1.868$$

The refractive index of silicon nitride (Si_3N_4) is ~2, so it can be used. Note that the refractive index of SiO_2 is ~1.54, so it is too low. The required thickness is

$$d_{AR} = \frac{600\,\text{nm}}{4 \times 2} = 75\,\text{nm}$$

Note that if SiO_2 was used as the AR coating, then there would be more re ections, thus the amount of light absorbed into the photodetector would be less and its sensitivity would be reduced.

8.3.3 Bandwidth

The time delay for the conversion from incident optical radiation to an electrical signal should be as small as possible if the photodetector is to operate with a wide bandwidth or at a high speed. The speed of a photodetector is a measure of how fast the photodetector can respond to a series of light pulses. It is mainly limited by two time constants. First is the transit time that is needed by the photogenerated carriers to travel through the active layers of the photodetector and be collected by the contacts of the device. If the speed or bandwidth is mainly limited by this time constant, it is called the *transit time–limited bandwidth* and a short response time is required to increase the bandwidth. The second time constant that limits the speed of the photodetector is the $R_{Tot}C_D$ time constant where C_D is the photodetector junction capacitance (and other parasitic capacitances) and R_{Tot} is the equivalent parallel resistance that includes the parasitic resistance of the photodetector. Since this $R_{Tot}C_D$ time constant depends on the dimensions of the photodetector, then it can be minimized by the reduction of the area of the photodetector that reduces its capacitance, but this increases the detector's parasitic resistance. Further, a reduction in detector area makes it more dif cult to collect the incident light, with this resulting in reduced quantum ef ciency. The overall time constant τ is approximated by the following expression:

$$\tau = \sqrt{\tau^2_{\text{transit}\ time} + \tau^2_{R_{Tot}C_D}}\tau. \tag{8.9}$$

For optimal quantum ef ciency and speed, the dimensions of the photodetector must be optimized to solve this trade-off issue.

8.3.4 Gain

In addition to high responsivity or quantum ef ciency and low-noise characteristics, in some situations, it is desirable for the photodetector itself to have internal gain. This is particularly important in situations where weak optical signals must be detected. The photodetector should be able to detect weak optical signals at a speci c wavelength. The gain of the photodetector is de ned as the ratio of the number of collected electron–hole pairs to the number of primary photogenerated pairs at the operating wavelength. Examples of photodetectors with internal gain are avalanche photodiodes or phototransistors. However, in general, too high an internal gain also results in increased noise. Therefore, designers usually make a compromise between gain and noise when designing and operating photodetectors with internal gain.

8.3.5 Noise and Noise-Equivalent Power

Electrical noise is normally de ned as random uctuations of an electrical signal, either current or voltage uctuations. The noise is usually evaluated as a root mean square (RMS) value of the current or the voltage. In a photodetector, the main sources of noise include shot noise from the photocurrent due to incident radiation (described in this subsection as *photon noise*) and, from background sources, icker noise (also known as *pink noise* or *1/f noise*); noise from the thermal agitation of carriers, especially in the parasitic resistances; dark current noise due to generation–recombination and leakage currents typically due to surface defects; and also multiplication noise in the case of avalanche photodiodes. The dark noise due to the thermally generated electron–hole pairs follows a Poisson relationship to dark current, and it is proportional to the square root of the number of thermal electrons generated within the exposure time and highly dependent on the operating temperature. Low operating temperatures dramatically reduce the dark current and hence the dark noise.

The shot noise I_{shot} is due to the random uctuation of the dark current I_{dark} that follows Poisson statistics, and it is given by

$$I_{shot} = \sqrt{2qI_{dark} \cdot \Delta f} \text{ in Amperes,} \tag{8.10}$$

and thermal (or Johnson) noise ($I_{thermal}$) due to the shunt resistance (R_{shunt}) is

$$I_{thermal} = \sqrt{\frac{4kT}{R_{shunt}} \cdot \Delta f} \text{ in Amperes.} \tag{8.11}$$

The total noise current is given by

$$I_{noise} = \sqrt{I_{thermal}^2 + I_{shot}^2}. \tag{8.12}$$

where q is the electronic charge (C), k is Boltzmann's constant (J/K), T is the absolute temperature (K), R_{shunt} is the value of resistance (Ω), and Δf is the bandwidth (Hz).

Typically, noise power scales with the area of the detector *and* its operating bandwidth. To decrease noise, lower bandwidth and smaller areas are better, but this results in lower quantum ef ciency and responsivity.

Example 8.8

Given a resistor that is 1 GΩ, determine its thermal noise current in a 1 Hz bandwidth at 300K. (Answer: 4 fA.)

Example 8.9

If a photodiode has noise currents due to shot noise in the incident radiation and other noise sources in the absorption region amounting to 1 nA, what is the total noise current in a 1 Hz bandwidth? Note that a leakage current of 1 nA gives the same shot noise current in a 1 Hz bandwidth. (Answer: 18 fA.)

In addition to the electrical noise described above, there is also photon noise that is inherent in the statistical nature of the photon arrival rate in the semiconductor. The electron–hole pairs generated in the semiconductor active region follow the Poisson statistical distribution of the phonon arrival rate in the semiconductor. This gives rise to a photon noise that is proportional to the square root of the photocurrent.

For some applications, it is required to know the minimum detectable light (power) of the photodiode. In this case, we de ne the noise-equivalent power (NEP′) as the minimum power in a 1 Hz bandwidth incident on a photodiode such that the photocurrent is equal to the total noise currents in the photodiode. In equation form, we express NEP′ as

$$NEP' = \frac{\text{Total Noise Currents } (A)}{\text{Responsivity } (A/W)} = \frac{I_{noise}}{\Re} \text{ in } W. \tag{8.13}$$

The lower the NEP, the better would the photodetector be able to detect low levels of incident radiation. Also, note that because we compute the noise in a 1 Hz bandwidth, the different frequency dependences are included; to remove these dependences, the NEP′ is divided by the square root of the bandwidth. If we did not include the 1 Hz bandwidth, then the resulting NEP would be expressed in W/$\sqrt{}$Hz. Another way of stating this is that the total noise current is expressed as the square root of total noise current power spectral density (A^2/Hz). So now, NEP becomes

$$NEP = \frac{\sqrt{\text{Total Noise Currents Power Spectral Density } (A^2/Hz)}}{\text{Responsivity } (A/W)} \left(W/\sqrt{Hz}\right). \quad (8.14)$$

Example 8.10

Calculate the NEP′ for a photodiode with a dark current of 2 nA, a shunt resistance of 500 MΩ, and a responsivity of 0.45 A/W (at a wavelength $\lambda = 0.6\,\mu$m). Assume that the bandwidth is 1 Hz.

The shot noise is $I_{shot} = \sqrt{2qI_{dark} \cdot \Delta f} = 25fA$.

The thermal noise is $I_{thermal} = \sqrt{\frac{4kT}{R_{shunt}} \cdot \Delta f} = 5.75fA$ at 300K.

The total noise current is $I_{noise} = \sqrt{I_{thermal}^2 + I_{shot}^2} = 26fA$.

The $NEP' = \frac{I_{noise}}{\Re} = \frac{26fA}{0.45A/W} \simeq 58fW$.

Note that if the bandwidth increases, then the NEP will also increase as $\sqrt{\Delta f}$.

A related performance measure is the detectivity *D*, de ned as the reciprocal of NEP. However, a more useful measure is the speci c detectivity *D**, that is, the detectivity normalized to the area of the photodetector. So we can write

$$D* = D\sqrt{Area \times \Delta f} = \frac{\sqrt{Area \times \Delta f}}{NEP}$$

$$= \sqrt{\frac{\text{Area}(\text{cm}^2) \times \Delta f(Hz)}{\text{Total noise current power spectral density } (A^2/Hz)}} \times \text{Responsivity } (A/W). \quad (8.15)$$

And the units of D^* are cm $\sqrt{Hz}/W$ or *Jones*. Note that D^* is independent of the photodetector area and a high D^* also implies a high sensitivity. Note that we can also express D^* as

$$D* = \sqrt{\frac{\text{Area}(\text{cm}^2) \times \Delta f(Hz)}{\text{Total noise voltage power spectral density } (V^2/Hz)}} \times \text{Responsivity } (V/W). \quad (8.16)$$

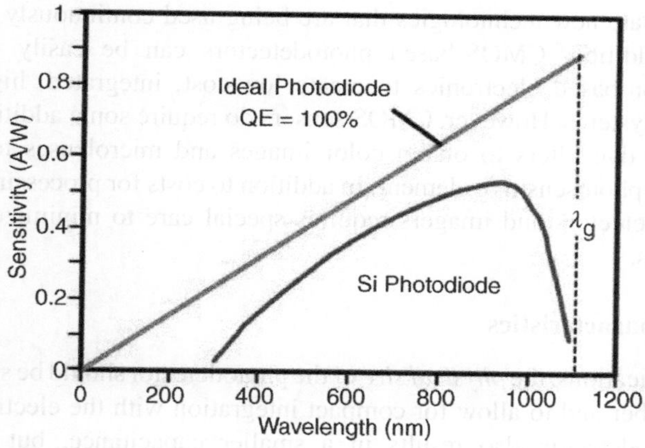

Figure 8.5 *Example of the sensitivity (i.e., responsivity) of a silicon photodiode.*

Note that the units would still be cm $\sqrt{Hz}/W$ because we are computing a voltage noise and expressing the responsivity as a voltage signal per unit incident illumination power.

8.3.6 Wavelength Sensitivity Range

The spectral range of the photodetector is the range of wavelengths that the absorption layer of the semiconductor material of the photodetector is sensitive to. The *spectral range* of a photodetector is de ned as the cutoff wavelength that is determined by the material of the absorption layer. Since the photodetector is sensitive only to the photons with energy larger than the band gap energy of the material of the absorption layer, there is a cutoff wavelength λ_{CO} for the sensitivity of a photodetector, and it has been de ned in Eq. (8.1). At the short-wavelength end, for example in the short-ultraviolet (UV) wavelength range, light is absorbed in a very thin layer (~100 nm at ~400 nm) of silicon, so the responsivity drops off. In addition, some of the electron–hole pairs that are generated will be lost due to surface recombinations that always occur due to surface damage of silicon. To minimize surface recombinations, a surface coating may be used, but this could result in re ection of part of the incident light. To minimize surface re ections, typically an antire ection coating may be used to enhance the ef ciency of light collection at a desired wavelength. Figure 8.5 shows a schematic representation of the responsivity or sensitivity (A/W) versus wavelength of a typical silicon photodiode. The straight line is for an ideal photodiode with a quantum ef ciency of 100% – in this case, R increases linearly with wavelength up to the cutoff wavelength. In a real diode, the quantum ef ciency is usually less than 100% and the limited absorption or surface recombinations cause R to also roll off at the shorter wavelengths.

8.3.7 Cost and Yield

A prime consideration in the development of commercial photodetectors and optical detection systems is cost. This is the major motivation for concentrating exclusively on silicon-based photodetectors. In fact, CMOS image sensors are typically fabricated in the same semiconductor plants as digital or mixed-signal chips, resulting in low costs and the ability to incorporate new technologies that are being used continuously for mainstream electronics. In addition, CMOS-based photodetectors can be easily integrated with mainstream silicon-based electronics to create low-cost, integrated, high-functionality optical detection systems. However, CMOS sensors do require some additional processing steps such as for dye lters to obtain color images and microlenses to improve light collection of each photosensitive element. In addition to costs for processing, the assembly of CMOS photodetectors and imagers requires special care to minimize contamination with dust particles.

8.3.8 Other Characteristics

In ber-optic applications, the *physical size* of the photodetector should be small for ef cient coupling to the ber and to allow for compact integration with the electronics. Having a small-sized photodetector also results in a smaller capacitance, but larger parasitic resistance.

Fidelity is a measure of how well the photodetector can replicate the received signal waveform. This is very important for analog transmission, and the response of the photodetector must be linear with respect to the incident optical signal over a wide range of input optical powers. In this way, the photodetector is able to recover the received optical pulses over a wide range of optical power. In digital systems, delity would not be critical because the digital applications are less sensitive to the uctuations in optical power. However, for gigabit systems, the delity becomes critical to preserve the signal integrity and avoid spurious bit detection.

Linearity refers to the output of a photodiode being linear with respect to the light intensity (also called *illuminance*) applied to the absorption region of the photodiode. At the low end of illuminance, the response is limited by the noise characteristics of the photodiode, and at the high end, by saturation of the photodiode. The range of responsivity versus illuminance over which the photodiode is linear is its *dynamic range*.

Low operating or supply voltages are important for the photodetector to be compatible with the supplies used for the CMOS electronics that follow. Low supply voltages are also required to reduce the power consumption of the photoreceivers, which is of signi cant concern for portable systems operating under a battery supply.

Stability is also important as the photodetector must be capable of continuous stable operation over the commercial temperature range for long times without signi cant degradation or change in its performance. However, the characteristics of photodetectors such as sensitivity, noise, internal gain, and bandwidth are usually temperature dependent. Therefore, for temperature-stable operation, additional circuits are required to change operating conditions so that the photodetector and associated electronics have acceptable performance as the ambient temperature changes. Also, good designs, proper process development, and tight control of fabrication technologies are required to produce stable photodetectors with desired performance and improved reliability and repeatability.

8.4 Examples of Types of Photodetectors

Semiconductor photodetectors can be broadly classi ed into two categories – those without internal gain and those with internal gain. Photodetectors without internal include *p–n photodiodes*, *p–i–n photodetectors*, *Schottky barrier photodetectors*, and *metal–semiconductor–metal photodetectors* (*MSM-PDs*). Photodetectors with internal gain include *photoconductors*, *phototransistors*, and *avalanche photodetectors* (*APDs*). The latter types of photodetectors are used in optical ber communications systems to improve the overall sensitivity of the front-end photodetector in the optical receiver. In this section, we discuss several of these types of photodetectors.

A *p–n photodiode* is a p–n junction diode operating under reverse-bias conditions. The incident photons are absorbed in both the depletion and the neutral regions, where the number of the generated electron–hole (e–h) pairs is proportional to the optical power. The e–h pairs get separated in the depletion region and drift under the effect of the applied electric eld. In general, the depletion region should be wide enough to achieve high conversion of light to e–h pairs – or to have a high quantum ef ciency. However, the thickness of the absorption layer should not be so large because the drift time through the depletion region is long, resulting in a low bandwidth of the photodiode. This illustrates one of the compromises between quantum ef ciency and speed for many photodetector structures.

An example of the cross-section and energy band diagram of a simple p–n photodiode is shown in Figure 8.6.

The *p–i–n photodetector* (PIN-PD) is one of the more popular types used in ber-optic communications. The basic PIN-PD consists of three regions, heavily doped p+ and n+ layers and an intrinsic or very lowly n- or p-doped layer that is sandwiched between them. The photon absorption takes place mainly in the intrinsic region that is depleted when a reverse-bias voltage is applied to the PD's terminals. The collection process of the photogenerated carriers is usually fast and ef cient, resulting in a high intrinsic bandwidth. Just like the p–n photodiode, there is a compromise between the quantum ef ciency and the bandwidth for a PIN-PD. But for an optimized design, the quantum ef ciency–bandwidth

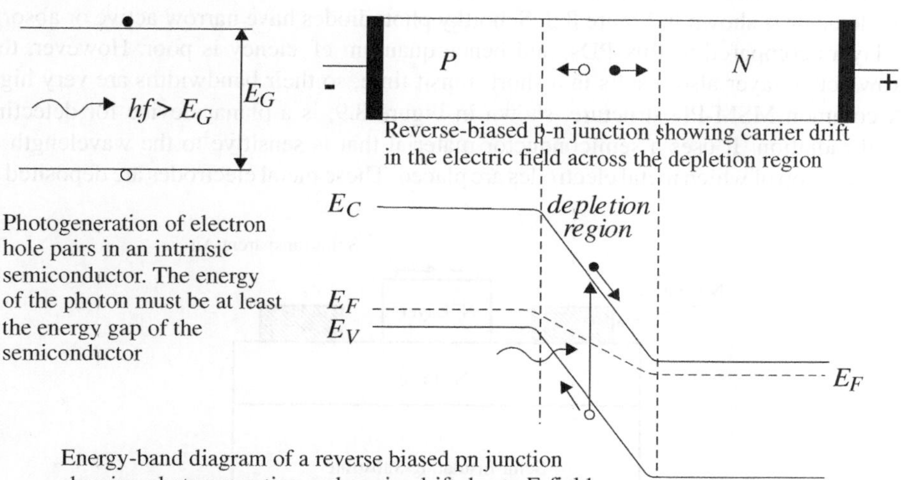

Figure 8.6 *Schematic representation of the energy band diagram of a semiconductor absorbing light, cross-section of a simple p–n junction, and photodetection in a p–n diode.*

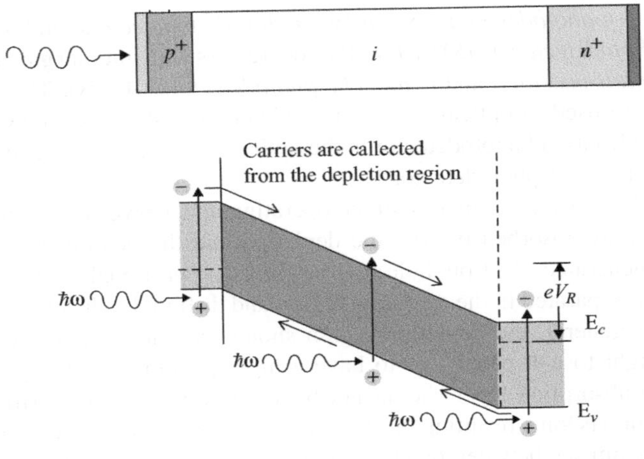

Figure 8.7 Cross-section and energy band diagram of a simple p–i–n photodiode.

product (η-BW) is nearly constant. Later, we will explain how this compromise can be broken by having the light incident perpendicular to the transport of charges across the depleted region or with resonant cavity structures. The fabrication of PIN-PDs is relatively easy with well-established processes, and the fabricated devices are very reliable and have low noise. However, this option may not be possible in a standard CMOS process because an intrinsic or lowly doped layer may not be available.

The cross-sectional view and energy band diagram of a simple p–i–n photodiode are shown in Figure 8.7.

A *Schottky barrier photodiode* is a metal–semiconductor–metal (MSM) rectifying junction rather than a p–n semiconductor junction. A schematic structure of a Schottky photodetector is shown in Figure 8.8. Schottky photodiodes have narrow active or absorption layers compared to PIN-PDs, and hence quantum ef ciency is poor. However, this narrow active layer also results in a short transit time, so their bandwidths are very high.

A common MSM-PD structure, shown in Figure 8.9, is a planar device for detecting optical radiation. It uses a semiconductor material that is sensitive to the wavelength of interest, on top of which metal electrodes are placed. These metal electrodes are deposited as

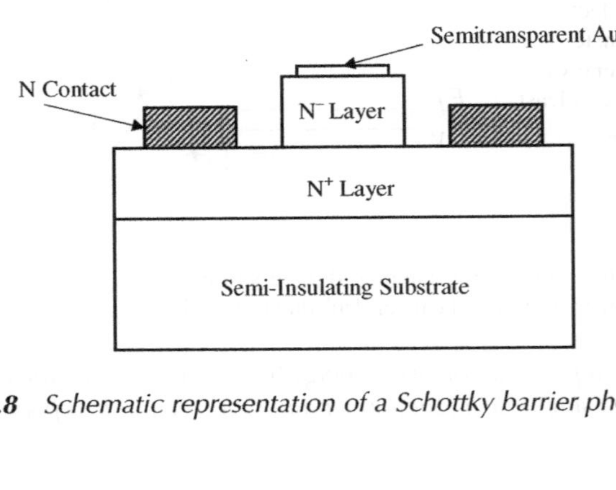

Figure 8.8 Schematic representation of a Schottky barrier photodetector.

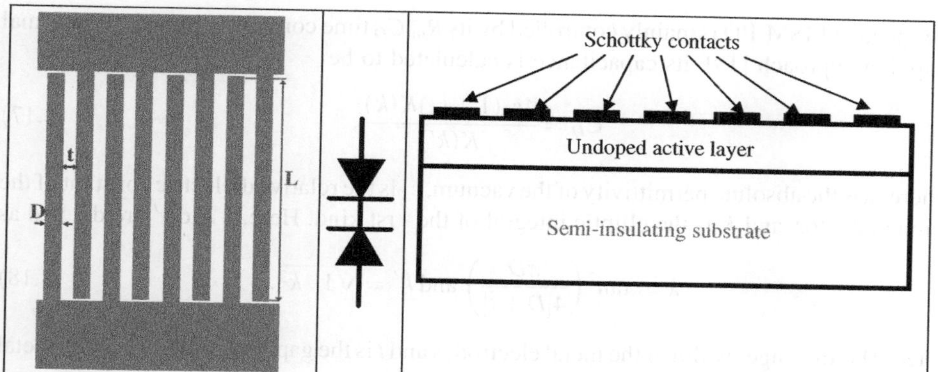

(a) (b) (c)

Figure 8.9 *(a) Top view of a MSM-PD; (b) a circuit schematic of a MSM-PD; and (c) a schematic structure of a MSM-PD.*

interdigitated ngers to form back-to-back Schottky diodes. Each set of electrodes forms a Schottky barrier contact with the semiconductor, and is connected to a large pad for connection to the external circuit. Therefore, a MSM-PD is a variation of a Schottky barrier photodetector with both contacts made on the same side of the substrate. MSM photodetectors have very high speeds due to their low capacitance. However, their responsivity is low compared to p–n or p–i–n photodiodes because of re ections from the metal and semiconductor surfaces, a nite carrier lifetime as they transit the gap between the electrodes before being collected, and recombination currents due to surface defects and deep traps within the semiconductor material. Their performance can be improved with a suitable antire ection coating on the surface and passivation of surface defects.

Like a p–n junction, a Schottky diode exhibits recti ed I–V characteristics. However, a Schottky barrier occurs only for certain combinations of metal–semiconductor junctions. The semiconductor doping level controls the potential barrier of a p–n junction. However, in a Schottky diode, the potential barrier (φ_b) is controlled by the metal and semiconductor work functions. The key difference between a p–n and a Schottky diode is that p–n junctions allow both electrons and holes to ow under forward bias, while a Schottky diode is a majority-carrier device.

The MSM-PD also has the issue of compromise between quantum ef ciency and bandwidth. Its quantum ef ciency can be improved by using a multi nger arrangement. However, its bandwidth can be increased only by reducing the effective absorption layer thickness. One solution is to restrict the electric eld within a certain absorption layer near the surface by introducing a highly doped layer at a certain depth. Another problem of the MSM-PD is that its impulse response shows a tail due to the photogenerated carrier distribution away from the surface. But the slow tail response can be reduced by tailoring the carrier lifetime in the material so that the lifetime is almost equal to the transit time between the electrodes. However, this calls for a specialized process and so may not be possible in standard CMOS technology.

Current MSM-PD technology can provide very thin ngers and narrow nger spacing. Thus, the transit time can be made very small, but the limiting factor is its capacitance and the speed of MSM-PD is mainly controlled by its $R_{tot}C_D$ time constant. Using the conformal mapping approach [19], its capacitance is calculated to be

$$C_D = \frac{4\varepsilon_0(1+\varepsilon_S)K(k)}{K(k')}. \tag{8.17}$$

where ε_0 is the absolute permittivity of the vacuum, ε_S is the relative dielectric constant of the semiconductor, and K is the elliptic integral of the rst kind. Here, k and k' are de ned as

$$k = \tan^2\left(\frac{\pi D}{4[D+t]}\right) \text{ and } k' = \sqrt{1-k^2}. \tag{8.18}$$

where D is the nger width of the metal electrodes and t is the gap spacing between the metal ngers.

A simple analytic expression for the capacitance of a MSM photodiode was provided in [20], which is accurate to within a small percentage of the more complicated Eq. (8.17), is

$$C_D = 0.226NL\varepsilon_0(\varepsilon_S+1)(6.5\theta^2+1.08\theta+2.37), \tag{8.19}$$

where L is the length of a nger. The parameter $\theta = (D+D/t)$ is the nger-to-period ratio. Now, the MSM photodiode's external quantum ef ciency is given by

$$\eta = (1-R)\frac{t}{t+D}(1-e^{-\alpha d}), \tag{8.20}$$

where R is the re ection coef cient, α is the absorption coef cient, and d is the thickness of the active or absorption layer of the semiconductor. Note that one method to increase η is to replace the metal ngers with transparent metal ngers such as indium tin oxide (ITO). Also, η can be increased by reducing the nger width, but then the resistance would increase, so some optimization is required.

Example 8.11

Given an MSM photodetector with $W_r = t = 0.5\,\mu\text{m}$, and 40 ngers, each being $10\,\mu\text{m}$ long (L). Its capacitance can be calculated to be 24fF, which is a very small value in comparison to other conventional photodetectors. From a design perspective, this value can be further reduced by using a series-parallel con guration of the ngers, and it can become as low as 6fF [19].

Avalanche photodiodes (APDs) are the most common photodetectors with internal gain that are used in optical ber communications systems. Because of their internal gain they are commonly used for the detection of optical radiation of extremely low intensities [21–25]. Many variations of APDs using Si, Ge, or III–V semiconductor materials are possible. Its internal gain arises from an avalanche multiplication process caused by impact ionization events. The impact ionization phenomenon has been extensively investigated both theoretically and experimentally [26–30]. APDs are typically operated under a suf ciently high reverse voltage to generate highly energized carriers. Under a high electric eld in the conduction band, the high-energy electrons initially scatter with a bound electron in the valence band and ionize it (or knock it out into the conduction band), resulting in the creation of an electron–hole pair. Now, these two electrons plus the hole, again travelling under the influence of the electric field, can impact ionization to create two additional electrons and a hole. Continuing this impact ionization process, a significant multiplication in the number of current-carrying charges through the avalanche process is eventually obtained. This process terminates when all the free carriers are swept out of the high–electric field region. The same phenomenon could also happen to high-energy holes that excite extra holes in the valence band. In the end, one initial electron or hole generates M extra e–h pairs, where M is called the *multiplication gain* of the photodetector.

Note that impact ionization is not the only energy loss mechanism for high-energy carriers. Electrons and holes can lose energy in nonionizing collision processes such as phonon scattering. The required minimum carrier energy for causing impact ionization is the ionization threshold energy that should be larger than the band gap energy. Also, the ionization coefficients of electrons α_i and holes β_i, defined as the inverse of the mean distance between ionization collisions, are two other important parameters [31].

In an APD, the statistical variation of the multiplication process is responsible for multiplication excess noise. In fact, the multiplication M which increases the photocurrent also gives rise to excess multiplication noise F that is dependent on the ratio k between the ionization coefficients of both photogenerated carriers ($k = \alpha_i/\beta_i$). The excess multiplication noise F is given by

$$F = kM + (1-k)\left(2 - \frac{1}{M}\right) = M\left\{1 - \left[(1-k)\left(\frac{M-1}{M}\right)^2\right]\right\}. \tag{8.21}$$

Note that for gains >10 and low k-values <0.1, Eq. (8.21) can be written as

$$F \simeq 2 + kM. \tag{8.22}$$

The total dark current in an APD is

$$I_d = I_s + I_{dm}M. \tag{8.23}$$

where I_s is the unmultiplied surface current and I_{dm} is the multiplied component of the dark current.

If the primary DC photocurrent in an APD is I_{ph0}, the photo-induced background current is I_{bk}, and the bandwidth is B, then the total mean square current noise spectral density is given by

$$\langle I_M \rangle^2 = 2q\left[I_s + \left(I_{dm} + I_{bk} + I_{ph0}\right)M_0^2 F\right]B. \tag{8.24}$$

Therefore, for low excess noise, the carrier with a higher ionization rate should be injected into the multiplication region. Thus, using one carrier with a much lower ionization rate compared to the other is preferred. With the definition of $k = \alpha_i/\beta_i$, injection of electrons and a lower hole ionization rate are preferred. Also, note that the ratio k of ionization coefficients is chosen so that it is less than 1. Now we can write an expression for the signal-to-noise ratio (SNR) of APDs as

$$SNR = \frac{M^2 I_{ph0}^2}{\langle I_M \rangle^2} = \frac{M^2 I_{ph0}^2}{2q\left[I_s + \left(I_{dm} + I_{bk} + I_{ph0}\right)M^2 F\right]B}. \tag{8.25}$$

Note that this expression neglects the thermal noise of parasitic resistance(s) R which would add a term $(4kT/R)$ to the denominator.

For an APB with a breakdown voltage of V_B and a total effective resistance of R biased at a voltage V_a, the multiplication M is given by the empirical relation

$$M \simeq \frac{1}{\left\{1-\left(\frac{V_a - I_{ph0}R}{V_B}\right)^n\right\}}. \tag{8.26}$$

where n is a constant that depends on the design of the device.

An APD provides gain without the need of an ampli er, that is, it functions as both a photodetector and an ampli er combined into one device. An important limitation of an APD is its bandwidth. Because of the long avalanche buildup time, that is, the time for the carriers to gain suf cient energy in the electric eld to impact ionize, the bandwidth of APDs is small. However, they have important photodetectors because their internal gain decreases the number of repeaters used in long-haul communications systems or in dense wavelength-division multiplexing systems. To improve the high-frequency performance of APDs, some special structures will be described later.

A *bipolar junction transistor* can be used as a *phototransistor.* Its structure is similar to a normal n–p–n or p–n–p bipolar transistor. However, in this case, only two terminals have electrical contacts – the collector and the emitter. The base is left oating. The base and the base–collector junction regions are used as the absorption layer for the incident light. For an n–p–n bipolar transistor, the photogenerated holes in the absorption region accumulate in the base. This excessive charge results in electrons injected from the emitter, and the current gain mechanism is the same as in a normal bipolar transistor.

The *MOS phototransistor* is sometimes also called a *photogate transistor.* Generally, in photogate devices, the incident light changes the surface potential under the MOS gate oxide, which in turn modulates the current owing between the source–drain terminals. Another name used to describe the photogate or MOS phototransistors is a charge modulation device (CMD).

8.5 Examples of Photodiodes in Standard Silicon Technology

We begin by illustrating the richness in possibilities for the types of photodiodes that can be made with standard or mainstream CMOS technology. In particular, we discuss several possibilities in a standard 0.18 μm CMOS technology as illustrations of what may be possible. Typically, this technology offers many layers such as n+, p+, n-well, p-well, and p-substrate that could be suitably combined to make photodiodes. Figure 8.10 illustrates several of these photodiode options.

In general, the responsivity of these diodes varies because the depletion regions of the junctions are at different distances from the surface of the silicon substrate and the doping concentrations are different. Deeper junctions are more responsive to longer wavelengths of incident radiation. Also, to a rst order, the responsivity depends on the volume of the depletion region which also depends on the doping concentrations and doping pro les. The photocurrent I_{PC} in a junction photodiode can be written as

$$I_{PC} = G_{OP}\left(qA\left[W_D + L_n + L_p\right]\right)\ (in\ A). \tag{8.27}$$

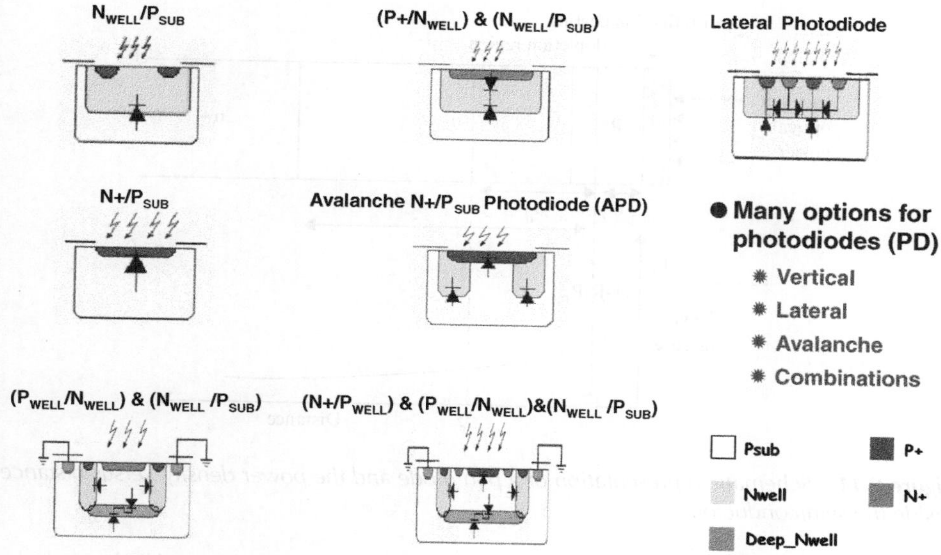

Figure 8.10 *Some of the possibilities of photodiodes in standard CMOS technology. Note the differences in doping concentrations as well as junction depths which would contribute to varying optimum wavelength responses and quantum efficiencies.*

In Eq. (8.27), G_{OP} is the generation rate of electron–hole pairs from the absorbed light, A is the photosensitive area of the photodiode, and W_D is its depletion width. Note that L_n and L_p are the electrons' and holes' diffusion lengths in the p-type and n-type layers, respectively (these are termed the *minority-carrier diffusion lengths*); they are included because the excess carriers in the neutral regions will have a diffusion current component and contribute to the total photocurrent. Note that the generation rate is given by

$$G_{OP} = \frac{1}{A}\frac{dP}{dx}\frac{1}{h\nu}, \tag{8.28}$$

where P is the incident optical power which changes with distance into the semiconductor because of the absorption coef cient changes and $h\nu$ is the energy of a photon.

A similar expression to Eq. (8.27) can be written for the dark current due to thermal generation of electron–hole pairs with G_{OP} replaced with G_{TH}, the thermal generation rate. Note that Eq. (8.27) assumes a constant photocurrent generation rate, which is only an approximation. If we consider the p–n diode shown in Figure 8.11, then the power per unit time, entering x_1 and leaving x_2, is given by

$$G_{OP} = (1-R)\left(\int_{x_1}^{x_2} G_0 e^{-\alpha x}\cdot dx\right) = (1-R)G_0(e^{-\alpha x_1} - e^{-\alpha x_2}). \tag{8.29}$$

where G_0 is constant and α is the absorption coef cient. For simplicity, we can assume a constant generation rate in the photodiode, and hence the photocurrent from the depletion region $I_{PC} = eG_{OP}W_D$.

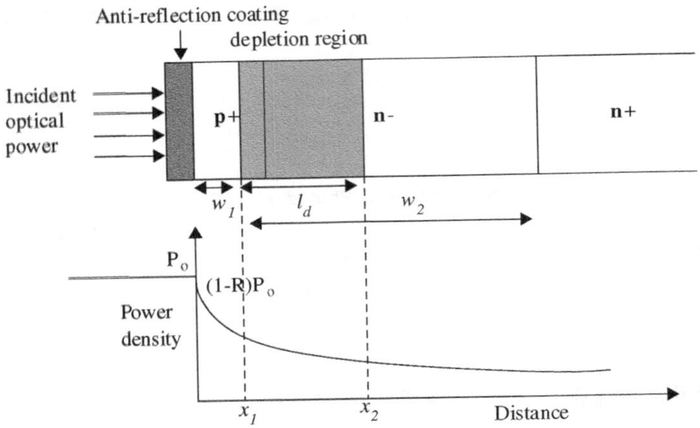

Figure 8.11 *Schematic representation of a p–n diode and the power density versus distance inside the semiconductor.*

As mentioned earlier in this chapter, the excess minority carriers generated in the neutral n and p regions will have a diffusion current component and contribute to the total photocurrent. The diffusion current can be obtained by solving the ambipolar transport equation to get the excess carriers' distribution. The excess-minority carrier electron distribution in the p-region can be obtained from

$$D_n \frac{\partial^2(\delta n_p)}{\partial x^2} + G_{OP} - \frac{\delta n_p}{\tau_{n0}} = \frac{\partial(\delta n_p)}{\partial t} \tag{8.30}$$

where the electric eld is assumed to be zero in the neutral region, D_n is the diffusion coef cient of electrons in the p-region, τ_{n0} is the minority carrier (electron) lifetime, and δn_p comprises the excess-minority carriers (electrons) in the p-region due to illumination. Equation (8.29) can be solved at steady state and assuming uniform generation rate, as in [32], and hence the excess electrons in the p-region are given by

$$\delta n_p = G_L \tau_{n0} - (G_L \tau_{n0} + n_{p0}) e^{-x/L_n} \tag{8.31}$$

where $L_n = \sqrt{D_n \tau_{n0}}$ is the minority-carrier diffusion length (electrons in the p-region) and n_{p0} is the number of electrons in the p-region in equilibrium. The gradient in the minority concentration will create diffusion current density J_n in the p–n junction, which is calculated from

$$J_n = qD_n \frac{d(\delta n_p)}{dx}\bigg|_{x=0} = qG_{OP}L_n + \frac{qD_n n_{p0}}{L_n}. \tag{8.32}$$

The rst term is the diffusion component of the photocurrent, while the second term is the normal reverse-saturation current which is called the *dark current*. The same approach can be taken to get the diffusion current component of the excess hole concentration in the n-region. The total photocurrent is the sum of the space charge region drift current and the diffusion currents of the excess electrons and holes in the neutral n- and p- regions. The total photocurrent is then given by

$$J_{PC} = qG_L W_D + qG_{OP} L_n + qG_{OP} L_p = qG_{OP}(W_D + L_n + L_p). \quad (8.33)$$

The dark-current density J_{dark}, the current under zero illumination, which is the reverse-saturation current, is given by

$$J_{dark} = \frac{qD_n n_{p0}}{L_n} + \frac{qD_p p_{n0}}{L_p}, \quad (8.34)$$

where p_{n0} is the number of holes in the n-region in equilibrium, and $L_p = \sqrt{D_p \tau_{p0}}$ is the minority–carrier diffusion length (holes in the n-region).

We notice from the physical operation of the p–n photodiode that a large part of the optical power will be lost in the neutral n and p regions. The excess carriers generated will recombine after diffusing a distance that equals a diffusion length and hence do not contribute to the photocurrent. The excess carriers that do contribute to the photocurrent are the generated carriers, mostly within one diffusion length of the depletion region edges next to the neutral regions. Therefore, carrier recombination effectively decreases the sensitivity of the device and also reduces the quantum ef ciency. Furthermore, carrier diffusion in the neutral region is slower than carrier drift current through the depletion region. The effect of these different carrier transport times is that the diffusion component of the current affects the frequency response and limits the maximum allowable modulation frequency that can be used. To get around this problem, usually, the p–n photodiode is replaced with a p–i–n photodiode because it has higher sensitivity and better frequency response than the p–n photodiode.

Example 8.12

Calculate G_{OP} for a light source illuminating a p–n diode of area $(100\,\mu\text{m})^2$ with $1\,\mu$W of light. Assume that the light is ltered, its wavelength is 800 nm, and all the incident photons are converted into electron–hole pairs.

The light intensity is the light power per unity area, $1\,\mu W/(100\,\mu\text{m})^2 = 10^{-2}\,W/\text{cm}^2$. The photon energy is given by

$$hv = hc/\lambda = (6.62 \times 10^{-34}\,\text{Js} \times 3 \times 10^{10}\,\text{cm/s})/800 \times 10^{-7}\,\text{cm} = 2.48 \times 10^{-19}\,\text{J}.$$

$$\text{The photon ux } \Phi = \frac{\text{Light Intensity}}{\text{Photon Energy}} = \frac{10^{-2}\,W\,\text{cm}^{-2}}{2.48 \times 10^{-19}\,\text{J}} = 4.03 \times 10^{16}\,\text{Photons}/(\text{cm}^2 \cdot \text{s}).$$

Since each photon generates one electron–hole pair (EHP), then $G_{OP} = 4.03 \times 10^{16}\,\text{EHP}/(\text{cm}^2 \cdot \text{s})$.

For a p–n junction with abrupt doping pro les, the depletion width W_D is given by

$$W_D = \sqrt{\frac{2\varepsilon_0 \varepsilon_{Si}(V_{BI} - V)}{q}\left(\frac{1}{N_A} + \frac{1}{N_D}\right)}\ (in\ \text{m}). \quad (8.35)$$

where, with ε_0 being the permittivity of free space (8.854×10^{-12} F/m) and ε_{Si} the relative permittivity (11.9) of silicon, V_{BI} is the built-in voltage of the junction, V is the applied bias (it would be negative for a reverse bias), and N_A and N_D are the doping concentrations of the p- and n-type materials in the junction. The built-in voltage is given by

$$V_{BI} = \frac{kT}{q}\ln\left(\frac{N_A N_D}{n_i^2}\right) \ (in\ V). \tag{8.36}$$

where kT/q is the thermal voltage, k is Boltzmann's constant (1.38×10^{-23} J/K), T is the absolute temperature, and n_i is the intrinsic carrier concentration ($\sim 1.4 \times 10^{10}\ \text{cm}^{-3}$ at room temperature). Finally, the capacitance of the photodiode C_J (operated with a reverse-bias voltage) is given by

$$C_J = \frac{\varepsilon_0 \varepsilon_{Si} A}{W_D} \ (in\ F). \tag{8.37}$$

From Eqs. (8.17), (8.19), (8.27), (8.35) and (8.36), we note that the characteristics of the photodiode are very dependent on the depletion width, with a larger depletion width due to lower doping concentrations or higher reverse biases resulting in higher photocurrents because of an increased light-capturing area. Also, the lower junction capacitance results in a lower *RC* time constant. This may seem like a win–win situation, but the wider depletion width also results in higher *carrier transit times*, de ned as the time it takes the electrons and holes to drift across the depletion region. Thus, there is a compromise between the responsivity given by Eqs. (8.7), (8.17) and (8.19). The carrier transit time τ given by

$$\tau = \frac{W_D}{v_{drift}} = \frac{W_D}{\mu E} \simeq \frac{W_D^2}{\mu V} \ (in\ s), \tag{8.38}$$

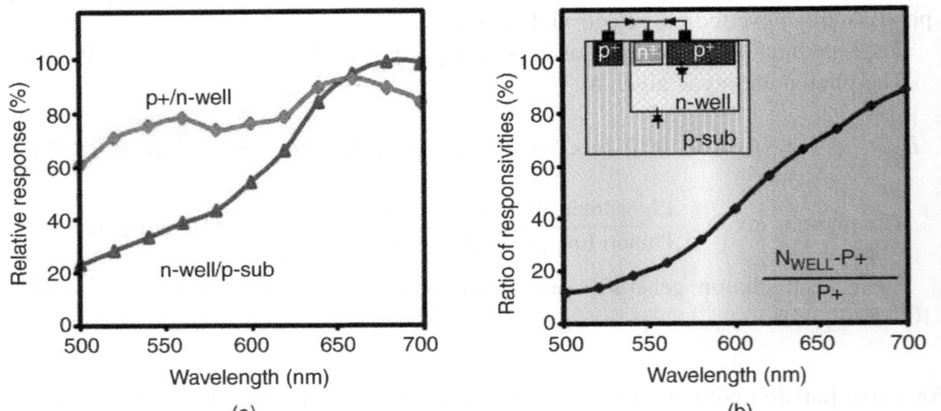

__Figure 8.12__ Wavelength detection using two photodiodes with different junction depths. (a) The responsivity of the two photodiodes (shown in the inset of (b)); and (b) a ratio of the difference in responsivities divided by the responsivity of the shallow photodiode.

where v_{drift} is the drift velocity of the carrier, μ is its mobility, E is the electric eld across the depletion region, and $E \sim V/W_D$ where V is the reverse-bias voltage on the diode. From Eq. (8.38), to achieve a short carrier transit time, the depletion width should be narrow and the reverse voltage high. As a rough guide, the depletion width is usually selected according to

$$\frac{1}{\alpha} < W_D < \frac{2}{\alpha}. \tag{8.39}$$

For a combinations of diodes, for example a "deep" diode (one in which the junction depth is far below the surface) combined with a "shallow" diode can be used as a wavelength detector by taking a suitable ratio of the responsivities, if it is calibrated at one wavelength. This is illustrated in Figure 8.12 which shows the measured relative responsivities of a shallow p + /n-well diode and a deep n-well/p-subdiode. By simultaneously measuring the photocurrents from both diodes, and then taking an appropriate mathematical combination of the two photoresponses, a ratio can be obtained to indicate the wavelength of incident light. In this case, the ratio is the difference in the two photocurrents normalized to the p + /n-well responsivity.

8.6 Phototransistors in Standard Silicon Technology

By using the p + /n-well p-substrate, the p + /n-well polysilicon gate, or the n + /p-well p+ vertical or combined vertical-lateral phototransistors that are only bipolar, metal oxide semiconductors (MOS) or a combination of bipolar and MOS transistors can be made. Three examples are illustrated later for a vertical bipolar phototransistor, a combined vertical-lateral bipolar phototransistor, and a MOS phototransistor. Generally, bipolar transistors offer better photoresponse characteristics because they are bulk-transport rather than surface-controlled devices as in the case of a eld effect transistor.

In a popular bipolar phototransistor structure, the base is left oating and the collector–base junction has a large window through which the light is incident. In this con guration and if it is a p–n–p transistor (collector (p), base (n), and emitter (p)), then the photogenerated holes in the reverse-biased collector–base will be swept to the collector terminal, giving rise to the photocurrent. The electrons from the photogenerated electron–hole pairs, on the other hand, will be swept into the base, causing the emitter–base potential to rise. This increase in emitter–base potential, in turn, causes more holes to be injected from the emitter into the base, where they diffuse to the collector, thus causing an increase in the collector current. Note also that since the base terminal is oating, current conservation requires that the emitter and collector currents be equal. If β is the common-emitter current gain, then the gain of this phototransistor will be $(1 + \beta)$.

In a typical CMOS technology, the availability of thin- and thick-oxide options allows for the design of bipolar and MOS phototransistors with different responsivities because of the differing electrical characteristics between these transistors using different gate oxide thicknesses. Examples of the opto-electronic responses of these different types of phototransistors are illustrated in Figure 8.13.

A summary of typical performance characteristics in a 0.18 μm CMOS technology is shown in Figure 8.14 [33].

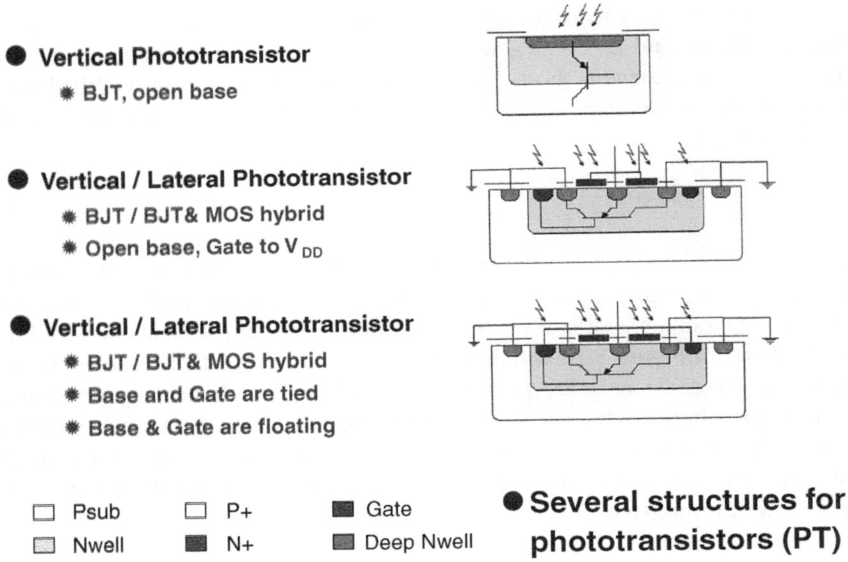

Figure 8.13 *Examples of possible phototransistor structures in a standard CMOS technology.*

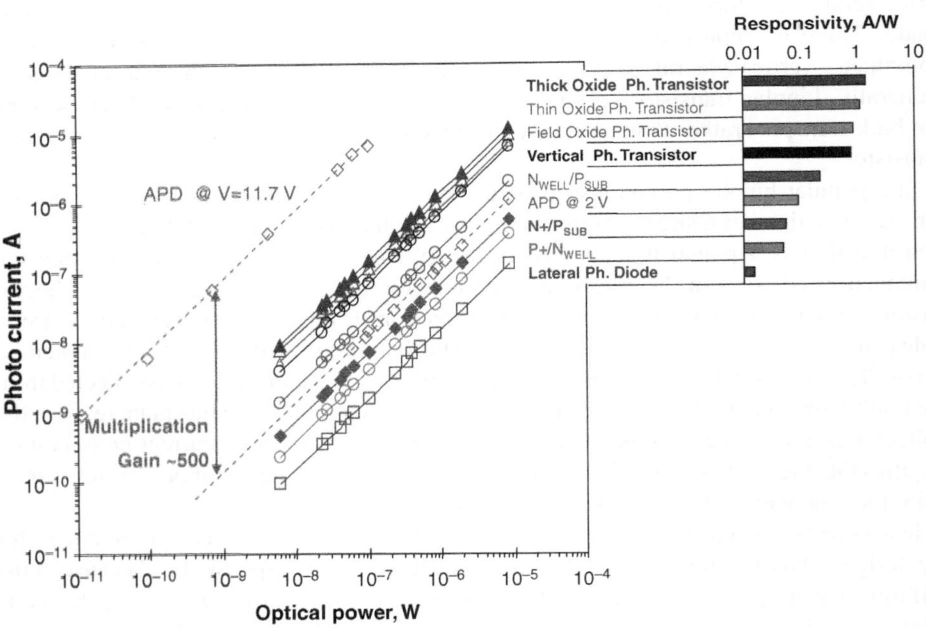

Figure 8.14 *Photocurrent versus optical power for different photodetector structures in a 0.18 μm standard CMOS technology. Also shown is the responsivity values at the top-right side of the figure.*

8.7 CMOS and BiCMOS

The ability to realize a high-performance silicon-based photodetector in a production CMOS process is an attractive option for systems designers, but it is not easy to accomplish. This fact, in combination with the costs associated with optical packaging and the challenges of making high-performance CMOS receiver circuits, has challenged researchers and systems developers working to make fully integrated CMOS photoreceivers. The motivation and potential to produce such elements in CMOS have increased markedly as the speed of CMOS circuits has made them viable candidates for gigabit rate data communication. The integration of gigabit receivers became feasible in submicrometer CMOS [34].

Considering the CMOS vertical structure, three factors determine the responsivity. These factors are in the three square brackets in the following equation:

$$\frac{I_{ph}}{P_{opt}} = \left[\prod A(1-R)\right] \times \left[1 - \frac{\exp(-\alpha W_D)}{1+\alpha L_{diff}}\right] \times \left[\frac{q\eta}{hv}\right] \tag{8.40}$$

The rst square bracket denotes optical attenuations $A \approx \exp(-\alpha_{SiO2} W_{SiO2})$ and re ections R by the layers above silicon, where α_{SiO2} and W_{SiO2} are the absorption coef cient and thicknesses of insulating silicon dioxide layers that hold the metallization in CMOS.

The second brackets denote the optical absorption in a silicon-depleted layer of thickness W_D, where the Beer–Lambert law is assumed for the absorption coef cient $\alpha(\lambda)$ of silicon and L_{diff} is the diffusion length of minority carriers.

The last bracket denotes the responsivity of silicon material, where h is Plank's constant, hv is the photon energy, q is the electron charge, and $\eta(\lambda)$ is the quantum ef ciency of silicon that is also a function of wavelength.

By the downscaling of CMOS, the number of metallization layers and their thicknesses increases W_{SiO2}, which attenuates the optical signals prior to reaching the photodetector active region. As the absorption coef cient of silicon $\alpha(\lambda)$ is material property, then one desires suf cient depletion widths W_D, so that $(\alpha \times W_D) > 3$. However, W_D is determined by the doping of wells, contact, and other junctions, as optimized for anticipating the electrical parameters of MOS transistors. In general, the doping is high in downscaled CMOS technologies to remedy effects such as drain-induced barrier lowering and maintain the threshold voltages in proportion with the supply. Thus, W_D is in the range of fractions of micrometers for well junctions and less than 100 nm for contact junctions that are used for the drain or source terminals. Since $\alpha \sim (10^{-3}$ to $1)\ \mu m^{-1}$, then the product $(\alpha \times W_D) < 1$, and the design of a photodetector in CMOS is quite dif cult, since the exponential term in the middle square brackets in Eq. (8.40) is not negligible.

As discussed in [35], the absorption coef cient of silicon $\alpha(\lambda)$ is given as a function of carrier concentration. A comparison of the feasibility of different diode and transistor structures for photo-detection is given in [36] in relation to materials, device properties, and speed.

8.8 Silicon-on-Insulator (SOI)

SOI-CMOS has advantages producing high-speed and low-power circuits that allow for extensive signal processing in miniature systems on a chip. The integration of photodetectors in SOI is desired for several reasons. One reason is to build imager arrays with in-pixel processing. Another reason is to replace electrical interconnections with optical

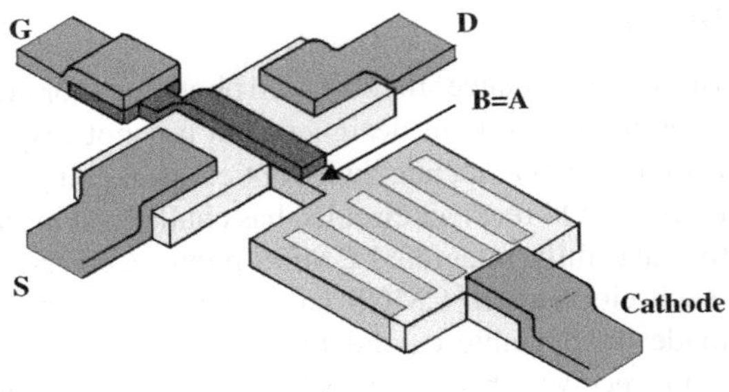

Figure 8.15 Partially depleted SOI structure consisting of an nMOS transistor with an extension of the transistor body as the anode of a lateral photodiode [37]. The photodiode is arranged as interdigitized (comb-like) fingers of anodes and cathodes, and operates in the photovoltaic regime, with the photovoltage controlling the body potential of the MOS transistor and the photocurrent serving as a base current for the parasitic lateral bipolar transistor (source–body–drain). Reproduced with permission from [37]. Copyright (2002) Japan Society of Applied Physics.

ones, since the parasitic capacitances, inductances, and resistance of electrical interconnections are presently major performance limiters at the chip level.

The semiconductor layers in SOI are thin, normally in the range of 20–200 nm for fully or partially depleted SOI. Therefore, vertical p–n junction photodiodes have low responsivity, since $\alpha W_D << 1$, and the term in the middle brackets of Eq. (8.40), is a small fraction of 1. To remedy the low responsivity, one employs the advantage of SOI to build ampli cation structures.

One option is electrical ampli cation, which is implemented in several ways. One way is to tie a relatively large-area photodiode with the body of partially depleted SOI [37], as shown for example in Figure 8.15. Even at low responsivity, since the leakages and capacitances of the structure are relatively small, the photo generation creates signi cant photovoltage (0.3–0.5 V) at the anode (A) of the photodiode, which is applied to the body of the negative-channel MOS (nMOS) transistor and decreases the threshold voltage approximately with the same value and, consequently, the nMOS transistors' turns conducting. In addition, since the gate length is small, then the body also serves as the base of the lateral BJT (S–B–D = n–p–n), and the small photocurrent is ampli ed by the current gain β of the BJT. This approach of controlling the body of the MOS transistor has extraordinarily high effective current gain ($\sim 10^4$ and more) at low photocurrents and logarithmic-like decrease of the gain at high photocurrents.

Further increase of the gain can be achieved by connecting the oating gate of the MOS transistor to the body [38, 39]. Figure 8.16a illustrates the logarithmic dependence of the body potential as function of the illumination, which is a result of the voltage drop of the primary photocurrent on the body–source junction. This variation of the voltage drop, together with the current gain of the lateral BJT, causes larger variations in the drain current on the order of several decades, when compared with the primary body photocurrent, as seen in Figure 8.16b.

In fully depleted SOI, the thickness of silicon is low, for example 50 nm, and one does not have an opportunity to contact the body of the MOS transistor. Consequently, a

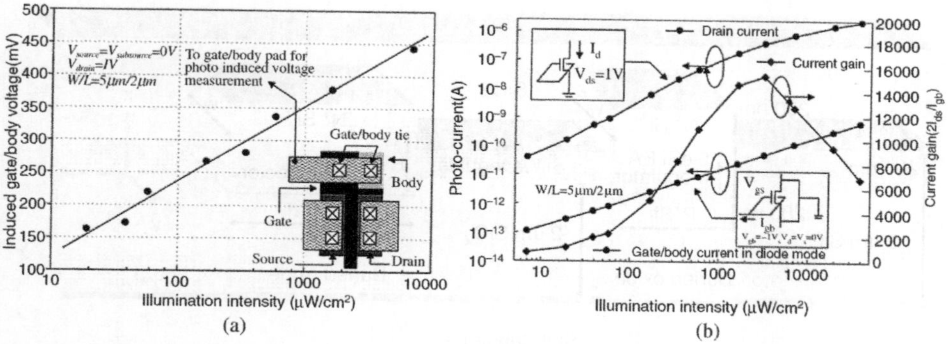

Figure 8.16 Performance of a gate-body tied partially depleted SOI MOSFET (W/L = 5 μm/2 μm) as a photodetector. (a) Drain and body currents, and their ratio as the gain; and (b) a shift of body and gate potentials as a function of illumination. Reproduced with permission from [39]. Copyright (2000) Elsevier Ltd.

photo-generated majority for the body carriers, holes for nMOS, are trapped in the body, causing threshold shift and activation of the lateral bipolar transistor, resulting in high dark current. To remove these majority carriers from the body, it is suggested in [40] that one can use charge pumping to bring the majority carriers to the gate oxide interface and recombine with surface states (see Figure 8.17). The procedure is not simple, as it requires one to

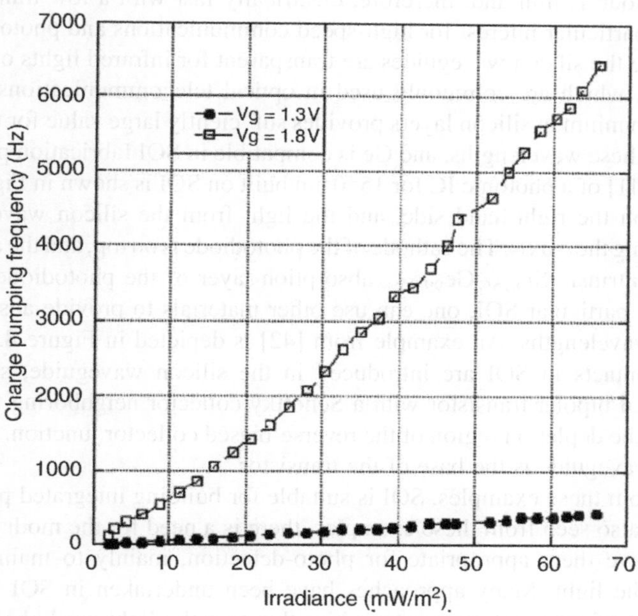

Figure 8.17 Charge-pumping frequencies that compensate for the photo-generated majority carriers in the body of a fully depleted SOI, and bring the transistor current as a dark condition [40]. Reproduced with permission from [40]. Copyright (2008) Elsevier Ltd.

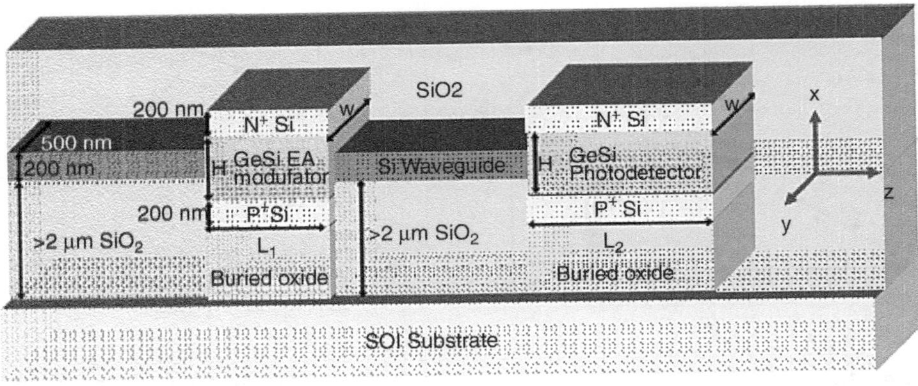

Figure 8.18 *An optical modulator connected with a silicon waveguide to a p–i–n photodiode with 99.25% germanium content in the intrinsic region [41]. Reproduced with permission from [41]. Copyright (2007) The Optical Society.*

increase the charge pumping frequency at higher illuminations and gate biases, but the procedure allows one to extend the detectivity of fully depleted SOI-MOS almost a decade toward lower illuminations and to apply readout techniques with conversion to frequency.

A second option for an increase in the responsivity of SOI photodetectors is to pass the light laterally, so that the lateral photodiode has large width for light absorption, while vertically the diode is thin and, therefore, electrically fast with a low transit time. This approach is of particular interest for high-speed communications and photonic integrated circuits, because the silicon waveguides are transparent for infrared lights of wavelengths 1.3 and 1.5 µm, which are commonly used in optical telecommunications, whereas the addition of germanium in silicon layers provides suf ciently large value for the absorption coef cient α at these wavelengths, and Ge is compatible in SOI fabrication processes. One example from [41] of a photonic IC for 1550 nm built on SOI is shown in Figure 8.18. The photodiode is on the right-hand side, and the light from the silicon waveguide travels horizontally along the z-axis. The cathode of the photodiode is on top, and the anode is on the bottom of the intrinsic $Si_{0.75\%}Ge_{99.25\%}$ absorption layer of the photodiode. If Ge is not available in the particular SOI, one can use other materials to provide absorption of the telecom light wavelengths. An example from [42] is depicted in Figure 8.19, where the silicides for contacts in SOI are introduced in the silicon waveguide, simultaneously building a lateral bipolar transistor with a Schottky collector neighboring the absorption $NiSi_2$ island in the depletion region of the reverse-biased collector junction. The extension of the silicon waveguide is the base of the transistor.

As is seen from these examples, SOI is suitable for building integrated photodetectors. However, as is also seen from these examples, there is a need for the modi cation of SOI structures to have them appropriate for photo-detection, mainly to maintain suf cient absorption of the light. Many approaches have been undertaken in SOI with this goal including Bragg mirrors and resonant cavities that trap the light until absorbed [44, 45], vertical metal–semiconductor–metal diodes and transistors [43, 46], and chip ipping [47]. Recent SOI technologies stack wafers, using through-silicon vias (TSVs) for interconnections, as shown in Figure 8.20a, and allow them to place several photodiodes on top of each

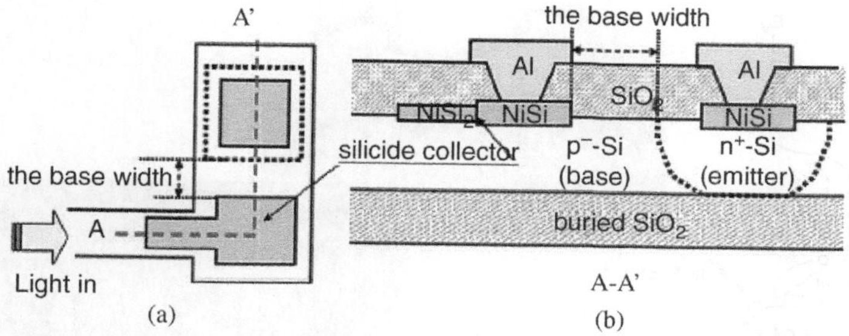

Figure 8.19 Layout (a) and (b) cross-section of a SOI waveguide-based two-terminal Ni-silicide bipolar phototransistor [42]. $NiSi_2$ neighboring the Schottky collector is in the depletion region of the collector and serves as an absorber of the infrared light ($\lambda \sim 1.5 - 1.6\,\mu m$), which is delivered from the left through the Si waveguide. Reproduced with permission from [42]. Copyright (2008) American Institute of Physics.

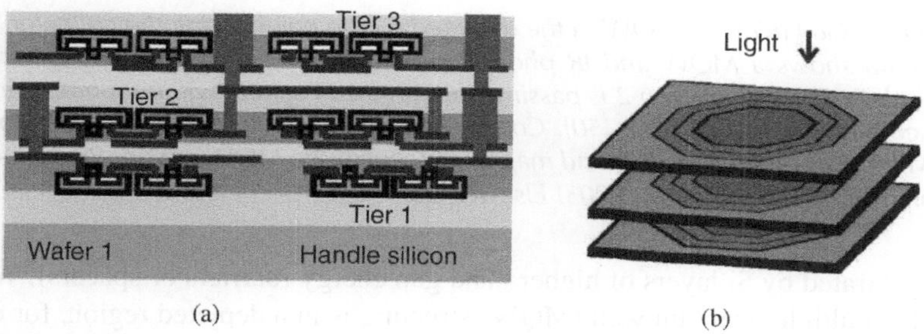

Figure 8.20 Vertical stacking of SOI wafers and photodiodes [48]. (a) Cross-section of the MIT Lincoln Laboratories 3D integration of three SOI dies; and (b) their photodiodes stacked vertically. Reproduced with permission from [48]. Copyright (2007) IEEE.

other [48] to increase the absorption width, as illustrated in Figure 8.20b. One repeatedly observe the use of germanium in SOI to provide light absorption at the telecom wavelengths of 1.3 and 1.5 μm.

8.9 Photodetectors Using Heteroepitaxy

Heteroepitaxy in silicon is used to fabricate silicon photodetectors for an extended range of wavelengths and to increase the absorption of physically thin silicon layers. Owing to material properties and existing technologies, the heteroepitaxy in silicon usually considers germanium as the second material. The SiGe structures provide photodetection of infrared light for the widely used telecom wavelengths of 1.3 and 1.5 μm, for which the silicon itself is transparent and unsuitable due to negligible light absorption.

8.9.1 Si and SiGe Multiple Quantum Wells

The alternation (or undulation) of several nm-thin layers of silicon and germanium creates a quantum well (QW) superlatice with lower band gap energy (well) where Ge is present,

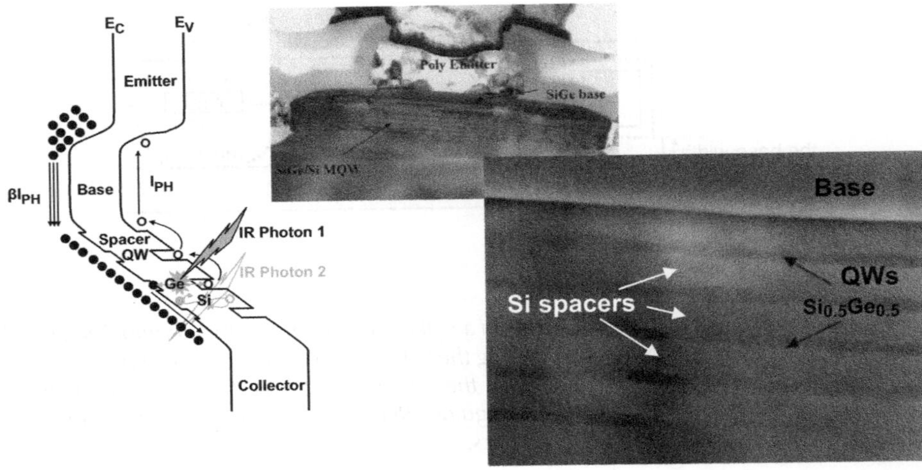

Figure 8.21 SiGe HBT with MQW in the depleted region neighboring the collector [49, 50]. Band diagram shows a MQW and IR photon 1 incident on Ge in the QW, generating an electron–hole pair, while photon 2 is passing through the Si spacer without photo-generation. Reproduced with permission from [50]. Copyright (2003) IEEE. Transmission electron microscopy (TEM) images of the transistor and magnified view of the MQW region. Reproduced with permission from [49]. Copyright (2003) Elsevier Ltd.

which is separated by Si layers of higher band gap energy (barriers or spacers). When the stack of the multiple quantum well (MQW) structure is in a depleted region, for example near the collector of a bipolar transistor, as illustrated in Figure 8.21 with an example from [49, 50], then the photogenerated charges are normally con ned in the Ge quantum wells, but do not freely propagate through MQW, since there are energy barriers that the charges in the quantum well need to overcome in order to reach the "continuum" of the silicon-separating layers. In this way, Ge QWs provide photogeneration for IR, while Si separators prevent large dark currents.

While the incorporation of Ge in Si is a solution for IR photodetectors, the incorporation of large amount of Ge meets has constraints, owing mostly to the ~4.2% mismatch in the lattice constants of Si and Ge, as heuristically depicted in Figure 8.22. The lattice mismatch causes compressive strain in SiGe layers grown on Si, which is also accompanied with the Stransky–Krastanov growth of Ge semispherical islands on Si and quantum–dot like layers, instead of QW layers with uniform thickness. Several monolayers of Ge usually can follow the Si lattice, as shown on the left in Figure 8.23 for MQW with thin and low-Ge-content SiGe layers in a side-illuminated photodector [51], but a 5–10 nm Ge layer relaxes the strain, causing dislocations, as depicted on the right of Figures 8.22 and 8.23 for photodetectors with thicker and high-Ge-content SiGe layers in MQW [52]. The dislocations are defects, which assist the carrier recombination, and are undesired, because the recombination "annihilates" the photo-generated electron–hole pairs.

Various efforts were made to increase the Ge content in MQW photodetectors, but the face dif culties, owing to the lattice mismatch between Si and Ge. The requirement for uniform MQW was relaxed in [53], allowing for embedding Ge islands in the MQW

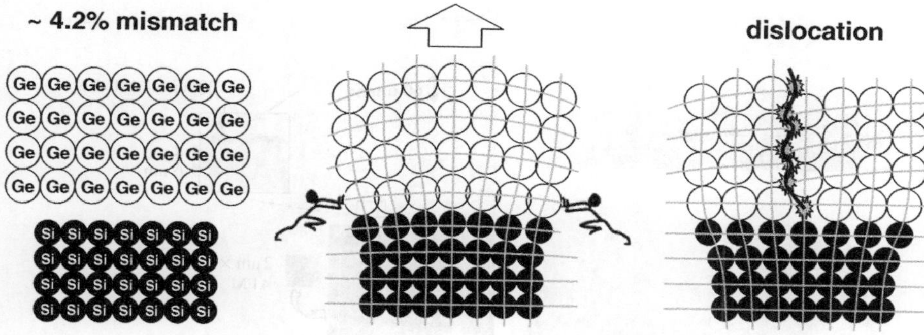

Figure 8.22 (Left) A attice mismatch; (middle) a compressively strained Stransky–Krastanov growth of Ge on Si; and (right) a Ge layer relaxed by a missing Ge atom that causes threading dislocation extending to upper layers.

structure of p–i–n waveguide photodiodes for SOI and SiGe technologies, as given in Figure 8.24a and Figure 8.24b, respectively. The superlattice in Figure 8.24a is of ve periods with thick 65 nm Si barrier spacers and $Si_{0.5}Ge_{0.5}$ islands that are 85 nm in diameter and either 11 nm or 17 nm high, with density 3×10^9 islands/cm^2. The islands are spatially uncorrelated. In contrast, the thinner 50 nm $Si_{0.98}Ge_{0.02}$ spacer in the 20-period MQW in Figure 8.24b has forwarded the strain from lower layers to upper layers during the growth, which aligned vertically the $Si_{0.5}Ge_{0.5}$ islands, increasing the diameter of the islands from bottom to top, and changing the dome shape of the bottom islands toward the arc shape for the upper. The average diameter of the islands in Figure 8.24b is 100 nm and the average height is 20 nm, with an areal density of the islands (2.6×10^9 islands/cm^2 reported earlier by the authors in [54]) similar to that in Figure 8.24a. Interestingly, although the MQWs in Figure 8.24a and Figure 8.24b are apparently different in structure and thickness, the responsivities are similar within a factor of two, 25 mA/W for (a) versus 55 mA/V for (b) at $\lambda = 1.3$ μm and 0.25 mA/W for (a) versus 0.15 mA/V for (b) at $\lambda = 1.55$ μm. One may think that a factor of 2 is signi cant. However, it is stated in [53] that the results are representative

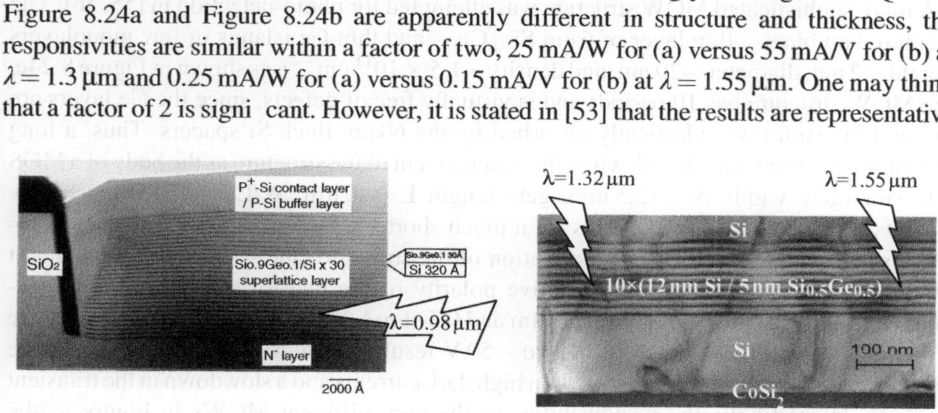

Figure 8.23 (Left) A defect-free MQW of planar photodector with side illumination [51] when Ge content is low (10%), the SiGe layer is thin (3 nm), and the spacer is thick (32 nm). Reproduced with permission from [51]. Copyright (1997) IEEE. (Right) A vertical MSM MQW photodetector [52], with threading dislocations and dome-shaped islands, when the Ge content is high (50%), the SiGe layer is thicker (5 nm), and the Si buffer is thin (12 nm). Reproduced with permission from [52]. Copyright (2002) Elsevier Ltd.

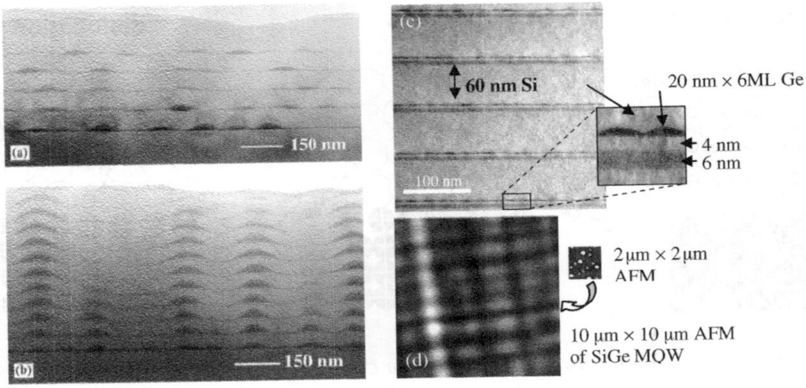

Figure 8.24 *Attempts to utilize a higher content of Ge in MQW. TEM of Ge islands in MQW structure of p–i–n waveguide photodiodes [53] for (a) SOI and (b) SiGe technologies. Reproduced with permission from [53]. Copyright (2003) Elsevier Ltd; (c) TEM of MQW combining the SiGe layer and Ge islands [55]. Reproduced from [55]. Copyright (2006) American Physical Society and (d) AFM of cross-hatch pattern of a SiGe MQW surface caused by underlying Ge islands [56]. Reproduced with permission from [56]. Copyright (2006) Linkoping University, Sweden.*

for lengths 3–7 mm of the waveguide of the photodiodes, whereas for samples similar to that shown in Figure 8.24b, the same authors had reported 210 mA/W at $\lambda = 1.3$ µm and 3 mA/W at $\lambda = 1.5$ µm earlier in [54], that is, 4–20 times larger values. This gives an estimate for the repeatability that one normally observes between different trials of photodetector characterization. Thus, a factor of 2 is small.

A more sophisticated MQW structure was attempted for photo-detection in [55, 56]. The structure combines a thin layer of 6 nm $Si_{0.8}Ge_{0.2}$ and thin Ge islands of few monolayers (height <2 nm, diameter ~20 nm, and density $\sim 1.5 \times 10^{11}$ cm^{-2}), as shown in Figure 8.24c. The MQW structure has 10 periods and is virtually free of defects, since the Ge layers are thin and the strain was elastically absorbed by the 60 nm thick Si spacers. Thus, a long recombination time is achieved, while the arrangement of the structure as the body of a MOS transistor (gate width W = 12.5 mm, gate length L = 40 µm, W/L ~ 60) with Schottky contacts for drain and source allows for a much shorter transit time ~100 ns. The photoresponse was measured by side illumination of the structure, and was bias-dependent. At low drain bias ($V_D \sim 6$–8 V) and positive polarity of the gate bias ($V_G \geq 0$V), the responsivity was 8–10 mA/W for $\lambda = 1.31$ µm and 0.5–1 mA/W for $\lambda = 1.55$ µm. The increase of V_D to 12 V and the decrease of V_G to -50 V resulted in a 20–50-fold increase of the responsivity, but it was accompanied with high dark currents and a slowdown in the transient response. Interestingly, the responsivities of the very different MQWs in Figure 8.24a, 8.24b, and 8.24c are similar: several tens of mA/W at $\lambda = 1.31$ µm and a fraction of 1 mA/W for $\lambda = 1.55$ µm. In addition, irrespectively of the fabrication approach, the surface above the SiGe MQW becomes "wavy" with a cross-hatch pattern typical for Si–Ge devices, as illustrated in Figure 8.24d with an example from [56]. Such waviness of surfaces is not favorable when resonant cavity enchantment is undertaken, since the top mirror of the resonant cavity will be not at.

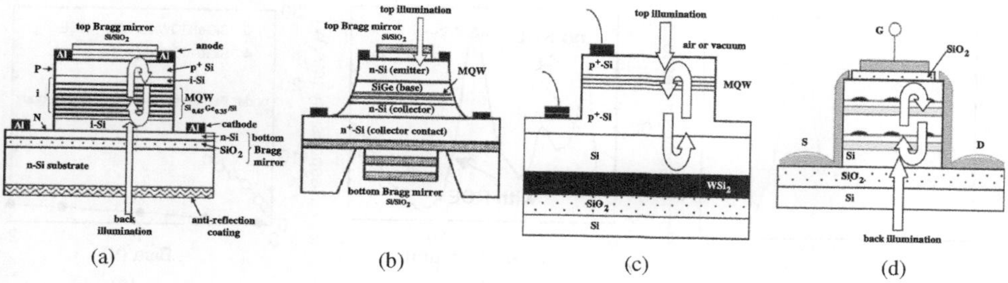

Figure 8.25 Typical arrangements of MQW in Fabry–Perot resonators in vertical or back-illuminated photodetectors based on (a) a p–i–n diode [57]. Reproduced with permission from [57]. Copyright (2000) IEEE; (b) a junction transistor [44]. Reproduced with permission from [44]. Copyright (2000) IEEE; and (c) a MSM diode [45]. Reproduced with permission from [45]. Copyright (2000) IEEE; and (d) in the body of MOS structure [55].

A second useful property of alternating materials by heteroepitaxy (or other means) is that the materials have different refraction indices, which result in re ection at the interfaces. This is exploited to create optical Fabry–Perot resonators, which re ect the light back to the optical absorption layer, thus effectively increasing the width of the otherwise physically thin absorption layer. Owing to the electromagnetic wave property of the light, the "cavity" of mirrors causes constructive summation of the waves of interest, although destructive subtraction can occur, depending of the ratio between the distance between the mirrors and the wavelength of the light. As depicted in Figure 8.25, Fabry–Perot resonators have been used to boost the responsivity of MQWs placed in the intrinsic region of p–i–n diodes [57], or in the depleted region of a base near the collector junction in SiGe heterojunction bipolar transistor (HBT) [44], in the semiconductor layer of metal–semiconductor–metal (MSM) diodes [45], or in the body of a MOS structure [55]. The sketches in Figure 8.25 are adapted from the publications. These RCE–MQW structures have different goals.

The p–i–n diode and SiGe HBT in Figure 8.25a and 8.25b were for near-infrared illuminations ($\lambda = 1.3\,\mu$m), using a re ection from Si–SiO_2 Bragg re ectors, whereas the MSM in Figure 8.25c, in which the doping is high, is for far-infrared ($\lambda = 6$–$12\,\mu$m); it operates at cryogenic temperature (T = 25 K), using a Si–vacuum interface as the top mirror and the metallic (tungsten) silicide WSi_2 as the bottom mirror. The effect of RCE is illustrated in Figure 8.26a and 8.26b, where RCE causes peak-to-valley waviness on the order of 5–10 times, and overall responsivity enhancement by about 5 times, as seen in Figure 8.26c. Other arrangements of MQW in optical resonators are also possible, for example for waveguide MQW photodetectors in ring resonators or photonic crystals. Indeed, since the photodetectors are surrounded by insulating materials of low refractive indices (oxides, air gaps, etc.), then RCE (resonant cavity enhancement) effects always accompany the photodetectors and special care is taken in reference photodetectors in order to reduce RCE and the associated waviness in the responsivity.

One observation from the literature is that SiGe heteroepitaxy is not used for the creation of Bragg mirrors, since Si and Ge have similar refraction indices. Instead, alternating Si and SiO_2 layers are used, owing to the higher optical contrast and compatibility with silicon

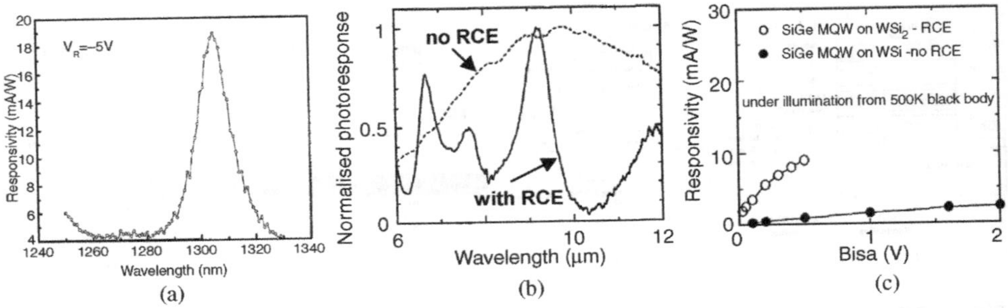

Figure 8.26 Responsivity of SiGe MQW photodetectors with RCE for (a) a p–i–n diode [57] with structure shown in Figure 8.25a and 8.25b for a MSM diode [45], shown in Figure 8.25c, at 25 K. Reproduced with permission from [57]. Copyright (2000) IEEE. The data for the latter are normalized, with absolute values of responsivity as shown in (c), measured under illumination from a black-body source with temperature of 500 K. Reproduced with permission from [45]. Copyright (2000) IEEE.

technologies. A second observation is that SiGe MQW photodetectors have similar responsivities for various MQW structures and RCE arrangements. Perhaps the limitation is due to the problems with the growth of Ge-rich layers in MQWs without defects.

8.9.2 Ge Detectors on Si

The problems associated with the defects in SiGe MQWs led to a search for other methods to incorporate germanium in silicon detectors. The Ge-rich layers are usually deposited by heteroepitaxy, but are not alternating with silicon spacers as in the multiple quantum well (MQW) structures.

Ge-on-Si photodetectors are fabricated in various structures, and the Ge-rich layer might be even amorphous. Typical arrangements as vertical or lateral p–i–n diodes or metal–semiconductor–metal (MSM) diodes are shown in Figure 8.27. Usually, the vertical arrangement in Figure 8.27a is a mesa diode. Typically, the lateral arrangement uses a pair of comb-shaped contacts. Depending on the contacts and doping of the semiconductor under

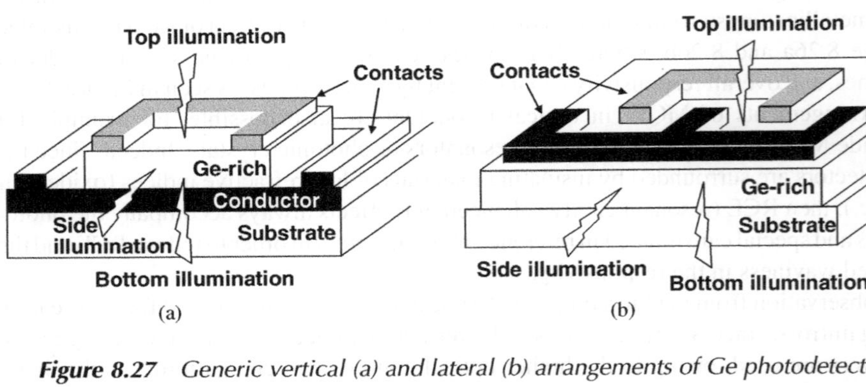

Figure 8.27 Generic vertical (a) and lateral (b) arrangements of Ge photodetectors.

or above the contacts, a large variety of photo-sensing devices are proposed, since one can arrange the structures as

- a p–i–n diode (one of the contacts is with n^+ doping, the other p^+, with the Ge-rich layer being the undoped intrinsic layer),
- a MSM diode (both contacts are either n^+ or p^+, and the diode behaves as an optically controlled resistor), or
- a junction transistor (typically, the Ge-rich region is p-type, n^+ serves as an emitter at one of the contacts, with a Schottky or n^- as collector at the other contact).

Depending on the growth of the Ge-rich layer, there are metamorphic and amorphous combinations with MQW. Furthermore, depending on the coupling of the light, we have top-, bottom-, or side-illuminated structures, with and without resonant cavity enhancement. Last but not least is a structure in which the photo-sensing Ge-rich layer can be part of an electrical device, for example as the base of a junction transistor, the body of a eld-effect transistor, or the absorption layer in an avalanche diode. Or the Ge-rich layer can be a structural extension of the device (e.g., an extension of the base of SiGe HBT). Or the Ge-rich layer can be on a separate structure or a discrete device (or a matrix called a *focal plane array*) electrically connected by wiring to the electrical input of photo-receivers.

The variations in structures and applications of Ge-on-Si photodetectors do not allow for a fair comparison, since the requirements for building these detectors are varied. The comparison requirements range from compatibility with standard Si processes and access for modi cation of these processes, availability of equipment, different goals in the wavelength–sensitivity–speed trade-offs, requirements for fabricating imager arrays, coupling to bers, sensors for measurement instrumentation, to being part of photonic systems-on-chip. Nevertheless, with the several examples given in this section, one can outline approaches and problems related with Ge-on-Si photodetectors.

From the perspective of infrastructural investment and manufacturing cost, a highly desirable approach is to utilize existing standard SiGe technology for photo-sensing applications. Without modifying the fabrication process, and with proper device design and layout, this approach is feasible. In the past, SiGe technologies were mostly developed for vertical junction transistors in a bipolar junction transistor–MOS (BiCMOS) process. An example of a vertical junction transistor in 0.35 μm SiGe BiCMOS [58, 59] is shown in Figure 8.28. The surface photodiode (SPD) is an extension of the 100-nm thin $Si_{0.78}Ge_{0.12}$ base of the HBT, contacted with a polysilicon layer for the emitter, instead of a non-transparent silicide normally used for base contacts. The layout of SPD replicates a second large-area emitter with silicide only at the edge.

The responsivity of the surface photodiode is shown in Figure 8.28. This photoresponse is good for visible light detection (5–9 A/W) up to wavelengths of $\lambda = 850$ nm, which corresponds to some ber communication applications. The responsivity of the diode itself is about 30 times lower (0.2–0.4 A/W) when the emitter is not connected, corresponding to the responsivity of standard silicon photodiodes. However, several oxide layers above the photodiode cause resonance effects. Note that the reponsivity vanishes for telecom wavelengths of $\lambda = 1.3$ and 1.55 μm. This is because of the thin absorption layer (100 nm) and the low content of germanium (12% on average, 14% maximum). Therefore, this approach of using standard SiGe technologies by proper layout is useful for visible light, and there may be an improvement for ultraviolet wavelength detection. However, the

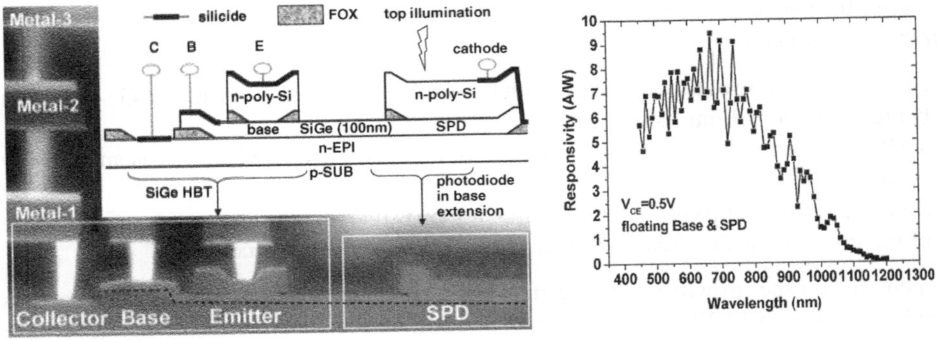

Figure 8.28 *Utilization of the layout of the SiGe HBT in standard 0.35 μm BiCMOS technology as a photodiode in [58, 59]. Reproduced with permission from [58]. Copyright (2007) IEEE. The SiGe layer is denoted with a dashed line due to the low contrast in the SEM image. FOX = field oxide for insulation between contacts. Reproduced with permission from [59]. Copyright (2008) IEEE.*

approach is very limited for IR photo-sensing, because the SiGe layers are thin and with a low content of germanium.

Standard SiGe technologies do not provide infrared (IR) photosensitivity, so Ge-rich layers have to be grown by additional steps in the manufacturing process. A desirable option is for this extra processing to be postfabrication after a standard CMOS process. This is usually done by postprocessing, depositing Ge on top of the silicon chip (or wafer) that was fabricated in a standard or mainstream silicon technology.

The examples shown in Figure 8.29 were reviewed in [60]. A polycrystalline Ge layer was deposited on Si in windows that were opened in the insulation layers above silicon. The structures are arranged as a linear array of photodetectors for top illumination, contacting to the photoreceiver's electrical front end by metal strips on both sides around a meandering common electrode that was used for biasing. The poly-Ge photodetector is shown in Figure 8.29a. Two types of photodiodes for linear arrays were fabricated. One type was a

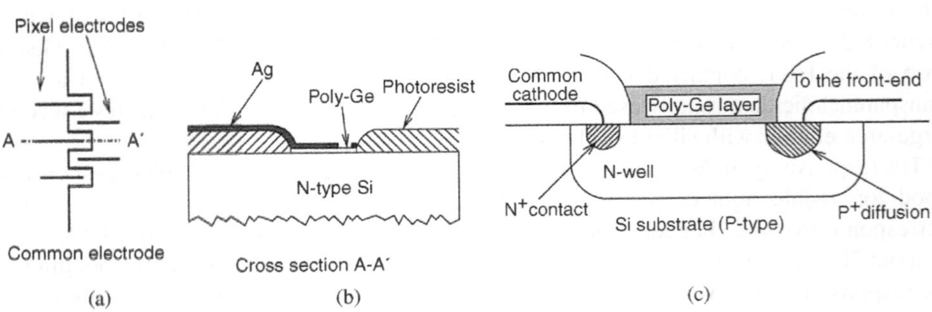

Figure 8.29 *Utilization of a noncrystalline Ge-rich layer by postprocessing of Si wafers, as reviewed in [60]. (a) A linear array of (b) MSM poly-Ge photodiodes. Reproduced with permission from [60]. Copyright (2006) Springer Science + Business Media; or (c) a poly-Ge hetero-junction on a Si p–i–n diode proposed in [61]. Reproduced with permission from [61]. Copyright (2004) IEEE.*

MSM diode with metallization contacting the Ge-rich layer, the latter grown on weakly conducting n-type Si at 300°C, as shown in Figure 8.29b. The area of a pixel in the array is $100 \times 100\,\mu m^2$, and the responsivity was 16 mA/W at $\lambda = 1.32\,\mu m$. The second type, shown in Figure 8.29c, is with metallization contacting a Si lateral p–i–n diode, on top of which the poly-Ge layer was grown, overlapping the p^+ diffusion. Thus, poly-Ge is the anode of the heterojunction diode.

In this Si lateral p–i–n diode, the n-well serves as the intrinsic region that transfers photo-electrons from the Ge layer to the n^+ contact of the cathode. This con guration was proposed in [61]. The photo-holes ow from Ge directly to the p^+ diffusion. The area of the hetero-p–i–n diode pixel is $66 \times 66\,\mu m^2$, and its responsivity was 0.9 mA/W at $\lambda = 1.3\,\mu m$ and 1 V bias. We note the much lower responsivity of the poly-Ge p–i–n diode, as compared with the MSM diode. However, the comparison is not fair, because the thicknesses of the Ge layer and other relevant details were not provided.

While a deposition of poly-Ge lms is relatively simple and cost-effective, the responsivity of poly-Ge photodiodes is lower than the SiGe MQWs described in Section 8.9.1. This reduction in responsivity is from the low level of crystallinity in polycrystalline and amorphous lms, which causes short recombination times (few ns) and low mobility ($<1200\,cm^2/Vs$, which is less than one third of the mobility in crystalline Ge), owing to a very short diffusion length of $\sim$20–30 nm [60]. The low-temperature deposition of Ge results in low-quality lms, as shown in Figure 8.30a from [62]. A large number of mis t dislocations is present at the Ge–Si interface, resulting in signi cant surface roughness in the range of hundreds of nanometers.

The quality of the Ge lm can be signi cantly improved after annealing in a hydrogen ambient and at a high temperature (see Figure 8.30b). However, an hour of annealing at 825°C will ruin the contrast of implantations in the underlying Si structures, smearing the base regions in junction transistors and short channels in CMOS. Thus, high temperature annealing is not an acceptable solution for postprocessing. In addition, the Si–Ge interface still has many dislocations. This interface is important for photodetection, since the depletion region and charge photo-generation are around the Si–Ge junction. The photo-generated electron–hole pairs in the better quality bulk of the Ge lm are electrically not accessible, because the electric eld is low in the bulk. Indeed, one needs multiple hydrogen annealing for heteroepitaxy (MHAH) in order to reduce the dislocation density from $10^9\,cm^{-2}$ to

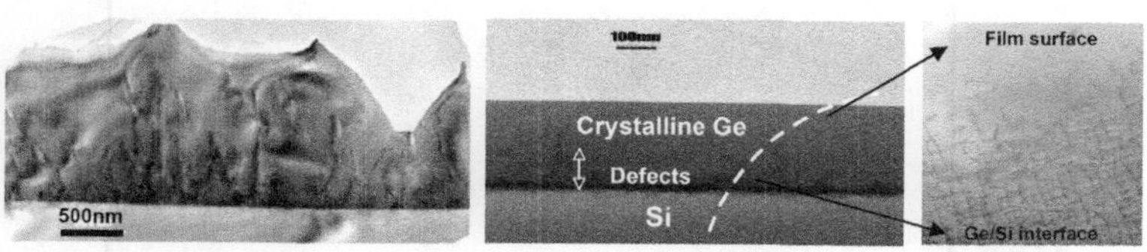

(a) (b) (c)

Figure 8.30 *TEM images from [62] of Ge films (a) as deposited at 400°C. Reproduced with permission from [62]. Copyright (2007) A. Okyay; and (b) after one-hour hydrogen anneal at 825°C. (c) A plan view of annealed film of an arc cross-section as depicted in (b). Reproduced with permission from Opt. Lett., High-efficiency metal-semiconductor-metal photodetectors on heteroepitaxially grown Ge on Si, by A. Okyay, A. Nayfeh, K. Saraswat, T. Yonehara, A. Marshall and P. MacIntyre,* ***31****, 17, 2565–2567 Copyright (2006) The Optical Society.*

$10^7\,cm^{-2}$ and to achieve a responsivity ~0.8 A/W at $\lambda = 1.55\,\mu m$ and 2 V reverse bias from TiGeTi lateral MSM photodiodes on a Si substrate. The MSM structure on a Si substrate replicates the lateral arrangement in Figure 8.27b with a spacing of 5 µm between the comb ngers that are of width 5 µm.

Calculations in [62] for bulk Ge and the experiments with TiGeTi lateral MSM of MHAH Ge lms, baked with XRD and theoretical analyses, indicated a 24 meV reduction of the Ge band gap. This reduction was due to tensile strain after cooling from the processing temperature (850°C, as mentioned) to room temperature, which causes a 47 nm red shift of Ge absorption toward longer wavelengths, as shown in Figure 8.31. This wavelength shift is helpful when utilizing Ge for the longer telecom wavelength $\lambda = 1.55\,\mu m$.

The lattice defects in poly-Ge layers can be reduced if metamorphic SiGe lattices grown by heteroepitaxy are used. This growth is more expensive, but it is a silicon-compatible technique. In addition, the processing can be carried out at lower temperatures, rather than using 800°C for annealing, which severely degrades the doping contrast in preprocessed Si wafer. A low-temperature growth at 380°C of photosensitive Ge epitaxial layer on silicon substrate was reported in [63]. The photodetector is a lateral MSM structure with comb-shaped silver (Ag) electrodes, as shown in Figure 8.32a. The thin amorphous-Ge (α–Ge) layer shown in Figure 8.32b increases the Schottky barrier height between Ag and the Ge epitaxial layer, reducing the dark current of the photodetector. These dark-current results are shown in Figure 8.32c. The external quantum ef ciency was above 10% at $\lambda < 1.4\,\mu m$ and 10 times lower at $\lambda = 1.55\,\mu m$, as illustrated in Figure 8.32d. The resulting responsivity was 0.15 A/W at $\lambda = 1.3\,\mu m$, and the electrical bandwidth was from 1.5 to 4.3 GHz at biasing voltages between 1 and 4 V. These respectable results indicate a high-quality Ge epitaxial layer, in which the surface was relatively smooth with a small roughness (0.9 nm). While a low-temperature process was used, one notes the use of strong acids $H_2O_2 + H_2SO_4$ (1:2) and HF in the precleaning process. Therefore, because of this cleaning method, the α-Ge layer cannot be used on prefabricated Si chips, since metal pads and elds oxides will be damaged or dissolved.

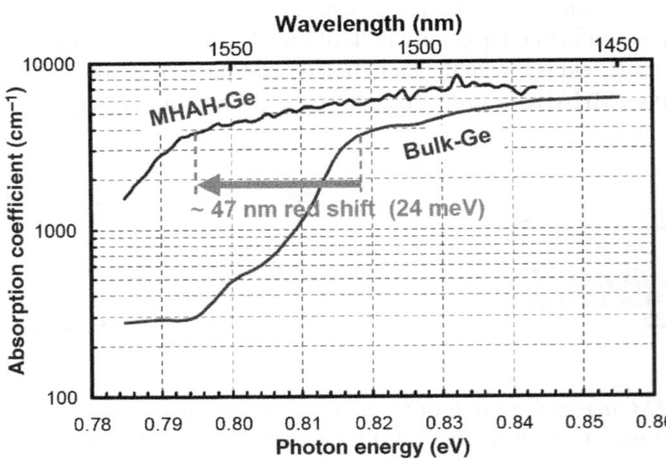

Figure 8.31 *Red shift of Ge absorption toward longer wavelengths by tensile strain in MHAH Ge layers, as deduced in [62]. Reproduced with permission from [62]. Copyright (2007) A. Okyay.*

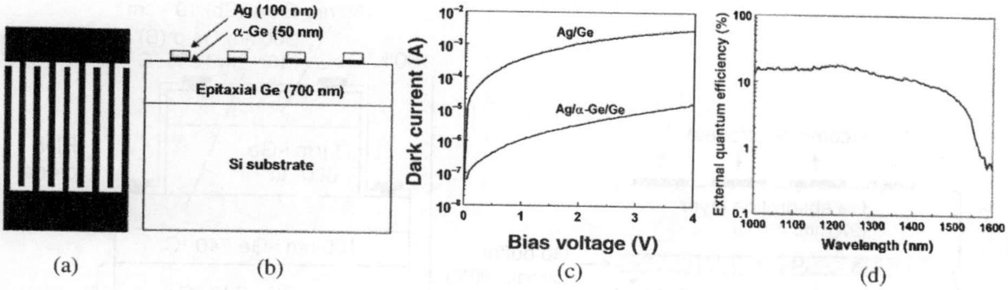

Figure 8.32 *MSM Ge-on-Si photodetector with a low-temperature grown Ge epitaxial layer [63]. (a) Comb-shaped silver (Ag) electrodes with 1 μm finger width and 2 μm spacing in an area of 25 × 50 μm; and (b) deposited on amorphous α–Ge for higher built-in voltage between Ag and the epitaxial Ge layer, the latter grown on Si substrate at 380° C. (c) Two decades of reduction of dark current by using α–Ge; and (d) External quantum efficiency, being about one third of the internal quantum efficiency, without antireflection coating and at low bias of 0.2 V. Reproduced with permission from [63]. Copyright (2004) IEEE.*

The lattice defects in Ge layers can be further reduced (or at least controlled) by using a buffer layer and a gradual transition from Si toward Ge-rich layers in the epitaxially grown metamorphic SiGe lattices. These growths are more complicated than the deposition of Ge lms, and thus are more expensive. But they allow for building advanced vertical structures that can operate as active devices, for example vertical phototransistors and avalanche diodes, or for having a Ge layer virtually free of defects in the active depleted region of the photodetector.

There are two approaches in using buffer layers to obtain Ge layers of good quality on Si. These approaches are outlined using the examples in Figure 8.33. One approach is to grow the Ge buffer layer at low temperature, for example 50 nm at 400°C as shown in Figure 8.33a. In this way, the mismatch between Si and Ge lattices with many defects is relaxed. Then, the temperature is increased to grow a Ge layer that is almost monocrystalline. This Ge layer is the active absorption layer of the photodetector, since the electrical contacts are connected to the high-quality layer, while the defective buffer is in passive regions of low electrical eld far from the contacts. Using this approach, a lateral MSM photodetector with a bandwidth of 35 GHz was achieved in [64] when the spacing between the metal (Ti–Au) contacts is 0.5 μm. These results are suitable for 40 Gbit/s data transfer, appropriate for both telecom wavelengths of $\lambda = 1.31$ μm and $\lambda = 1.55$ μm.

The other approach of using buffer layers to obtain Ge layers with good quality on Si is to grow the buffer layer with a gradual transition from Si toward the Ge-rich layer. This is illustrated in Figure 8.33b for a Ge mesa p–i–n diode [65]. The buffer layer is pure Si at the bottom, and gradually, in steps of 10% Ge per micrometer, it becomes pure Ge at the top. The mismatches in the lattice are relaxed by defects in the graded buffer. Then, a 100 nm intrinsic Ge layer is grown to achieve monocrystalline Ge. The p^+, intrinsic, and n^+ layers of the p–i–n photodiode are then grown on a high-quality base, while the defective buffer is in the electrically inactive region below the anode of the diode. These defective buffer layers do not interfere with photodetection.

The growth of the graded buffer usually results in a tall structure, which is not very easy to implement in standard SiGe technology. This is because the metallization from the top of the

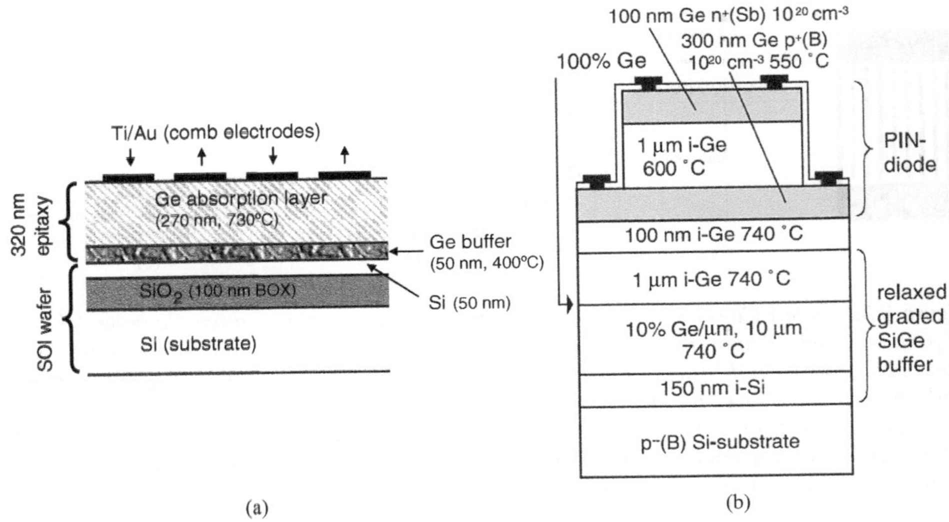

Figure 8.33 Two methods reducing defects in a photo-active Ge layer. (a) Thin low-temperature grown Ge buffer layer accumulates defects due to lattice mismatch, and the upper high-temperature Ge layer is with low defect density [64]. Reproduced with permission from [64]. Copyright (2005) American Institute of Physics; and (b) a gradual transition from Si to Ge in the buffer layer with a rate of 10% Ge per micrometer [65], resulting in much taller epitaxial growth (~12 µm). Reproduced with permission from [65]. Copyright (2004) Elsevier Ltd.

structure needs to go back to the silicon substrate for the electrical input of the photoreceiver. The distance is more than 12 µm vertically, and metal lines and vias will have resistances, inductances, and capacitances, which reduce the bandwidth of the photo-receiver and cause signi cant losses at radiofrequencies. Therefore, the approach from Figure 8.33a for the low-temperature buffer was utilized in recent high-speed photodetectors [66], as is shown in Figure 8.34. The fabricated mesa photodiode with a radius of 5 µm had a bandwidth of

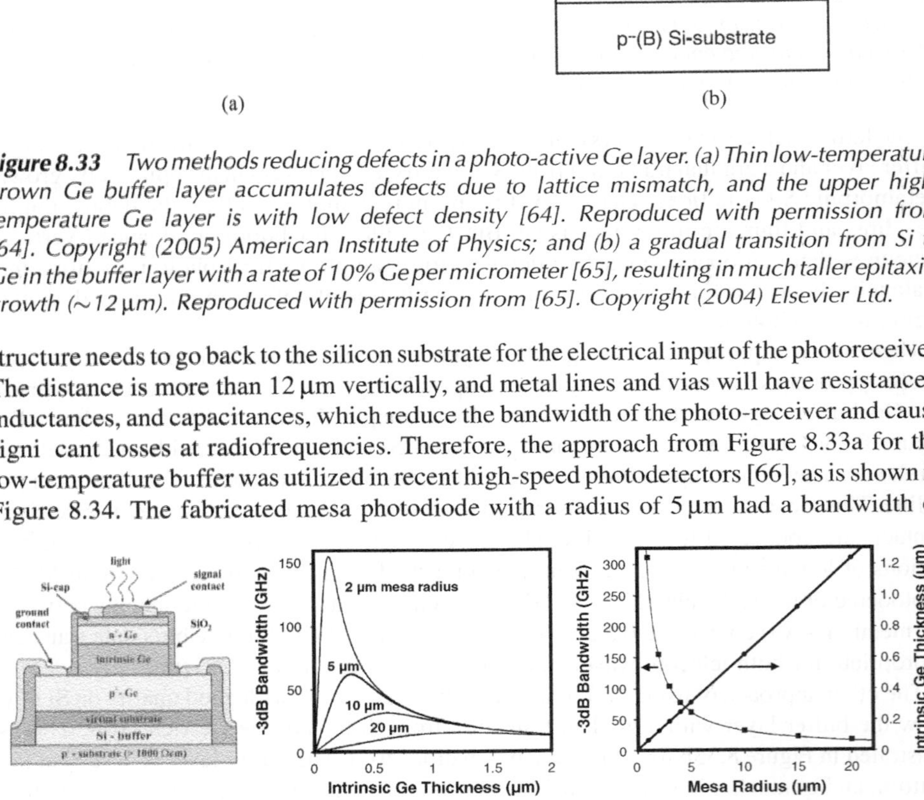

Figure 8.34 High-speed vertical mesa diode [66], using the approach from Figure 8.33a of a low-temperature Ge buffer. Virtual substrate stands for the 50–100 nm Ge buffer grown at low temperature. The middle graph illustrates the trade-off between sizes and speed in round mesa Ge diodes. The right-hand graph illustrates optimum designs in terms of speed. Note that high speed requires a small radius of the mesa and a thinner absorption layer, which reduce the responsivity of the photodiode. Reproduced with permission from [66]. Copyright (2006) American Institute of Physics.

39 GHz at $\lambda = 1.552\,\mu$m using a laser source of 6 mW power. The responsivity was not stated, perhaps because it is low, and the thickness of the absorption layer was not stated, perhaps because is not exactly at an optimum value.

8.9.3 Related Theoretical Discussion

The equations used in computing the graphs in Figure 8.34 are described in this subsection. The RC time constant is

$$\tau_{RC} = RC = R\frac{\varepsilon_{Ge}A}{W_{Ge}} \tag{8.41}$$

where $R \sim 50\Omega$ is the load resistance of the photodiode, since the contact resistances of the structure are expected to be less than a small fraction of the load resistance. Also, W_{Ge} is the thickness of the intrinsic Ge layer, $\varepsilon_{Ge} \approx 1.4 \times 10^{-12}$ F/cm is the permittivity of Ge, and $A = \pi r^2$ is the area of the mesa diode of radius r. The transit time of the carriers traversing the intrinsic Ge layer is

$$\tau_T = \frac{\sqrt{2}}{4}\frac{W_{Ge}}{\upsilon_{sat}} = 0.35\frac{W_{Ge}}{\upsilon_{sat}} \tag{8.42}$$

where $\upsilon_{sat} \sim 0.6 \times 10^7$ cm/s is the saturation velocity of carriers in Ge, since the electric eld $V_{bias}/W_{Ge} \sim$ few V/few tenths of a μm ~ several kV/cm is larger than the critical electric eld for velocity saturation (a few tens of kV/cm). Then, the combined time constant of the structure is

$$\tau = \sqrt{\tau_{RC}^2 + \tau_T^2} = \frac{1}{2\pi f_{3dB}} \tag{8.43}$$

Substituting (8.35) and (8.36) in (8.37), we obtain

$$\begin{aligned} f_{3dB} &= \frac{1}{2\pi\tau} = \frac{1}{2\pi\sqrt{\tau_{RC}^2+\tau_T^2}} = \frac{1}{2\pi\sqrt{\left(R\frac{\varepsilon_{Ge}A}{W_{Ge}}\right)^2 + \left(\frac{\sqrt{2}}{4}\frac{W_{Ge}}{\upsilon_{sat}}\right)^2}} \\ &= \frac{\sqrt{2}}{\pi}\upsilon_{sat}\frac{W_{Ge}}{\sqrt{8\pi^2(R\upsilon_{sat}\varepsilon_{Ge})^2 r^4 + W_{Ge}^4}} = \frac{a \times W_{Ge}}{\sqrt{b \times r^4 + W_{Ge}^4}} \end{aligned} \tag{8.44}$$

$$a = \frac{\sqrt{2}}{\pi}\upsilon_{sat} \approx \frac{1.4}{3.14}0.6 \times 10^7[\text{cm/s}] \sim 2.7\,\text{MHz cm}$$

$$b = 8\pi^2(R\upsilon_{sat}\varepsilon_{Ge})^2 \sim 1.4 \times 10^{-5}$$

The function for f_{3dB} is the reciprocal function of the mesa radius r and is a peaking function of depletion thickness W_{Ge}, having a maximum

$$\max(f_{3dB}) = \frac{a}{r\sqrt{2}\sqrt[4]{b}} = \frac{\sqrt[4]{2}}{2\pi r}\sqrt{\frac{\upsilon_{sat}}{\pi R\varepsilon_{Ge}}} \sim \frac{30\,\text{MHz} \times \text{cm}}{r} \quad at$$

$$\frac{\partial f_{3dB}}{\partial W_{Ge}} = 0 \Rightarrow W_{Ge,\max(f_{3dB})} = r \times \sqrt[4]{b} = r\sqrt{2\pi R\upsilon_{sat}\varepsilon_{Ge}} \times \sqrt[4]{2} \sim r \times 6\% \tag{8.45}$$

Note that the equations and the values of the parameters *a* and *b* are approximate, since many details in the structure are neglected in these equations. Nevertheless, one can use these equations as a rst-order approximation for optimizing the speed of a photodetector. Also, using the absorption coef cient of Ge, one can also estimate the maximum responsivity of the detector from

$$\frac{I_{ph}}{P_{opt}} \leq \{1-\exp[-\alpha(\lambda) \times W_{Ge,\max(f_{3dB})}]\} \frac{\lambda[\text{nm}]}{1240},$$
$$\textit{with} \quad \alpha(\lambda) \approx \begin{cases} 3.6\,\mu\text{m}^{-1} \; at \; \lambda = 0.85\,\mu\text{m} \\ 0.74\,\mu\text{m}^{-1} \; at \; \lambda = 1.31\,\mu\text{m} \\ 0.046\,\mu\text{m}^{-1} \; at \; \lambda = 1.55\,\mu\text{m} \end{cases} \tag{8.46}$$

More conveniently, one can rewrite Eq. (8.45) as function of the desired bandwidth f_{3dB}, and then calculate the optimum mesa radius *r* and the optimum width W_{Ge} of the Ge absorption layer, as shown with the top two lines in Figure 8.35. Then, one calculates the responsivity at different wavelengths using Eq. (8.46), and draws the curves in the

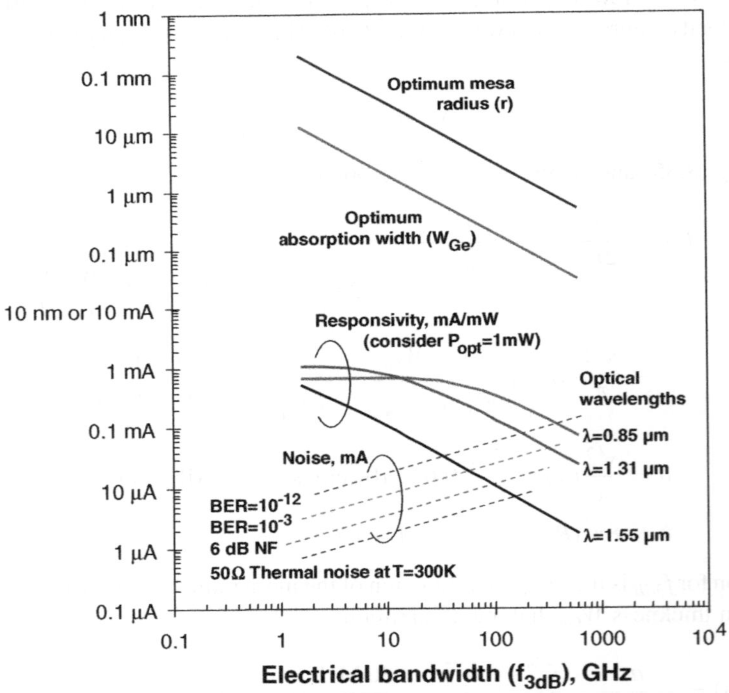

Figure 8.35 Chart for estimation of a Ge mesa photodiode. The assumptions are that 1mW optical power reaches the Ge absorption layer, the photodiode load is 50Ω at temperature 300 K, the noise figure of the system is NF = 6dB, and the shot noise is included in NF.

middle of Figure 8.35 in mA/mW, assuming that 1 mW optical power is absorbed in the absorption layer, despite the small area of the photodiode and re ections from surfaces and other loss in the optical path.

Next, one calculates the thermal noise of the load ($R = 50\Omega$) and the added noise from the system (e.g., the photodetector's ampli er, the latter with noise gure $NF = 6\text{dB}$), according to

$$\text{Thermal Noise [mA]} = 1000\sqrt{\frac{4\,\text{kT}}{50\Omega} f_{3dB}}, \quad T = 300\,\text{K} \tag{8.47}$$
$$\text{Noise} = (\text{Thermal Noise}) \times 10^{NF/20dB}$$

where the noise current is calculated in mA in order to compare with responsivity at $P_{opt} = 1$ mW. The calculation is illustrated with the bottom two dashed lines in Figure 8.35 From the requirement for an allowable bit error rate ($BER = 10^{-3}$ to 10^{-12}), one nds the signal-to-noise ratio (SNR) by equalizing

$$\frac{I_{ph}}{Noise} \geq SNR, \; where\; BER = \frac{1}{2} erfc(SNR) \tag{8.48}$$

assuming nonreturn-to-zero (*NRZ*) signals. Approximate values for *SNR* are {2.2, 3.4, 4.3, and 5} for $BER = \{10^{-3}, 10^{-6}, 10^{-9}$, and $10^{-12}\}$, respectively. The minimum I_{ph} for $\text{BER} = 10^{-3}$ and 10^{-12} is shown in Figure 8.35 with the remaining two dashed lines.

The chart in Figure 8.35 indicates dif culties in building high-speed and long-wavelength Ge photodetectors. For example, a 3 GHz photodetector can have a large area $30\,000\,\mu\text{m}^2 \approx \pi(100\,\mu\text{m})^2 \approx (170\,\mu\text{m})^2$, a thick absorption layer $\sim 6\,\mu$m, and a high responsivity $\sim 0.3 - 1$ mA/mW at any of the wavelengths $\lambda = (0.85, 1.31, \text{or } 1.55)\,\mu$m. The requirement for low $\text{BER} = 10^{-12}$ is $I_{ph} > 10\,\mu$A, and it is easy to meet even a low optical power of 10–30 μW. Moving to a 10 GHz bandwidth, the area reduces to $2800\,\mu\text{m}^2 \approx \pi(30\,\mu\text{m})^2 \approx (50\,\mu\text{m})^2$, the absorption layer thins to $\sim 2\,\mu$m, and the responsivity at long-wavelength $\lambda = 1.55\,\mu$m drops to 0.1 mA/mW, while the requirement for low $\text{BER} = 10^{-12}$ increases to $I_{\text{ph}} > 20\,\mu$A; thus, one has to provide optical power $P_{opt} > 200\,\mu$W in the absorption layer. Nevertheless, the values do not dramatically change for shorter wavelengths. The two examples below further illustrate some of the technical challenges for higher speed Ge photodiodes.

Example 8.13

Using Figure 8.35, determine the requirements for a 40 GHz Ge photodiode.

In this case, the requirements for 40 GHz bandwidth are tight. The area shrinks to $200\,\mu\text{m}^2 \approx \pi(8\,\mu\text{m})^2 \approx (14\,\mu\text{m})^2$, which requires alignment of a light beam with micrometer accuracy, thus making it expensive to package the photodetector.

The absorption layer thins further to $\sim 0.45\,\mu$m, degrading the responsivity at long-wavelength $\lambda = 1.55\,\mu$m to 0.02 mA/mW, and also reducing the responsivity to 0.3 mA/mW at the medium-wavelength $\lambda = 1.31\,\mu$m.

At the same time, the requirement for low $BER = 10^{-12}$ increases further to $I_{ph} > 40\,\mu$A, which still can be met with an optical power of ~ 0.1 mW at $\lambda = 1.31\,\mu$m, but much larger power $P_{opt} > 2\,\text{mW}$ is necessary for $\lambda = 1.55\,\mu\text{m}$. Indeed, the power density at $\lambda = 1.55\,\mu\text{m}$ will become $2\,\text{mW}/200\,\mu\text{m}^2 = 1\,\text{kW/cm}^2$ (and perhaps 10 times larger at the photodetector surface), which will raise the temperature and leakage characteristics of the Ge photodetector.

As a calibration of this high light intensity, the sun's irradiation on Earth is maximum $0.14\,\text{W/cm}^2$.

Example 8.14

Using Figure 8.35, determine the requirements for a Ge photodiode to reach 100 GHz.

This example illustrates some of the technical limitations of using a Ge photodetector for long wavelengths from a simple analysis.

Now, the area should be $3\,\mu\text{m}^2 \approx \pi(1\,\mu\text{m})^2 \approx (1.7\,\mu\text{m})^2$, the absorption layer should be as thin as $0.2\,\mu\text{m}$, and the responsivity falls to $10\,\mu\text{A/mW}$ at $\lambda = 1.55\,\mu\text{m}$.

For a $BER = 10^{-3}$, the photodiode's photocurrent $I_{ph} > 25\mu\text{A}$, which requires $2.5\,\text{mW}/3\,\mu\text{m}^2 \sim 0.1\,\text{MW/cm}^2$. Note that laser metal cutters usually use more than $1\text{–}10\,\text{MW/cm}^2$ beams to cut steel that is a centimeter or so thick with a speed of cm/s. Certainly, $0.1\,\text{MW/cm}^2$ will destroy the photodetector in a minute or so.

On the other hand, one can use the shortest wavelength $\lambda = 0.85\,\mu\text{m}$ at an optical power of 0.1 mW or less, but Si itself is absorbing lights of this wavelength. Therefore, in this case, it is dif cult to justify the expensive heteroepitaxial Ge.

From the discussion in this subsection, it becomes clear that devices with internal gain should be utilized for high-speed detectors, for example avalanche photodiodes or bipolar transistors, which can compensate for the low responsivity in the primary opto-electrical conversion.

Problems

8.1 Show that a photon with a wavelength λ (μm) has energy given by

$$Energy(Joules) = \frac{1.985 \times 10^{-19}}{\lambda(\mu\text{m})}$$

8.2 Using Eq. (8.2), calculate the cutoff wavelength of $Si_{0.2}Ge_{0.8}$. Then, using Eq. (8.3) and the parameters in Table 8.2, determine the absorption coef cient of this binary Si–Ge material.

8.3 A light source at a wavelength of 500 nm illuminates a photodiode of area $(20\,\mu\text{m})^2$ with $10\,\mu\text{W}$ of light. Compute the incident ux of photons on the photodiode surface. Assume that the re ection coef cient is 0 (no re ection).

8.4 A photodiode absorbs 1 nW of incident power over its photoactive area. If the incident light is of wavelength 500 nm, how many photons per second are incident on the photodiode? If the quantum ef ciency of the photodiode is 0.8, how much current ows through the photodiode?

8.5 If the ux of photons is at a wavelength of 600 nm, incident on a photodiode is 10^{12} photons per second, and a current of 10 nA ows through the photodiode, determine the quantum ef ciency of the photodiode. Then determine its responsivity in A/W.

8.6 A photodetector with a responsivity of 0.4 A/W is illuminated with light of wavelength 700 nm at an intensity of 100 μW/cm^2. If the photodiode has an area of 0.1 cm^2, what is its photocurrent?

8.7 Assume that a silicon photodiode is biased so that the electric eld in the depletion region results in carriers traveling at 5×10^6 cm/s. The depletion region is 1 μm thick. If the photodiode's resistance–capacitance time constant is 50 ps, determine the overall time constant of the photodiode.

8.8 Consider a silicon photodiode with the following parameters: $N_D = 5 \times 10^{16}$ cm^{-3}, $N_A = 10^{16}$ cm^{-3}, $D_p = 10$ cm^2/s, $D_n = 25$cm^2/s, $\tau_{po} = 0.1$ μs, and $\tau_{no} = 0.5$ μs. A reverse bias of 5 V is applied, and the generation rate of excess carriers is 10^{21} per cm^3 per second. (a) Calculate the electron and hole diffusion lengths. (b) Determine the built-in voltage of the photodiode. (c) Compute the total depletion width. (d) Find the steady-state photocurrent if the active area of the photodiode is a circle of area 1 cm^2.

8.9 Consider a silicon photodiode with the following parameters: area = (100 μm)^2, $N_D = 10^{16}$ cm^{-3}, $N_A = 5 \times 10^{16}$ cm^{-3}, $D_p = 12$ cm^2/s, $D_n = 20$ cm^2/s, $\tau_{po} = 0.01$ μs, and $\tau_{no} = 0.01$ μs. A reverse bias of 2 V is applied and the generation rate of excess carriers is 5×10^{21} per cm^3 per second. (a) Calculate the electron and hole diffusion lengths. (b) Determine the built-in voltage of the photodiode. (c) Compute the total depletion width. (d) Find the steadyistate photocurrent. (e) What is the minimum length of the photodiode?

8.10 Consider a p–i–n photodiode in which the width of the intrinsic region (*i*) is 1 μm. Assume that the absorption coef cient $\alpha = 10^3$ cm^{-1} and the incident photon ux is 10^{18} per cm^2 per second. If the quantum ef ciency is unity, determine the generation rate of electron–hole pairs at the front of the *i*-region and then at the back of the *i*-region. Then determine the photocurrent per cm^2 through in the photodiode.

8.11 A photodiode has a photocurrent of 20 μA. The total shot noise in the photodiode is 5 pA and the thermal noise is 1 pA. Compute the signal-to-noise ratio of this photodiode. If the responsivity of this photodiode is 0.4 A/W, determine its noise-equivalent power (NEP) in watts. State two ways in which the noise-equivalent power may be improved.

8.12 For a typical receiver systems application, an avalanche photodiode with the lowest possible excess noise is preferable. Using Eq. (8.21), make a plot of the calculated excess noise factor as a function of multiplication gain for various *k* values from 0.1 to 1.0 and for multiplication values *M* from 1 to 100.

8.13 Discuss the conditions under which Eq. (8.25) will result in a higher signal-to-noise ratio compared to a photodiode with no multiplication. That is, discuss the role of *F*, *k*, and *M* in your answer.

8.14 An avalanche photodiode operating at 700 nm has a signal power of 1 nW, a *k*-factor of 0.01, a quantum ef ciency of 70%, a negligible bulk current, and a multiplied dark current of 10 pA. The DC photocurrent I_{pho} is 0.36 nA and the resistance is 100 kΩ.

Make a plot of the signal, noise, and signal-to-noise ratio as a function of M between 1 and 1000. Comment on the characteristics of your graphs plotted.

8.15 Using Figure 8.35, determine the requirements for a Germanium photodiode to operate at 10 GHz.

References

1. Bandyopadhyay, A. and Deen, M.J. (2001) Photodetectors for optical ber communications, in *Photodetectors and Fiber Optics* (ed H.S. Nalwa), Academic Press, San Diego, CA, pp. 307–368.
2. Ma, C.L.F., Deen, M.J., and Tarof, L. (1998) in *Characterization and Modeling of SAGCM InP/InGaAs Avalanche Photodiodes for Multigigabit Optical Fiber Communications, Advances in Imaging and Electron Physics*, vol. **99** (ed P. Hawkes), Academic Press, San Diego, CA, pp. 65–170.
3. Kfouri, M., Marinov, O., Quevedo, P. *et al.* (2008) Towards a miniaturized wireless uorescence-based diagnostic imaging system. *IEEE J. Sel. Top. Quant. (special issue on biophotonics)*, **14** (10), 226–234.
4. Faramarzpour, N., El-Desouki, M.M., Deen, M.J. *et al.* (2009) CMOS photodetector systems for low-level light applications. *J. Mater. Sci.-Mater. El. (special issue)*, **20**(Suppl. 1), S87–S93.
5. El-Desouki, M.M., Palubiak, D., Jamal Deen, M. *et al.* (2010) A novel, high dynamic range, high speed and high sensitivity CMOS imager using time-domain single photon counting and avalanche photodiodes. *IEEE Sens. J.*, **11**(4), 1078–1083.
6. El-Desouki, M., Deen, M.J., Fang, Q. *et al.* (2009) CMOS image sensors for high speed applications. *Sensors*, **9**(1), 430–444.
7. Campos, F.S., Marinov, O., Faramarzpour, N. *et al.* (2008) A multisampling time-domain CMOS imager with synchronous readout circuit. *Analog Integr. Circ. S.*, **57**(1–2), 151–159.
8. Faramarzpour, N., El-Desouki, M.M., Deen, M.J. *et al.* (2008) CMOS imaging for biomedical applications. *IEEE P.*, **27**(3), 31–36.
9. Faramarzpour, N., Deen, M.J., Shirani, S., and Fang, Q. (2008) Fully integrated single photon avalanche diode detector in standard CMOS 0.18 μm technology. *IEEE T. Electron. Dev.*, **55**(3), 760–767.
10. Faramarzpour, N., Deen, M.J., Shirani, S. *et al.* (2007) CMOS based active pixel for low-light-level detection: analysis and measurements. *IEEE T. Electron. Dev.*, **54**(12), 3229–3237.
11. Faramarzpour, N., Deen, M.J., and Shirani, S. (2006) An approach to improve the signal-to-noise ratio of active pixel sensor for low-light-level applications. *IEEE T. Electron. Dev.*, **53**(9), 2384–2391.
12. Ardeshirpour, Y., Deen, M.J., and Shirani, S. (2004) 2-D CMOS based image sensor system for uorescent detection. *Can. J. Elect. Comput. E.*, **29**(4), 231–235.
13. Ardeshirpour, Y., Deen, M.J., and Shirani, S. (2006) Evaluation of CMOS based photodetectors for low-level light detection. *J. Vac. Sci. Technol. A (special issue – CSTC 2005)*, **A24**(3), 860–865.
14. Deen, J., Mehdi, M., and Kazemeini, H. (2005) Photosensitive polymer thin- lm eld-effect transistors based on poly(3-octylthiophene). *P. IEEE (special issue on flexible electronics technology, Part 1: Systems and Applications)*, **93**(7), 1312–1320.
15. Polleux, J.L. and Rumelhard, C. (2000) Optical absorption coef cient and physical modeling of strained SiGe/Si photodetectors. *Proc. 8th IEEE International Symposium on High Performance Electron Devices for Microwave and Optoelectronic Applications, Glasgow, UK*, pp. 167–172.
16. Dash, W.C. and Newman, R. (1955) Intrinsic optical absorption in single crystal germanium and silicon at 77 K and 300 K. *Phys. Rev.*, **99**(4), 1151–1155.

17. *Handbook of Optical Constants*, 1st and 2nd edn. Silvaco Atlas User's Manual. Silvaco, Santa Clara, CA.
18. Lang, D.V., People, R., Bean, J.C., and Sergent, A.M. (1985) Measurement of the band gap GeSi/Si strained layer heterostructure. *App. Phys. Lett.*, **47**(12), 1333–1335.
19. Lim, Y.C. and Moore, R.A. (1968) Properties of charged coplanar parallel strips by conformal mapping. *IEEE T. Electron. Dev.*, **15**, 173–180.
20. Averine, S.V., Chen, Y.C., and Lan, Y.L. (2001) Geometry optimization of interdigitated Schottky-barrier metal-semiconductor-metal photodiode structures. *Solid State Electron.*, **45**, 441–446.
21. Woodward, T. and Krishnamoorthy, A. (1999) 1-Gb/s integrated optical detectors and receivers in commercial CMOS technologies. *IEEE J. Sel. Top. Quant.*, **5**(2), 146–156.
22. Bielecki, Z. (2000) Photoreceiver with avalanche C-30645 E photodiode. *IEEE Proc. – Optoelectron.*, **147**, 234-236.
23. Itzler, M.A., Loi, K.K., McCoy, S., and Codd, N. (2000) High-performance, manufacturable avalanche photodiodes for 10 Gb/s optical receivers. *Opt. Fiber Commun. Conf.*, **4**, 126–128.
24. Gnauck, A.H., Burrus, C.A., and Ekholm, D.T. (1992) A transimpedance APD optical receiver operating at 10 Gb/s. *IEEE Photonics Tech. L.*, **4**, 468–470.
25. Yun, T.Y., Park, M.S., Han, J.H. *et al.* (1996) 10-Gigabit-per-second high-sensitivity and wide-dynamic-range APD-HEMT optical receiver. *IEEE Photonics Tech. L.*, **8**, 1232–1234.
26. Torres, A. and Gutiérrez, E.A. (1997) A planar amorphous Si1-xGex separated-absorption-multiplication avalanche photodiode. *IEEE Electr. Dev. L.*, **18**, 568–570.
27. Chynoweth, A.G. (1968) Charge multiplication phenomena, in *Semiconductor and Semimetals*, vol. **4** (eds R.K. Willardsonand A.C. Beer), Academic Press, San Diego, CA.
28. Stillman, G.E. and Wolfe, C.M. (1977) Avalanche photodiodes, in *Semiconductor and Semimetals*, vol. **12** (eds R.K. Willardsonand A.C. Beer), Academic Press, San Diego, CA.
29. Okuto, Y. and Crowell, C.R. (1972) Energy-conservation considerations in the characterization of impact ionization in semiconductor. *Physics Rev. B*, **6**, 3076–3081.
30. Capasso, F. (1985) Physics of avalanche photodiodes, in *Semiconductors and Semimetals*, vol. **22D** (ed W.T. Tsang), Academic Press, San Diego, CA.
31. Okuto, Y. and Crowell, C.R. (1974) Ionization coef cients in semiconductors: a nonlocalized property. *Physics Rev. B*, **10**, 4284–4296.
32. Neamen, A.D. (1992) *Semiconductor Physics and Devices*, Irwin, Homewood, IL.
33. Ardeshirpour, Y. and Deen, M.J. (2005) *CMOS Photodetectors*, unpublished report, McMaster University, Hamiton, ON.
34. Pavesi, L. (2003) Will silicon be the photonic material of the third millenium? *J. Phys.: Condens. Matter*, **15**(26), R1169–R1196.
35. Schroder, D., Thomas, R., and Swartz, J. (1978) Free carrier absorption in silicon. *IEEE J. Solid-State Circuits*, **13**(1), 180–187.
36. Beneking, H. (1982) Gain and bandwidth of fast near-infrared photodetectors: a comparison of diodes, phototransistors, and photoconductive devices. *IEEE Trans. Electron. Dev.*, **29**(9), 1420–1431.
37. Uryu, Y. and Asano, T. (2002) CMOS image sensor using SOI-MOS/photodiode composite photodetector device. *Jpn. J. Appl. Phys.*, **41**(4B), 2620–2624.
38. Yamamoto, H., Taniguchi, K., and Hamaguchi, C. (1996) High-sensitivity SOI MOS photodetector with self-ampli cation. *Jpn. J. Appl. Phys.*, **35**(2B), 1382–1386.
39. Zhang, W., Chan, M., Huang, R., and Ko, P. (2000) High gain gate/body tied NMOSFET photodetector on SOI substrate for low power applications. *Solid-State Electron.*, **44**(3), 535–540.
40. Harik, L., Sallese, J., and Kayal, M. (2008) Transient charge pumping as an ef cient technique to measure low light intensity with PD SOI MOSFET. *Solid-State Electron.*, **52**(5), 597–605.

41. Liu, J., Pan, D., Jongthammanurak, S. *et al.* (2007) Design of monolithically integrated GeSi electroabsorption modulators and photodetectors on an SOI platform. *Opt. Express*, **15**(2), 623–628.
42. Zhu, S., Lo, G., Yu, M., and Kwong, D. (2008) Low-cost and high-gain silicide Schottky-barrier collector phototransistor integrated on Si waveguide for infrared detection. *Appl. Phys. Lett.*, **93**(7), 071108-1–071108-3.
43. Colace, L., Masini, G., and Assanto, G. (1999) Ge-on-Si approaches to the detection of near-infrared light. *IEEE J. Quant. Electron.*, **35**(12), 1843–1852.
44. Zhu, Y., Yang, Q., and Wang, Q. (1997) Resonant cavity SiGe/Si MQW heterojunction phototransistor grown on the SIMOX substrate for 1.3 μm operation. *IEEE Electron. Comp. Technol. Conf.*, **47**, 1199–1204.
45. Carline, R., Nayar, V., Robbins, D., and Stanaway, M. (1998) Resonant cavity longwave SiGe-Si photodetector using a buried silicide mirror. *IEEE Photon. Tech. Lett.*, **10**(12), 1775–1777.
46. Kimura, S., Maio, K., Doi, T. *et al.* (2002) Photodetectors monolithically integrated on SOI substrate for optical pickup using blue or near-infrared semiconductor laser. *IEEE Trans. Electron Devices*, **49**(6), 997–1004.
47. Krishnamoorthy, A. and Goossen, K. (1998) Optoelectronic-VLSI: photonics integrated with VLSI circuits. *IEEE J. Sel. Top. Quant. Electron.*, **4**(6), 899–912.
48. Culurciello, E. and Weerakoon, P. (2007) Three-dimensional photodetectors in 3-D silicon-on-insulator technology. *IEEE Electron. Dev. Lett.*, **28**(2), 117–119.
49. Pei, Z., Lai, L., Hwang, H. *et al.* (2003) $Si_{1-x}Ge_x$/Si multi-quantum well phototransistor for near-infrared operation. *Physica E: Low-Dim. Sys. Nanostruc.*, **16**(3–4), 554–557.
50. Pei, Z., Liang, C., Lai, L. *et al.* (2003) A high-performance SiGe–Si multiple-quantum-well heterojunction phototransistor. *IEEE Electron. Dev. Lett.*, **24**(10), 643–645.
51. Tashiro, T., Tatsumi, T., Sugiyama, M. *et al.* (1997) A selective epitaxial SiGe/Si planar photodetector for Si-based OEIC's. *IEEE Trans. Electron. Dev.*, **44**(4), 545–550.
52. Winnerl, S., Buca, D., Lenk, S. *et al.* (2002) Fast IR Si/SiGe superlattice MSM photodetectors with buried $CoSi_2$ contacts. *Microelectron. Eng.*, **64**(1–4), 205–209.
53. El Kurdi, M., Boucaud, P., Sauvage, S. *et al.* (2003) Silicon-on-insulator and SiGe waveguide photodetectors with Ge/Si Self-assembled islands. *Physica E: Low-Dim. Sys. Nanostruc.*, **16** (3–4), 523–527.
54. El Kurdi, M., Boucaud, P., Sauvage, S. *et al.* (2002) Near-infrared waveguide photodetector with Ge/Si self-assembled quantum dots. *Appl. Phys. Lett.*, **80**(3), 509–511.
55. Elfving, A., Karim, A., Hansson, G., and Ni, W.-X. (2006) Three-terminal Ge Dot/SiGe quantum-well photodetectors for near-infrared light detection. *Appl. Phys. Lett.*, **89**(8), pp. 083510-1 to -3.
56. Elfving, A. (2006) *Near-Infrared Photodetectors Based on Si/SiGe Nanostructures*, Dissertation No. 1003, Linköping University, Linköping, Sweden.
57. Li, C., Yang, Q., Wang, H. *et al.* (2000) Back-incident SiGe-Si multiple quantum-well resonant-cavity-enhanced photodetectors for 1.3-μm operation. *IEEE Photon. Tech. J.*, **12**(10), 1373–1375.
58. Lai, K-S., Huang, J-C., and Hsu, K. (2007) High-responsivity photodetector in standard SiGe BiCMOS technology. *IEEE Electron. Dev. Lett.*, **28**(9), 800–802.
59. Lai, K-S., Huang, J-C., and Hsu, K. (2008) Design and properties of phototransistor photodetector in standard 0.35-μm SiGe BiCMOS technology. *IEEE Trans. Electron. Dev.*, **55**(3), 744–781.
60. Schneider, K. and Zimmerman, H. (2006) Photodetectors, in *Highly Sensitive Optical Receivers*, Springer, Berlin.
61. Masini, G., Cencelli, V., Colace, L., and Notaristefani, F. (2004) Linear array of Si-Ge hetero-junction photodetectors monolithically integrated with silicon CMOS readout electronics. *IEEE J. Sel. Top. Quant. Electron.*, **10**(4), 811–815.

62. Okyay, A. (2007) *Si-Ge Photodetection Technologies for Integrated Optoelectronics*, PhD dissertation, Stanford University, Palo Alto, CA.
63. Oh, J., Banerjee, S., and Campbell, J. (2004) Metal-germanium-metal photodetectors on heteroepitaxial Ge-on-Si with amorphous Ge Schottky barrier enhancement layers. *IEEE Photon. Tech. Lett.*, **16**(2), 581–583.
64. Rouviere, M., Vivien, L., Le Roux, X. *et al.* (2005) Ultrahigh speed germanium-on-silicon-on-insulator photodetectors for 1.31 and 1.55 μm operation. *Appl. Phys. Lett.*, **87**(23), pp. 231109-1 to -3.
65. Jutzi, M., Berroth, M., Wohl, G. *et al.* (2004) SiGe PIN photodetector for infrared optical ber links operating at 1.25 Gbit/s. *Appl. Surf. Sci.*, **224**(1–4), 170–174.
66. Oehme, M., Werner, J., Kasper, E. *et al.* (2006) High bandwidth Ge p-i-n photodetector integrated on Si. *Appl. Phys. Lett.*, **89**(7), 071117-1–071117-3.

9

Raman Lasers

9.1 Introduction

Light emitters in the form of light-emitting diodes (LEDs) and laser diodes using silicon and related materials have been discussed in Chapter 6. The present chapter describes yet another form of light emitter, the operation of which depends on stimulated Raman emission.

The Raman effect was discovered in 1928 by Raman and Krishnan [1] in some organic liquids. Subsequently, the effect was observed in crystalline solids. Basically the effect is the scattering of photons by vibrational modes in materials or by optical phonons in crystalline solids. As a consequence of scattering, the frequency of the incoming radiation is downshifted (Stokes line) or upshifted (anti-Stokes line). The effect has been employed as a powerful analytical tool in spectroscopy. Simulated Raman scattering, in which a strong pump power causes a growth of the Stokes radiation, has been the subject of intensive research in materials since the 1960s. The possibility of Raman amplification in optical fibers was investigated as early as 1973 [2]. Recent years have witnessed a spectacular growth of investigations and practical applications of Raman amplifiers, in particular in optical fiber communication. Fiber Raman amplifiers in conjunction with Er-doped fiber amplifiers (EDFAs) are essential components in long-haul communications networks [3].

The Raman effect in crystalline silicon received attention from workers as early as 1973 [4]. The possibilities of stimulated Raman scattering (SRS) and consequent amplification of Stokes radiation were studied by a number of workers. Only very recently, a few groups announced Raman lasers based on silicon.

The Raman effect is a nonlinear phenomenon. There exist other nonlinear optical effects, notably two-photon absorption (TPA) in silicon. TPA tends to reduce Raman gain in silicon. On the other hand, nonlinearity in the refractive index can be exploited to produce photonic devices like wavelength converters and so on, which are essential parts in all optical networks.

In this chapter, we shall first discuss the general principle of Raman scattering: spontaneous and stimulated. The characteristics of emission in crystal silicon will then be presented. The structures and experimental arrangements necessary for observing stimulated Raman scattering in silicon will then be described. The results obtained so far by different groups will be presented next, and some of the application areas will be pointed out. Other nonlinear effects expected to play significant roles in photonic devices will be discussed after this, and the device application aspects will be considered.

9.2 Raman Scattering: Basic Concepts

9.2.1 Stokes and Anti-Stokes Lines

When a beam of monochromatic light of frequency ω_0 passes through a transparent medium (solid, liquid, or gas), the light scattered at right angles to the incident beam is found to contain, in addition to the original frequency, other frequency components. The spectrum of the scattered light shows a number of new lines situated symmetrically on both sides of the original line. This type of incoherent scattering was first observed in liquid by Raman in 1928 and is known as the *Raman effect*. Subsequently, the phenomenon was also observed in solids and gaseous media.

The satellite lines on the low-frequency (high-wavelength) side are called the *Stokes lines*, and those on the high-frequency (low-wavelength) side are known as *anti-Stokes lines*. Taken together, all these satellite lines are usually referred to as *Raman lines*.

The origin of the Stokes and anti-Stokes lines lies in the interaction between the incident photon of energy $\hbar\omega_l$ and quantized vibrational modes of energy $\hbar\omega_v$ in liquids or optical phonons of energy $\hbar\omega_0$ in crystalline solids.

Assume that a plane electromagnetic field described by

$$\mathbf{F}(\mathbf{r}, t) = \mathbf{F}_0(\mathbf{k}, \omega)\exp[j(\mathbf{k}\cdot\mathbf{r}-\omega t) + cc], \tag{9.1}$$

is present in the medium, inducing a sinusoidal polarization of the form

$$\mathbf{P}(\mathbf{r}, t) = \mathbf{P}(\mathbf{k}, \omega)\exp[j(\mathbf{k}\cdot\mathbf{r}-\omega t) + cc]. \tag{9.2}$$

The term *cc* means *complex conjugate*. The polarization and field are related by

$$\mathbf{P} = \chi\mathbf{F}. \tag{9.3}$$

At finite temperatures, excited atomic vibrations will cause a fluctuation in susceptibility χ. The normal modes of atomic vibrations are quantized into phonons, and displacements related to phonons are expressed as

$$\mathbf{Q}(\mathbf{r}, t) = \mathbf{Q}(\mathbf{q}, \omega_0)\exp[j(\mathbf{q}\cdot\mathbf{r}-\omega_0 t] + cc, \tag{9.4}$$

with wave vector $\mathbf{q}$ and frequency ω_0. Under quasi-static or adiabatic approximation, electronic frequencies determining χ are large compared to ω_0, and hence χ is assumed to be a function of $\mathbf{Q}$. One may expand χ as a Taylor series and thus write

$$\chi(\mathbf{k}, \omega, \mathbf{Q}) = x_0(\mathbf{k}, \omega) + (\partial\chi/\partial\mathbf{Q})_0\mathbf{Q}(\mathbf{r}, t), \tag{9.5}$$

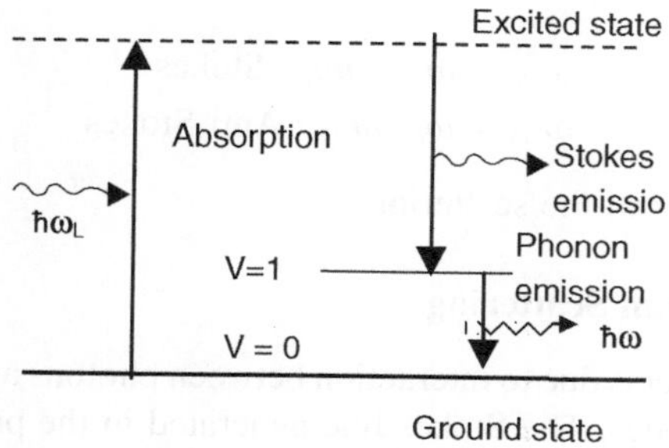

Figure 9.1 *Illustration of the creation of Stokes photons by molecules.*

where the first term represents the susceptibility of the medium without any fluctuation. The second term is induced by lattice waves. The corresponding nonlinear polarization is

$$\mathbf{P_{NL}}(\mathbf{r}, t, \mathbf{Q}) = (\partial\chi/\partial\mathbf{Q})_0\mathbf{Q}(\mathbf{r}, t)\mathbf{F}(\mathbf{r}, t). \tag{9.6}$$

Substituting the forms of **Q** and **F** as given by Eqs. (9.4) and (9.1), we rewrite the above as

$$\begin{aligned} P_{NL}(\mathbf{r}, t, \mathbf{Q}) &= (\partial\chi/\partial\mathbf{Q})_0\mathbf{Q}(\mathbf{q}, \omega_0)\cos(\mathbf{q}\cdot\mathbf{r}-\omega_0 t) \times \mathbf{F_0}(\mathbf{k}, \omega)\cos(\mathbf{k}\cdot\mathbf{r}-\omega t) \\ &= (1/2)(\partial\chi/\partial\mathbf{Q})_0\mathbf{Q}(\mathbf{q}, \omega_0)\mathbf{F_0}(\mathbf{k}, \omega) \qquad (9.7) \\ &\quad \times\{\cos[(\mathbf{k}+\mathbf{q})\cdot\mathbf{r}-(\omega+\omega_0)t] + \cos[(\mathbf{k}-\mathbf{q})\cdot\mathbf{r}-(\omega-\omega_0)t]\} \end{aligned}$$

It is easy to conclude that the polarization consists of two sinusoidal waves: a *Stokes* wave with frequency $\omega_S = (\omega - \omega_0)$ and wavevector $\mathbf{k}_S = \mathbf{k} - \mathbf{q}$, and an *anti-Stokes* wave with frequency $\omega_{AS} = (\omega + \omega_0)$ and wavevector $\mathbf{k}_{AS} = \mathbf{k} + \mathbf{q}$.

We illustrate the basic transitions involved in Raman scattering by considering the molecular energy levels shown in Figure 9.1.

In Figure 9.1, the different molecular energy levels are denoted by the symbol v. In Figure 9.1, the molecule is initially in the ground state, $v = 0$. The molecules absorb the incident photon of energy $\hbar\omega_l$, and simultaneously a Stokes photon of frequency $\omega_{S=}\omega_l - \omega_v$ is emitted. To conserve energy, the molecule is excited to the vibrational level $v = 1$ of energy $\hbar\omega_v$. The emission of anti-Stokes radiation may be understood by using a diagram like Figure 9.1. In this case, the molecule is initially in the excited state ($v = 1$). Absorption of the incident light takes it to a higher level (dashed line). The downward transition to the ground level gives rise to spontaneous emission of the anti-Stokes photon of frequency $\omega_{AS} = \omega_l + \omega_v$. The stimulated absorption of a Stokes photon in the presence of incident light may be illustrated in a similar fashion.

It follows easily from this discussion that the occurrence of anti-Stokes scattering depends on the number of molecules in the excited state ($v = 1$), the number of which is proportional to $\exp(-\hbar\omega_v/k_BT)$. Therefore the intensity of the anti-Stokes line is weaker than that of the Stokes line.

In the discussion in this subsection, only the first-order Stokes and anti-Stokes processes are mentioned. In general, however, multiple absorption and emission of vibrational modes are possible. The expressions for the frequencies of Stokes and anti-Stokes lines may be expressed generally as

$$\begin{aligned} \omega_S &= \omega_l + n\omega_v : \text{Stokes} \\ \omega_{AS} &= \omega_l - n\omega_v : \text{Anti Stokes} \end{aligned} \tag{9.8}$$

where n denotes the order of the scattering.

9.2.2 Stimulated Raman Scattering

As noted, the Raman effect is due to interaction between photons and phonons (in solids) or quantized vibrational modes. The Stokes line generated in the process is spontaneous in nature. In the scattering event, energy gets transferred from one lightwave to another wave of lower frequency (or energy). The lost energy is absorbed by the molecular vibrations or phonons. The wave of lower energy is the Stokes wave. The first wave may be considered as a "pump" wave that amplifies the Stokes wave. As the pump wave propagates through the medium, it continuously depletes by giving energy to the Stokes wave which in turn grows with distance. In spontaneous scattering, there is no self-sustained growth of the Stokes wave. However, in the presence of a sufficiently strong pump wave, the Stokes wave gains enough energy to overcome the loss and as a consequence there is self-sustained growth or oscillation of the Stokes wave. This is the stimulated Raman scattering. In the following, a simplified model will be developed to demonstrate that growth of the Stokes wave is indeed possible.

9.3 Simplified Theory of Raman Scattering

The theory presented in this section is based on electromagnetic (EM) theory and closely follows the treatment given by Wang [5].

In the classical explanation of Raman scattering, a lightwave with an electric field $\mathbf{E} = \mathbf{E}_0 \cos \omega t$ falling on a molecule induces a dipole moment given by

$$\mathbf{p} = \alpha \mathbf{E}_0 \cos \omega t \tag{9.9}$$

In the absence of vibration or rotation of molecules, the polarizability α is a scalar; however, molecules do have vibration or rotation, and consequently the induced polarization changes with direction leading to the tensorial nature of α. The phenomenological expression for α may be written as

$$\alpha = \alpha_0 + \alpha_1 \cos(\omega_v t + \delta) \tag{9.10}$$

where ω_v denotes the angular frequency of one of the Fourier components of the vibration spectra and δ is the random phase angle for spontaneous process. Substituting Eq. (9.10) in Eq. (9.9) shows that the induced dipole moment and hence the scattered light have three components: the original wave with angular frequencies ω and two additional waves with angular frequencies $\omega \pm \omega_v$, representing Stokes and anti-Stokes waves.

To understand SRS in which the phase angle δ appearing in Eq. (9.10) is no longer random, the nonlinear polarization caused by a high field associated with a pump wave must be considered. The electric displacement contains both the linear and nonlinear polarizations and is expressed by

$$\mathbf{D} = \varepsilon \mathbf{E} + \mathbf{P}_{NL} \tag{9.11}$$

The linear component is included in the first term. The nonlinear polarization $\mathbf{P_{NL}}$ is expressed in a power series as

$$P_{NL,a} = \sum_{\alpha,\beta} \chi_{a\alpha\beta} E_\alpha E_\beta + \sum_{\alpha,\beta,\gamma} \chi_{a\alpha\beta\gamma} E_\alpha E_\beta E_\gamma \tag{9.12}$$

where $a = \alpha, \beta$, and $\gamma = x,y,z$. The susceptibility χ is tensor, and SRS arises due to the presence of the cubic polarization, that is, due to $\chi_{a\alpha\beta}$. In view of Eq. (9.11), Maxwell's equations are written as

$$\nabla \times \mathbf{E} = -j\omega\mu_0\mathbf{H} \tag{9.13a}$$

$$\nabla \times \mathbf{H} = j\omega\varepsilon\mathbf{E} + j\omega\mathbf{P_{NL}} \tag{9.13b}$$

Assuming the plane wave nature, that is, an $\exp[j(\omega t - \mathbf{k}\cdot\mathbf{r})]$-like variation, the following wave equation results:

$$\nabla \times \nabla \times \mathbf{E} - \omega^2\mu_0\varepsilon\mathbf{E} = \omega^2\mu_0\mathbf{P_{NL}} \tag{9.14}$$

For solids, the motion of nuclei is to be considered and the electronic polarizability depends parametrically on the displacement $\mathbf{R}$ of the nucleus from its position of equilibrium in the following way:

$$\alpha = \alpha_0 + \frac{\partial\alpha}{\partial R}R + \ldots \tag{9.15}$$

Since R is related to optical vibrational waves, one may write,

$$R = 2\mathrm{Re}\,A_v \exp[j(\omega_v t - \mathbf{k}\cdot\mathbf{r})], \tag{9.16}$$

where A_v, ω_v and $\mathbf{k_v}$ denote, respectively, the amplitude, angular frequency, and wave vector of the vibrational wave; and Re denotes the *real part*. Assume that two lightwaves are present so that the total electric field is

$$\mathbf{E} = \mathbf{E_0}\exp[j(\omega_0 t - \mathbf{k_0}\cdot\mathbf{r})] + \mathbf{E_1}\exp[j(\omega_1 t - \mathbf{k_1}\cdot\mathbf{r})]. \tag{9.17}$$

The component of the nonlinear polarization at frequency ω_1 becomes

$$P_{NL}(\omega_1) = N\frac{\partial\alpha}{\partial R}A_v^* E_0 \exp[j(\omega_1 t - \mathbf{k_1}\cdot\mathbf{r})], \tag{9.18}$$

with the condition that $\omega_0 = \omega_1 + \omega_v$, $\mathbf{k_0} = \mathbf{k_1} + \mathbf{k_v}$. In Eq. (9.18), N is the number of molecules per unit volume. The component for nonlinear polarization at frequency ω_0 may be expressed by a similar equation.

The next task is to find the amplitudes E_1 and A_v. Substituting Eqs. (9.17) and (9.18) in Eq. (9.14) and assuming that the variation of E_1 over a wavelength is small, that is, $|\partial^2 E_1/\partial z^2| \ll k_1|\partial E_1/\partial z|$, one obtains

$$2jk_1\frac{\partial E_1}{\partial z} = N\frac{\partial\alpha}{\partial R}\omega_1^2\mu_0 E_0 A_v^*. \tag{9.19}$$

To obtain an equation for A_v, the free energy U for the medium containing N molecules per unit volume is given according to Eq. (9.15) as

$$U = \frac{N}{2}\alpha_0 E^2 + \frac{N}{2}\frac{\partial\alpha}{\partial R} R E^2. \tag{9.20}$$

The first term in above stands for the ordinary polarization energy, while the second one arising due to nonlinear polarization is the additional free energy due to *photoelastic coupling*. The quantity R is the displacement of the nucleus from its equilibrium position and is proportional to the strain S. The additional stress T due to photoelastic coupling is then written as

$$T = \frac{\partial U}{\partial S} = \frac{1}{B}\frac{N}{2}\frac{\partial\alpha}{\partial R} E^2, \tag{9.21}$$

where B is a proportionality constant such that $S = BR$.

Equation (9.21) gives the stress component at frequency ω_v as

$$T(\omega_v) = \frac{1}{B}\frac{N}{2}\frac{\partial\alpha}{\partial R} E_0 E_1^* \exp\{j[\omega_v t - (\mathbf{k_0} - \mathbf{k_1}) \cdot \mathbf{r}]\}. \tag{9.22}$$

The differential equation relating displacement and stress in a piezoelectric solid is

$$\rho \frac{d^2 \varsigma}{dt^2} = \frac{\partial T}{\partial z}. \tag{9.23}$$

Adding the component $T(\omega_v)$ and noting that $S = \partial\varsigma/\partial z$, one obtains

$$\rho \frac{\partial^2 S}{\partial t^2} - \frac{1}{s}\frac{\partial^2 S}{\partial z^2} = \frac{\partial^2 T(\omega_v)}{\partial z^2}. \tag{9.24}$$

Here ρ is the mass density and the elastic constant s relates the stress and strain components by the relation

$$S_i = \sum_j s_{ij} T_j, \quad i,j = 1 \text{ to } 6. \tag{9.25}$$

The vibrational mode is assumed to be a longitudinal wave propagating along the z-direction. Substituting Eq. (9.22) in Eq. (9.24) and using $S = BR$, the following differential equation obtains:

$$\rho \frac{\partial^2 R}{\partial t^2} - \frac{1}{s}\frac{\partial^2 R}{\partial z^2} = k_{vg}^2 \frac{E_0 E_1^*}{B^2}\frac{N}{2}\frac{\partial\alpha}{\partial R} \exp[j(\omega_v t - \mathbf{k_v} \cdot \mathbf{r})]. \tag{9.26}$$

Using the expression for R (Eq. (9.16)) and following the same procedure as used in deriving Eq. (9.19) and putting $B = k_v$, we find

$$2jk_{vg}\left(\frac{dA_v}{dz} + \frac{A_v}{v_g \tau}\right) = N \frac{\partial\alpha}{\partial R}\frac{E_0 E_1^*}{2\rho\omega_v^2} k_{vg}^2 \tag{9.27}$$

In Eq. (9.27), v_g is the group velocity of the vibrational (optical) wave, $k_{vg} = \omega_v/v_g = \omega_v(\partial k_v/\partial\omega_v)$, and the term $A_v/(v_g\tau)$ represents the loss associated with the vibration wave, with τ denoting the characteristic damping time. Equation (9.27) can be derived by using the normalized vibrational coordinates instead of R, and $B = k_v$ is used to make the present result agree with that.

From Eq. (9.27), it appears that a coherent generation of vibrational waves can occur if the power received from the pump laser beam can overcome the loss. The condition for SRS is therefore

$$\frac{2\pi}{\rho\omega_v^2}\left(N\frac{\partial\alpha}{\partial R}\right)^2\left(\frac{\omega_1}{c}\right)^2\left|\frac{E_0}{2}\right|^2\frac{k_{vg}}{k_1} \geq \frac{1}{(v_g\tau)^2}. \tag{9.28}$$

In the above derivation, the two light beams differ in frequency by $\omega_0 = \omega_1 + \omega_v$. Thus the secondary beam of frequency $\omega_1 = \omega_0 - \omega_v$ represents the Stokes wave. Once the first Stokes line grows coherently, its intensity builds up to serve as the pump wave for the second Stokes line of frequency $\omega_2 = \omega_0 - 2\omega_v$. In this way, more and more higher order Stokes lines are coherently generated.

9.4 Raman Effect in Silicon

Research on Raman scattering in bulk silicon was initiated in the early 1970s, to study optical phonons in the material (see Temple and Hathaway in [4]). The zone center optical phonon in crystalline Si (c-Si) is threefold degenerate and is responsible for a Stokes shift in the material. Figure 9.2 gives the observed Raman spectra. The scattered Stokes radiation is downshifted from the pump frequency by an amount equal to the zone center optical phonon frequency, and the shift is shown as 15.6 THz in Figure 9.2. The full width at half maximum (FWHM) of the spectra is ~105 GHz, which corresponds to a response time of ~ 10 ps. The second-order Stokes line is also observed, as indicated in Figure 9.2. However, its intensity is more than an order of magnitude lower.

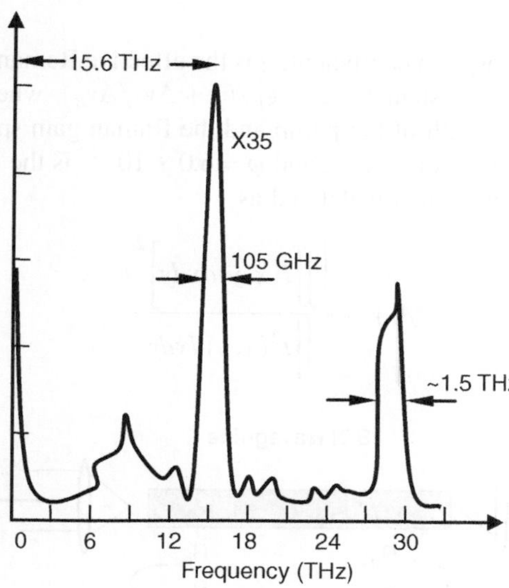

Figure 9.2 *Raman spectra in crystalline silicon. Reproduced with permission from [4]. Copyright (1973) American Physcial Society.*

9.5 Raman Gain Coefficient

In this section, a numerical analysis made by Krause *et al.* [6] of Raman lasing characteristics in silicon-on-insulator (SOI) waveguides will be presented. The analysis shows that in spite of the presence of detrimental-effect two-photon absorption and free-carrier absorption (FCA), continuous-wave (CW) Raman lasing is possible.

9.5.1 Mathematical Model

The analysis assumes the presence of a SOI waveguide of length L, as shown in Figure 9.3. A pump laser light of wavelength λ_p and power P_O is coupled to the left-hand side of the waveguide with coupling efficiency T_p. The Stokes wave generated inside the waveguide due to Raman scattering has a wavelength λ_s. The end surfaces of the waveguide are coated with reflective materials. The reflectivities are denoted by R, and the suffixes l, r, p, and s refer, respectively. to the left mirror, right mirror, pump, and Stokes wave. For example, $R_{p,r}$ denotes the reflectivity of the pump at the left-hand mirror.

A simple incoherent model that describes the laser in terms of longitudinally varying optical powers of forward- and backward-propagating waves, but neglects phase-related effects, will be used. The model is similar to that using fiber Raman lasers but includes TPA and FCA.

Considering the depletion of pump due to SRS, and including backward propagation, the differential equations for forward- and backward-propagating pump waves (P_p^+ and P_p^-, respectively) are

$$\pm\frac{dP_p^{\pm}}{dz} = \left\{-\alpha - \frac{g}{A_{eff}}\frac{\lambda_s}{\lambda_p}(P_s^+ + P_s^-) - \frac{\beta}{A_{eff}}\left[P_p^{\pm} + 2\left(P_p^{\mp} + P_s^+ + P_s^-\right)\right] - \bar{\varphi}\lambda_p{}^2\bar{N}_{eff}\right\}P_p^{\pm} \tag{9.29}$$

In Eq. (9.29), α is the absorption coefficient; g is the effective Raman gain constant related to the peak Raman gain constant by $g = g_R/(1 + \Delta v_p/\Delta v_R)$, where Δv_p and Δv_R are, respectively, the spectral width of the pump and the Raman gain spectrum; β is the two-photon absorption coefficient of silicon; and $\bar{\varphi} = 6.0 \times 10^{-10}$ is the efficiency of the FCA process. The modal effective area is defined as

$$A_{eff} = \frac{\left[\iint I(x,y)dxdy\right]^2}{\iint I^2(x,y)dxdy}, \tag{9.30}$$

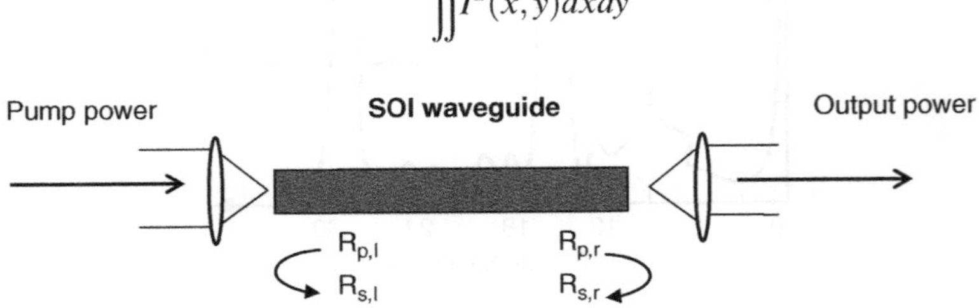

Figure 9.3 Structure of the Raman laser used for analysis.

where *I(x,y)* is the intensity profile of the fundamental waveguide mode in the plane transverse to the direction of propagation. The effective charge carrier density is given by

$$\bar{N}_{eff}(z) = \frac{\beta\tau_{eff}}{2h\nu_p A_{eff}^2}\left\{\begin{array}{l} P_p^{+2} + P_p^{-2} + P_s^{+2} + P_s^{-2} + \\ 4\left[P_p^+ P_p^- + P_s^+ P_s^- + (P_p^+ + P_p^-)(P_s^+ + P_s^-)\right] \end{array}\right\} \quad (9.31)$$

where $\nu_s \approx \nu_p$ is used, h denotes the Planck constant, and τ_{eff} is an effective charge carrier lifetime, whose typical values are in the range of 0.7–100 ns.

The variation of Stokes power $P_s^{\pm}$ can be expressed as

$$\pm\frac{dP_s^{\pm}}{dz} = \left\{-\alpha + \frac{g}{A_{eff}}(P_p^+ + P_p^-) - \frac{\beta}{A_{eff}}\left[P_s^{\pm} + 2(P_s^{\mp} + P_p^+ + P_p^-\right] - \bar{\varphi}\lambda_s^2\bar{N}_{eff}\right\}P_s^{\pm}. \quad (9.32)$$

The absorption coefficients at pump and Stokes wavelengths are assumed equal. The boundary conditions at the two ends $z = 0$ and L are

$$P_p^+(0) = T_p P_0 + R_{p,l}P_p^-(0), \quad P_p^-(L) = R_{p,r}P_p^+(L), \quad (9.33)$$

$$P_s^+(0) = R_{s,l}P_s^-(0), \quad P_s^-(L) = R_{s,r}P_s^+(L). \quad (9.34)$$

9.5.2 Simulation Parameters

First of all, reflectivities are assumed due to the Si–air interface, for which $R = 30\%$. The transmittivity of pump power is $T_p = 1 - R_{p,l}$. Assuming peak Raman gain $g_R = 30$ cm/GW, $\Delta\nu_p = 50$ GHz, and $\Delta\nu_R \approx 100$ GHz, effective Raman gain constant $g = 20$ cm/GW. A large effective modal area $= 5\ \mu m^2$, $\lambda_p = 1427$ nm, and $\lambda_s = 1542$ nm are chosen so that the Stokes shift is 15.6 THz.

9.5.3 Threshold Power

Around threshold, the Stokes power is much smaller than the pump power, and Eqs. (9.29), (9.32), and (9.31) are simplified as

$$\pm\frac{dP_p^{\pm}}{dz} = \left\{-\alpha - \bar{\varphi}\lambda_p^2\bar{N}_{eff} - \frac{\beta}{A_{eff}}\left[P_p^{\pm} + 2P_p^{\mp}\right]\right\}P_p^{\pm}, \quad (9.35)$$

$$\pm\frac{dP_s^{\pm}}{dz} = \left\{-\alpha - \bar{\varphi}\lambda_s^2\bar{N}_{eff} + \frac{g - 2\beta}{A_{eff}}\left(P_p^+ + P_p^-\right)\right\}P_s^{\pm}, \quad (9.36)$$

$$\bar{N}_{eff}(z) = \frac{\beta\tau_{eff}}{2h\nu_p A_{eff}^2}\left(P_p^{+2} + P_p^{-2} + 4P_p^+ P_p^-\right). \quad (9.37)$$

The laser is at threshold when the round-trip gain of the Stokes signal equals the losses due to reflectivities at the left- and right-hand end mirrors with reflectivities $R_{s,l}$ and $R_{s,r}$. From

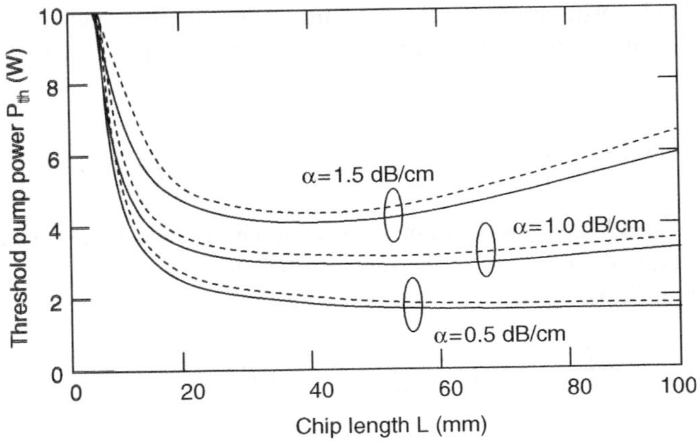

Figure 9.4 Calculated threshold pump power versus chip length neglecting FCA. Solid and dashed curves are obtained with $\beta = 0$ (absence of TPA) and with $\beta = 0.7$ cm/GW. Reproduced with permission from [6]. Copyright (2004) The Optical Society.

Eqs. (9.36) and (9.34), we end up with threshold condition

$$R_{s,l}R_{s,r}\exp\left\{2\int_0^L \left[-\alpha - \bar{\varphi}\lambda_s^2 \bar{N}_{eff}(z) + \frac{g-2\beta}{A_{eff}}\left[P_p^+(z) + P_p^-(z)\right]dz\right\}\right. = 1. \quad (9.38)$$

The threshold pump power, the pump power distribution along z, is obtained numerically by using Eqs. (9.35) and (9.33) for different pump powers P_0 until $P_p^+(z)$ and $P_p^-(z)$ satisfy Eq. (9.38). The corresponding pump power is then the threshold pump power P_{th}.

The threshold pump power of the SOI Raman laser for different values of chip length L is plotted in Figure 9.4. In Figure 9.4, the solid lines represent the variation when both FCA and TPA are neglected, that is, $\beta = 0$ and $\tau_{eff} = 0$. The results for three values of α indicate that it is possible to obtain laser action by using a few watts of pump laser output. The dashed curves in Figure 9.4 are calculated, including TPA with value $\beta = 0.7$ cm/GW, but still assuming FCA to be absent. TPA is seen to increase the threshold but only weakly.

The influence of FCA on the threshold pump power is illustrated in Figure 9.5 for a fixed waveguide loss $\alpha = 1.0$ dB/cm. The solid and dashed lines are once again obtained with and without TPA. Different curves in Figure 9.5 correspond to different values of charge carrier lifetimes (τ_{eff}), the increase of which increases the threshold. There exists a limiting value of chip length L above which there is no threshold at all. For example, for $L = 80$ mm and $\tau_{eff} > 3.0$ ns, the device has no lasing threshold. For larger values of τ_{eff} the usable range becomes smaller, and at τ_{eff} 3.2 ns the range vanishes. The dotted curves show the shutdown thresholds.

The variation of output power with input pump power is shown in Figure 9.6, with effective carrier lifetimes τ_{eff} as the parameter. The dashed curve corresponds to an absence of TPA and FCA. The curves clearly indicate the existence of a threshold pump power.

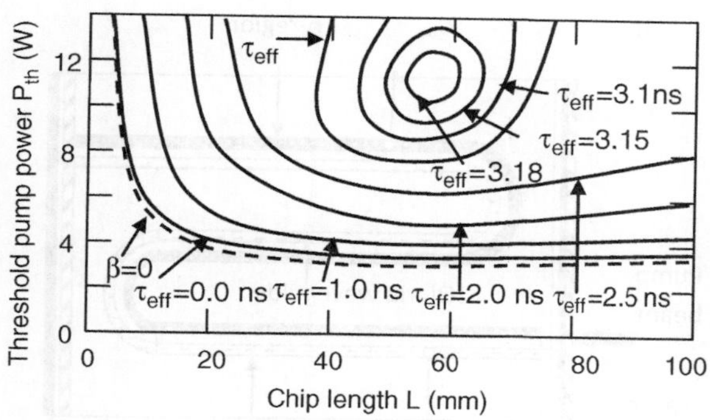

Figure 9.5 *Threshold pump power versus chip length for SOI Raman lasers. Reproduced with permission from [6]. Copyright (2004) The Optical Society.*

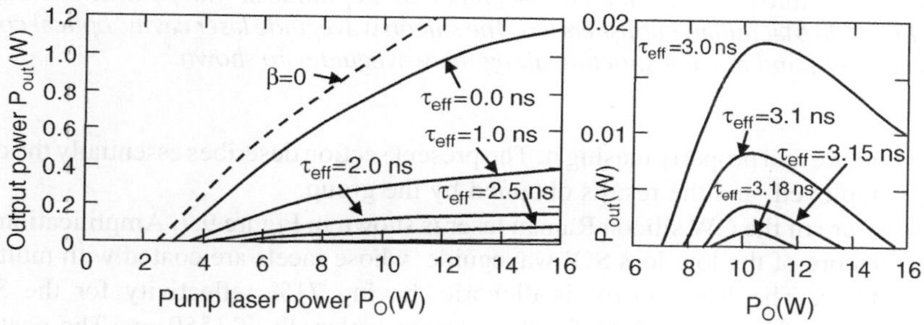

Figure 9.6 *Input–output characteristics of SOI Raman lasers with $L = 55$ mm and for several values of τ_{eff}. The curves for large τ_{eff} are shown in the right-hand side. Reproduced with permission from [6]. Copyright (2004) The Optical Society.*

The output power increases first with pump power. However, increasing the pump power also increases the losses for both the pump and Stokes waves via nonlinear loss mechanisms like FCA and TPA. This leads to the breakdown of lasing operations at the shutdown threshold. The effect of large τ_{eff} is shown in the right, and the curves indicate that the conversion efficiency dramatically decreases and the lasing and shutdown thresholds come closer to each other.

9.6 Continuous-Wave Raman Laser

We now present some experimental results related to continuous-wave Raman lasing. Several groups reported Raman lasing in SOI waveguides; however, the operation was in pulsed mode [7–13]. As may be inferred from discussions in this chapter, the main disadvantage in having proper gain is the absorption by free carriers generated by TPA. The detrimental role of TPA was pointed out by Liang and Tsang [14].

The CW operation of the Raman laser was first reported by the workers at Intel Corporation [15, 16]. They circumvented the problem by wiping out the electrons by using

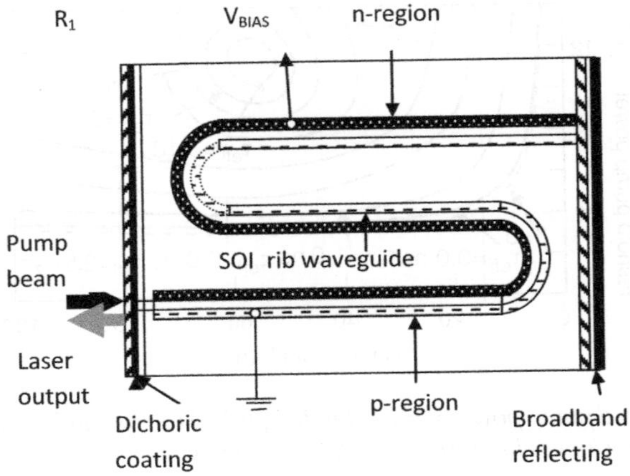

Figure 9.7 Structure of the CW Raman laser reported. Reproduced with permission from [15]. Copyright (2005) Macmillan Publishers Ltd. The silicon waveguide laser cavity, optical coatings applied to facets, and a p–i–n structure along the waveguide are shown.

a p–i–n structure and properly biasing it. The present section describes essentially the device structure employed, and the results obtained by the group.

The structure of the CW silicon Raman laser is shown in Figure 9.7. Amplification takes place in the core of the low-loss SOI waveguide, whose facets are coated with multilayer dielectric films. The front mirror is dichoric, having 71% reflectivity for the Stokes wavelength of 1686 nm, and 24% for the pump wavelength of 1550 nm. The coating at the back facet has a reflectivity of 90% for both the pump and Stokes wavelengths.

The waveguide, the schematic of which is shown in Figure 9.8, has a rib structure and is fabricated on the (100) surface of an undoped SOI substrate using standard photolithography and RIE. The waveguide dimensions were so chosen that a small threshold pump power is ensured and at the same time high transmission loss is avoided. The dimensions of

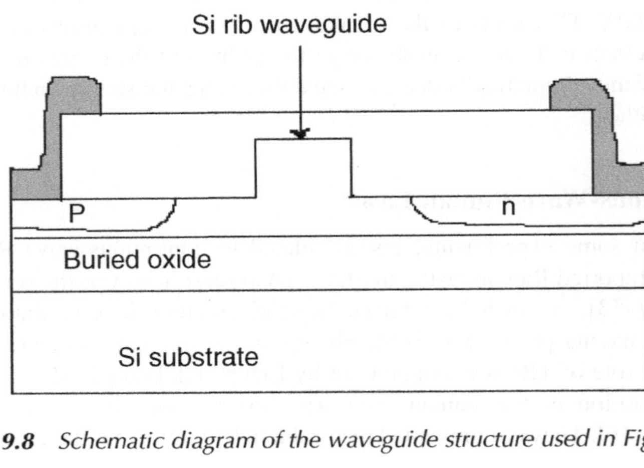

Figure 9.8 Schematic diagram of the waveguide structure used in Figure 9.7.

the structure are given in the reference. The waveguide is in the form of an S-shaped curve with a total length of 4.8 cm, and the straight section of the waveguide is oriented along the [011] direction.

As mentioned already, a pin diode structure was used to reduce the TPA-induced free-carrier absorption loss. The p and n regions are obtained by implanting B and P in the slab on either side of the rib waveguide, as illustrated in Figure 9.8.

The TPA process creates EHPs that contribute to the overall loss and increase in the threshold pump power. The reverse bias between the p- and n-doped regions of the pin diode sweeps these pairs, thus reducing the lifetime of the free carriers. The measured photocurrent in the diode is found to increase with the square of the light power inside the waveguide, indicating thereby that the charge carriers are generated by the TPA process. At a reverse bias of 25 V, the effective carrier lifetime reduces to only 1 ns, while the value is as large as a few tens of nanosecons in ordinary Si rib waveguides.

The experimental arrangement for the measurement of Raman gain consists of an external cavity diode laser operating at 1550 nm, amplified by an EDFA system to produce a pump beam of power up to 3 W. The pump beam is then allowed to pass through a polarization controller and a thin film–based wavelength demultiplexer (demux). The beam is then launched into the waveguide cavity by a lensed fiber through the dichoric mirror at the front facet. The Raman laser output and the reflected pump beam are coupled back into the lensed fiber. The Raman beam, separated by a demux, is further filtered by a low-pass filter and finally analyzed by a spectrum analyzer or detected by a power meter.

The dichoric mirror at the front facet and the high-reflectivity mirror at the back facet form a low-finesse cavity at the pump wavelength. When the pump laser is tuned to the resonant wavelength of the cavity, the circulating power inside the cavity is enhanced due to the resonant cavity effect. This ensures lowering of the lasing threshold.

The measured Raman laser output power is plotted in Figure 9.9 against the input pump power for two different voltages, 5 and 25 V, applied to the pin diode. As is seen from

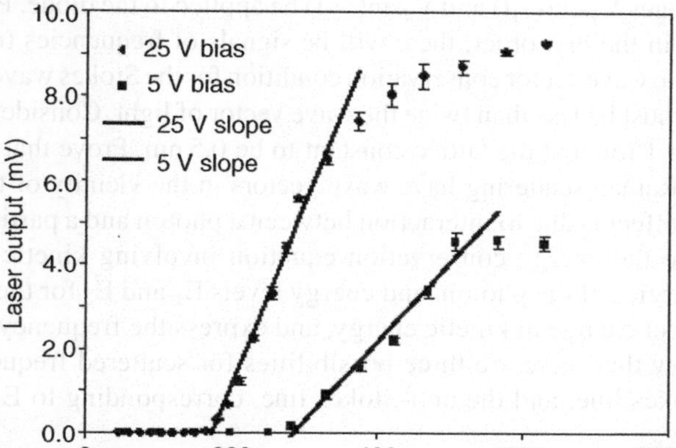

Figure 9.9 *Raman laser output power versus input pump power for 5 and 25 V bias voltages applied to the pin diode. Reproduced with permission from [15]. Copyright (2005) Macmillan Publishers Ltd.*

Figure 9.9, the lasing threshold is about 180 mW for 25 V reverse bias, whereas it is 280 mW at 5 V. The lower threshold and higher laser power at higher reverse bias are due to a shorter effective carrier lifetime, resulting in lower nonlinear loss and higher gain.

The laser output tends to saturate at higher pump power levels. Since TPA-induced FCA increases at higher pump power, the net gain is reduced. In addition, the cavity enhancement factor for the pump reduces with increasing nonlinear loss. These factors make the output power to saturate.

The spectrum of the laser output was measured, and a very narrow linewidth ~80 MHz was obtained for single-mode output.

9.7 Further Developments

Further improvement of the performance of the laser may be made by (1) optimizing the cavity mirror and cavity length design, (2) reducing the threshold power by using a waveguide of smaller cross-sections and/or by introducing a larger cavity enhancement for the pump beam, (3) improving the fiber-to-waveguide coupling efficiency by adding a mode converter, (4) reducing the effective carrier lifetime below 1 ns by optimizing the p–i–n diode design, and (5) replacing the multilayer mirrors with waveguide Bragg reflectors, rings, or microdisk resonators.

It is to be mentioned that Raman lasing has since been observed by other groups and in other material systems. Attempts are continuously being made to improve its performance. However, inclusion of all these recent works is beyond the scope of the book. The reader is referred to some of the recent publications [17–20].

Problems

9.1 Consider the expression for the current–voltage characteristic of a p–n junction diode. Let two voltages $V_1\sin(\omega_1 t)$ and $V_2\sin(\omega_2 t)$ be applied to the diode. Prove that, due to nonlinearity in the first order, there will be signals of frequencies $(\omega_1 \pm \omega_2)$.

9.2 Show from the wave vector conservation condition for the Stokes wave that the phonon wavevector must be less than twice the wave vector of light. Consider the wavelength of light to be 1 μm and the lattice constant to be 0.5 nm. Prove that optical phonons involved in Raman scattering have wave vectors in the vicinity of the zone center.

9.3 The Raman effect is due to interaction between a photon and a particle (molecule or atom). Write the energy conservation equation involving kinetic energies of the particle, energies of the photon, and energy levels E_1 and E_2 for the initial and final states. Neglect change in kinetic energy, and express the frequency of the scattered photon. Show that there are three possibilities for scattered frequencies: unmodified, the Stokes line, and the anti-Stokes line, corresponding to $E_1 = E_2$, $E_1 > E_2$, and $E_1 < E_2$.

9.4 The intensity of spontaneous emission is proportional to the number of particles in the energy level. Using Boltzmann statistics, prove that the intensity of the anti-Stokes line is much less than that for the Stokes line.

9.5 Calculate the loss due to TPA in Si by using $I_p = 4\ \mathrm{W/cm^2}$, $I_s = 0.2\ \mathrm{W/cm^2}$, $\beta = 0.5$ cm/GW, and $\tau_{eff} = 2.0$ ns Assume that the Stokes wavelength is 1550 nm.

References

1. Raman, C.V. and Krishnan, K.S. (1928) A new type of secondary radiation. *Nature*, **121**, 501–502.
2. Stolen, R.H. and Ippen, E.P. (1973) Raman gain in glass optical fiber waveguides. *Appl. Phys. Lett.*, **22**, 276–278.
3. Islam, M.N. (2003) *Raman Amplifiers for Telecommunications: Sub-systems and Systems*, Springer,
Berlin.
4. Temple, P.A. and Hathaway, C.E. (1973) Multiphonon Raman spectrum of silicon. *Phys. Rev. B*, **7**, 3685–3697.
5. Wang, S. (1989) *Fundamentals of Semiconductor Theory and Device Physics*, Prentice-Hall, Englewood Cliffs, NJ.
6. Krause, M., Renner, H., and Brinkmeyer, E. (2004) Analysis of Raman lasing characteristics in silicon-on-insulator waveguides. *Opt. Express*, **12**, 5703–5710.
7. Claps, R., Dimitropoulos, D., Han, Y., and Jalali, B. (2002) Observation of Raman emission in silicon waveguides at 1.54 μm. *Opt. Express*, **10**, 1305–1313.
8. Claps, R., Dimitropoulos, D., Raghunathan, V. *et al.* (2003) Observation of stimulated Raman amplification in silicon waveguides. *Opt. Express*, **11**, 1731–1739.
9. Espinola, R.L., Dadap, J.I., Osgood, R.M. Jr. *et al.* (2004) Raman amplification in ultrasmall silicon-on-insulator wire waveguides. *Opt. Express*, **12**, 3713–3718.
10. Rong, H., Liu, A., Nicolaescu, R. *et al.* (2004) Raman gain and nonlinear optical absorption measurement in a low loss silicon waveguide. *Appl. Phys. Lett.*, **85**, 2196–2198.
11. Boyraz, O. and Jalali, B. (2004) Demonstration of a silicon Raman laser. *Opt. Express*, **12**, 5269–5273.
12. Liu, A., Rong, H., Paniccia, M. *et al.* (2004) Net optical gain in a low loss silicon-on-insulator waveguide by stimulated Raman scattering. *Opt. Express*, **12**, 4261–4267.
13. Xu, Q., Almeida, V., and Lipson, M. (2004) Time-resolved study of Raman gain in highly confined silicon-on-insulator waveguides. *Opt. Express*, **12**, 4437–4442.
14. Liang, T.K. and Tsang, H.K. (2004) Role of free carriers from two-photon absorption in Raman amplification in silicon-on-insulator waveguides. *Appl. Phys. Lett.*, **84**, 2745–2747.
15. Rong, H., Jones, R., Liu, A. *et al.* (2005) A continuous wave Raman silicon laser. *Nature*, **433**, 725–728.
16. Rong, H., Kuo, Y-H., Liu, A. *et al.* (2006) Monolithic integrated Raman silicon laser. *Opt. Express*, **14**, 6705–6712.
17. Sih, V., Xu, S., Kuo, Y-H. *et al.* (2007) Raman amplification of 40 Gb/s data in low-loss silicon waveguides. *Opt. Express*, **15**, 357–362.
18. Rong, H., Xu, S., Kuo, Y.-H. *et al.* (2007) Low threshold continuous-wave Raman silicon laser. *Nature Photonics*, **1**, 232–237.
19. Krause, M., Renner, H., Fathpour, S. *et al.* (2008) Gain enhancement in cladding-pumped silicon Raman amplifiers. *IEEE J. Quant. Electron.*, **44**, 692–704.
20. Jalali, B., Raghunathan, V., Shori, P. *et al.* (2006) Prospects for silicon mid-IR Raman lasers. *IEEE J. Sel. Top. Quant. Electron.*, **12**, 1618–1627.

10

Guided Lightwaves: Introduction

10.1 Introduction

Earlier chapters in this volume discussed the principle of operation of various active devices and problems and the prospects of using Si and its alloys to fabricate the devices. While the field for active devices is not yet well developed, there exists an important area covering passive devices in which Si-based devices are finding widespread use. Mature Si microfabrication facilities are used to fabricate these devices as well as to make optical integrated circuits (OICs).

The basic element in all the passive devices or passive lightwave circuits is an optical waveguide. To understand the light-guiding properties in optical waveguides, one needs to solve the wave equation subject to appropriate boundary conditions. There exists another approach, known as the *ray optic theory*, which proves to be very useful in understanding the basic principle of how light is confined in the structure. The theory has its limitations, of course, but it forms a good starting point.

In this chapter, the ray optic theory of guidance of light as well as the wave theory will be presented for a simple three-layer planar light guide structure. The different practical planar structures, however, deviate considerably from the ideal structure. The guiding mechanism in these structures will be discussed qualitatively, mentioning how the ideal theory may be made use of in explaining their mode behavior.

The present chapter will also discuss the loss mechanisms in the guides and methods of coupling the waveguides to optical fibers.

10.2 Ray Optic Theory for Light Guidance

The condition for light guiding may be established by using Snell's law of refraction. Consider that a light ray E_i propagating in a medium of refractive index (RI) n_1 is incident at the interface between two media at an angle θ_1. The ray will be partially reflected (E_r) with angle θ_1 and partially refracted (E_t) with angle θ_2 into the other medium of RI n_2 (Fig. 10.1).

Silicon Photonics: Fundamentals and Devices, First Edition. M. Jamal Deen and P. K. Basu.

Snell's law states that

$$n_1 \sin\theta_1 = n_2 \sin\theta_2. \tag{10.1}$$

When $n_1 > n_2$, there exists an incident angle θ_c, known as the critical angle, for which the refracted angle is 90°, and the above relation reduces to

$$\sin\theta_c = n_2/n_1. \tag{10.2}$$

For angles of incidence greater than θ_c, there is total internal reflection and no light is transmitted into medium 2.

Consider now the three-layer structure in which medium 1 having RI n_1, called the *core layer*, is sandwiched between two media of RI's n_2, called the *cladding layers*, as shown in Figure 10.2. A light ray in a medium (say, air with RI n_0) is incident at the air–core interface with an angle of incidence α. The ray entering into the core is then incident at the core–upper cladding interface. If the angle exceeds the critical angle, the ray will be totally internally reflected, strike the core–lower cladding layer, get reflected once again, and propagate through the core layer by following the zigzag path as shown. This is light guidance by the core layer. A simple geometrical argument will reveal that there exists a cone of maximum half angle α_{max} and if the light rays in air are confined within this cone, they will be guided by the core by total internal reflection. The numerical aperture of the guide is defined as

$$n_0 \sin\alpha_{max} = n_1 \cos\theta_c = \sqrt{n_1^2 - n_2^2} \approx n_1\sqrt{2\Delta}, \tag{10.3}$$

where $\Delta = (n_1 - n_2)/n_1$ is the fractional RI difference.

This simplified approach might tempt anybody to conclude that the waveguide will guide light when the incidence angle may have any value exceeding the critical angle. This is not true, however, as will be shown in Section 10.3.

10.3 Reflection Coefficients

The complex amplitudes of the incident and reflected rays, denoted by E_i and E_r respectively in Figure 10.1, are related to each other by

$$E_r = rE_i \tag{10.4}$$

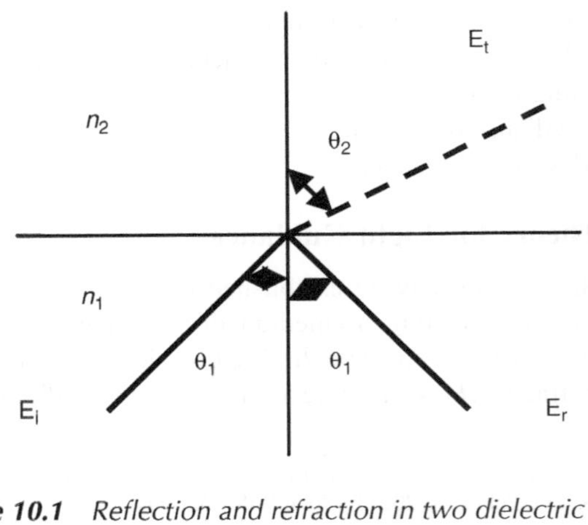

Figure 10.1 *Reflection and refraction in two dielectric media.*

where r is the complex reflection coefficient dependent on both the angle of incidence and the polarization of the light ray. The polarization of a light ray is related to the direction of the electric field vector associated with the electromagnetic wave. The transverse electric (TE) condition refers to the situation when the electric fields of the waves are perpendicular to the plane of incidence, that is, the plane containing the wave normal and normal to the interface. Similarly, the transverse magnetic (TM) case occurs when the magnetic field of the wave is normal to the plane of incidence. The reflection coefficients r for the two cases, r_{TE} and r_{TM}, are expressed by the following two Fresnel formulae:

$$r_{TE} = \frac{n_1 \cos\theta_1 - n_2 \cos\theta_2}{n_1 \cos\theta_1 + n_2 \cos\theta_2} \tag{10.5a}$$

$$r_{TM} = \frac{n_2 \cos\theta_1 - n_1 \cos\theta_2}{n_2 \cos\theta_1 + n_1 \cos\theta_2} \tag{10.5b}$$

These two equations may be rewritten, by using Snell's law, Eq. (10.1), as

$$r_{TE} = \frac{n_1 \cos\theta_1 - \sqrt{n_2^2 - n_1^2 \sin^2\theta_1}}{n_1 \cos\theta_1 + \sqrt{n_2^2 - n_1^2 \sin^2\theta_1}} \tag{10.6a}$$

$$r_{TM} = \frac{n_2^2 \cos\theta_1 - n_1 \sqrt{n_2^2 - n_1^2 \sin^2\theta_1}}{n_2^2 \cos\theta_1 + n_1 \sqrt{n_2^2 - n_1^2 \sin^2\theta_1}} \tag{10.6b}$$

For any angle of incidence less than the critical angle, only partial reflection occurs and the reflection coefficient is real. However, when the critical angle is exceeded, the terms inside the square root in Eqs. (10.6a) and (10.6b) become negative. As a result, the reflection coefficient becomes complex with magnitude unity and a phase shift is imposed on the reflected wave. The phase shifts related to TE and TM waves may be expressed as

$$\phi_{TE} = 2\tan^{-1}[\sqrt{\sin^2\theta_1 - (n_2/n_1)^2}/\cos\theta_1] \tag{10.7a}$$

$$\phi_{TM} = 2\tan^{-1}[\sqrt{(n_1/n_2)^2 \sin^2\theta_1 - 1}/(n_2/n_1)\cos\theta_1] \tag{10.7b}$$

The reflection coefficients r given in Eqs. (10.6a) and (10.6b) are the field reflection coefficients, that is, the ratio between the electric fields related to incident and reflected waves. The power reflection coefficient, by using the definition of Poynting vector, is simply given by

$$R = \frac{S_r}{S_i} = \frac{E_r^2}{E_i^2} = r^2 \tag{10.8}$$

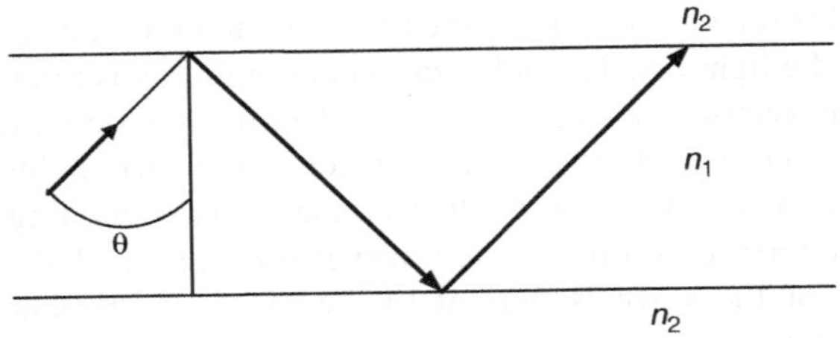

Figure 10.2 Ray path showing total internal reflection and guidance of mode in a symmetric slab waveguide.

10.4 Modes of a Planar Waveguide

The simplest optical waveguide is a planar waveguide, shown in Figure 10.2. We modify the structure slightly to consider a more general structure, in which the refractive indices of the upper and lower cladding layers are different and are, respectively, n_3 and n_2 as shown in left of Figure 10.3. The Cartesian coordinate system used to define the geometry is also shown.

The thickness of the guide layer along the y direction along which the light is confined is a and the direction of propagation of the electromagnetic wave is z. The zigzag path denotes the direction of wave normal as the waves propagate through the waveguide with wave vector k ($= k_0 n_1$), where k_0 is the free space wave vector ($k_0 = 2\pi/\lambda_0$, where λ_0 is the free space wavelength). The components of k along the y- and z-directions are

$$\begin{aligned} k_z &= n_1 k_0 \sin\theta_1 \\ k_y &= n_1 k_0 \cos\theta_1 \end{aligned}. \tag{10.9}$$

The relation between the propagation constants and the wave normal is shown in Figure 10.3 (right).

As the light is guided along the y-direction, the corresponding electromagnetic wave will form a standing wave pattern along this direction. We consider the total phase shifts introduced in making a round trip along the y-direction covering a distance $2a$. This can be written as

$$\phi_t = 2k_0 n_1 a \cos\theta_1 - \phi_u - \phi_l. \tag{10.10}$$

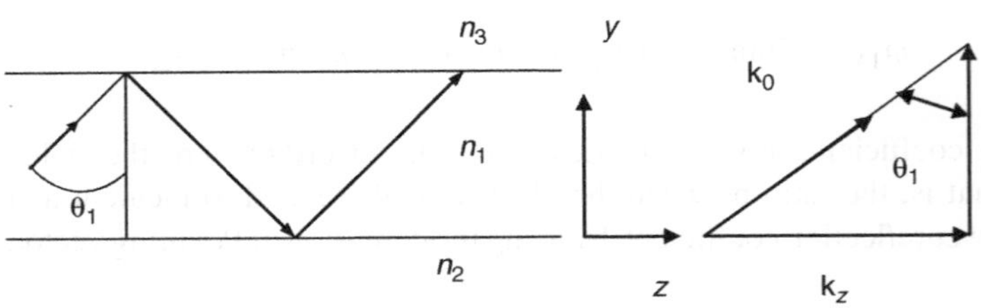

Figure 10.3 (Left) Ray path and propagation in planar waveguide; and (right) propagation constants in y- and z-directions.

The first term in the right-hand side of Eq. (10.10) is the phase change in traversing a distance $2a$ across the y direction. The other two terms denote the phase shifts due to reflection at the core–upper cladding (ϕ_u) and core–lower cladding (ϕ_l) interfaces, respectively. These two terms may be obtained from Eq. (10.7a) after proper modification. In order that the wave is sustained along the y-direction, this total phase shift must be an integral multiple of 2π, so that

$$2k_0 n_1 a \cos\theta_1 - \phi_u - \phi_l = 2m\pi, \tag{10.11}$$

where m is an integer. Since the phase changes ϕ_u and ϕ_l are functions of θ_1, Eq. (10.11) is an implicit equation of θ_1, and, when solved, it gives the value of θ_1 for a given integral value of m.

We now notice that only discrete values of θ_1 are permitted, corresponding to integral values of m. For each solution of Eq. (10.11), there will be corresponding propagation constants along the y- and z-directions, for each polarization. The conclusion is then that light cannot propagate with any arbitrary value of the angle of incidence, but only discrete angles are allowed. Each allowed solution is referred to as a *mode of propagation*, and the mode number is given by the value of the integer m. Each mode is characterized by the polarization and the mode number. For example, in the first TE mode, the fundamental mode is denoted as TE_0. Higher order modes are correspondingly described by using the appropriate value for m.

10.4.1 The Symmetrical Planar Waveguide

Let us first consider the symmetrical planar waveguide structure shown in Figure 10.2 in which both the upper and lower cladding layers have the same RI n_2. In this case, the phase shift due to reflections at the two interfaces will be equal, $\phi_u = \phi_l$. Using Eq. (10.7a) for TE polarization, Eq. (10.11) becomes

$$2k_0 n_1 a \cos\theta_1 - 4\tan^{-1}\left[\sqrt{\sin^2\theta_1 - (n_2/n_1)^2}/\cos\theta_1\right] = 2m\pi. \tag{10.12}$$

This equation may be rearranged to give

$$\tan[(k_0 n_1 a \cos\theta_1 - m\pi)/2] = \left[\sqrt{\sin^2\theta_1 - (n_2/n_1)^2}/\cos\theta_1\right]. \tag{10.13}$$

The only variable in this equation is θ_1, which may be obtained by solving the equation. The corresponding equation for the TM mode may be written in a similar fashion by using Eq. (10.7b).

Number of modes supported: to have an estimate of the total number of modes supported by the symmetric guide, we note that the minimum value of the angle of incidence θ_1 is the critical angle θ_c, corresponding to the highest possible order mode. Putting $\theta_1 = \theta_c$ in Eq. (10.13) makes the right-hand side zero and thus

$$(k_0 n_1 a \cos\theta_c - m_{max}\pi)/2 = 0 \tag{10.14}$$

The maximum allowed mode number m_{max} is therefore

$$m_{max} = \frac{k_0 n_1 a \cos\theta_c}{\pi} \tag{10.15}$$

In an actual calculation, one should consider the nearest integer $[m_{max}]_{int}$ that is less than the computed value of m_{max}. The total number of modes will actually be $[m_{max}]_{int} + 1$, since the lowest order mode has a mode number $m = 0$.

Example 10.1

Let $\lambda_0 = 1.0\,\mu m$, $n_1 = 1.5$, and $a = 5\,\mu m$. The value of m_{max} turns out to be 7.5. The maximum number of modes to be supported is $[7 + 1] = 8$.

10.4.2 The Asymmetric Waveguide

The asymmetric planar waveguide structure has different RIs for the upper and lower cladding layers, as shown in Figure 10.3. In this case, the phase changes due to reflections at the upper and lower interfaces are unequal. The eigenvalue equation for the TE mode now reads

$$[k_0 n_1 a \cos\theta_1 - m\pi] = \tan^{-1}[\sqrt{\sin^2\theta_1 - (n_2/n_1)^2}/\cos\theta_1] + \tan^{-1}[\sqrt{\sin^2\theta_1 - (n_3/n_1)^2}/\cos\theta_1] \quad (10.16)$$

For a given mode number m, this equation may be solved numerically or graphically to find the propagation angle θ_1. Note, however, that the critical angles for the two interfaces are different in this case. For total internal reflections at both the boundaries, the angle θ_1 must exceed the larger of these two. It is possible that when the guide thickness a is small, the solution to Eq. (10.16) ceases to exist for a fundamental mode with number $m = 0$ (see Problem 10.5).

10.4.3 Single-Mode Condition

Depending on the fractional change in RIs and the guide thickness, sometimes it is possible that the guide supports only a single mode. The condition for such a monomode operation will now be derived for a symmetric waveguide using Eq. (10.13).

Consider the second mode with mode number $m = 1$. In the limiting condition, $\theta_1 = \theta_c$, the critical angle. Since the incident angle for the second mode is less than that for the fundamental mode, then for all angles greater than this critical angle, the waveguide will be monomode. We may write Eq. (10.13) for this situation as

$$\tan[(k_0 n_1 a \cos\theta_c - \pi)/2] = 0. \quad (10.17)$$

This yields

$$\cos\theta_c = \frac{\pi}{k_0 n_1 a} = \frac{\lambda_0}{2 n_1 a}. \quad (10.18)$$

Hence, for monomode conditions,

$$\theta_c \le \cos^{-1}\left(\frac{\lambda_0}{2 n_1 a}\right). \quad (10.19)$$

10.4.4 Effective Index of a Mode

The propagation constants of the ray along the y- and z-directions have been defined in Eq. (10.9). The constant k_z is often denoted by the variable β.

We may now define a parameter N, called the *effective index of the mode*, in the following way:

$$N = n_1 \sin \theta_1. \tag{10.20}$$

Then Eq. (10.9) becomes

$$k_z = \beta = N k_0. \tag{10.21}$$

We may now state that the ray propagates straight along the z-direction with RI N, *without* following the zigzag path shown in Figures 10.2 and 10.3.

The lower bound on β is determined by the critical angles at the interfaces. As noted already for the guidance of light in the asymmetric guide, the propagation angle must be greater than the larger of the two critical angles. Since, in most cases, the upper cladding layer is air, the larger critical angle corresponds to the lower cladding. This means that $\theta_1 \geq \theta_{\text{lower}}$. The lower bound on β is given by

$$\beta \geq n_1 k_0 \sin \theta_l = k_0 n_2. \tag{10.22}$$

The upper bound on β is determined by the maximum value of θ, that is, 90° and then $\beta = k = n_1 k_0$. The final result is

$$k_0 n_1 \geq \beta \geq k_0 n_2. \tag{10.23}$$

Using Eq. (10.21), it may be written as

$$n_1 \geq N \geq n_2. \tag{10.24}$$

10.5 Wave Theory of Light Guides

We now wish to develop the theory of optical waveguides and of the modes by using the EM wave equation.

10.5.1 Wave Equation in a Dielectric

Like all other EM phenomena, the propagation of EM waves along a dielectric waveguide is governed by Maxwell's equations, given by Eqs. (4.1a)—(4.1d). For most of the guiding materials, $\mathbf{M} = 0$; further, the current density and free charge densities are assumed to be absent. The electric polarization is related to the electric field through a susceptibility χ, which is taken to be constant in most of the situations.

By taking curl of Eq. (4.1a) and using (4.1b)–(4.1d), the wave equation is obtained in the following form:

$$\nabla \times \nabla \times \mathbf{E} = -\mu_0 \varepsilon \varepsilon_0 \frac{\partial^2}{\partial t^2} \mathbf{E} \tag{10.25}$$

where $\varepsilon = 1 + \chi$ is the relative permittivity of the medium. Now assuming that **E** varies sinusoidally as $\mathbf{E} = \mathbf{E}\exp(j\omega t)$, we may write

$$\nabla \times \nabla \times \mathbf{E} = -\varepsilon(\omega^2/c^2)\mathbf{E}, \tag{10.26}$$

where $c^2 = 1/\sqrt{\varepsilon_0\mu_0}$ is the velocity of light in free space. In general, ε is complex and its real and imaginary parts are related to the refractive index n and absorption coefficient α (see Chapter 4). We shall consider now that ε is real and equals n^2 because of low optical loss in the guides and write $\nabla \times \nabla \times \mathbf{E} = \nabla(\nabla \cdot \mathbf{E}) - \nabla^2\mathbf{E} = -\nabla^2\mathbf{E}$, since $\nabla \cdot \mathbf{D} = 0$. We get finally

$$\nabla^2\mathbf{E} + n^2 {k_0}^2\mathbf{E} = 0. \tag{10.27}$$

The symbol k_0 is defined in terms of free space wavelength λ_0 by the relation

$$k_0 = \omega/c = 2\pi/\lambda_0. \tag{10.28}$$

10.5.2 The Ideal Slab Waveguide

The planar slab waveguide structure forms the starting point of all theoretical developments of wave propagation in waveguides. We shall first discuss the origin of different modes in this structure.

10.5.2.1 Modes in Slab Waveguide

A simple three-layer planar waveguide structure is shown in Figure 10.4. The structure consists of a dielectric layer, denoted as layer 1 having RI n_1, sandwiched between the upper cladding layer 3 of RI n_3 and lower cladding layer 2 having RI's n_2. All the layers are assumed to extend to infinity along the x- and z-directions, and layers 2 and 3 are semi-infinite also along the y-direction. If the radiation propagates along the z-direction, then a

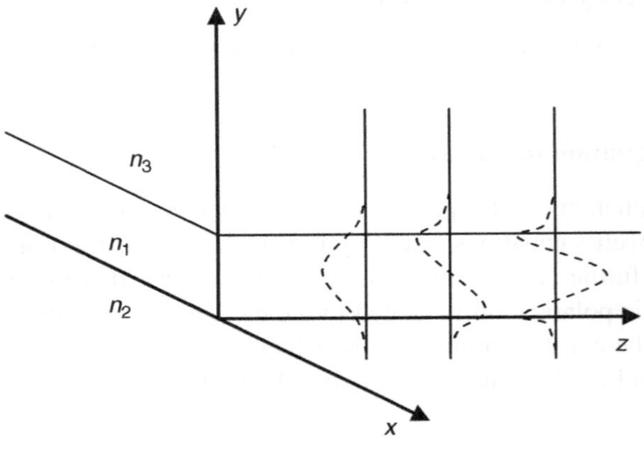

Figure 10.4 *Three-layer slab waveguide. The electric field distribution of the first three modes is shown.*

mode should be a solution of the following wave equation, obtained by modifying Eq. (10.27) to account for spatial variation of the RI:

$$\nabla^2\mathbf{E}(\mathbf{r}) + n^2(\mathbf{r})k_0{}^2\mathbf{E}(\mathbf{r}) = 0. \tag{10.29}$$

If we now consider the propagation of a uniform plane wave along the z-direction, that is, $\mathbf{E}(\mathbf{r}) = \mathbf{E}(x,y)\exp(-j\beta z)$, with β being the propagation constant, Eq. (10.29) then becomes

$$\partial^2\mathbf{E}(x,y)/\partial x^2 + \partial^2\mathbf{E}(x,y)/\partial y^2 + [k_0^2 n^2(r) - \beta^2]\mathbf{E}(x,y) = 0. \tag{10.30}$$

Since the guide extends to infinity along x, we may write Eq. (10.30) as

$$\partial^2\mathbf{E}(x,y)/\partial x^2 + \partial^2\mathbf{E}(x,y)/\partial y^2 + [k_0^2 n_i^2(\mathbf{r}) - \beta^2]\mathbf{E}(x,y) = 0, i = 1,2,3. \tag{10.31}$$

The solutions of Eq. (10.31) are either sinusoidal or exponential depending on whether $[k_0^2 n_i{}^2 - \beta^2]$ is greater or less than 0. The $E(x,y)$ and its derivative $\partial E/\partial y$ must be continuous at the interfaces between the dielectric layers. The possible modes are shown in Figure 10.4.

Some insight of the mode shape may be obtained from Eq. (10.31) for constant frequency ω and $n_1 > n_2 > n_3$. For $\beta > kn_1$, $(1/E)\partial^2 E/\partial y^2$ is positive in all three regions, leading to exponential solutions. The resulting field distribution denoted as (a) in Figure 10.5 is the only solution compatible with continuity of field and its derivative. The monotonically increasing wave amplitudes do not represent real waves. For values of β satisfying the condition, $kn_3 < \beta < kn_2$, $(1/E)\partial^2 E/\partial y^2$ is positive in regions 2 and 3 and negative in region 1, leading to sinusoidal solution in region 1 and decaying exponentials in regions 2 and 3. We now have mode confinement and guiding in region 1. The exponential solutions in regions 2 and 3 do not give propagating modes, but rather they represent evanescent waves. It follows therefore that for waveguiding, the condition $kn_2 < \beta < kn_1$ must be satisfied. Mode (b) indicated in Figure 10.5 belongs to the family of

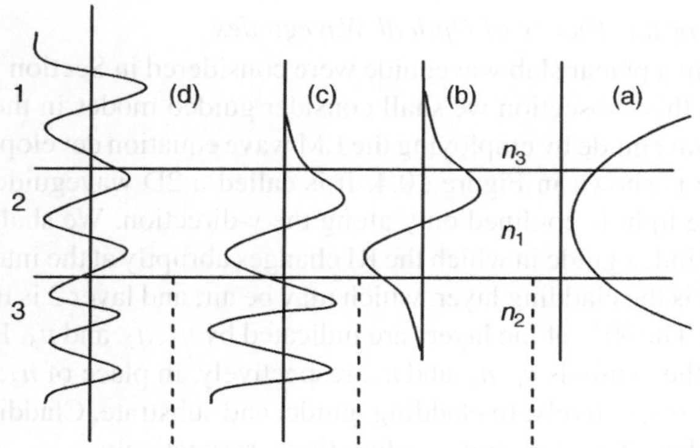

Figure 10.5 *Possible modes in a planar waveguide. Reproduced with permission from Quantum Electronics by A. Yariv, John Wiley NY Copyright (1993) John Wiley and Sons, Inc.*

well-confined guided modes generally referred to as *transverse electric modes*. The mode shown in Figure 10.5 is the TE_1 mode.

If $kn_2 > \beta > kn_3$, a mode as shown in Figure 10.5c will result. The mode is confined in an upper cladding layer, for example air, but is sinusoidal in the lower substrate layer. This is called the *substrate radiation mode*. However, as the mode loses energy from the guiding region to the substrate, it dampens out over a short distance. The mode is not useful in signal transmission, but finds application in the coupler structures as discussed in this chapter.

If $\beta < kn_3$, the solution for *E(x, y)* is oscillatory in all three regions as indicated by Figure 4.2d. In this case, energy propagates freely out of the guiding layer and the modes are called *radiation modes*.

An interesting parameter for guided mode propagation is the cutoff parameter. We shall see in Section 10.5.2.2 that β can have any value when it is less than kn_2 but only discrete values of β are allowed when $kn_1 > \beta > kn_2$. These discrete values of β correspond to various modes, TE_j, $j = 0,1,2\ldots$ (also TM_k, $k = 0,1,2\ldots$). The number of modes that can be supported depends on the thickness a of the guiding layer, frequency ω, and the RI's. For given a and n_i's, there is a cutoff frequency ω_c below which no waveguiding can occur. This frequency leads to a long-wavelength cutoff λ_c. It may be shown by solving the wave equation that for a given mode number m to be supported by an asymmetric guide shown in Figure 10.4, the following condition applies:

$$\Delta n = n_2 - n_1 \geq (2m+1)^2 \lambda_0^2/(32 n_2 a^2), m = 0, 1, 2, \ldots \qquad (10.32)$$

where λ_0 is the wavelength in free space.

Example 10.2

For Si, $n_2 = 3.5$; for $m = 0, a = \lambda_0, \Delta n \approx 9 \times 10^{-3}$ for propagation of the TE_0 mode.

10.5.2.2 Outline of the Theory of Optical Waveguides

The guided modes in a planar slab waveguide were considered in Section 10.4 by using the ray optic theory. In this subsection we shall consider guided modes in the step index two-dimensional (2D) waveguide by employing the EM wave equation developed in Eq. (10.27). The basic structure is shown in Figure 10.4. It is called a 2D waveguide (or *slab optical waveguide*) because light is confined only along the y-direction. We shall assume that the waveguide is a step index guide in which the RI changes abruptly at the interfaces of the two dielectrics. Layer 3 is the cladding layer, which may be air, and layer 2 is the other cladding (or substrate) layer. The RI's of the layers are indicated by $n_1, n_2,$ and n_3. However, in some cases we shall use the symbols $n_c, n_g,$ and n_S, respectively, in place of $n_3, n_1,$ and n_2 where the subscripts refer, respectively, to cladding, guide, and substrate. Cladding layers 1 and 3 extend to infinity along the $+y$- and $-y$-directions, respectively.

We consider plane waves propagating along the z-direction with a propagation constant β. The EM fields vary as

$$E = E(x, y) \exp j(\omega t - \beta z), \qquad (10.33a)$$

$$H = H(x, y) \exp j(\omega t - \beta z), \qquad (10.33b)$$

where ω is the angular frequency, $\omega = 2\pi\lambda/c$, and $c = 1/\sqrt{\varepsilon_0\mu_0}$ is the velocity of light in free space. In the step index guide, the fields are independent of x. Therefore since $\partial/\partial t = j\omega$, $\partial/\partial z = -j\beta$ and $\partial/\partial x = 0$, Eq. (10.31) yields two different modes with mutually orthogonal polarization states. One is the TE mode having only E_x as the electric field component. Using Eq. (4.1), the magnetic field components are H_y and H_z. The other is the TM mode which has components E_y, H_x and E_z. Wave equations for the TE and TM modes are

$$\text{TE}: \frac{d^2E_x}{dy^2} + ({k_0}^2n^2 - \beta^2)E_x = 0, \tag{10.34a}$$

$$H_y = -\frac{\beta}{\omega\mu_0}E_x, H_z = -\frac{1}{j\omega\mu_0}\frac{dE_x}{dy}, \tag{10.34b}$$

$$\text{TM}: \frac{d^2H_x}{dy^2} + ({k_0}^2n^2 - \beta^2)H_x = 0, \tag{10.35a}$$

$$E_z = \frac{\beta}{\omega\varepsilon_0 n^2}H_x, E_y = \frac{1}{j\omega\varepsilon_0 n^2}\frac{dH_x}{dy}. \tag{10.35b}$$

Solving for the fields with the boundary conditions at the interface $y = -a$ and $y = 0$, one obtains eigenvalue equations that determine the propagation characteristics of the TE and TM modes.

Dispersion of the guided modes: we consider here TE modes only. The corresponding expressions for TM modes may be obtained in a similar fashion. From Eq. (10.31), the field solutions may be written as follows:

$$\begin{aligned} E_x &= E_c \exp(-\gamma_c y); y > 0, \text{ in the upper cladding,} \\ E_x &= E_g \cos(k_y y + \phi_1); -a < y < 0, \text{ in the guiding film,} \\ E_x &= E_s \exp\{\gamma_s(y + a)\}; y < -a, \text{ in the substrate.} \end{aligned} \tag{10.36}$$

The propagation constants in the y-direction are expressed in terms of the following symbols:

$$\gamma_c = k_0\sqrt{N^2 - n_c^2}, k_y = k_0\sqrt{n_g^2 - N^2}, \gamma_s = k_0\sqrt{N^2 - n_s^2}. \tag{10.37}$$

where $N = \beta/k_0$ is the effective index. Since the tangential field components E_x and H_z are continuous at the interface $y = 0$, one obtains

$$\begin{aligned} E_c &= E_g \cos\phi_1 \\ \tan\phi_1 &= \gamma_c/k_y \end{aligned} \tag{10.38}$$

Similarly,

$$\begin{aligned} E_s &= E_g \cos(k_y a - \phi_1), \\ \tan(k_y a - \phi_1) &= \gamma_s/k_y. \end{aligned} \tag{10.39}$$

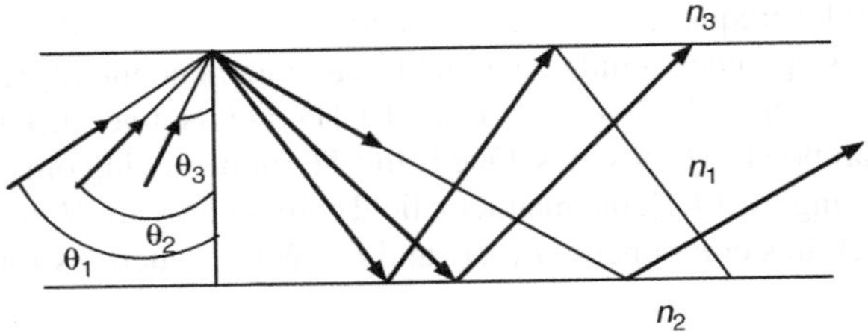

Figure 10.6 *Different modes incident at different angles at the core–cladding interface.*

at $y = -a$. Eliminating arbitrary coefficients in Eqs. (10.38) and (10.39) yields the following eigenvalue equation:

$$k_y a = (m+1)\pi - \tan^{-1}(k_y/\gamma_s) - \tan^{-1}(k_y/\gamma_c). \tag{10.40}$$

where $m = 0,1,2\ldots.$. denotes the *mode number*. The value of k_x may be obtained from the values of RI's and the thickness of the guide. When this value is substituted in Eq. (10.37), the value of effective index N may be obtained. Since the mode number is a positive integer, N must have discrete values lying in the range $n_s < N < n_g$. This means that zigzag rays with certain incident angles can propagate as guiding modes along the guide as shown in Figure 10.6. It is to be noted that the above eigenvalue Eq. (10.40) is identical with Eq. (10.14) if the RI's for upper cladding and substrate are assumed equal.

The above transcendental Eq. (10.40) may be solved numerically to evaluate the dispersion characteristics of the guided modes. For this purpose one introduces the following two normalized parameters: *normalized frequency V* and the *normalized guide index b*, defined as

$$V = k_0 a \sqrt{n_g^2 - n_s^2}, \tag{10.41}$$

and

$$b = (N^2 - n_s^2)/(n_g^2 - n_s^2). \tag{10.42}$$

The measure of the asymmetry of the guide is defined by

$$\alpha = (n_s^2 - n_c^2)/(n_g^2 - n_s^2). \tag{10.43}$$

Thus, when $n_c = n_s$, which refers to a symmetrical waveguide, the asymmetry parameter $\alpha = 0$. In general, waveguides are asymmetrical. Using Eqs. (10.41)—(10.43), Eq. (10.40) may be rewritten as

$$V\sqrt{1-b} = (m+1)\pi - \tan^{-1}\sqrt{(1-b)/b} - \tan^{-1}\sqrt{(1-b)/(b+\alpha)}. \tag{10.44}$$

Solving this equation numerically, a plot may be obtained which is the normalized dispersion curve with m and α as parameters. When material RI's and guide thickness are given, the effective index of the guided mode may be obtained from the graph.

When the incident angle equals the critical angle related to layers 1 and 2 (i.e., the guide–substrate interface), the light is no longer confined in the guide layer, but begins to leak into the substrate at the interface $y = -a$. This situation is called the *cutoff of the guided mode*. In this case, the effective index $N = n_s, b = 0$. From Eq. (10.44), the value of V_m at the cutoff condition is given by

$$V_m = V_0 + m\pi, V_0 = \tan^{-1}\sqrt{\alpha} \tag{10.45}$$

where V_0 is the cutoff value for the fundamental mode. When the normalized frequency V lies in between V_m and V_{m+1}, $V_m < V < V_{m+1}$, the TE_0, TE_1, ... and TE_m modes are supported, the number of guided modes being $m + 1$. For symmetrical waveguides, $n_c = n_s$ and $V_0 = 0$. Thus the fundamental mode is not cut off in a symmetrical waveguide.

For a symmetrical guide, use of Eqs. (10.34a) and (10.37) indicated that for waveguiding of a given mode to occur, one should have

$$\Delta n = n_g - n_c > \frac{m^2 \lambda_0^2}{4(n_c + n_g)a^2}. \tag{10.46}$$

A family of curves showing b parameters against V parameters for mode numbers 0, 1, and 2 and for different values of a is given in Kogelnik and Ramaswamy [1].

10.5.2.3 Mode Profiles and Confinement Factor

Eq. (10.36) allows us to plot the electric field distribution $E_x(y)$ or the intensity distribution $|E_x(y)|^2$. For the fundamental mode the electric field is maximum at the center of the guide as shown in Figure 10.4. It then decreases following a cosine distribution. At the two interfaces the field does not decay to zero but penetrates into the upper and lower claddings following the exponential nature. The field distribution for the higher order mode, for example $m = 2$, may be qualitatively depicted and the corresponding intensity profile is also shown in Figure 10.4. The number of minima in this profile equals the mode number.

As seen from the field profiles, the power propagating along the guide is not totally confined inside the core of the guide, but rather a part of the power leaks into the two cladding layers. A useful parameter describing the light-guiding property of the waveguide is the *mode confinement factor* defined as follows:

$$\Gamma = \frac{\int_0^a E_x^2(y)dy}{\int_{-\infty}^{\infty} E_x^2(y)dy} \tag{10.47}$$

This factor quantifies how much fraction of the total power belonging to the particular mode is confined within the guide layer. It is a function of polarization, of the refractive index difference, of the thickness of the guide, and of the mode number.

10.5.2.4 Goos–Hänchen Shift

It has been noted in Section 10.3 that a phase shift is introduced when a ray suffers total internal reflection at the core–cladding interface. The EM theory of guided waves developed in this chapter indicates that the electric field penetrates into the cladding layer. There is complete consistency about these two approaches as described in this subsection.

Consider the ray model as depicted in Figure 10.7. The field penetrates into both the upper and lower claddings to a depth equal to the inverse of the decay constants γ_c and γ_s. It may now be assumed that the ray gets reflected from a plane which is displaced by $1/\gamma_c$ from the

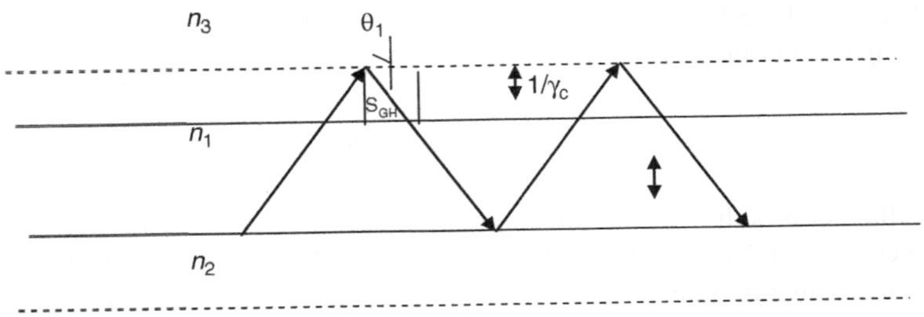

Figure 10.7 Propagation in an asymmetrical guide, showing ray penetration in the cladding layers and showing Goos–Hänchen shifts.

core–upper cladding interface. Clearly the ray suffers additional phase shift due to additional path length, and the phase shift may be calculated from simple geometry as

$$\tan\theta_1 = \frac{S_{GH}/2}{1/\gamma_c} = S_{GH}\gamma_c/2 \tag{10.48}$$

The phase shift as well as the lateral shift of the reflecting plane are known as the Goos–Hänchen shift and are denoted by S_{GH}. The amount of shift may be expressed from Eq. (10.48) as

$$S_{GH} = \frac{2\tan\theta_1}{\gamma_c} = \frac{2}{\gamma_c}\frac{\beta}{k_y} \tag{10.49}$$

The shift can amount to significant distance and may exceed even the guide thickness. The field outside the core or guide layer is known as the *evanescent field*. We shall see in later chapters that this leaky field when coupled into an adjacent guide may give rise to interesting device applications.

10.6 3D Optical Waveguides

We now extend the theory for practical waveguides, which are basically three-dimensional (3D) in structure.

10.6.1 Practical Waveguiding Geometries

The symmetric and asymmetric slab waveguide structures discussed in Section 10.5.2 form a good starting point for a discussion of waveguiding properties in dielectric waveguides. A guiding layer of silica grown on silicon substrate having air as the upper cladding layer may serve as an example of the asymmetric guide. The silica guiding layer may have finite thickness but may extend to infinity in the other two dimensions. In accordance with the earlier discussions in this chapter, the RI profile makes step changes at the air–silica and silica–silicon interfaces.

A number of techniques that are employed presently for the fabrication of practical IO circuits, however, lead to significantly different types of guiding structures. When ions or molecules are driven into a substrate by diffusion or ion implantation, a guiding layer is

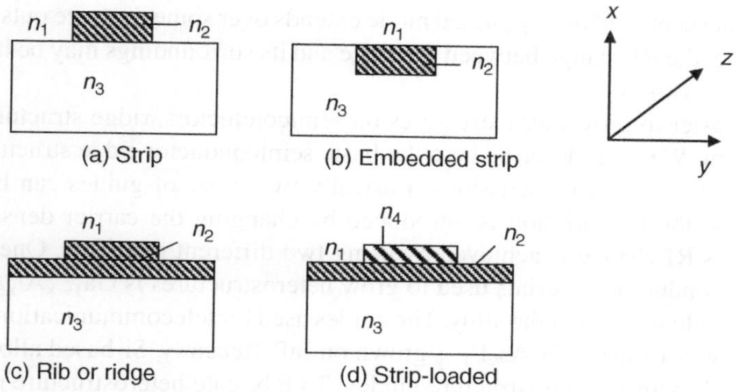

Figure 10.8 Various 3D waveguides.

produced at the surface of the substrate. However, the RI profile in the guide layer is far from being constant; rather, the RI decreases exponentially or nonlinearly from the value at the surface. Such structures are referred to as *graded index waveguides*.

Another important feature in practical waveguides is that optical energy is confined along both the *x*- and *y*-directions, unlike in the earlier case where confinement occurred only along the *y*-direction due to RI index steps. The size of the IO circuits is then reduced and a number of such circuits may then be placed in close proximity on the same substrate. The structures are called *three-dimensional waveguides*, because in contrast to the two-dimensional structures considered so far, field variation in three rather that two directions must be considered now. A few 3D waveguide structures are shown in Figure 10.8. They are called the *buried channel*, *rib* or *ridge*, and *strip-loaded* waveguides. We shall now describe these in some detail, together with a summary of typical fabrication methods.

The *strip guide* is formed by growing a layer on the substrate. It is surrounded by air. The *embedded strip* or *buried channel guide* shown in Figure 10.8b is made by modifying the properties of the substrate material so that a higher RI region is obtained locally. The buried layer beneath the surface in most situations is a graded index weakly guiding region and does not have a well-defined cross-sectional shape under a normal fabrication scheme.

Diffusion is normally used to fabricate this type of waveguide. For example, Ti can be diffused into $LiNbO_3$ or $LiTaO_3$ substrates, by first depositing the metal in patterned strips of about 1000 A thicknesses and then carrying out an in-diffusion at higher temperature (~1000 C) for 3–9 hours. The additional impurities produce a RI change $\Delta n \approx 0.01$. In Si-based structures, such kind of waveguides may be realized by diffusing Ge into a selective region on the Si substrate. The RI of $Si_{1-x}Ge_x$ alloy is larger than the surrounding Si and light is guided in the alloy channel.

Ridge waveguide: Figure 10.8c shows a ridge waveguide. This is basically a step index structure; a planar guide is patterned and then etched to leave a ridge. Total internal reflection takes place at the sides of the guide, as well as at the top and bottom faces. Often the etching is designed to follow specific crystal planes, so the edges of the ridge may be sloped. In a different fabrication scheme, materials are often regrown around the sides of the guide, as in laser structures. Much higher field confinement is possible when using a ridge rather than a buried guide. Since the index change induced by material modification is small, the evanescent field of any guided mode extends over some distance outside the core. In a ridge guide, the RI change between the ridge and its surroundings may be larger. Bends can be tighter in ridges.

Since it is easier to grow mesa structures on semiconductors, ridge structures are often made with them. We shall describe here the basic semiconductor ridge structures, leaving for the present the detailed discussions. Basically two types of guides can be grown. In homostructures, the RI variation is introduced by changing the carrier density, while in heterostructures RI change is achieved by using two different materials. One of the most common semiconductor materials used to grow heterostructures is $Ga_{1-x}Al_xAs$ on GaAs, where x is the mole fraction of the alloy. The guides used for telecommunication wavelength 1.55 μm are layers of $In_{1-x}Ga_xAs_yP_{1-y}$ grown on InP. Recently, Si-based alloys grown on Si are also used to grow heterostructure guides. To fabricate heterostructure guides in the GaAs–GaAlAs systems, successive layers of material are first grown on a GaAs substrate in differing compositions. Because GaAs and GaAlAs have similar lattice parameters, these additional layers are lattice matched to the substrate and the composite crystal could grow without strain. In GaAlAs the RI ranges roughly linearly from $n \approx 3.57$ for $x = 0$ to $n \approx 3.36$ for $x = 0.35$ (at near-infrared). The variation of the alloy content between the layers then yields the RI difference necessary for waveguiding. For example, a planar guide might be fabricated using a layer of high-index GaAs grown epitaxially on top of a thick layer of GaAlAs which acts as a low-index isolation layer separating the guide from the high-index GaAs substrate. The guiding GaAs layer may then be etched down to a narrow rib to form a ridge waveguide. More complicated layer structures like a double heterostructure waveguide have the low-index GaAlAs cladding layers on both sides of the guide. This structure is quite often used in double heterostructure (DH) lasers to provide effective mode confinement. The band gap difference between GaAs and GaAlAs provides carrier confinement. A similar ridge structure may be realized by growing alternate layers of $Si_{1-x}Ge_x$ and Si on a Si substrate.

Instead of using compositional variation, waveguides can also be formed by reducing the RI by the introduction of free carriers. The theory developed in earlier chapters of this volume indicates that the change in RI $\Delta n \propto N$, where N is the carrier density. For example, undoped GaAs can be grown on top of a doped (n-type) GaAs substrate to form a planar homostructure guide. Alternatively, the guide can be formed by reducing the carrier density by proton bombardment. A problem in homostructure guides is that the increased carrier density in the substrate causes a rise in absorption which increases the propagation loss.

Ridge guides can also be made by growing amorphous layers on semiconducting substrates. One common system involves silica on Si or Si oxy-nitride (SiON) layers grown on a silica buffer layer on Si substrate. SiON has high RI but less than Si, for which $n \approx 3.5$. A buffer layer of silica ($n \approx 1.47$) separates SiON from Si substrate. Recently waveguide structures are also being fabricated using a silicon-on-insulator (SOI) structure.

Strip-loaded waveguides: In a strip-loaded guide shown in Figure 10.8d, a three-layer structure has been used, which has a substrate, a planar layer, and then a ridge. The planar layer is arranged as a guide but one which is just cut off. However, the addition of a high-index overlay is sufficient to induce guiding in a localized region near the ridge. Confinement of the optical field in the y-direction is obtained because the RI when averaged in the x-direction is higher in the region of the ridge. The additional ridge overlay may have a different RI from that of a planar guide, or, as in Figure 10.8c, the same index.

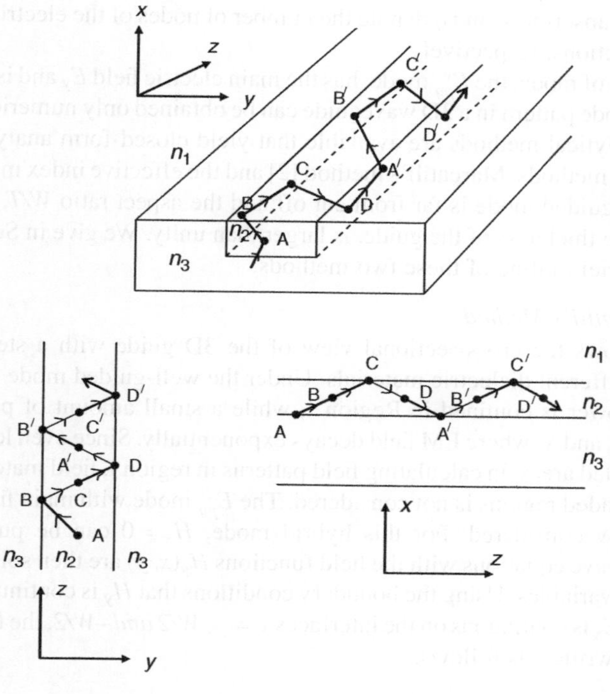

Figure 10.9 *Ray path in a 3D waveguide.*

10.6.2 Ray Optic Approach for 3D Guides

It is easy to understand guiding action in a strip or embedded strip waveguide. In both, a high-RI (n_2) core region is surrounded by regions of lower RI (n_1 and n_3). We may think that the light ray propagates in a zigzag fashion within the high-RI region, suffering total internal reflections at the n_2/n_1 and n_2/n_3 boundaries. The path followed by these rays is shown in Figure 10.9.

It is difficult to employ the above line of thinking to explain the guidance in rib or strip-loaded waveguides, in which no central region of higher RI capable of confinement in both the x- and y-directions can be identified. An alternative method, known as the *effective index method*, is usually employed to explain the confinement of light rays along two dimensions. The method is discussed in Section 10.6.3.2.

A powerful numerical method, known as the *beam propagation method* (BPM), has been developed which is used extensively for obtaining the field profile, the mode behavior of the complex structures that are fabricated practically.

10.6.3 Approximate Analyses of Guided Modes

Pure TE and TM modes are not supported in optical 3D waveguides surrounded by different dielectrics. Instead, two types of hybrid modes exist that are essentially TEM modes polarized along the x- and y-directions. The mode having the main electric field E_x is called the E_{pq}^x *mode* or *TM-like mode*, that is, the mode resembles the TM mode in a slab waveguide. The subscripts p and q denote the number of nodes of the electric field E_x along the x- and y-directions, respectively.

The other class of mode, the E^y_{pq} mode, has the main electric field E_y and is TE-like mode. In general, the mode pattern in a 3D waveguide can be obtained only numerically. However, approximate analytical methods are available that yield closed-form analytical solutions. The approximate methods, Marcatili's method [2] and the effective index method [3, 4], are useful when the guided mode is far from cut off and the aspect ratio W/T, where W is the width and T is the thickness of the guide, is larger than unity. We give in Sections 10.6.3.1 and 10.6.3.2 a brief outline of these two methods.

10.6.3.1 Marcatili's Method

Figure 10.10 shows the cross-sectional view of the 3D guide with a step index profile surrounded by different dielectric materials. Under the well-guided mode condition, most of the optical power is confined in Region I, while a small amount of power leaks into regions II, III, IV, and V, where EM field decays exponentially. Since even less power enters into the four shaded areas, in calculating field patterns in region I, field matching at the four corners of the shaded regions is not considered. The E^x_{pq} mode with main field components E_x and H_y is now considered. For this hybrid mode, $H_x = 0$ can be put in Maxwell's equations. The wave equations with the field functions $H_y(x, y)$ are then solved analytically by separation of variables. Using the boundary conditions that H_y is continuous at $x = 0$ and $x = -T$ and that E_x is continuous on the interfaces $y = +W/2$ *and* $-W/2$, the field distribution $H_y(x,y)$ may be written as follows:

$$\begin{aligned}
&Region\ I: H_1 \cos(k_x x + \phi_1)\cos(k_y y + \phi_2),\\
&Region\ II: H_1 \cos(k_x T - \phi_1)\cos(k_y y + \phi_2)\exp\{\gamma_{x2}(x + T)\},\\
&Region\ III: H_1 \cos\phi_1 \cos(k_y y + \phi_2)\exp(-\gamma_{x3} x),\\
&Region\ IV: H_1 \cos(k_y W/2 + \phi_2)\cos(k_y x + \phi_1)\exp\{-\gamma_{y4}(y - W/2)\},\\
&Region\ V: H_1 \cos(k_y W/2 + \phi_2)\cos(k_x x + \phi_1)\exp\{\gamma_{y4}(y + W/2)\}.
\end{aligned} \tag{10.50}$$

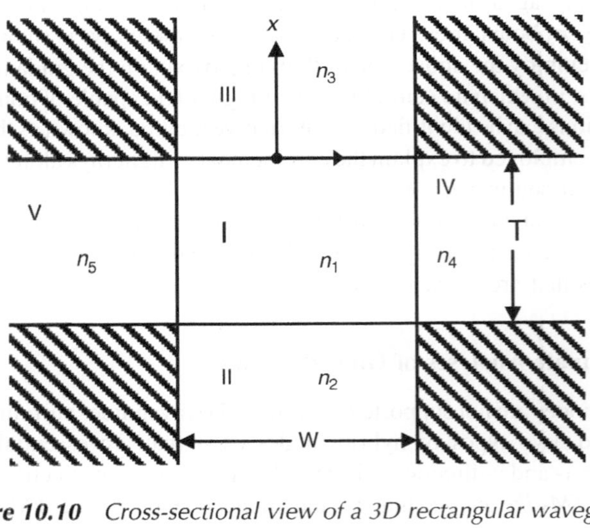

Figure 10.10 *Cross-sectional view of a 3D rectangular waveguide.*

The waveguide is assumed to be symmetrical along the y direction, that is, $n_4 = n_5$. The phase $\phi_2 = 0$ refers to symmetrical modes whose fields vary as cos $(k_y y)$. The condition $\phi_2 = \pi/2$ refers to the existence of antisymmetrical modes having field variation in the form sin $(k_y y)$. The propagation constants included in Eq. (10.50) are expressed as

$$\begin{aligned} \beta^2 &= k_0^2 n_1^2 - k_x^2 - k_y^2, \\ \gamma_{xi}^2 &= (\beta^2 + k_y^2) - k_0^2 n_i^2. \end{aligned} \tag{10.51}$$

where $i = 2,3,4$. Continuity of E_z on $x = 0$ and $-T$ yields the phase shift

$$\phi_1 = \frac{\pi}{2} - \tan^{-1}\left(\frac{n_3}{n_1}\right)^2 \left(\frac{k_x}{\gamma_{x3}}\right). \tag{10.52}$$

and the eigenvalue equation

$$k_x T = (p+1)\pi - \tan^{-1}\left(\frac{n_3}{n_1}\right)^2 \left(\frac{k_x}{\gamma_{x3}}\right) - \tan^{-1}\left(\frac{n_2}{n_1}\right)^2 \left(\frac{k_x}{\gamma_{x2}}\right), \tag{10.53}$$

where $p = 0,1,2\ldots$. Considering that H_z is continuous at $y = W/2$ and $-W/2$, the following equation obtains

$$k_y W = (q+1)\pi - 2\tan^{-1}\left(\frac{k_y}{\gamma_{y4}}\right) \tag{10.54}$$

where $q = 0,2,4\ldots$ or $1,3,5\ldots$, corresponding to symmetrical or asymmetrical modes, respectively. Solutions of the transcendental Eqs. (10.51) and (10.54) lead to the value of propagation constant β of the hybrid modes. The field distributions $H_y(x, y)$ of the fundamental E_{00}^x mode and the first-order E_{01}^x modes are shown in Figure 10.11 by solid and dotted lines, respectively. The E_{pq}^y modes may be analyzed in a similar manner.

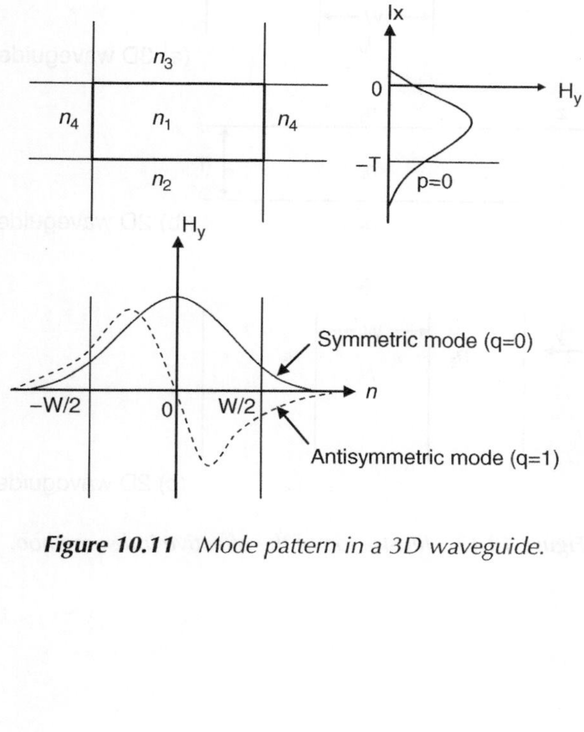

Figure 10.11 *Mode pattern in a 3D waveguide.*

10.6.3.2 Effective Index Method

In order to introduce this method, let us choose the embedded-strip 3D waveguide shown in Figure 10.8b. This guide consists of a rectangular core of high RI n_f, embedded in a substrate of lower RI n_s. The surrounding cladding region above has a RI n_c.

Guidance in this structure may be viewed as total internal reflections at the core–cladding and the core–substrate interfaces. In almost all practical situations, the core RI is slightly larger than the substrate RI, and therefore the reflected waves propagate nearly parallel to the z-direction for total internal reflections to occur. The propagation wave vector for these waves is essentially directed along z. Thus from Maxwell's equation the electric and magnetic field vectors for each mode are perpendicular to **k** and therefore nearly transverse to the z-direction. Assuming two polarizations, the possible solutions are E_y, H_x, and E_x,H_y.

As seen from Figure 10.12, in the region $|y| \leq W$, the structure is an asymmetric waveguide, having a RI, n_c, and thickness T. One cannot, however, directly use results in Section 10.4 to obtain the propagation constant k_z from the normalized b–V plot, because there is also a field variation along the y-direction.

At the present moment, the field variation along the y-direction is ignored. This means that W is assumed to be large. The structure we need to analyze is then as shown in Figure 10.12b. Assuming that the electric field is polarized along the y-direction, then the solutions correspond to modes which are TE with respect to the xz-plane. Alternatively, if the electric field is along x, then the magnetic field, **H**, is along y and the solutions are TM with respect to this plane. The normalized dispersion curves discussed in Section 10.4 may

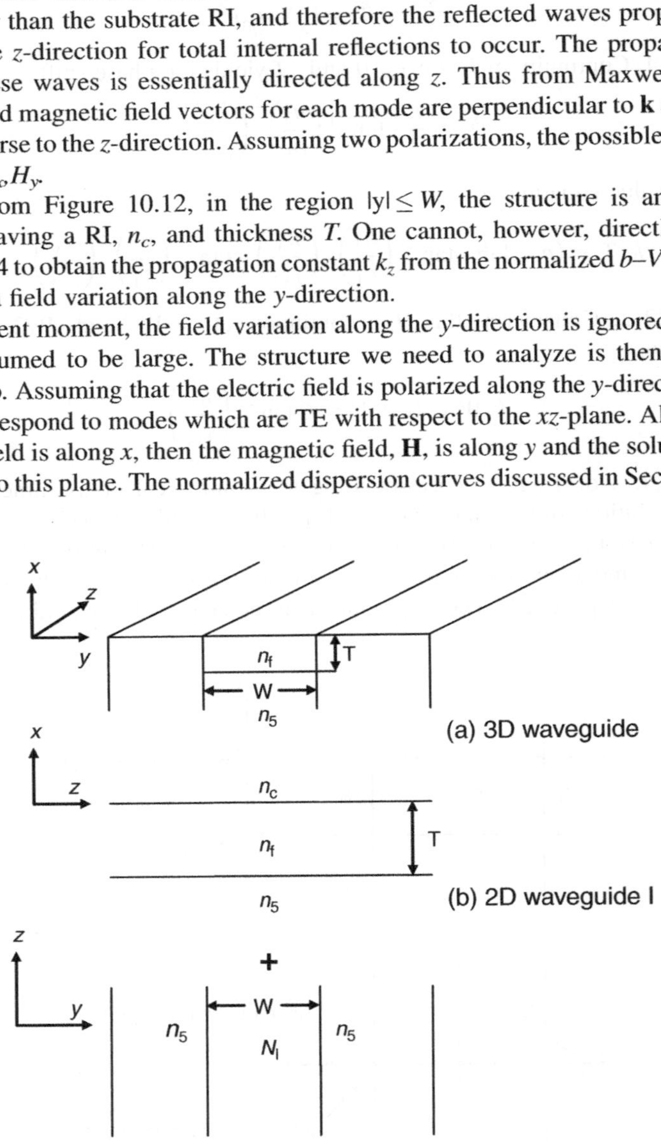

Figure 10.12 *Illustration of the effective index method.*

now be utilized to find the propagation constant for a given frequency. The normalized frequency parameter may be expressed as follows:

$$v = k_0 T(n_f^2 - n_s^2), \tag{10.55}$$

The asymmetry parameter is α_{TE} or α_{TM}. The symbols n_f and n_s denote, respectively, the RI's for the film (guide) and the substrate. Usually $(n_f - n_s)/n_f$ is small and as discussed in this chapter the same normalization curves may be used for both TE and TM modes. For each allowed mode p, the corresponding parameter b_p may be defined as

$$b_p = \left[(n_{eff})_p^2 - n_s^2\right] / (n_f^2 - n_s^2). \tag{10.56}$$

The index, $p = 0, 1, \ldots.,$ indicates the number of half cycles of field variation along x in the core region. An effective RI $(n_{eff})_p$ may now be assigned to each value of b_p. The propagation constant associated with each solution may now be expressed as

$$(k_z)_p = k_0 (n_{eff})_p. \tag{10.57}$$

This propagation constant is valid for the limit that W is infinite.

It is now easy to find out the corresponding transverse wave vectors in the three regions. The expressions are

$$(\gamma_{cx})_p = \sqrt{k_z^2 - k_c^2} = k_0 \sqrt{(n_{eff})_p^2 - n_c^2}, \tag{10.58a}$$

$$(\beta_{fx})_p = \sqrt{\beta_f^2 - k_z^2} = k_0 \sqrt{n_f^2 - (n_{eff})_p^2}, \tag{10.58b}$$

$$(\gamma_{sx})_p = \sqrt{k_z^2 - k_s^2} = k_0 \sqrt{(n_{eff})_p^2 - n_s^2}, \tag{10.58c}$$

The entire information about guidance in the xz-plane is contained in the effective index $(n_{eff})_p$. Thus, in the region $|y| \le W/2$, the entire asymmetric slab waveguide could be replaced by an equivalent uniform material of RI $(n_{eff})_p$ or N_I, as shown in Figure 10.12c. The presence of subscript p indicates that different modes need different values of the effective index.

Using the effective index determined by the method given in this subsection, the guidance along the yz plane may now be analyzed. As shown in Figure 10.12c, we need to consider a symmetric waveguide having core RI $(n_{eff})_p$ or N_I of width W surrounded by a material of RI n_s. We are now in a position to calculate a new parameter v' analogous to v:

$$v' = k_0 W \left[(n_{eff})_p^2 - n_s^2\right]^{1/2} = v b_p^{1/2} W/T, \tag{10.59}$$

As we consider a symmetric guide, the asymmetry parameter α' is equal to zero. Again using the weak guiding approximation, we can use the normalized dispersion curve for the TE modes to calculate a parameter b' from v':

$$(b')_{pq} = \left[(n'_{eff})_p^2 - n_s^2\right] / \left[(n_{eff})_p^2 - n_3^2\right]. \tag{10.60}$$

The presence of an additional subscript q in Eq. (10.60) signifies that for each value of index p, there are q solutions, $q = 0, 1, \ldots$, describing the number of half-cycle variations along y within the core. The effective index $(n'_{eff})_{pq}$ can be expressed in terms of b_p and b_{pq} by using Eqs. (10.56) and (10.60) as

$$(n'_{eff})_{pq} = \left[n_s^2 + b_p b'_{pq}(n_f^2 - n_s^2)\right]^{1/2}. \tag{10.61}$$

n_s^2 Usually the difference in RI, Δn, between RI's of the core and substrate is small, so that $n_f^2 - n_s^2 \cong 2\Delta n n_s$, and therefore $(n'_{eff})_{pq} \cong n_s + \Delta n b_p b'_{pq}$. The associated propagation constant for each mode $(k_z)_{pq}$ is given by

$$(k_z)_{pq} = k_0 (n'_{eff})_{pq}. \tag{10.62}$$

The corresponding transverse wave vectors along the y-direction for regions $|y| < W/2$ and $|y| > W/2$ take the following respective forms:

$$(\gamma_y)_{pq} = \sqrt{(k_z)_{pq}^2 - \gamma_s^2} = k_0 \sqrt{(n'_{eff})_{pq}^2 - n_s^2} \cong k_0 (2\Delta n\, n_s b_p b'_{pq})^{1/2}, \tag{10.63}$$

$$(k_y)_{pq} = \sqrt{k_0^2 (n_{eff})_p^2 - (k_z)_{pq}^2} = k_0 \sqrt{(n_{eff})_p^2 - (n'_{eff})_{pq}^2} \cong k_0 \left[2\Delta n\, n_s b_p (1 - b'_{pq})\right]^{1/2}. \tag{10.64}$$

A comparison between $b'_{pq} - V$ diagrams obtained by the effective index method and computer calculation shows that when the modes are far from cutoff, the approximate solutions agree with the exact numerical solution.

Example 10.3

This example illustrates the calculation of the effective refractive index for the fundamental y-polarized mode on an embedded-strip 3D waveguide having the following parameters: $n_c = 1.0$, $n_f = 2.234$, $n_s = 2.214$, $W = 2.5\ \lambda_0$, and $T = 1.5\ \lambda_0$.

Solution: The mode is y-polarized, or it is TE with respect to the xz-plane. The asymmetry parameter is now

$$\alpha^{TE} = (n_s^2 - n_c^2)/(n_f^2 - n_s^2) = 43.9$$

$$v = k_0 T (n_f^2 - n_c^2)^{1/2} = 2.81.$$

We use the normalized curve given in Kogelnik and Ramaswamy [1], and obtain $b_0 = 0.44$ and $b_p = 0$, for $p > 0$. From Eq. (10.56),

$$(n_{eff})_o = \left[(n_f^2 - n_s^2) b_0 + n_s^2\right]^{1/2} = 2.223.$$

For the yz-plane, we use Eq. (10.59) to obtain

$$v' = k_0 W \left[(n_{eff})_0^2 - n_s^2\right]^{1/2} = 3.11.$$

Although the mode appears TM with respect to the *yz*-plane, due to small differences between (n_{eff}) and n_s, we may use the normalized dispersion curve for the TE modes. Thus,

$$b'_{00} = 0.64, b'_{pq} = 0, p \neq q \neq 0.$$

From Eq. (10.61),

$$(n_{eff})'_{00} \cong n_s + (n_f - n_s) b_0 b'_{00} = 2.220.$$

10.7 Loss Mechanisms in Waveguides

Attenuation or loss suffered by a lightwave, as it travels the guide, is an important parameter that should be minimized for a good device.

There are three contributions to the loss: from scattering, absorption, and radiation. Scattering loss is usually important in glass or dielectric waveguides, while absorption loss is most important in semiconductor guides. Radiation loss becomes important when the guides are bent through a curve. We shall discuss briefly the various loss mechanisms at play within a guide.

10.7.1 Scattering Loss

There are two types of scattering loss in optical waveguides: volume scattering and surface scattering.

10.7.1.1 Volume Scattering

Volume scattering is caused by imperfections present in the guide material, such as impurities, defects, voids, and so on. The loss per unit length is proportional to the number of imperfections per unit volume of the medium. Also the size of the imperfections relative to the wavelength of radiation is another factor determining the loss. In a well-prepared guide material, the density of imperfections should be minimized leading to a negligible loss due to volume imperfections.

There are situations in which volume imperfections may play a dominant role. For example, an ion implantation process may introduce a sufficient number of defects in the waveguide material. Also, the size of the defects with respect to the wavelength of propagation, and the correlation length along the guide, are important. In bulk media, Rayleigh scattering is the dominant loss mechanism exhibiting λ^{-4} dependence. The dependence follows λ^{-3} variation for confined waves, for correlation length of the defects shorter than or of the order of the wavelength. For long correlation lengths compared to the wavelength, radiation loss is dominant and the loss varies as λ^{-3}.

10.7.1.2 Surface Scattering

Loss due to surface imperfections is more dominant due to limitations in processing technology and therefore merits attention. Surface scattering occurs at the core–cladding interfaces and its origin lies in the interfaces having fluctuations from an atomically smooth perfect plane. Various authors have treated surface roughness scattering in different ways. We describe in this subsection a simple and instructive theory developed by Tien [5] based on the ray optic theory.

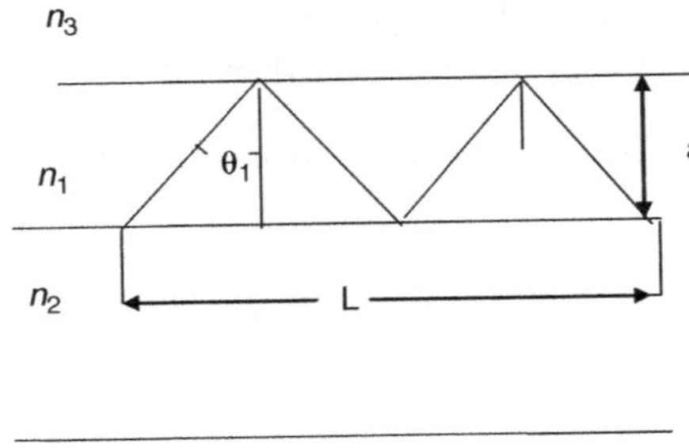

Figure 10.13 *Ray optic approach to determine the surface-scattering loss.*

Consider again the asymmetric slab guide redrawn in Figure 10.13. The light ray bounces back and forth in the guide due to total internal reflections. It follows from the geometry that the number of reflections encountered by the ray from each interface while covering an axial distance L may be expressed as

$$N = \frac{L}{2a \tan \theta_1}, \tag{10.65}$$

where θ_1 is the angle of incidence at the core–upper cladding interface.

Example 10.4

Let the thickness of SiO_2 layer be 1 μm, $n_1 = 1.5$, and $\theta_1 = 60°$. Then the number of reflections from the upper interface is 2887 for 1 cm length of the guide.

The propagating waves interact very strongly with the interfaces. Higher order modes undergo more reflections because of the larger angle of incidence and therefore suffer higher loss.

Tien's analysis is based upon the specular reflections from the surface and on Rayleigh criteria. It states that if the incident power at the surface is P_i, the specular reflected power P_r from the surface is

$$P_r = P_i \exp\left[-\left(\frac{4\pi\sigma n_1}{\lambda_0}\cos\theta_1\right)^2\right], \tag{10.66}$$

where σ is the variance of the surface roughness. In order to obtain the expression for the loss coefficient, both the surfaces and the distance covered must be considered. The final expression given by Tien takes the following form:

$$\alpha_s = \frac{\cos^3\theta}{2\sin\theta}\left(\frac{4\pi n_1(\sigma_u^2 + \sigma_l^2)^{1/2}}{\lambda_0}\right)^2\left(\frac{1}{a + (1/\gamma_c) + (1/\gamma_s)}\right), \tag{10.67}$$

where γ's are the decay constants for the fields penetrating into the cladding layers, as defined by Eq. (10.37).

A number of conclusions may be derived from Eq. (10.67). First the loss coefficient is proportional to the square of the ratio between roughness and wavelength in the material.

It is inversely proportional to the effective width of the guide that contains the actual width and the penetration tails in the claddings (GH shift). Thus, well-confined modes will be more affected than modes with large evanescent tails. The scattering is reduced if the inverse decay constants are larger than the guide thickness. In other words, loss will be higher if the difference in RI's is larger. Physically, deeper penetration of waves into claddings makes them less sensitive to the effect of roughness. Since θ_1 is larger for higher order modes, the factor ($\cos^3\theta/\sin\theta$) accounts for greater loss for higher order modes. This is expected as higher order modes suffer larger number of reflections at the interfaces for a given distance.

Equation (10.67), developed by Tien, though approximate, works well for long correlation lengths of surface roughness. A more rigorous theory was developed by Marcuse, who treated surface scattering as a form of radiation loss in which irregularities at the interface were assumed to couple energy from the propagating modes to the radiation modes. However, under suitable approximation and in the limit of long correlation length, the result of Marcuse agrees well with that predicted by Eq. (10.67).

10.7.1.3 Sidewall Loss

Tien's theory works well for slab waveguides in which the correlation length connected with irregularities is large. However, most of the practical waveguides are in the form of a rib or ridge, in which the width of the guide is made small by some form of etching: dry or wet. The scattering by the sidewalls of the guide introduces a good amount of loss [6, 7]. Payne and Lacey [6] derived a relation for the upper bound of the scattering loss coefficient for a waveguide of core index n_1 and width d given by

$$\alpha \leq \frac{\sigma^2}{k_0 d^4 n_1}\kappa, \qquad (10.68)$$

where κ depends on the waveguide geometry and the statistical distribution of the roughness. It seems from Eq. (10.68) that the loss would increase with decreasing width of the rib. The sidewall roughness is described by a root mean square roughness σ, a measure of the mean height of the asperity, and a correlation length l_c. Lee *et al.* [8] assumed an exponential autocorrelation for the sidewall roughness,

$$C(u) = \sigma^2 \int_{-\infty}^{\infty} \exp(-|u|/l_c)du. \qquad (10.69)$$

The function κ in Eq. (10.68) was then calculated by using the exponential function given here. The authors found good agreement between their calculated values with the experimental values for Si waveguides buried in SiO_2 as given in Figure 10.14. It appears that when the width of the rib is less than 2 μm, sidewall loss assumes a significant role. Using a TEM picture, the authors obtained $\sigma = 9$ nm and $l_c = 50$ nm.

10.7.2 Absorption Loss

While the surface-scattering loss may be present in all kinds of waveguides, however small it might be, the important loss mechanism in semiconductors is the absorption: both interband and intraband free-carrier absorption. We discuss the two loss processes in Sections 10.7.2.1 and 10.7.2.2.

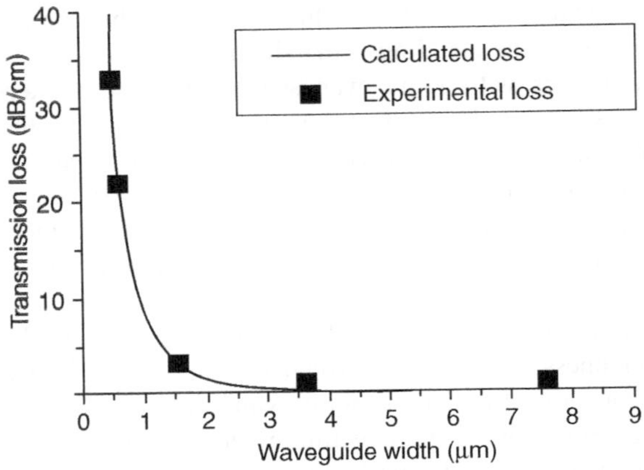

Figure 10.14 *Transmission loss for a Si strip waveguide embedded in silica as a function of width of the Si strip. Reproduced with permission from Appl. Phys. Lett., Effect of size and roughness on light transmission in a Si/SiO2 waveguide: Experiments and model by K. Lee et al., 77, 11, 1617–1620. Copyright (2000) American Institute of Physics.*

10.7.2.1 Interband Absorption

The interband absorption, alternatively known as the *fundamental absorption*, occurs when photons having energy exceeding the fundamental band gap energy of a semiconductor or a crystalline material are strongly absorbed. There is very little absorption when the photon energy is smaller than the band gap energy, E_g.

The fundamental absorption coefficient is quite large in III–V compounds, more than $10^4\ cm^{-1}$. The value is, however, small in indirect-gap materials like silicon. To avoid the interband absorption, one should use a wavelength that is significantly longer than the absorption edge wavelength of the guide material. The cutoff wavelength for Si is approximately 1.11 μm. Therefore, the loss is negligible in Si waveguides at the standard telecommunication wavelengths of 1.3 and 1.55 μms.

10.7.2.2 Free Carrier Absorption

This process has been discussed in Chapters 4 and 7, where the expressions for the absorption coefficient in terms of injected carriers are given.

10.7.3 Radiation Loss

We now consider the radiation-bending losses. In an optical integrated circuit (OIC), a number of components are interconnected. The dielectric pathways interconnecting them often contain bends as shown in Figure 10.15. As may also be seen from Figure 10.15, the direction of propagation of light in the bend region changes, as a result of which power may be radiated away from the guide.

The dielectric waveguide shown in Figure 10.15 has a curved section of radius R. The propagation direction z is along an axis of the waveguide. The geometry shown may correspond to the top view of a 3D waveguide with n_1 and n_2 signifying the waveguide effective refractive indices.

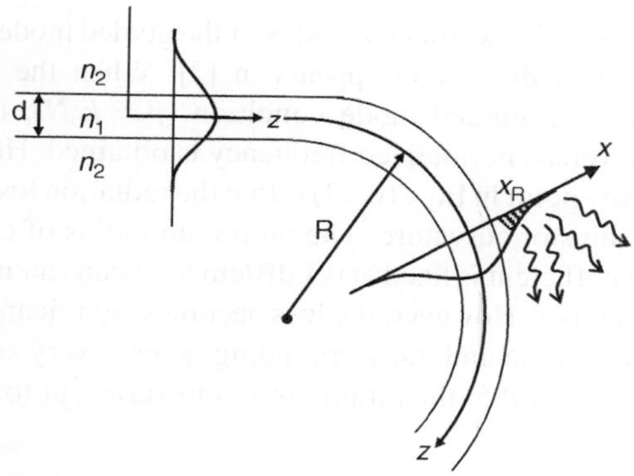

Figure 10.15 *Section of a curved slab waveguide illustrating radiation loss due to bending of the guide.*

Let us consider that only the fundamental mode is above the cutoff frequency. If the radius of curvature of the bend is large, then it may be concluded that the properties of the mode for any point z along the bend would not differ significantly from the straight waveguide limit ($R \to \infty$). The constant phase fronts are normal to z, that is, in the radial direction. The field distribution measured along such a phase front should be almost identical with that of the mode in the straight guide. Further the phase velocity, $v_p = \omega/\beta_z$, measured along the axis of the bend should be approximately equal to that for $R \to \infty$. This implies

$$\frac{\omega}{\beta_z} = R\frac{d\theta}{dt}. \tag{10.70}$$

Now consider the tangential velocity of points on the constant phase front which are farther from the center of curvature. Since these points lie on the same phase front, they all have the same angular velocity $d\theta/dt$. Their tangential velocity $v_{\tan} = (x+R)d\theta/dt$ increases with x and at some point x_r, shown in Figure 10.15, this velocity may exceed the velocity of light in the surrounding medium. We may define this transition point by the expression

$$\frac{\omega}{\beta_0} = (x_r + R)\frac{d\theta}{dt}. \tag{10.71}$$

Combining the above two expressions, one obtains

$$x_r = \frac{\beta_z - \beta_0}{\beta_0} R \tag{10.72}$$

The radiation process can be explained in the following terms. Photons located at radii greater than $R + x_r$ cannot travel fast enough to keep up with the rest of the mode. They therefore split away and are radiated into layer 2. An estimate of the attenuation coefficient has been made by some workers [2], which shall not be reproduced here. The attenuation constant α of the guided mode may be expressed in the following approximate form by Nishihara *et al* [1989]:

$$\alpha = \frac{2\gamma_s^2 k_y^2 \exp(2\gamma_s a)}{(N_2 - n_s^2)k_0^2 \beta(2+\gamma_s a)} \exp\left(-\frac{2\gamma_s^3}{3\beta^2}R\right). \tag{10.73}$$

if $\alpha R << 1$. In Eq. (10.73), N is the effective index of the guided mode, and a is the width of the guide. A more detailed discussion appears in [2]. When the change in RI $\Delta N = N - n_s << n_s$ under the well-guided mode condition ($\beta \approx k_0 N$), the relation between normalized attenuation versus normalized frequency is obtained. Here n_s is the RI of the substrate. The noteworthy point in Eq. (10.73) is that the radiation loss coefficient depends exponentially on the radius of curvature. The minimum radius of curvature that may be allowed when the loss coefficient is fixed at 0.1 dB/cm has been calculated by Goell [9] for several dielectric waveguides. However, the loss becomes significant when the difference between RI's of the waveguide and the surrounding layer is very small. As an example, when the RI's are 1.5 and 1.4985, the minimum radius turns out to be 10.4 cm for a loss of 0.1 dB/cm.

10.7.4 Coupling Loss

The waveguide components and devices are in general coupled to input or output fibers. The coupling between the fiber and the device should be perfect; otherwise, an insertion loss appears. The principal characteristics of any coupler are its efficiency and its mode selectivity. Coupling efficiency gives the measure of the fraction of total power in the beam which is coupled into (or out of) the waveguide. It is given approximately by the square of the overlap of the normalized field distribution function of the fiber and that of the waveguide. Alternatively, it may be expressed in terms of the coupling loss in dB, defined as

$$L_{Cm} = 10 \log \frac{\text{total power in optical beam prior to coupling}}{\text{power coupled into (out of)the } m\text{th order mode}}. \tag{10.74}$$

The coupling loss arises due to mode mismatch between the waveguide and the fiber and the misalignment of the fiber relative to the waveguide. The mode mismatch results due to the differences in the RI and size between the cores of waveguide and fiber. A higher value of Δ permits higher integration density but makes the core size smaller than that of the fiber. A coupling loss is therefore unavoidable. The mismatch and misalignment losses can be calculated approximately by assuming the optical fields to be Gaussian beams. Simple analytical expressions are obtained for losses in that situation. Most practical waveguides have rectangular cross-sections and the coupling efficiency is calculated numerically. The loss depends on the value of Δ, the height and width of the waveguide.

Various methods exist for coupling fiber with waveguides, namely, focusing, end-butt coupling, prism coupling, and grating coupling, or matching by index matching fluid. Each method has its own advantages and disadvantages.

10.8 Coupling to Optical Devices

One of the challenging tasks in optical systems is efficiently coupling light into and out of the OEIC. In general, light from a standard optical fiber or an external light source needs to be coupled to a silicon waveguide. A single-mode fiber core ($n = 1.5$) usually has a diameter of about 8 μm with a symmetric mode, while a silicon waveguide with RI = 3.45 is typically a few μms wide with an asymmetrical mode. To overcome these large differences in effective index, core size, and symmetry, various techniques are developed for efficiently coupling light from fiber to the guide. In the end-fire and butt-coupling schemes, light is

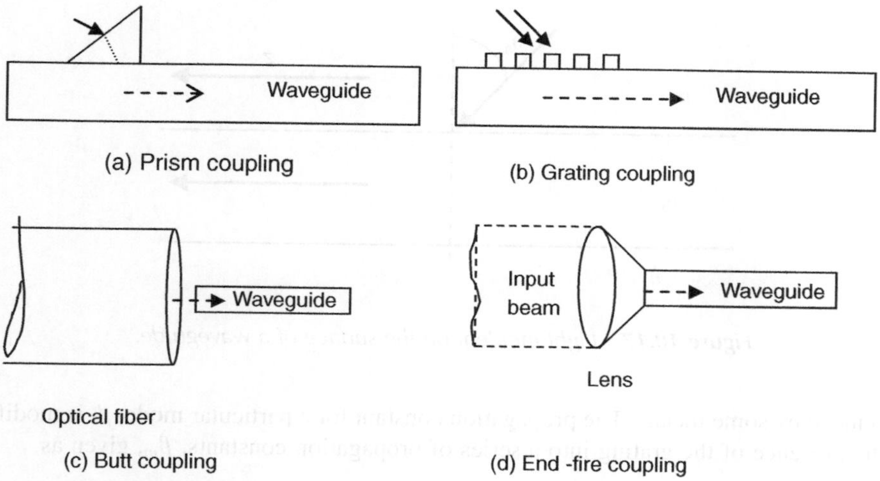

Figure 10.16 *Various methods of coupling light from the fiber to waveguides: (a) prism coupling; (b) grating coupling; (c) butt coupling; and (d) end-fire coupling.*

simply allowed to shine onto the end faces of the waveguide. These techniques are shown schematically in Figure 10.16. In butt coupling, the two devices are simply butted, such that the mode field of the transmitting device falls onto the end face of the receiving device. In end-fire coupling, a lens is employed to focus the light beam onto the end face of the collecting device. In these two schemes, the input light beam can excite all the modes of the waveguide.

In the prism coupling and grating coupling, the input beam impinges on the guide at a specific angle, as shown in Figure 10.16. Due to several disadvantages, prism coupling is still not very popular. In Sections 10.8.1 and 10.8.2, the three other methods that work well for silicon-based technology will be described.

10.8.1 Grating Couplers

In order to couple light into the waveguide mode, the phase velocity of the coupled wave in the direction of propagation (the z-direction) must be the same as that in the waveguide. This is referred to as the *phase match condition*. Let the ray be incident on the surface of the waveguide at an angle θ_a, as shown in Figure 10.17. The propagation constant of the ray in medium 3 (RI $= n_3$) is $k_0 n_3$ and the same along the z-direction is $k_0 n_3 \sin\theta_a$. This requires the following phase match condition:

$$\beta = k_z = k_0 n_3 \sin \theta_a, \tag{10.75}$$

where β is the waveguide propagation constant, which is $\geq k_0 n_3$. The above phase match condition can never be met unless some other technique is employed. This is accomplished by a grating or a prism, as explained below.

The structure and detailed working principle of a grating will be described in Chapter 11. It is sufficient at this moment to note that in a grating, a periodic modulation of RI is

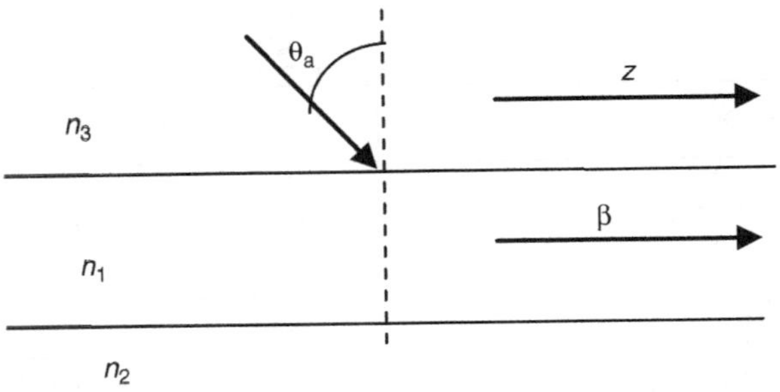

Figure 10.17 *Light incident on the surface of a waveguide.*

introduced by some means. The propagation constant for a particular mode, β, is modified by the presence of the grating into a series of propagation constants, β_m, given as

$$\beta_m = \beta + \frac{2m\pi}{\Lambda}, m = \pm 1, \pm 2, \pm 3, \text{etc.} \tag{10.76}$$

and Λ is the period of the grating. For positive values of m, β_m is still less than $k_0 n_3$. Therefore negative values of m can ensure phase mismatch. In most cases, the grating is fabricated in such a way that with only $p = -1$, the phase match results. Therefore, the waveguide propagation constant becomes

$$\beta_m = \beta - \frac{2\pi}{\Lambda}. \tag{10.77}$$

The phase match condition now reads

$$\beta - \frac{2\pi}{\Lambda} = k_0 n_3 \sin\theta_a. \tag{10.78}$$

Writing $\beta = k_0 N$, where N is the effective index and $k_0 = 2\pi/\lambda$, the condition becomes

$$\Lambda = \frac{\lambda}{N - n_3 \sin\theta_a}. \tag{10.79}$$

Example 10.5

Let $\lambda = 1.55\ \mu\text{m}, N = 3.5$ for Si, and the angle of incidence $= 45°$. Then $\Lambda = 555$ nm.

There are not many reports about the grating couplers using silicon, presumably due to the small size of the grating period. Ang *et al.* [10, 11] achieved an output coupling efficiency of about 70% for rectangular gratings, and about 84% with a nonsymmetrical profile.

10.8.2 Butt Coupling and End-fire Coupling

When a light beam falls upon the end face of a waveguide, the efficiency with which the light beam couples with the waveguide depends on four factors: (1) matching between modes of the light beam and waveguide, (2) the degree of reflection from the waveguide facet, (3) the quality of the waveguide facet, and (4) the spatial misalignment of the excitation and waveguide fields. In Sections 10.8.2.1, 10.8.2.2, 10.8.2.3, and 10.8.2.4, we will consider these factors.

10.8.2.1 Overlap of Excitation and Waveguide Fields

We consider a simple situation when both the fields are single moded. The overlap integral between the exciting field E_{ex} and the guide field in the guide E_g is expressed as

$$\Omega = \frac{\int_{-\infty}^{\infty} dy \int_{-\infty}^{\infty} dx E_{ex} E_g}{\left(\int_{-\infty}^{\infty}\int_{-\infty}^{\infty} E_{ex}^2 dxdy \cdot \int_{-\infty}^{\infty}\int_{-\infty}^{\infty} E_g^2 dxdy \right)^{1/2}}, \tag{10.80}$$

where the denominator is a normalizing factor that ensures that the value of Ω lies between 0 and 1. Eq. (10.80) greatly simplifies this, and the result gives a good idea of the values of overlap, if the fields for both the excitation beam and the waveguide mode are assumed to have the Gaussian shapes, described by the following equations:

$$E_g = \exp\left[-\left(\frac{x^2}{d_x^2} + \frac{y^2}{d_y^2} \right) \right], \tag{10.81}$$

and

$$E_{ex} = \exp\left[-\frac{x^2 + y^2}{d_0^2} \right]. \tag{10.82}$$

Here different d's represent the half widths of the beams along the respective directions. Using the identity $\int_0^{\infty} \exp[-r^2x^2]dx = \sqrt{\pi}/2r$, Eq. (10.80) may now be evaluated analytically to yield

$$\Omega = \frac{\frac{2}{d_0}\left[\frac{1}{d_x d_y}\right]^{1/2}}{\left[\frac{1}{d_x^2} + \frac{1}{d_0^2}\right]^{1/2}\left[\frac{1}{d_y^2} + \frac{1}{d_0^2}\right]^{1/2}}. \tag{10.83}$$

The power-coupling efficiency is Ω^2.

Example 10.6

When all d's are equal, $\Omega = 1.0$ and $\Omega^2 = 1.0$, and the loss is 0 dB. When $d_O = d_x = 10\ \mu m$, and $d_y = 5\ \mu m$ $\Omega = 0.894$ and $\Omega^2 = 0.8$, making the loss $= 0.97$ dB.

10.8.2.2 Reflections from the Waveguide Facet

Equations (10.6a) and (10.6b) for the reflection coefficients for TE- and TM-polarized rays may be modified as follows using Snell's law:

$$r_{TE} = \frac{-\sin(\theta_1 - \theta_2)}{\sin(\theta_1 + \theta_2)}. \tag{10.84}$$

Hence the power reflectivity, $R = r^2$, is expressed as

$$R_{TE} = \frac{\sin^2(\theta_1 - \theta_2)}{\sin^2(\theta_1 + \theta_2)}. \tag{10.85}$$

Similarly, for the TM ray,

$$R_{TM} = \frac{\tan^2(\theta_1 - \theta_2)}{\tan^2(\theta_1 + \theta_2)}. \tag{10.86}$$

At normal incidence, $\theta_1 = 0$, both the reflectivities are the same. Furthermore, end-fire coupling allows light to be incident nearly normal to the face. In that case,

$$R_{TE} = R_{TM} = R = \left| \frac{n_1 - n_2}{n_1 + n_2} \right|^2. \tag{10.87}$$

For the interface between silicon (RI = 3.5) and air (RI = 1.0), $R = 31\%$, leading to a loss of about 1.6 dB/facet. Use of an antireflection coating between the two media eliminates the loss. The thickness of the antireflection coating is $\lambda/4$, so the rays reflected from the front face of the coating and the back face, which is the end face of the guide, are in antiphase. This eliminates the reflection. It may be shown that for normal incidence, the reflectivity is given by

$$R = \left| \frac{n_1 n_2 - n_{ar}^2}{n_1 n_2 + n_{ar}^2} \right|^2, \tag{10.88}$$

where n_{ar} is the RI of the antireflection coating. When $n_1 n_2 = n_{ar}^2$, the reflectivity is zero.

10.8.2.3 The Quality of the Waveguide End Face

The end face of the waveguide is prepared as sufficiently smooth to avoid scattering from it. Usually three methods are employed for this purpose. A smooth surface may be prepared by cleaving the facet, that is, applying a mechanical pressure to the sample, so that it cracks along a crystallographic plane. Though the result is good, but the method is difficult to master, and is hardly used in commercial applications. Polishing is the most preferred way of preparing the end face. The facet is polished by lapping with abrasive materials, decreasing the grain size of the abrasive in steps. Chemical and dry etching is the third method employed to prepare a polished end face.

10.8.2.4 Spatial Misalignment

The overlap between the exciting and receiving fields may alter when there is slight misalignment of the two fields. The effect of spatial misalignment is estimated by calculating the overlap integral defined by Eq. (10.80), when the exciting and waveguide fields are displaced by an amount a, relative to each other. Let us assume that the waveguide field is still described by Eq. (10.81), but the exciting field is offset by a, so that

$$E_{ex} = \exp\left[-\frac{x^2+(y-a)^2}{d_0^2}\right]. \quad (10.89)$$

The offset is assumed to be along the y-direction, to simplify calculation. The overlap integral, in the presence of misalignment, may now be written as

$$\Omega_{misal} = \exp\left[-\frac{a^2}{d_y^2+d_0^2}\right]\Omega = A\Omega. \quad (10.90)$$

Example 10.7

We calculate the additional loss due to misalignment given by the square of the exponential factor, A^2, in Eq. (10.90). Let $a = 2\,\mu$m, and all d's are $5\,\mu$m. Then $A^2 = 0.85$, which accounts for an additional loss of about 0.7 dB.

Equation (10.90) is obtained if the offset is along the y-direction only. There would be additional loss if the offset is along the x-direction also.

10.9 Tapers

All the methods of coupling, described in Section 10.8, do not work when the guide dimension is reduced. The frequently used method to overcome the large differences in effective index, core size, and symmetry between the fiber and the waveguide is to employ a waveguide taper.

As mentioned in this chapter, some loss is incurred while coupling light from fiber to waveguides, because there is mismatch between the mode field diameters in the two media. The loss is not appreciable for coupling of fibers to silica-based waveguides, since the RIs have very close match. However, when a SOI waveguide is employed, the mode field dimension in high-index Si is considerably smaller than that in the core of the fiber. For example, in a SOI rib guide with a 4 μm thick epilayer, 3 μm rib width, and 2 μm etch depth, the mode field radius is 1.6 μm, whereas the fiber mode field radius is 5.2 μm. The coupling loss in this case is more than 4.7 dB. To reduce this coupling loss, it is necessary to use a mode transformer or a taper. Tapers reduce the coupling loss by an adiabatic modal transformation. Essentially there is a gradual, smooth, and lossless transition from a large cross-sectional waveguide area to a smaller one. The angle of the taper should be very small to achieve the smooth transition, and the surface roughness should also be as small as possible. Tapers may be of different forms. Lateral tapers are easily fabricated, since they involve an etching process from the top of the wafer. To produce a vertical taper, a differential etch rate is needed along the length of the taper. Various structures of the tapers are given in the review by Moerman *et al.* [12].

The simplest transformer expands the waveguide dimension horizontally, which expands the mode field in one dimension only. The coupling loss is not reduced significantly, however.

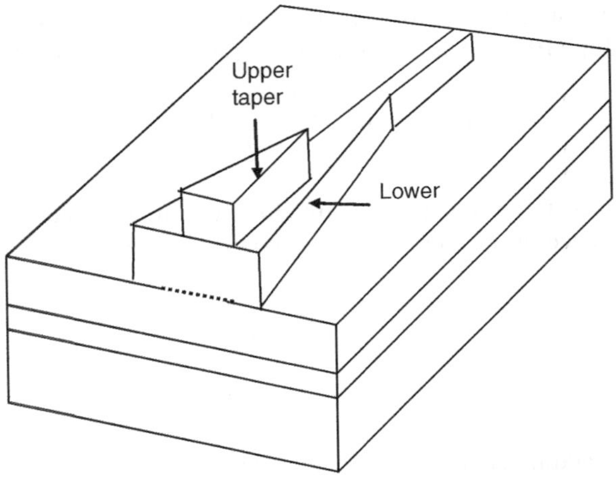

Figure 10.18 Schematic of SOI taper for low loss coupling between a waveguide and single-mode fiber. Reproduced with permission from [13]. Copyright (2004) Springer Science + Business Media.

Further improvement in coupling is achieved when the height of the guide is also expanded. An SOI guide with a 12.5 μm thick epilayer, 12.5 μm rib width, and 10 μm etch depth has a mode field radius of 4.8 μm, and only 0.24 dB coupling loss to fiber. A schematic diagram of a mode transformer or taper, consisting of two regions, is shown in Figure 10.18 [13]. The lower taper is about 20 μm wide, produced in a 4 μm thick silicon etched to a depth of 2 μm. The upper taper on it is 12.5 μm high. Initially the light is confined in the upper taper, but after a distance of 6 μm, the light is transferred to the lower rib guide. The lower guide is tapered beyond the end of the upper taper to produce the final single-mode waveguide with a 3 μm width. The total length of the taper is about 3 mm.

The taper transforms the fundamental mode of the optical fiber to that of the single-moded SOI waveguide. It is necessary that there should not be any coupling between the fundamental mode and higher order modes, if any, and between the propagating fundamental mode and the radiation modes into the substrate. The usual design approach is to obtain the pattern of the propagating fundamental mode as a function of the taper length L using a beam propagation tool. The analysis gives the adiabatic limit to the taper length L. However, this analysis does not give the minimum length for a given loss for various reasons, and ones need to solve coupled-mode equations along the taper.

A simple, empirical approach for determining the length for an adiabatic taper is given by Love [14]. For the taper to be adiabatic, the length scale over which its width varies should be long compared to the beat length between the fundamental mode and the nearest mode in the effective index, to which the fundamental mode can couple. The condition can be derived by considering Figure 10.19.

The waveguide half width at a point z along the taper is denoted by $\rho(z)$; the propagation constants for the fundamental and unwanted modes, that may be coupled, are respectively,

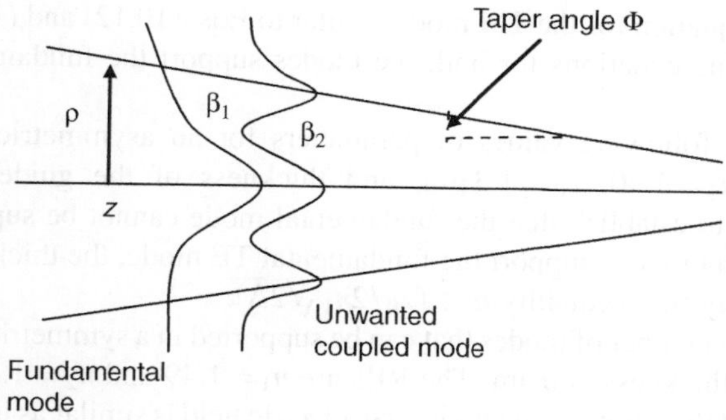

Figure 10.19 Taper with coupling between a fundamental mode and unwanted mode.

β_1 and β_2. The beat length between the modes is given by

$$L_B = \frac{2\pi}{\beta_1 - \beta_2}. \tag{10.91}$$

The length scale over which the taper width is changing is $[\rho/(\partial\rho/\partial z)]$, and to satisfy the condition for achieving an adiabatic taper, this length scale should be much larger than the beat length. Therefore

$$\rho/(\partial\rho/\partial z) >> L_B. \tag{10.92}$$

The local taper angle must satisfy the relation

$$\Phi = \frac{d\rho}{dz} << \frac{\rho}{L_B}. \tag{10.93}$$

Several taper methods have been proposed and fabricated that demonstrate efficient coupling from a relatively large silicon waveguide into an optical fiber. An inverted nano-taper has been proposed and fabricated by Almeida *et al.* [15]. The reader is referred to the literature cited to get an idea about the new developments [16].

Problems

10.1 Establish Eq. (10.3), the expression for the numerical aperture for a light ray in air falling on the core of the guide.

10.2 Find the value of numerical aperture and maximum acceptance angle in degrees for a core layer of RI = 1.5 and cladding layer RI = 1.48.

10.3 Show by using Snell's law that the angle of refraction becomes imaginary for the critical angle of incidence.

10.4 Derive Eq. (10.7a), the expressions for the phase shift for the totally reflected rays.

10.5 Using the definition of the Poynting vector, derive Eq. (10.8).

10.6 Use $n_1 = 1.5$, $n_2 = 1.49$, $\lambda_0 = 1.3\ \mu m$, and $a = 5\ \mu m$. Show by plotting two sides of Eq. (10.13) that the angle of incidence is maximum for the fundamental TE mode.

10.7 Set up the equation for the TM mode similar to Eqs. (10.12) and (10.13). Prove that the eigenvalue equations for both the modes support the fundamental mode with $m = 0$.

10.8 Assume the following values of parameters for an asymmetric guide: $n_1 = 1.5$, $n_2 = 1.49$, $n_3 = 1.40$, $\lambda_0 = 1.3\,\mu m$, and thickness of the guide $= 0.3\,\mu m$. Solve Eq. (10.16) to establish that the fundamental mode cannot be supported.

10.9 Prove that in order to support the fundamental TE mode, the thickness of the guide should satisfy the inequality $a < (\lambda_0/2n_1\sqrt{2\Delta})$.

10.10 Estimate the number of modes that can be supported in a symmetric planar dielectric guide with thickness 100 μm. The RI's are $n_1 = 1.49$ and $n_2 = 1.47$.

10.11 Assume that the nature of the transverse electric field is similar as in Eq. (3.21): show that the eigenvalue equation for a symmetric planar waveguide is $\gamma = k\tan(ka/2)$.

10.12 Using the electric field variation assumed in Problem 10.11, show that the mode confinement factor may be expressed as $\Gamma = \frac{1 + 2\gamma a/D^2}{1 + (2/\gamma D)}$, where $D = k_0(n_1{}^2 - n_2{}^2)^{1/2}a$.

10.13 Calculate the mode confinement factor for Si–SiGeC waveguides by solving a wave equation. Also calculate the approximate value by using the simplified formula $\Gamma = D^2/(2 + D^2)$ [17]. and compare the two values ($n(Si_{1\text{-}x\text{-}y}Ge_xC_y) = n(Si) + [n(Ge) - n(Si)][x - 2.34y]$).

10.14 Calculate the overlap integral between excitation and waveguide modes and the loss in dB for the following combinations of (d_0, d_x, d_y): (5,5,5), (10,10,5), (20,1,1), (5,5,3), and (20,10,5). All the dimensions are in μms.

10.15 Calculate the facet loss when a layer of silicon nitride (RI = 2.05) is used as an antireflection coating.

10.16 The RI of SiO_xN_y varies from 1.46 (SiO_2) to 2.05 (Si_3N_4), by changing the O and N concentrations. Find the approximate composition of the oxynitride for a good AR coating for the silicon–air interface.

10.17 Derive Eq. (10.41), mentioning the approximation introduced.

10.18 Calculate the values of A, A^2, and excess loss in dB from Eq. (10.90), when the offset $a = 1$, 4, 5, and 10 μms.

10.19 The expression for Rayleigh scattering loss is $\alpha_R = \frac{8\pi^3}{3\lambda_0^4} n^8 p^2 \beta k_B T_F$. Calculate the loss in dB/km in silica at 1.0 and 1.55 μms. The RI n = 1.46, photoelastic constant p = 0.286, isothermal compressibility $\beta = 7 \times 10^{-11}\ m^2\ N^{-1}$, and the temperature at which imperfections are frozen $T_F = 1400$ K.

10.20 Calculate the scattering loss for a silica waveguide ($n_1 = 1.5$, upper cladding air $n_3 = 1.0$, and lower cladding silica $n_2 = 1.45$) for two modes TE_0 and TE_2. Assume $\sigma_1 = \sigma_2 = 1$ nm, and $\lambda_0 = 1.33\,\mu m$.

10.21 Work out the steps to arrive at Eqs. (10.83) and (10.90).

References

1. Kogelnik, H. and Ramaswamy, V. (1974) Scaling rules for thin film optical waveguides. *Appl. Opt.*, **13**, 1857–1862.
2. Marcatili, E.A.J. (1969) Dielectric rectangular waveguide and directional coupler for integrated optics. *Bell Syst. Tech. J.*, **48**, 2071–2102.

3. Burns, W.K. and Milton, A.F. (1975) Mode conversion in planar-dielectric separating waveguide. *IEEE J. Quant. Electron.*, **11**, 32–39.
4. Hocker, G.B. and Burns, W.K. (1977) Mode dispersion in diffused channel waveguides by an effective index method. *Appl. Opt.*, **16**, 113–118.
5. Tien, P.K. (1971) Light waves in thin films and integrated optics. *Appl. Opt.*, **10**, 2395–2413.
6. Payne, F.P. and Lacey, J.P.R. (1994) A theoretical analysis of scattering loss from planar optical waveguides. *Opt. Quant. Electron.*, **26**, 977–986.
7. Ladouceur, F., Love, J., and Senden, T. (1994) Effect of sidewall roughness in buried channel waveguides. *IEEE Proc. J.*, **141**, 242–248.
8. Lee, K.K., Lim, D.P., Luan, H-C. *et al.* (2000) Effect of size and roughness on light transmission in a Si/SiO_2 waveguide: experiments and model. *Appl. Phys. Lett.*, **77**, 1617–1619.
9. Goell, J. (1969) A circular harmonic computer analysis of rectangular dielectric waveguides. *Bell Syst. Tech. J.*, **48**, 2133–2160.
10. Ang, T.W. *et al.* (1999) Grating couplers using silicon on insulator. *Proc. SPIE*, **3620**, 79–86.
11. Ang, T.W. *et al.* (1999) Blazed grating coupler in unibond SOI. *Proc. SPIE*, **3896**, 360–368.
12. Moerman, I., Van Daele, P.P., and Demeester, P.M. (1997) A review on fabrication technologies for the monolithic integration of tapers with III–V semiconductor devices. *IEEE J. Sel. Top. Quant. Electron.*, **3**, 1308.
13. Bozeat, R.J., Day, S., Hopper, F. *et al.* (2004) Silicon based waveguides, in *Silicon Photonics* (eds L. Pavesi and D.J. Lockwood), Springer, Berlin, pp. 270–294.
14. Love, F.J. (1989) Application of low loss criterion to optical waveguides and devices. *IEE Proc. J.*, **136**, 225–228.
15. Almeida, V.R. *et al.* (2003) Nanotaper for compact mode conversion. *Opt. Lett.*, **28**, 1302–1304.
16. Salib, M. *et al.* (2004) Silicon photonics. *Intel. Technol. J.*, **08**, 142–160.
17. Botez, D.(1981) InGaAsP/InP double heterostructure lasers. *IEEE J. Quant. Electron.*, **QE-17**, 178.

Suggested Readings

Adams, M.J. (1981) *An Introduction to Optical Waveguides*, John Wiley & Sons, Inc., New York.
Bruce Buckman, A. (1992) *Guided-Wave Photonics*, Saunders College Publishing, Fort Worth, TX.
Ghatak, A.K. and Thyagarajan, K. (1989) *Optical Electronics*, Cambridge University Press, Cambridge, UK.
Haus, H. (1994) *Waves and Fields in Optoelectronics*, Prentice-Hall., Englewood Cliffs, NJ.
Hunsperger, R.G. (1993) *Integrated Optics: Theory and Technology*, 3rd edn, Springer-Verlag, Berlin.
Kasap, S.O. (2001) *Optoelectronics and Photonics: Principles and Practices*, Prentice Hall, Upper Saddle River, NJ.
Keiser, G. (1999) *Optical Fiber Communication*, 3rd edn, McGraw-Hill International, London.
Kogelnik, H. (1988) Theory of optical waveguides, in *Guided Wave Optoelectronics, Springer Ser. Electronics and Photonics*, vol. **26**(ed T. Tamir), Springer, Berlin.
Lee, D.L. (1986) *Electromagnetic Principles of Integrated Optics*, John Wiley & Sons, Inc., New York.
Marcuse, D. (1972) *Light Transmission Optics*, Van Nostrand-Reinhold, New York.
Nishihara, H., Haruna, M., and Suhara, T. (1989) *Optical Integrated Circuits*, McGraw-Hill, New York.
Okamoto, K. (2000) *Fundamentals of Optical Waveguides*, Academic Press, San Diego, CA.
Palais, J.C. (1998) *Fibre Optic Communication*, 4th edn, Prentice Hall International, London.
Reed, G.T. and Knights, A.P. (2004) *Si Photonics: An Introduction*, John Wiley & Sons, Ltd, Chicester, UK.

Saleh, B.E.A. and Teich, M.C. (1991) *Fundamentals of Photonics*, John Wiley & Sons, Inc., NY.
Syms, R. and Cozens, J. (1992) *Optical Guided Waves and Devices*, McGraw-Hill International, London.
Wilson, J. and Hawkes, J.F.B. (1996) *Optoelectronics: An Introduction*, 2nd edn, Prentice Hall International, London.
Yariv, A. (1993) *Quantum Electronics*, John Wiley & Sons, Inc., New York.
Yariv, A. and Yeh, P. (1984) *Optical Waves in Crystals*, John Wiley & Sons, Inc., New York.

11

Principle of Planar Waveguide Devices

11.1 Introduction

In the present chapter, we shall discuss the operating principle of some of the planar waveguide devices used in communication. We first develop the theory for mode coupling introduced by a small perturbation in the refractive index (RI) of the material in a waveguide. The coupled mode theory is useful in understanding the theory of two important classes of passive waveguide devices: the directional coupler and the Bragg reflector. As will be explained in this chapter as well as Chapters 12 and 13, both the devices play important roles in optical communication and networking. The general coupled mode theory will be developed first, and it will be applied to the two devices mentioned. A few more devices used in communication will then be introduced. The list includes a splitter, a directional coupler, a Mach–Zehnder type interferometer, filters, in particular a Fabry–Perot filter and a ring resonator, the Bragg grating, a dielectric mirror, and a variable optical attenuator [1–12].

11.2 Model for Mode Coupling

Electromagnetic energy propagating in one mode may be coupled into another mode with the introduction of a small perturbation in the RI. The purpose of this section is to develop a mathematical model for the mechanisms for the coupling [1, 6, 9].

Let us first consider a slab waveguide, the geometry of which is shown in Figure 11.1, in which an EM wave of angular frequency ω propagates along the z-direction. The RI distribution of the guide along the x-direction is denoted by $n(x)$. The general solution to the wave equation in the structure is a superposition of different modes with propagation constants β_m and an electric field profile $f_m(x)$, given as

$$E = \sum_m E_m(z) f_m(x) \exp(j\beta_m z). \qquad (11.1)$$

Silicon Photonics: Fundamentals and Devices, First Edition. M. Jamal Deen and P. K. Basu.

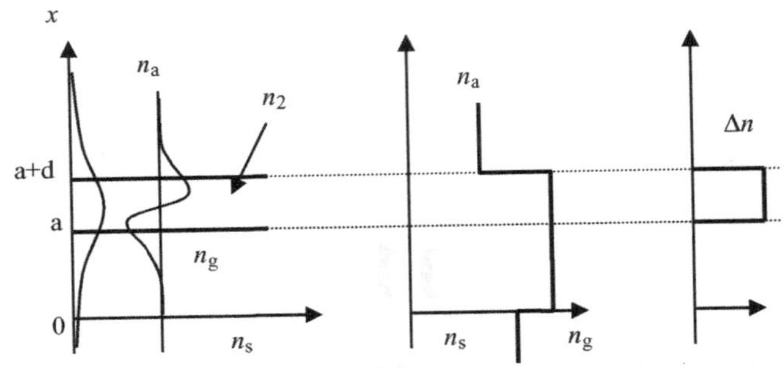

Figure 11.1 *Coupled mode model of a perturbed slab waveguide. (a) An unperturbed slab of RI n_g, thickness a, with an added perturbation layer of index n_2, thickness d, and unperturbed profile functions f_1 and f_2; (b) the unperturbed index profile; and (c) the index perturbation.*

The field profiles are assumed orthogonal, such that

$$\int f_m^*(x) f_n(x) dx = \delta_{mn}, \tag{11.2}$$

where δ_{mn} is the Kronecker delta. In Eq. (11.1) $E_m(z)$ is the z-dependent complex amplitude of the electric field of mode m.

Until now the different modes denoted by m have represented different independent solutions to the wave equation and there has been no interaction between these modes. Should there be any coupling between different modes, there must be present a perturbation of the RI distribution [9] of the following form:

$$n(x) = n_0(x) + \Delta n(x). \tag{11.3}$$

where $n(x)$ is the unperturbed RI distribution that leads to the set of modes characterized by β_m and $f_m(x)$, and $\Delta n(x)$ is a perturbation of the RI distribution.

A possible way of introducing coupling and hence power transfer between modes is shown in Figure 11.1. The unperturbed waveguide is a slab of thickness a, and RI n_g, sandwiched between a substrate of RI, n_s, and an upper cladding layer of RI, n_c. The perturbation may be introduced by inserting a thin layer of thickness d of RI n_2, so that the perturbation is $\Delta n = n_2 - n_c$. The solutions for the perturbed system may be built up as a linear combination of the modes for the unperturbed system.

The derivation of the coupled mode equations uses a spatial distribution of the square of the RI as follows:

$$n^2(x) = n_0^2(x) + 2n_0(x)\Delta n(x) + [\Delta n(x)]^2 = n_0^2(x) + \Delta n^2(x), \tag{11.4}$$

where $\Delta n^2(x)$ is the perturbation on the spatial distribution of n^2.

We now rewrite the wave equation as

$$\nabla^2 E - \mu_0 \frac{\partial^2}{\partial t^2}[\varepsilon_0 E + P] = 0. \tag{11.5}$$

The polarization, P, in Eq. (11.5) consists of two parts:

$$P = P_0 + \Delta P, \tag{11.6}$$

where the first term, P_0, is the contribution in the unperturbed system with RI, $n_0^2(x)$, and the perturbation ΔP has its origin in $\Delta n^2(x)$. Substituting Eq. (11.6) in Eq. (11.5) and rearranging terms, one obtains

$$\nabla^2 E + \omega^2 \mu_0 \varepsilon_0 n_0^2(x) E = \mu_0 \frac{\partial^2 \Delta P}{\partial t^2}. \tag{11.7}$$

As expected, this reduces to the usual wave equation in the absence of the perturbation. The Laplacian in Eq. (11.7) may now be expanded using Eq. (11.1) and the resulting equation is

$$\nabla^2 E = \left\{ \sum_m \left\{ E_m \frac{\partial^2 f_m}{\partial x^2} + \left[\frac{\partial^2 E_m}{\partial z^2} - \beta_m^2 E_m + 2j\beta_m \frac{\partial E_m}{\partial z} \right] f_m \right\} \exp(j\beta_m z). \right. \tag{11.8}$$

Since the perturbation in RI is small, mode coupling will be weak, leading to a slow variation of the mode field amplitude $E_m(z)$. This makes the $\partial^2 E_m / \partial z^2$ term negligible compared to other terms in the square bracket in Eq. (11.8). The other term on the left-hand side of Eq. (11.7) may be expressed as

$$k_0^2 n^2(x) E = \sum_m k_0^2 n^2(x) E_m f_m(x) \exp(j\beta_m z). \tag{11.9}$$

Adding Eqs. (11.8) and (11.9) and using an expanded form of the unperturbed wave equation,

$$\begin{aligned} \nabla^2 E + k_0^2 n^2(x) E &= \left[\frac{\partial^2 E}{dx^2} + \frac{\partial^2 E}{\partial z^2} + k_0^2 n_0^2(x) E \right] \\ &= \sum_m \left[E_m \frac{\partial^2 f_m}{\partial x^2} - \beta_m^2 f_m E_m + k_0^2 n_0^2(x) f_m E_m \right] \exp(j\beta_m z) = 0. \end{aligned} \tag{11.10}$$

Eq. (11.7) thus reduces to

$$\sum_m 2j\beta_m f_m \frac{dE_m}{dz} \exp(j\beta_m z) = \mu_0 \frac{\partial^2 \Delta P}{\partial t^2}. \tag{11.11}$$

We now use the relationship between displacement, electric field, and polarization given by

$$D = \varepsilon_0 E + P_0 + \Delta P = \varepsilon_0 n^2(x) E. \tag{11.12}$$

The total polarization is then

$$P_0 + \Delta P = \varepsilon_0 [(n_0^2(x) - 1) + \Delta n^2(x)] E, \tag{11.13}$$

from which the perturbation in polarization easily separates out to be

$$\Delta P = \varepsilon_0 \Delta n^2(x) E = \varepsilon_0 \Delta n^2(x) \sum_m E_m f_m \exp(j\beta_m z). \tag{11.14}$$

Now expressing the time variation of E_m as $\exp(j\omega t)$, in Eq. (11.11) and using the relation $k_0 = \omega/c$, $c = 1/\sqrt{\varepsilon_0\mu_0}$, we obtain

$$\sum_m 2j\beta_m f_m \frac{dE_m}{dz}\exp(j\beta_m z) = -k_0^2 \Delta n^2(x)\sum_m E_m f_m \exp(j\beta_m z). \tag{11.15}$$

Now multiplying Eq. (11.15) by f_n^*, integrating over x, and using the orthogonality property given by Eq. (11.2), we obtain

$$\frac{dE_n}{dz} - C_n E_n = j\sum_{m\neq n} E_m \kappa_{mn} \exp(j\Delta_{mn} z). \tag{11.16}$$

The parameters are defined by the following expressions:

$$C_n = \frac{-k_0^2}{2j\beta_n}\int f_n^* \Delta n^2 f_n dx, \tag{11.17}$$

$$\kappa_{mn} = \frac{-k_0^2}{2j\beta_n}\int f_m \Delta n^2 f_n^* dx, \tag{11.18}$$

and

$$\Delta_{mn} = \beta_m - \beta_n. \tag{11.19}$$

It is instructive to examine the physical significance of the parameters defined by Eqs. (11.17)–(11.19).

The parameter κ_{mn} is the coupling coefficient that links changes in E_n along z with the amplitude of another mode E_m. When this parameter is zero, the power in mode n remains unaltered.

The parameter C_n determines the change in propagation constant introduced by the RI perturbation Δn^2. Suppose all coupling coefficients, κ_{mn}, are zero, that is, the unperturbed structure supports only one mode, n. Eq. (11.16) then becomes

$$\frac{dE_n}{dz} - C_n E_n = 0. \tag{11.20}$$

The solution is

$$E_n(z) = A_n \exp(jC_n z), \tag{11.21}$$

where A_n is a constant. Substituting Eq. (11.21) in Eq. (11.1) we obtain for the resulting electric field

$$E = A_n f_n(x)\exp\{j(\beta_n + C_n)z\}. \tag{11.22}$$

The conclusion is that, even if no coupling between modes is present, the RI perturbation changes the propagation constant by an amount C_n, but no change occurs in the field profile.

The term Δ_{mn} is a measure of the mismatch in propagation constants between modes m and n. A later analysis will show that coupling of power between two modes is most efficient when the mismatch parameter is zero.

A set of coupled mode equations is now derived from Eq. (11.20), when amplitude variation is possible. Eq. (11.21) is now written as

$$E_n(z) = A_n(z)\exp(jC_n z), \tag{11.23}$$

where $A_n(z)$ is the z-dependent complex amplitude of the modified mode n, whose propagation constant has been modified to $\beta_n + C_n$. Eq. (11.23) is now substituted in Eq. (11.16) and the result is the following set of coupled mode equations for $A_n(z)$:

$$\frac{dA_n(z)}{dz} = \sum_{m \neq n} \kappa_{mn} A_m \exp(j\Delta'_{mn} z), \tag{11.24}$$

where Δ'_{mn} is the phase mismatch parameter for the pair of modified modes m and n, expressed as

$$\Delta'_{mn} = \beta_m + C_m - [\beta_n + C_n]. \tag{11.25}$$

Eq. (11.24) constitutes a model for calculating how power is coupled among the different modes in a waveguide structure. In the following, we shall consider the application of the coupled mode equation to different coupled structures.

11.3 Directional Coupler

As the first illustration of the use of the coupled mode theory, we shall consider the directional coupler, an important device used for the switching and splitting of power needed in optical communications [1–7]. In Section 11.3.1, we shall consider a simpler situation, in which the two coupled waveguides are identical, satisfying the phase-matched condition. In Section 11.3.2, the theory will be generalized for the non-phase-matched condition.

11.3.1 Phase-matched Directional Coupler

Figure 11.2a shows a schematic of a single-mode guided wave directional coupler, in which two identical waveguides of RI, n_g, surrounded by a dielectric of RI, n_s, are brought close to each other over a coupling distance L. As illustrated in Figure 11.2, the unperturbed field

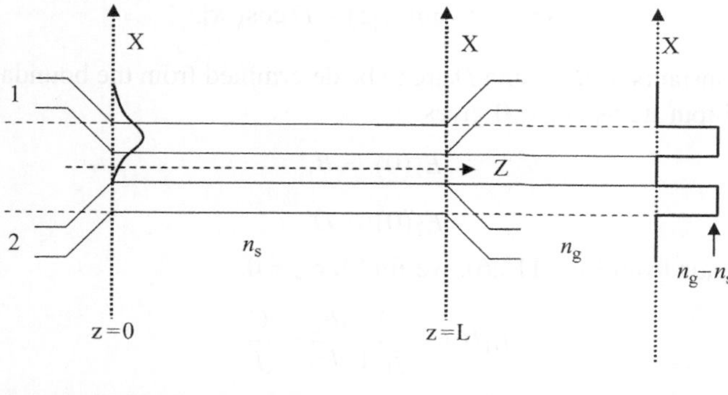

Figure 11.2 *(a) Schematic diagram of a directional coupler, showing the intensity profile of the radiation input into guide 1 at z = 0; and (b) plot of $n_0(x)$, $f_1(x)$, and $\Delta n(x)$ for the mode incident in waveguide 1 of (a).*

profile f1(x) associated with the mode incident in waveguide 1 indicates that there is negligible coupling between the waveguides for $z<0$ and $z>L$. In Figure 11.2b, the changes in RI in the guide regions are shown. When the mode is incident in waveguide 2, the distribution of the field and RI will have a similar variation.

Since we are assuming the two waveguides to be identical, their mode propagation constants are also identical, and the coupling is called *phase-matched*, with Δ' appearing in Eqs. (11.24) and (11.25) equal to zero. Also, since we are assuming weak mode coupling, the evanescent tails of the mode field shown in Figure 11.2(a) decay to very small values in the regions where $\Delta n(x)$ is nonzero.

As Eq. (11.21) indicates that E and A are identical within a phase shift, Eq. (11.24) is valid for both E and A. We consider a single-mode case and denote the fields in two waveguides by the respective subscripts. This yields the following two coupled equations:

$$\frac{dE_1}{dz} = \kappa_{21} E_2; \frac{dE_2}{dz} = \kappa_{12} E_1. \tag{11.26}$$

Assume that the waveguides are lossless; then the total power carried by both the modes must be constant with z. Thus,

$$\frac{d|E_1|^2}{dz} + \frac{d|E_2|^2}{dz} = 0. \tag{11.27}$$

It may be proved, using Eq. (11.26) and its complex conjugate, that

$$\kappa_{21} = -\kappa_{12}^{*}. \tag{11.28}$$

We differentiate one of the pairs of equations in Eq. (11.26) and then substitute the other one into it, to obtain

$$\frac{d^2 E_{1,2}}{dz^2} + |\kappa|^2 E_{1,2} = 0, \tag{11.29}$$

where $\kappa = \kappa_{12}$ or κ_{21}. The general solutions for Eq. (11.29) are

$$E_1(z) = A\sin(|\kappa|z) + B\cos(|\kappa|z), \tag{11.30}$$

and

$$E_2(z) = C\sin(|\kappa|z) + D\cos(|\kappa|z). \tag{11.31}$$

The unknown constants A, B, C, and D are to be determined from the boundary conditions. Evaluating the input fields at $z=0$ gives

$$E_1(0) = B, \tag{11.32a}$$

$$E_2(0) = D. \tag{11.32b}$$

Using now the pair from Eq. (11.26), we find for $z=0$,

$$E_1(0) = \frac{1}{j|\kappa|}\frac{dE_2}{dz} = \frac{C}{j}, \tag{11.33a}$$

$$E_2(0) = \frac{1}{j|\kappa|}\frac{dE_1}{dz} = \frac{A}{j}. \tag{11.33b}$$

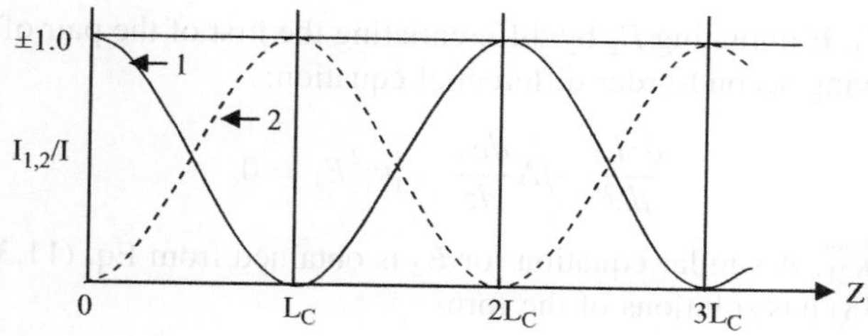

Figure 11.3 *Variation of intensity in the two guides as a function of the propagation distance. Complete power transfer to guide 2 occurs for the coupling length L_c. Reproduced with permission from [10]. Copyright (1997) American Institute of Physics.*

Using Eqs. (11.32a) and (11.33a) in Eqs. (11.30) and (11.31), the complete solution may be written as

$$E_1(z) = E_1(0)\cos(|\kappa|z) + jE_2(0)\sin(|\kappa|z),$$
$$E_2(z) = jE_1(0)\sin(|\kappa|z) + E_2(0)\cos(|k|z). \tag{11.34}$$

Let us consider a special case in which the EM power is launched into waveguide 1. In such situations, all the power $I = |E_1(z=0)|^2$ appears at waveguide 1 and $E_2(z=0) = 0$. Using these in Eq. (11.34), the variation of intensities in the two arms takes the form of

$$I_1(z) = |E_1(z)|^2 = I\cos^2(|\kappa|z),$$
$$I_2(z) = |E_2(z)|^2 = I\sin^2(|\kappa|z). \tag{11.35}$$

The intensity variation is plotted in Figure 11.3. It is noted that power is exchanged between the two guides and when the length

$$L = L_c = \pi/(2|\kappa|), \tag{11.36}$$

there is a complete transfer of power from guide 1 to guide 2. L_c is called the *coupling length* of the directional coupler. If the length of the coupler is such that $|k|L = \pi/4$, then, using Eq. (11.34), it follows that power is equally divided between the output of the two arms. Such a coupler is known as a 3 dB coupler. When the guides are excited by incoming fields of different amplitudes and phase, the power is divided between the two arms depending on the relative phase between the incoming fields.

11.3.2 Non-phase-matched Coupler

Under the phase-matched condition, the term $\Delta_{12} = 0$. When the two arms are no longer identical, the propagation constants β differ. The coupled mode equations are then written by using Eq. (11.24) as

$$\frac{dE_1}{dz} = \kappa_{21}E_2\exp(j\Delta z),$$
$$\frac{dE_2}{dz} = \kappa_{12}E_1\exp(-j\Delta z). \tag{11.37}$$

where $\Delta = \beta_1 - \beta_2$. Eliminating E_2 by differentiating the first of the pair of equations, one obtains the following second-order differential equation:

$$\frac{d^2E_1}{dz^2} - j\Delta\frac{dE_1}{dz} + |\kappa|^2 E_1 = 0, \quad (11.38)$$

where $\kappa = \sqrt{\kappa_{12}\kappa_{21}}$. A similar equation for E_2 is obtained from Eq. (11.37).

Equation (11.38) has solutions of the form

$$E_1 = (Ae^{jaz} + Be^{-jaz})e^{jbz}, \quad (11.39)$$

where

$$a = \frac{1}{2}\sqrt{4|\kappa|^2 + \Delta^2}, \quad \text{and} \quad b = \frac{\Delta}{2}. \quad (11.40)$$

Setting $E_1(z=0)=1$ and $E_2(z=0)=0$, then using Eq. (11.37) to evaluate dE_1/dz at $z=0$, gives two equations to solve the constants A and B. One then obtains

$$A = \frac{a-b}{2a}; B = \frac{a+b}{2a}. \quad (11.41)$$

If the power launched into the input of guide 1 is P_0, then the variation of power in the two guides as a function of distance may be expressed as

$$P_1(z) = P_0\left[\left(\frac{\Delta}{2a}\right)^2 \sin^2(az) + \cos^2(az)\right], \quad (11.42)$$

$$P_2(z) = P_0\frac{|\kappa_{12}|^2}{a^2}\sin^2(az). \quad (11.43)$$

For a phase-matched condition, $\Delta=0$, $a = |\kappa|$, and we recover Eq. (11.35).

It is to be noted from Eqs. (11.42) and (11.43) that in the non-phase matched condition, $\Delta \neq 0$, and the maximum power is transferred from guide 1 to 2 when $az = \pi/2$. Also, in this case, power cannot be transferred completely to guide 2.

11.4 Distributed Bragg Reflector

We now consider the second device, the distributed feedback or distributed Bragg reflector structure, in which coupling may exist between the forward- and backward-propagating modes [1–3]. The basic structure is shown in Figure 11.4. A periodic undulation has been introduced over the region $0 < z < L$. Such kind of undulation may be accomplished by etching the layer and then growing another layer on top of it. The DBR structure is introduced in narrow-width single-mode lasers used for telecommunication and the reader may consult the literature for more information.

11.4.1 Phase-Matched Grating

We first consider the phase-matched case. As stated already, the periodicity of perturbation along the z-direction allows coupling of modes propagating in opposite directions. The two

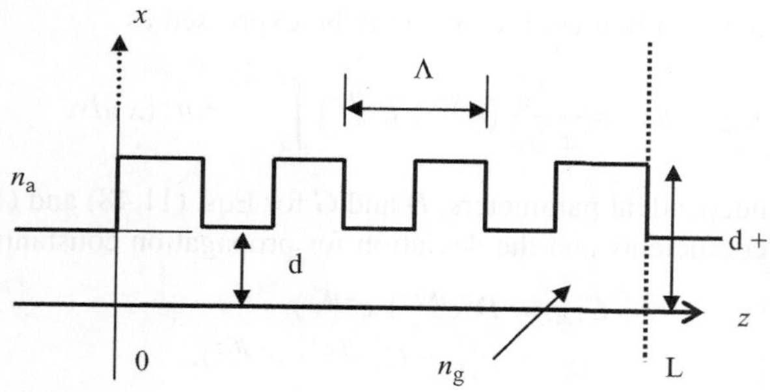

Figure 11.4 Schematic structure of a Bragg grating. The periodic perturbation in RI is introduced by changing the structure of width d_p along the x-direction.

modes of interest in this configuration have propagation constants $+\beta$ for the forward-traveling ($+z$) mode (mode 1) and $-\beta$ for the backward-traveling ($-z$) mode (mode 2). The field profiles $f(x)$'s remain the same for both the forward- and backward-propagating modes.

Assume that the structure is lossless. From energy conservation, it is required that the decrease of power along the positive z-direction in mode 1 must be exactly equal to the increase in power of back-propagating mode 2. Thus,

$$\frac{d}{dz}\left(|E_1|^2 - |E_2|^2\right) = 0. \tag{11.44}$$

Using the coupled mode Eq. (11.24), its complex conjugate, and Eq. (11.44) lead to the condition

$$\kappa_{21} = \kappa_{12}^*. \tag{11.45}$$

The ribbed, periodic structure on top of the waveguide causes a z-dependent perturbation of the RI in $n(x)$ in the region $d < x < d + d_p$. The perturbation is periodic in z and is expressed as

$$\begin{aligned}\Delta n(x,z) &= \delta n \cos(Kz), d < x < d + d_p \\ &= 0, \text{elsewhere.}\end{aligned} \tag{11.46}$$

The corresponding $\Delta n^2(x,z)$ is expressed as

$$\begin{aligned}\Delta n^2(x,z) &= [n(x) + \Delta n(x,z)]^2 - [n(x)]^2 \\ &= 2\Delta n(x,z) - [\Delta n(x,z)]^2\end{aligned} \tag{11.47a}$$

$$\cong 2\delta n \cos(Kz). \tag{11.47b}$$

The last term in Eq. (11.47a) has been neglected, as δn is small. Using Eqs. (11.46) and (11.47b) in Eq. (11.17), one obtains

$$C_1 = -C_2 = \frac{-k_0^2}{2j\beta}\left(e^{jKz} + e^{-jKz}\right)\int_d^{d+d_p} \delta n f^2(x)dx. \tag{11.48}$$

Using Eq. (11.18), the coupling coefficients may be expressed as

$$\kappa_{12} = \kappa_{21}^{*} = \frac{-k_0^2}{\pm 2j\beta}\left(e^{jKz} + e^{-jKz}\right)\int_{d}^{d+d_p} \delta n f^2(x)dx. \tag{11.49}$$

Introducing two z-independent parameters, D and G for Eqs. (11.48) and (11.49), we may write the coupling coefficients and the deviation for propagation constants as

$$\begin{aligned} C_{1,2} &= D(e^{jKz} + e^{-jKz}), \\ \kappa_{12} &= \kappa_{21}^{*} = G(e^{jKz} + e^{-jKz}). \end{aligned} \tag{11.50}$$

Substituting Eq. (11.50) in Eq. (11.25), one obtains for the phase-mismatch parameter as

$$\Delta'_{12} = 2\beta + 2D\left(e^{jKz} + e^{-jKz}\right) \cong 2\beta. \tag{11.51}$$

The approximate form, 2β, is valid only for weak coupling. Using Eq. (11.51) in Eq. (11.24), one obtains the following pair of coupled mode equations:

$$\frac{dE_1}{dz} = jGE_2[\exp\{j(2\beta - K)\} + \exp\{j(2\beta + K)z\}], \tag{11.52}$$

and

$$\frac{dE_2}{dz} = -jGE_1[\exp\{-j(2\beta - K)z\} + \exp\{-j(2\beta + K)z\}]. \tag{11.53}$$

Under weak coupling approximation, a significant exchange of power is likely to occur if the interaction length is long compared to the wavelength of an electromagnetic (EM) wave. The period of oscillation along the z-direction may be written from Eqs. (11.52) and (11.53) as

$$Z_{\pm} = \frac{2\pi}{2\beta \pm K} = \frac{2\pi}{2\left[\frac{2\pi}{\lambda_0}\bar{n}\right] \pm \frac{2\pi}{\Lambda}} = \left[\bar{n} \pm \frac{\lambda_0}{\Lambda}\right]^{-1}\lambda_0. \tag{11.54}$$

The second term in square brackets in Eqs. (11.52) and (11.53), corresponding to the + subscript in Eq. (11.54), always oscillates with a period that is short compared to an optical wavelength. Therefore, for short coupling, these terms may be dropped from Eqs. (11.52) and (11.53). Both these equations will have the same form as Eq. (11.26) for a phase-matched directional coupler, provided that

$$\begin{aligned} 2\beta - K &= 0, \\ \beta - K &= -\beta \end{aligned} \tag{11.55}$$

The last one in Eq. (11.55) is the momentum conservation condition. The initial photon momentum is β, the momentum change through interaction with the grating is $-K$, and the final photon momentum is $-\beta$. We now substitute Eq. (11.55) in Eq. (11.52) and (11.53), differentiate one of them, substitute the result in the other, and obtain finally

$$\frac{d^2E_{1,2}}{dz^2} - |\kappa|^2 E_{1,2} = 0, \tag{11.56}$$

where $|\kappa| = |\kappa_{12}| \text{or} |\kappa_{21}|$.Equation (11.56) has the following solutions:

$$E_i(z) = a_i e^{-|\kappa|z} + b_i e^{+|\kappa|z}, \quad i = 1, 2. \tag{11.57}$$

The unknown constants a and b in the above are determined from the boundary conditions, which are stated now. Let the coupling region length be L, and a wave is launched into the grating at $z=0$ from the left. The propagation constant of the incident wave is β, then $K=2\beta$, to satisfy momentum conservation between forward- and backward-traveling modes. At $z \geq L$ in Figure 11.3, there exists no mechanism to generate backward waves and thus $E_2(z \geq L)$ must be zero. Thus, using Eq. (11.26) and differentiating Eq. (11.57), one may write

$$\frac{dE_1}{dz}(z = L) = 0 = |\kappa|\left[-a_1 e^{-|\kappa|L} + b_1 e^{|\kappa|L}\right]. \tag{11.58}$$

Next consider that the incident power in mode 1 is unity, so that the incident field has a unity magnitude. Thus

$$E_1(z = 0) = a_1 + b_1 = 1. \tag{11.59}$$

Solving Eqs. (11.58) and (11.59) will completely specify the coefficients and thus

$$a_1 = \frac{1}{1 + e^{-2|k|L}}; \quad b_1 = \frac{e^{-2|k|L}}{1 + e^{-2|\kappa|L}}. \tag{11.60}$$

As there is no coupling other than in the region of $0 < z < L$, the z-dependence of the power in mode 1 is

$$|E_1(z)|^2 = 1, z < 0 \tag{11.61a}$$

$$|E_1(z)|^2 = \left|\frac{e^{-|\kappa|z} + e^{(|\kappa|z - 2|\kappa|L)}}{1 + e^{-2|\kappa|L}}\right|^2, \quad 0 < z < L, \tag{11.61b}$$

$$|E_1(z)|^2 = \left|\frac{2e^{-|k|L}}{1 + e^{-2|\kappa|L}}\right|^2, \quad \text{z} > \text{L}. \tag{11.61c}$$

Equation (11.61c) denotes the power transmittance, since the incident power is set at unity. It also indicates that for large L, there will be hardly any power transmitted into the region $z > L$.

In order to determine E_2, we use Eq. (11.26) and write

$$E_2 = \frac{1}{|\kappa|}\frac{dE_1}{dz}. \tag{11.62}$$

Using Eqs. (11.58), (11.60) and (11.62), one obtains

$$E_2(z) = \frac{e^{(|\kappa|z - 2|\kappa|L)} - e^{-|\kappa|z}}{1 + e^{-2|\kappa|L}}. \tag{11.63}$$

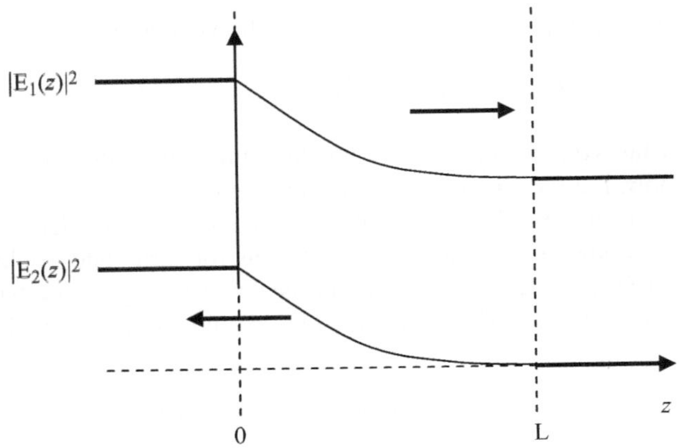

Figure 11.5 Incident and reflected intensities versus z, calculated with Eq. (11.64a), for the DBR structure.

The coupling is nonzero only in the region of $0 < z < L$. For $z > L$, $E_2 = 0$, and for $z < 0$, E_2 must have a value equal to $E_2\,(z = 0)$. The power in mode 2 is $|E_2(z)|^2$, and may be written as

$$|E_2(z)|^2 = \left|\frac{e^{-2|\kappa|L} - 1}{e^{-2|k|l} + 1}\right|^2 = \tanh^2(|\kappa|L), z < 0 \tag{11.64a}$$

$$= \left|\frac{e^{|\kappa|z-2|k|L} - e^{-|\kappa|z}}{1 + e^{-2|\kappa|L}}\right|^2, 0 < z < L \tag{11.64b}$$

$$= 0, z > L. \tag{11.64c}$$

The power variation in the two modes is shown in Figure 11.5.

11.4.2 Non-phase-matching

The analysis may be continued by using the equation developed for a directional coupler, Eq. (11.38). Here $\Delta = 2\beta - K$ and the starting equation is

$$\frac{d^2E_1}{dz^2} - j\Delta\frac{dE_1}{dz} - |\kappa|^2 E_1 = 0, \tag{11.65}$$

The solution is of the form

$$E_1 = (Ae^{az} + Be^{-az})e^{jbz}, \tag{11.66}$$

where

$$a = \frac{1}{2}\sqrt{4|\kappa|^2 - \Delta^2}, \quad \text{and} \quad b = \Delta/2. \tag{11.67}$$

The detailed solution will not be worked out. Assuming an input field E_{in}, evaluating the unknown constants A and B by applying the boundary conditions, the field variation $E_1(z)$ may be derived. The z-dependent intensity takes the form

$$I_1(z) = \frac{a^2 \cos \mathrm{h}^2[a(z-L)] + b^2 \sin \mathrm{h}^2[a(z-L)]}{a^2 \cos \mathrm{h}^2(aL) + b^2 \sin \mathrm{h}^2(aL)} |E_{in}|^2. \tag{11.68}$$

The reflectance is given by

$$R = 1 - \left| \frac{E_1(L)}{E_{in}} \right|^2 = 1 - \frac{a^2}{a^2 \cos \mathrm{h}^2(aL) + b^2 \sin \mathrm{h}^2(aL)}. \tag{11.69}$$

11.5 Some Useful Planar Devices

In this section we shall mention some basic devices which are used in communication and networks and which form the basic building blocks for more complex passive and active light waveguide components.

11.5.1 Splitters

Splitters are passive waveguides that are used in branching networks for integrated optics. These networks divide optical power among two or more outputs or combine power from two or more inputs. An important commonly used branching network is the waveguide *Y*-structure [5], shown in Figure 11.6 which can be used as a symmetric power divider or combiner.

The *Y*-structures are combined to yield 1 × N splitters. A 1 × 4 splitter is shown in Figure 11.7 using three *Y*-structures.

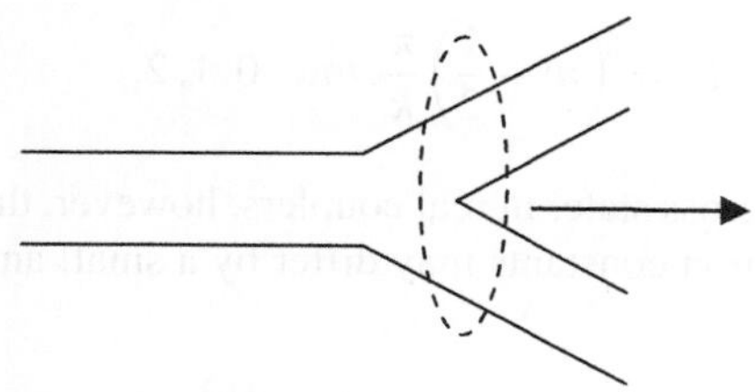

Figure 11.6 *Schematic diagram of a splitter or Y-coupler.*

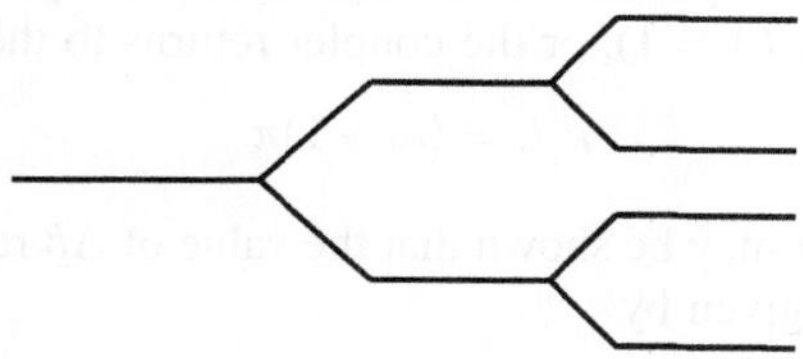

Figure 11.7 *1 × 4 splitter using three Y-structures.*

11.5.2 Directional Coupler

The dual-channel directional coupler is an important component that may work as both a passive and an active component in lightwave circuits. As shown already in Figure 11.2, it consists basically of parallel-channel optical waveguides sufficiently closely spaced so that optical energy can be transferred from one channel to the other. This energy transfer takes place by a process of synchronous coherent coupling between the overlapping evanescent tails of the modes guided in each channel. One channel in the structure is the driving channel, while the other one is the driven channel. Photons of the driving mode, say, in guide 1, transfer into the driven mode in guide 2, maintaining phase coherence. The fraction of the power coupled per unit length depends on the overlap of the modes in the separate channels, which in turn depends on the channel separation distance s, the interaction length L, and the mode penetration into the space between the channels, which are described by the extinction coefficients γ_1 and γ_3 introduced in Eq. (10.36).

In a dual-channel coupler, the energy transfers alternately from one guide to the other and then back again if the interaction length is sufficient. A concise theory of operation of the dual-channel coupler may be constructed by using the coupled mode theory developed by Yariv [9]. The energy exchange requires that the light propagating in both the guides have nearly the same phase velocity and propagation vector. When these parameters in the two guides are exactly identical, then the power propagating in the two guides is given by

$$\begin{aligned} P_1(z) &= \cos^2(\kappa z)e^{-\alpha z} \\ P_2(z) &= \sin^2(\kappa z)e^{-\alpha z} \end{aligned} \tag{11.70}$$

where z is the direction of propagation and κ is the coupling constant given by Eq. (11.18) and the attenuation factor has been included (compare with Eq. (11.35)).

The coupling length of the directional coupler L_c is defined to be the length at which total transfer of power takes place, and from Eq. (11.36) it is given by

$$L_c = \left(m + \frac{1}{2}\right)\frac{\pi}{\kappa}, m = 0, 1, 2, \ldots \tag{11.71}$$

The coupler is then in the cross state. In real couplers, however, the two guides may not be identical, and the propagation constants may differ by a small amount $\Delta\beta_z$. The coupling constant is then given by

$$\kappa_r^2 = \kappa^2 + (\Delta\beta_z/2)^2 \tag{11.72}$$

The power flow equations in the two guides are also modified. When an external agency like an electric field is employed to produce the change $\Delta\beta$, the coupling is completely canceled (i.e., $P_1(L) = 0$ and $P_0(L) = 1$), or the coupler returns to the bar state, if

$$\kappa_r L = (m + 1)\pi. \tag{11.73}$$

From the two conditions, it may be shown that the value of $\Delta\beta$ required for converting the coupler to the bar state is given by

$$\Delta\beta L = \sqrt{3}\pi. \tag{11.74}$$

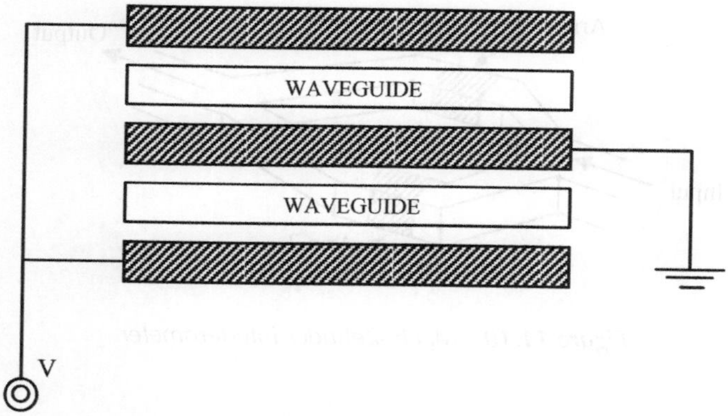

Figure 11.8 *Top view of a dual-channel directional coupler. The change in RI is due to EO effect and is introduced by applying voltage to the electrodes.*

Most of the directional couplers utilize the RI change or phase change $\Delta\beta$ mentioned in this subsection by the electro-optic (EO) effect, and are made into electro-optic modulators by merely adding two electrodes. A dual-channel modulator featuring strip-loaded guides on GaAs is shown in Figure 11.8. The three metal strips act as Schottky barrier electrodes of the modulator, and two are connected to an external voltage source V to create the electric field. In the absence of any field, the coupler is in a cross-state, that is, the power launched into port 1 totally comes out of port 3 (the tap port) (see Figure 11.9). However, when a field is applied, the resulting $\Delta\beta$ decreases the power in port 3, allowing some power to appear at port 2 (the throughput port). When the above condition $\Delta\beta L = \sqrt{3}\pi$ is satisfied, the whole power appears in port 2. Thus, by changing the RI with the field, the power in port 3 can be modulated.

The discussion in this subsection indicates that a predetermined amount of power may be available at the output ports 2 and 3 by choosing the length of the guides or the value of the applied electric field. The input power may be divided equally between ports 2 and 3; the coupler is then called a *3 dB directional coupler*. A 10 dB DC may also be realized in a similar way.

11.5.3 Mach–Zehnder Interferometer

The integrated optical components discussed so far are mainly passive and are used for routing optical signals. The DC may work as a passive device, although by application of an

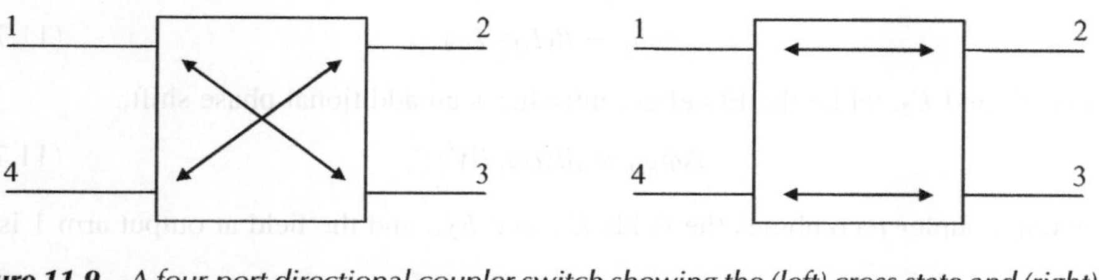

Figure 11.9 *A four-port directional coupler switch showing the (left) cross state and (right) bar state.*

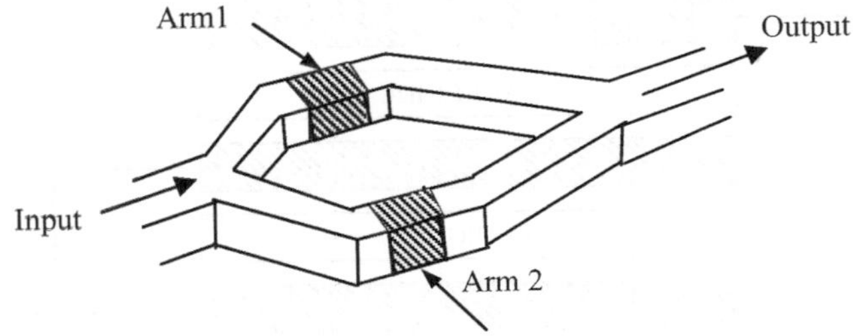

Figure 11.10 *Mach–Zehnder interferometer.*

external field it may be converted into an active optical device, as discussed in this chapter. There are also other active guided wave devices that may be used as modulators, interferometers, and filters.

We shall discuss in this subsection the working of an interferometer, named the Mach–Zehnder interferometer (MZI), that is used as a passive and also an active device by utilizing the EO effect. Its use as a light modulator has already been discussed in Chapter 7. The basic structure is shown in Figure 11.10. Assume that the input optical signal in the input waveguide is split equally by a *Y*-splitter. The input intensities in arm 1 and arm 2 are then equal, and the propagating electric fields are expressed as

$$E_1 = E_0 \sin(\omega t - \beta z) \tag{11.75a}$$

$$E_2 = E_0 \sin(\omega t - \beta z). \tag{11.75b}$$

At the input $z = 0$ and the two fields are identical. However, the two waves may travel with different propagation constants. Therefore, when the output *Y*-coupler recombines the two beams, a phase difference appears.

At the output terminal, the fields are phase-shifted due to propagation over different lengths in the two arms and may be expressed as

$$\begin{aligned} E_{o1} &= E_0 \sin(-\beta L_1) \\ E_{o2} &= E_0 \sin[-\{\beta L_2 + (\partial n/\partial V) l V\}] \end{aligned} \tag{11.76}$$

assuming that the two waves travel unequal physical lengths. In the second equation, the EO effect is introduced due to the bias V applied over a length l. Therefore, unequal lengths will introduce a phase difference

$$\Delta\phi_L = \beta(L_2 - L_1) \tag{11.77}$$

between E_1 and E_2, while the EO effect introduces an additional phase shift,

$$\Delta\phi_{EO} = \beta l(\partial n/\partial V)V \tag{11.78}$$

The output coupler recombines the fields E_{1o} and E_{2o}, and the field at output arm 1 is

$$E_{o1} = E_0 \exp(j\beta L_1)[1 + \exp\{j(\Delta\phi_L + \Delta\phi_{EO})\}] \tag{11.79}$$

The output intensity obtained by using the relation $S = [E \times H] = S_0(E_{01} + E_{02})^2$ can be expressed as

$$S = S_0\left\{\frac{E_0^2}{2} + \frac{E_0^2}{2} + E_0^2[\cos(\beta(L_2 - L_1) + \Delta\phi_{EO})]\right\}$$
$$= S_0\{E_0^2[1 + \cos(\beta(L_2 - L_1) + \Delta\phi_{EO})]\} \quad (11.80)$$

Note the similarity between Eqs. (11.80) and (7.27).

The guides are assumed lossless so that the sum of the output intensities equals the input intensity I_{in} which is taken to be unity. The normalized output intensity therefore varies sinusoidally between 0 and 1 depending on the magnitude of the cosine term. The device can be used as a modulator or a switch to transmit full power or no power to a particular output port. It may be mentioned that the phase difference $\Delta\phi_{EO}$, which in the present case is introduced via the EO effect, may well be induced by the thermo-optic effect also.

11.5.4 Fabry–Perot Resonators

Fast tunable filters are required for many WDM local area networks based on broadcast and select architectures. Usually Fabry–Perot (FP) etalons have been used for this purpose. The etalon is a cavity formed by two parallel mirrors. Light from an input fiber enters the cavity and suffers multiple reflections at the two mirrors. Depending on the distance between the mirrors, a particular wavelength is transmitted through the etalon, while all other wavelengths interfere destructively. The mirror separation may be altered mechanically or by changing the RI of the material within the cavity. The tuning range of the etalon specifies the range of wavelengths to be selected. Therefore, a wide tuning range allows a greater number of channels to be demultiplexed or selected. In etalons, the transfer function or the shape of the filter passband repeats itself after a certain period, referred to as the *free spectral range* (FSR). Another parameter of interest is finesse. The values of FSR and finesse determine the number of resolvable channels in WDM that can be resolved by the etalons. The tuning time of etalons is quite long, however. The etalons are now realized using Si-based waveguides.

Consider the schematic of a Fabry–Perot interferometer formed by two plane parallel highly reflecting mirrors, as shown in Figure 11.11. The two mirrors M_1 and M_2 have (field) reflection and transmission coefficients, denoted by (r_1, t_1) and (r_2, t_2), respectively. Let the electric field incident at the left-hand mirror be E_i. The field enters into the cavity and

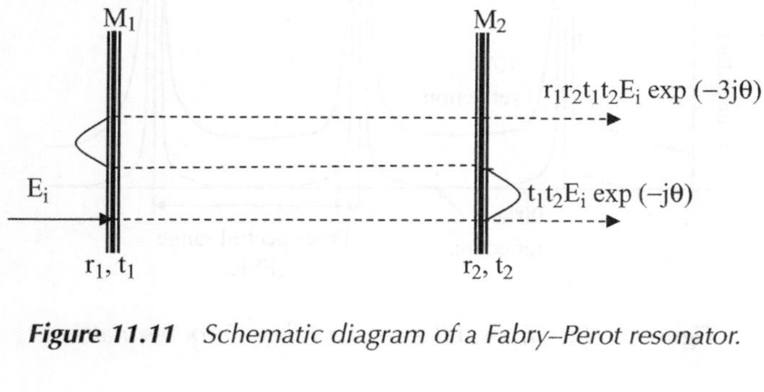

Figure 11.11 *Schematic diagram of a Fabry–Perot resonator.*

propagates to the second mirror, where it is partially transmitted and partially reflected. The fields of the various reflected and transmitted beams at the two mirrors are shown by parallel lines. Summing over the electric field of the transmitted rays coming out of mirror M_2, the total transmitted field E_t may be expressed as [8]

$$E_t = E_i t_1 t_2 e^{-j\theta}\left[1 + r_1 r_2 e^{-j2\theta} + (r_1 r_2)^2 e^{-j4\theta} + \ldots\right], \tag{11.81}$$

where $\theta = kL, k = \omega n_g/c.$ and L is the separation between the two mirrors.

The transmittivity of the FP resonator, that is, the ratio of output to input optical powers of the resonator, may be expressed in terms of the reflectivity of the mirrors as [8]

$$T = \frac{|E_t|^2}{|E_i|^2} = \frac{I_{out}}{I_{in}} = \frac{(1-R_1)(1-R_2)}{(1-\sqrt{R_1R_2})^2 + 4\sqrt{R_1R_2}\sin^2\theta}, \tag{11.82}$$
$$\theta = (2\pi f n_g/c)L, R_1 = r_1^2 = 1-t_1^2, R_2 = r_2^2 = 1-t_2^2.$$

where the R's are power reflection coefficients. The transmittivity of the etalon, plotted in Figure 11.12, shows different peaks, the sharpness of which depends on the reflectivities of the mirrors. The peaks occur at the successive longitudinal mode frequencies given by

$$f = f_m = mc/2n_gL, \tag{11.83}$$

where n_g is the group index, and m is an integer. The frequency spacing between two successive transmission peaks, or FSR, is given by

$$\Delta f = c/2n_gL \tag{11.84}$$

If the etalon is used to filter multichannel signals, the combined bandwidth should satisfy the condition

$$\Delta f_{sig} = NS_{ch}B \tag{11.85}$$

where N is the number of channels, S_{ch} is the normalized channel spacing $\Delta f_{ch}/B$, and B is the bit rate. The filter BW Δf_{FP}, the width of the transmission peak, should be large enough to

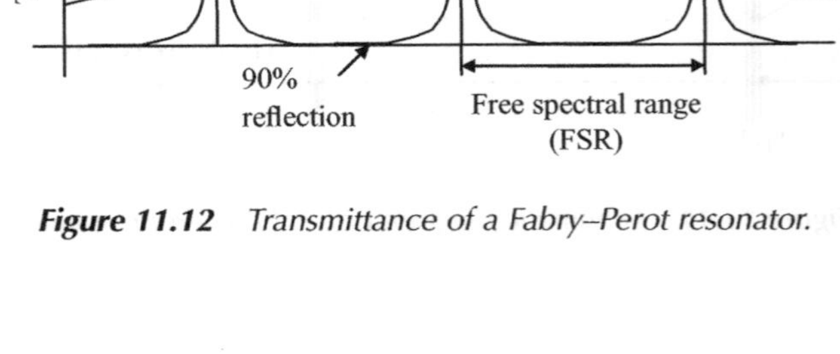

Figure 11.12 *Transmittance of a Fabry–Perot resonator.*

pass the entire frequency components of the selected channel. Usually, $\Delta f_{FP} = B$. It sets the limit to the total number of channels as

$$N < \frac{\Delta f}{S_{ch}\Delta f_{FP}} = \frac{F}{S_{ch}}, F = \Delta f/\Delta f_{FP}, \tag{11.86}$$

where F is known as the *finesse* of the FP filter, which is related to the reflectivities of the mirrors (assumed equal for both the mirrors) as

$$F = \frac{\pi\sqrt{R}}{1-R} \tag{11.87}$$

The selection of a particular channel is done by changing the filter length electronically. Usually the stress applied to a piezoelectric transducer changes the mirror separation by a small amount, and the channels are selected one by one. There are examples where the channel is selected by changing the group index of the material forming the cavity by, say, using the TO effect induced by heating.

The tuned frequency depends on the RI of the material filling the cavity, as seen from Eq. (11.83). A slight change in RI will therefore shift the transmittivity peak shown in Figure 7.6 to lower or higher wavelengths depending on the sign of RI change. This means that for a fixed wavelength, the amount of power coming out of the FP structure can be modulated by modulating the RI of the material. This technique has been employed to realize intensity modulators using Si-based reflectors, as discussed in Chapter 7.

11.5.5 Bragg Gratings

In the simplest form of Bragg grating, a periodic modulation of RI occurs along the z-direction, the direction of propagation of the electromagnetic wave [10–12]. The grating planes are of a constant period, and the phase fronts are perpendicular to z. The propagating light is scattered by each grating plane. If the Bragg condition is not satisfied, the reflected light from each of the planes becomes progressively out of phase and produces no net back reflection. When the Bragg condition is satisfied, the reflected lights from each grating plane add constructively in the backward direction and a reflected peak occurs.

The Bragg condition may be derived from the principles of energy and momentum conservation. The first one requires that the energies of forward- and back-reflected photons be equal, leading to the condition $\omega_f = \omega_b$. Momentum conservation requires that the incident wave vector β_f and the grating wave vector β_B should be equal to the back-reflected wave vector β_b. Thus

$$\beta_f + \beta_B = \beta_b. \tag{11.88}$$

The grating wave vector is along the direction normal to the grating planes, and it has a magnitude of $(2\pi/\Lambda)$, where Λ is the grating period shown in Figure 11.13. The back-reflected wave vector is equal in magnitude but opposite in direction to the incident wave vector. Therefore, Eq. (11.88) may be written as

$$2\left(\frac{2\pi n}{\lambda_B}\right) = \frac{2\pi}{\Lambda}, \tag{11.89}$$

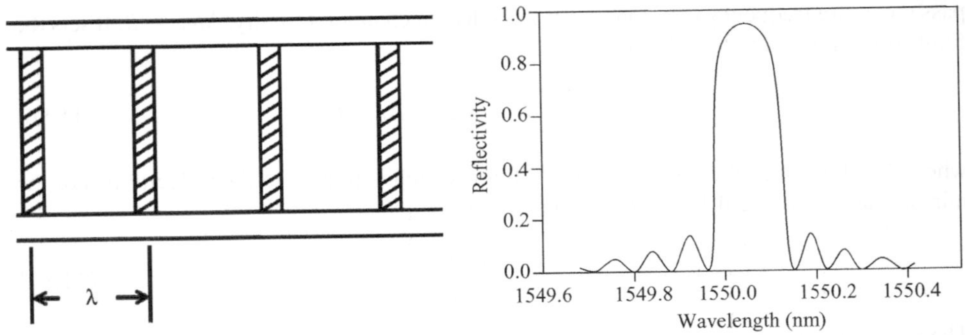

Figure 11.13 Typical reflectivity variation of a FBG. Reproduced with permission from [10]. Copyright (1997) American Institute of Physics.

from which the following first-order Bragg condition is obtained:

$$\lambda_B = 2n\Lambda. \tag{11.90}$$

The *Bragg wavelength* refers to free space, and n is the RI of the guide.

The coupled mode theory for both the phase-matched and phase-mismatched gratings has been developed. The reflectivity of the grating for mismatched conditions is given by Eq. (11.69) which reduces to Eq. (11.64a) for perfect matching.

The reflectivity increases with the increase of index perturbation Δn. The nature of the reflectivity spectrum is displayed in Figure 11.12 as a function of wavelength detuning. The bandwidth of the reflectivity spectrum is approximately given by

$$\Delta\lambda = \lambda_B \alpha \sqrt{\left(\frac{\Delta n}{2n_0}\right)^2 + \left(\frac{1}{N}\right)^2}, \tag{11.91}$$

where N is the number of grating planes. The parameter α is $\sim$1 for strong gratings and is $\sim$0.5 for weak gratings. The side lobes of the resonance shown in Figure 11.13 are due to multiple reflections to and from opposite ends of the grating region.

The Bragg resonance wavelength is affected by strain and temperature. Strain changes the grating spacing and, through the strain-optic effect, changes the mode index. Similar effects are also produced by temperature, the more dominant being the change of RI by the thermo-optic effect. The expected change of λ_B of a germanosilica fiber with temperature is 13.7 p.m./C at around 1550 nm.

The grating structures may be classified into three distinct types: (a) the common Bragg reflector, (b) the blazed Bragg grating, and (c) the chirped Bragg grating. A brief description of each type is given here.

The common Bragg reflector has a constant pitch and its reflectivity characteristics depend on the period of the grating and the amplitude of RI modulation. Its reflective character can be converted into a transmissive character when the gratings are inserted in the two arms of a MZI. A proper cascading of two Bragg gratings can result in a bandpass filter.

Tilting or blazing the Bragg grating planes at angles to the fiber (or propagation) axis will result in the light coupled out of the core into loosely guided cladding modes or into radiation modes outside the guide structure. This acts as a wavelength-selective tap which

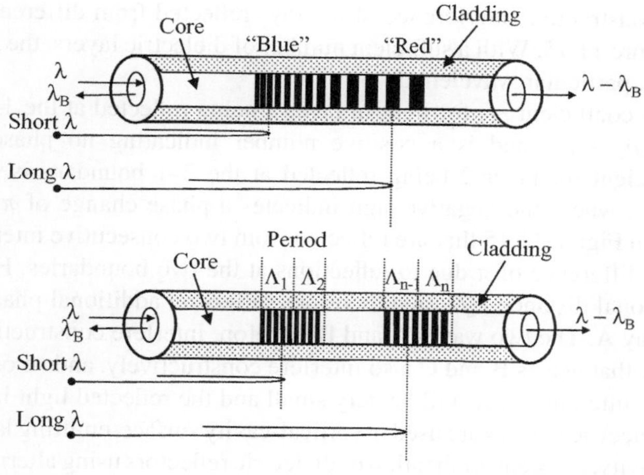

Figure 11.14 *(a) Schematic of a chirped FBG; and (b) a cascade of chirped FBGs having different periods.*

may form as an absorption-type filter that finds application as a gain equalizer of an erbium-doped fiber amplifier (EDFA).

A chirped Bragg grating has a monotonically varying grating period. It may be realized by axially varying the grating period Λ, or changing the RI of the core. By using the phase mask technique, a cascade of several gratings with increasing period may be produced. The step-like chirp, so produced, approximates the linear chirp. A characteristic of chirped grating is that different wavelengths are reflected from different regions of the grating as shown in Figure 11.14. A useful application of this type of grating is in dispersion compensation in fibers [12].

Although this discussion is related to optical fibers, similar structures have also been realized using a planar Si platform [13].

11.5.6 Dielectric Mirrors

A dielectric mirror is formed by a stack of dielectric layers of alternating RI, as illustrated in Figure 11.15. Here $n_1 < n_2$. The thickness of each layer is quarter wavelength. Mirror action

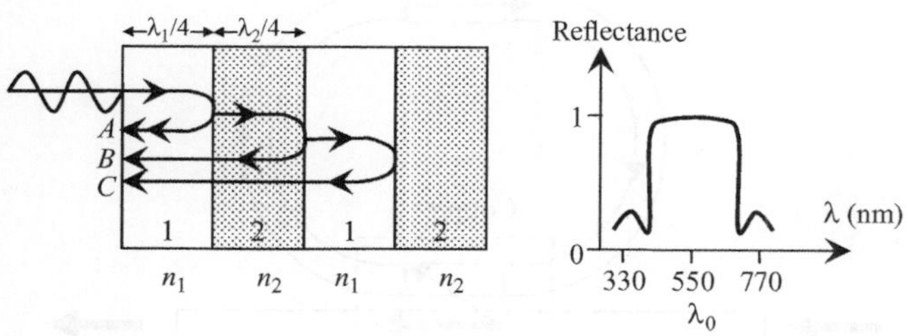

Figure 11.15 *Schematic illustration of the principle of a dielectric mirror with many low and high RI layers and its reflectance.*

is achieved by constructive interference of the rays reflected from different interfaces, as illustrated in Figure 11.15. With a sufficient number of dielectric layers, the reflectance can reach unity for a particular wavelength.

The reflection coefficient r_{12} for light in layer 1 being reflected at the 1–2 boundary is $r_{12} = (n_2 - n_1)/(n_2 + n_1)$ and is a positive number indicating no phase change. The reflection coefficient for layer 2 being reflected at the 2–1 boundary is $r_{21} = (n_1 - n_2)/(n_2 + n_1) = -r_{12}$, where the negative sign indicates a phase change of π. Consider two waves A and B in Figure 11.15 that are reflected from two consecutive interfaces. There is already a phase difference of π due to reflections at the two boundaries. Further, wave B travels an additional distance $\lambda_2/2$, and therefore suffers an additional phase change of π, while meeting ray A. The two waves A and B therefore interfere constructively. It may be proved similarly that waves B and C also interfere constructively, and so on. After several layers, the transmitted intensity will be very small and the reflected light intensity will be nearly unity. Dielectric mirrors are used in vertical-cavity surface-emitting lasers and also in realizing a FP cavity. A recent application of dielectric reflectors using alternate Si and SiO_2 layers may be found in Muscara *et al.* [14].

11.5.7 Ring Resonators

Another interesting device dependent on coupling between waveguides is a ring resonator [15], the schematic of which is shown in Figure 11.16. The resonator is in the form of a circular ring and it is excited by the straight guide, as a fraction of the input power is coupled to the ring. The light will circulate through the ring and suffer a phase shift $\Delta\varphi = \beta L$, where β is the propagation constant in the ring, and L is the optical path length. The device will act as a resonator if this phase shift is an integral multiple of 2π. The resonance condition is given by

$$\Delta\phi = \beta L = 2m\pi, \tag{11.92}$$

where m is an integer. Using $L = 2\pi R$, where R is the radius of the ring, and expressing β in terms of effective index N, the resonance condition may be written as

$$\lambda = \frac{2\pi NR}{m}. \tag{11.93}$$

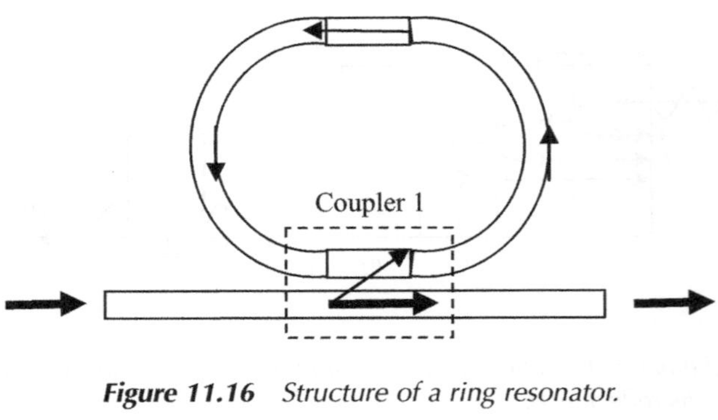

Figure 11.16 *Structure of a ring resonator.*

The analysis of the ring resonator may be performed by using coupled mode theory. The input and output electric fields of the straight guide are E_i and E_O, respectively. The electric field coupled into the ring is B_i and that after covering length L is B_O. The steady-state input and output fields are written as

$$E_\mathrm{O} = (1-\gamma)^{1/2}[E_i \cos(\kappa L_c) - jB_\mathrm{O} \sin(\kappa L_c)] \tag{11.94a}$$

$$B_i = (1-\gamma)^{1/2}[-jE_i \sin(\kappa L_c) + B_\mathrm{O} \cos(\kappa L_c)]. \tag{11.94b}$$

where κ is the mode-coupling coefficient, γ is the intensity insertion loss coefficient, and L_c is the coupling length. It is assumed that the propagation constants are the same, β, in straight and ring guides. If the intensity attenuation coefficient of the ring is denoted by α, B_O is expressed as

$$B_\mathrm{O} = B_i \exp[(-\alpha L/2) - j\beta L]. \tag{11.95}$$

The field transmittance of the ring resonator is then expressed by using Eqs. (11.94) and (11.95) as

$$\frac{E_\mathrm{O}}{E_i} = (1-\gamma)^{1/2}\left[\frac{\cos(\kappa L_c) - (1-\gamma)^{1/2}\exp[(-\alpha L/2) - j\beta L]}{1-(1-\gamma)^{1/2}\cos(\kappa L_c)\exp[(-\alpha L/2) - j\beta L]}\right]. \tag{11.96}$$

Introducing new parameters defined as

$$x = (1-\gamma)^{1/2}\exp(-\alpha L/2),\ y = \cos(\kappa L_c),\ \text{and}\ \phi = \beta L,$$

the intensity transmittance of the optical ring resonator may be expressed as

$$T(\phi) = \left|\frac{E_\mathrm{O}}{E_i}\right|^2 = (1-\gamma)\left[1 - \frac{(1-x^2)(1-y^2)}{(1-xy)^2 + 4xy\sin^2(\phi/2)}\right]. \tag{11.97}$$

The transmittance characteristics of the ring resonator show similar maxima and minima as in a FP resonator (see Figure 11.12). The minimum transmission occurs when

$$\varphi = \beta L = 2m\pi, \tag{11.98}$$

as noted before. The maximum and minimum transmittances are given by

$$T_{\max} = (1-\gamma)\frac{(x+y)^2}{(1+xy)^2}, \tag{11.99a}$$

$$T_{\min} = (1-\gamma)\frac{(x-y)^2}{(1-xy)^2}. \tag{11.99b}$$

It is noted from these equations that $x \cong y \cong 1$ should be satisfied in order to maximize $T_{\max}$ and to minimize $T_{\min}$. The full width at half maximum (FWHM) $\delta\varphi$ and finesse F are given by

$$\delta\phi = \frac{2(1-xy)}{\sqrt{xy}}, \tag{11.100a}$$

$$F = \frac{2\pi}{\delta\phi} = \frac{\pi\sqrt{xy}}{(1-xy)}. \tag{11.100b}$$

The value of $T_{\min}$ is zero when $x = y$, or when the condition

$$\cos(\kappa L_c) = (1-\gamma)^{1/2}\exp(-\alpha L/2), \tag{11.101}$$

is satisfied.

The FSR is determined by the spacing of two resonance peaks, for which φ differs by 2π. Writing the wavenumbers as k and $k + \Delta k$ respectively for $\varphi = 2m\pi$, and $= (2m + 1)\pi$ and assuming $\Delta k \ll k$, we obtain from Eq. (11.98)

$$\frac{d\beta}{dk}\Delta k = \frac{2\pi}{L}. \tag{11.102}$$

Since $\beta = kn$, where n is the effective index, and the frequency shift $\Delta f = (c/2\pi)\Delta k$, we may write

$$\Delta f = \frac{c}{NL}. \tag{11.103}$$

In Eq. (11.103), $N = n + k(dn/dk)$ is the group index. Using the relation $\delta\phi = \delta(\beta L) = (d\beta/dk)\delta k \cdot L = 2\pi/F$, the FWHM in terms of frequency may be expressed as

$$\delta f = \frac{c}{FNL}. \tag{11.104}$$

Ring resonators are useful as a wavelength selective filter, the operating principle of which is illustrated by Figure 11.17. The multiple-wavelength input into Terminal 1 will be partially coupled into the ring through Coupler 1. The optical wave in the ring will be partially coupled into the straight waveguide through Coupler 2 and outputs from Terminal 2, if the wavelength, for example, λ_1, satisfies the resonant condition. The coupling of the wave with wavelength λ_1 will be enhanced, and all others will be suppressed. As a result, only λ_1 will be dropped from Terminal 2, while the rest of the wavelengths will come out of terminal 4.

Since FSR is inversely proportional to the size of the ring resonator, the ring must be small in order to achieve a high FSR.

The finesse is another key specification of the ring resonator and is dependent on both the internal loss and the coupling (i.e., the external loss) of the resonator. The higher the total

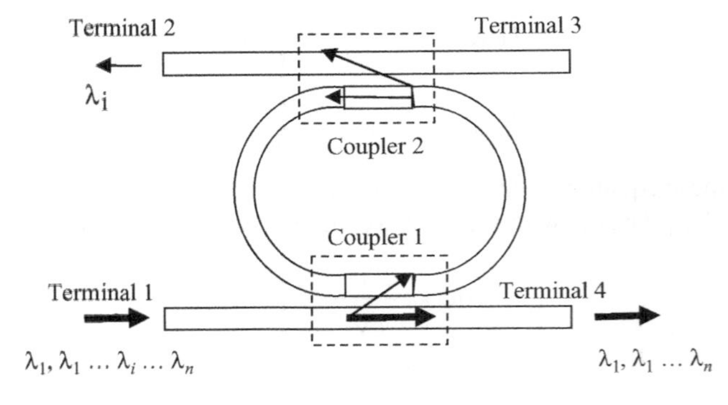

Figure 11.17 *Schematic diagram of a ring resonator used as a wavelength selective filter.*

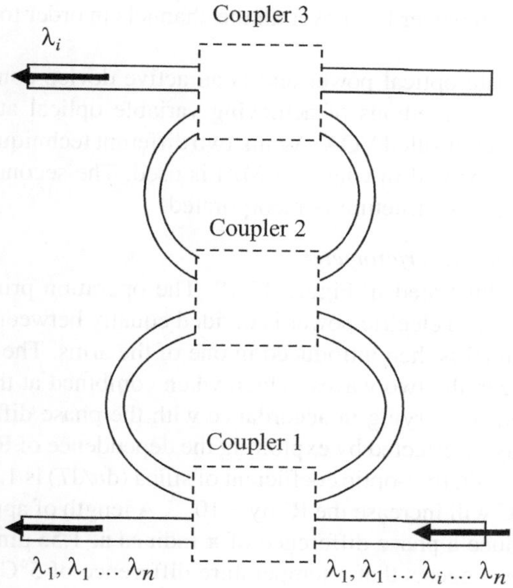

Figure 11.18 *Schematic diagram of a multiple ring resonator.*

losses are, the lower the finesse of the resonator, and it is advantageous to reduce both the internal and external losses in order to obtain higher finesse. However, the external loss due to coupling is necessary and cannot be too small for the resonator to operate as an optical filter. If the external loss is smaller than the internal loss, all the coupled power will be lost inside the cavity and no power will be coupled out. Because of these constraints, the ring resonator must use a strongly guided waveguide to minimize the bending loss for a curved waveguide with a very small radius.

11.5.8 Multiple-Ring Resonators

To increase the finesse, two or more rings could be used as shown in Figure 11.18. In this case, the combined FSR increases since $FSR = N.FSR_1 = M.FSR_2$ and the total FWHM in terms of wavelength, $\delta\lambda = \delta\lambda_1\delta\lambda_2/(\delta\lambda_1 + \delta\lambda_2)$, decreases, so there is a net increase of F. The two resonators have to be carefully designed to make sure that both N and M are integers. This way, the Nth peak of resonator 1 located at the same wavelength as the Mth peak of resonator 2 results in a sharper peak (high finesse). All the other peaks are blocked by each other.

11.5.9 Variable Optical Attenuator

Variable optical attenuators are used in many different applications. For example, in a DWDM communication, many wavelengths are transmitted through a single fiber. The power of different channels in the fiber may not be uniform due to variations in the characteristics of transmitters, of amplifiers, of attenuation in the network, and so on. It is important to have uniform power level over all the channels in order to improve the overall bit-error rate.

The VOA can attenuate optical power and is an active device controlled by a control signal. There are different methods of achieving variable optical attenuation. We shall discuss here electrically controlled VOA. Again, two different techniques will be discussed. First is the interference method in which an MZI is used. The second one uses a straight waveguide in which a p–i–n structure is incorporated.

11.5.9.1 Mach–Zehnder Interferometer

The basic structure is illustrated in Figure 11.19. The operation principle of a MZI has already been discussed. The electric power is divided equally between the two arms in the input side. A change in RI is then introduced in one of the arms. The change introduces a phase difference between the two waves, which when combined at the output produces a resultant beam of intensity varying in accordance with the phase difference.

The RI variation may be effected by exploiting the dependence of RI on heat, that is, the thermo-optic effect. The thermo-optic coefficient of silica (dn/dT) is 1.86×10^{-4}°C. Thus a temperature rise of 5°C will increase the RI by $\sim 10^{-3}$. A length of approximately 1 mm of the material may produce a phase difference of π radians at 1.55 μm.

The above example suggests that a temperature difference of 5°C must be maintained between two arms of the MZI, each having a length of 1 mm. The power needed to maintain this temperature difference is about 300 mW. The response time is in the range of a few tens of a millisecond.

The VOA may be realized by introducing free carriers. As mentioned in Chapter 4, the introduction or extraction of free carriers changes the absorption or RI in silicon. Both the changes may be exploited to have variable attenuation. The SOI waveguide is suited for this purpose.

In the first method, a p–i–n structure is incorporated in one arm of the MZI. The doping density in silicon is kept low in order to keep the attenuation in the arm as small as possible. Application of a forward bias injects both electrons and holes into the intrinsic layer incorporated in the waveguide. The excess carrier pairs introduce a change in RI and consequently a phase shift.

Alternatively, instead of using an interferometer, the p–i–n structure is incorporated directly into a straight waveguide. The injection of carriers then changes the absorption of

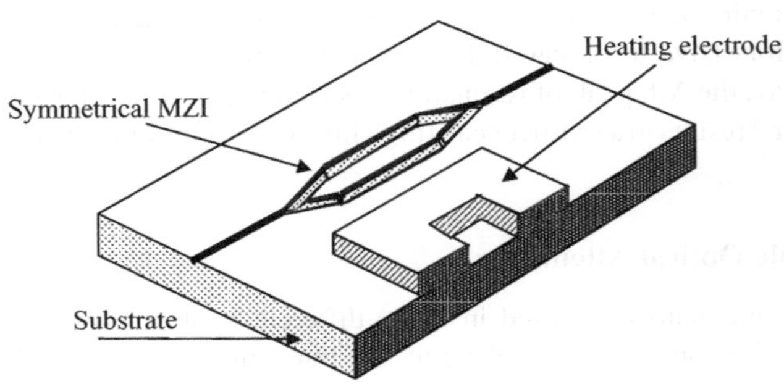

Figure 11.19 *Schematic of a VOA using MZI.*

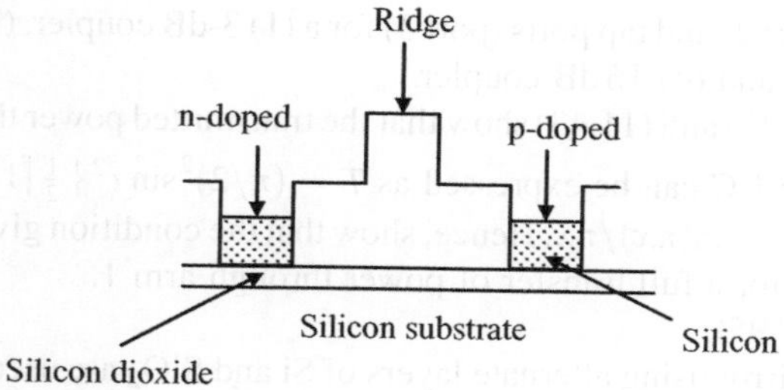

Figure 11.20 *Cross-section of a SOI waveguide attenuator.*

the medium, and the intensity of light coming out of the waveguide is changed according to the injection level or electrical bias. As mentioned, SOI structure is suited for the purpose. The injecting p and n regions are placed by the side of the silicon waveguide as shown in Figure 11.20.

Problems

11.1 Derive appropriate expressions for the output powers in the two arms, when the input fields E_1 and E_2 differ by a phase α.

11.2 Derive Eqs. (11.41)–(11.43).

11.3 Find out the condition for maximum power transfer in a non-phase-matched directional coupler.

11.4 Using Eqs. (11.42) and (11.43), discuss how the power in the two arms will vary with distance.

11.5 Verify the expected variation of powers in two arms from Problem 11.4 by plotting the variation of power for different values of coupling coefficients in a DC for the non-phase-matched condition. Also show that the sum of powers from the two arms is constant.

11.6 Derive Eqs. (11.68) and (11.69) by using boundary conditions stated in the text.

11.7 Prove that the ratio of the transmitted to the incident intensities in a FP resonator filled with a lossy material with loss coefficient α may be expressed as $\frac{I_t}{I_i} = \frac{(1-R)^2 e^{-\alpha L}}{(1-Re^{-\alpha L})^2 + 4Re^{-\alpha L}\sin^2\theta}$, where θ has been defined in Eq. (11.82) and the reflectivities of the two mirrors are assumed equal.

11.8 Express the loss in terms of the ratio of maximum-to-minimum transmitted intensities.

11.9 Show that the expression for transmittivity of a FP resonator, Eq. (11.83), reduces to the expression given by Eq. (7.28).

11.10 Show that the coupling constant is given by $\kappa = \frac{2\beta_y{}^2 b e^{-bd}}{\beta_z W(\beta_y{}^2 + \beta_z{}^2)}$, where b is the extinction coefficient, d is the separation between the guides, W is the width of each guide, and β_z and β_y are, respectively, the mode propagation constants in the propagation and transverse directions.

11.11 An X-dB coupler means that the power coming out from the tap port is X-dB below the input power launched into port 1. Find the ratio of powers coming out of the throughput (port 2) and tap ports (port 3) for a (1) 3-dB coupler, (2) 6-dB coupler, (3) 10 dB coupler, and (4) 16 dB coupler.

11.12 Using Eqs. (11.42) and (11.43) show that the transmitted power through the right end of arm 2 in the DC can be expressed as $T = (\pi/2)^2 \sin c^2\{\frac{1}{2}[1+(\Delta\beta L_0/\pi)^2]^{1/2}\}$, where $\sin c(x) = \sin(\pi x)/\pi x$. Hence, show that the condition given by Eq. (11.74) is to be satisfied for a full transfer of power through arm 1.

11.13 Derive Eq. (11.95).

11.14 A dielectric mirror using alternate layers of Si and SiO_2 are used to form a resonant cavity for a light-emitting device at 850 nm. Calculate the thicknesses of each layer ($n_{SiO2} = 1.5$ and $n_{Si} = 3.4$).

11.15 Prove that if the transmission of a FP resonator is represented by a Lorentzian, then the response is $g(f) = [(f - f_0)^2 + (f_0/2Q)^2]^{-1}$, where f is the frequency, f_0 is the resonant frequency, and Q is the quality factor of the resonator.

11.16 The cavity lifetime is defined as $dP/dt = -P/\tau_c$, where $P(t)$ is the power at time t. Prove that power decays as $P(t) = P(0)\exp(-t/\tau_c) = P(0)\exp(-\omega t/Q)$. From this, show that the linewidth is $\Delta f = (2\pi\tau_c)^{-1}$ and $\tau_c = \lambda^2/(2\pi c\Delta\lambda)$ where $\Delta\lambda$ is the FWHM.

11.17 Obtain an expression for a photon lifetime in a FP resonator enclosing a medium with loss $= \alpha$.

11.18 Draw different rays as shown in Figure 11.11 showing the field intensities coming out of mirror 2. The FP resonator, however, encloses a medium of gain $= g$ and loss coefficient $= \alpha$. By summing the field intensities, obtain the condition when $E_t/E_i \rightarrow \infty$. From this, prove that the self-sustained laser oscillation condition is $g = \alpha + (1/2L)\ln(1/R_1R_2)]$.

11.19 A 2 m long fiber Bragg grating is used to compensate the chromatic dispersion accumulated in a certain length (L) of fiber over the entire C-band (1530–1565 nm). The dispersion parameter is $D = 17$ ps/(nm.km). Calculate the length of the fiber; use $n = 1.5$ for the RI of the core. (Hint: delay by dispersion $= D\Delta\lambda L$, where $\Delta\lambda$is the bandwidth.)

References

1. Hunsperger, R.G. (1991) *Integrated Optics: Theory and Technology*, 3rd edn, Springer-Verlag, Berlin.
2. Yariv, A. (1993) *Quantum Electronics*, John Wiley, New York.
3. Bruce Buckman, A. (1992) *Guided-Wave Photonics*, Saunders College Publishing, Fort Worth, TX.
4. Lee, D.L. (1986) *Electromagnetic Principles of Integrated Optics*, John Wiley & Sons, Inc., New York.
5. Nishihara, H., Haruna, M., and Suhara, T. (1989) *Optical Integrated Circuits*, McGraw-Hill, New York.
6. Okamoto, K. (2000) *Fundamentals of Optical Waveguides*, Academic Press, San Diego, CA.
7. Agrawal, G.P. (2002) *Fiber Optic Communication Systems*, 3rd edn, John Wiley & Sons, Inc., Hoboken, NJ.
8. Verdeyen, J.T. (1995) *Laser Electronics*, 3rd edn, Prentice Hall, Englewood Cliffs, NJ.

9. Yariv, A. (1973) Coupled mode theory for guided wave optics. *IEEE J. Quant. Electron.*, **QE-9**, 919–933.
10. Orthonos, A. (1997) Fiber Bragg gratings. *Rev. Sci. Instr.*, **68**, 4309–4341.
11. Kashyap, R. (1999) *Fiber Bragg Gratings*, Academic Press, New York; and Kashyap, R., Maxwell, G.D., and Ainslie, B.J. (1993) Laser-trimmed four-port bandpass filter fabricated in single-mode photosensitive Ge-doped planar waveguide. *IEEE Photon. Technol. Lett.*, **5**, 191–194.
12. Ramaswami, R. and Sivarajan, K.N. (2002) *Optical Networks: A Practical Perspective*, 2nd edn, Morgan Kaufmann, Waltham, MA.
13. Saadany, B., Malak, M., Kubota, M. *et al.* (2006) Free-space tunable and drop optical filters using vertical bragg mirrors on silicon. *IEEE J. Sel. Top. Quant. Electron.*, **12**, 1480–1488.
14. Muscara, A., Castagna, M.E., Leonardi, S. *et al.* (2006) Design and electro-optical characterization of Si based resonant cavity light emitting devices at 850 nm. *J. Lumines.*, **121**, 293–297.
15. Lipson, M. (2006) Compact electro-optic modulators on a silicon chip. *IEEE J. Sel. Top. Quant. Electron.*, **12**, 1520–1526.

12

Waveguides for Dense Wavelength-Division Multiplexing (DWDM) Systems

12.1 Introduction

In the previous chapters we discussed the basic principle of waveguiding and the operation principles of a few simple passive waveguide devices like *Y*-switches, directional couplers, and Mach–Zehnder interferometers (MZIs). These simple devices nd applications in optical communications and networking systems.

With the advent of DWDM communications systems with the number of wavelengths exceeding 100, it becomes necessary to have, in addition to active devices, passive devices like multiplexers (muxes), demultiplexers (demuxes), and some of the sophisticated versions used in switching and routing. Demuliplexers require a wavelength-selective mechanism like ltering. In the literature there are extensive reports on different principles for the realization of demux operation as well as on their practical implementation. Among them, ber-optic and micro-optic technologies were in wider use. A comparison of these technologies with the integrated optic approach has been made by Pennings *et al.* [1]. Due to increased complexity of the current networks, the present trend shows a bias toward integrated optic (IO) components.

The IO components belong to two classes: grating based and interference based. Both are imaging devices in the sense that an image of the eld in one of the input guides is formed onto an array of output guides. The principle of operation of Bragg gratings has been studied in detail in Chapter 11. In this chapter, we focus our attention on PHASAR devices, which work on the principle of interference. They are so called because of the use of phased-array technique originally employed in the design of microwave antenna. Such phased arrays of optical waveguides, realized using the conventional waveguide formation technology, also act as a grating and are called arrayed waveguide gratings (AWGs).

Silicon Photonics: Fundamentals and Devices, First Edition. M. Jamal Deen and P. K. Basu.

PHASAR devices were proposed by Smit [2] and independently by Dragone [3]. The rst device was constructed by Vellekoop and Smit [4] to work at a shorter wavelength of 0.63 μm. Devices working at longer wavelengths were reported by Takahashi *et al.* [5]. Dragone [6] extended the concept of a 1 × N coupler to N × N devices, which are now called waveguide grating routers (WGRs).

Devices reported so far are mainly grown on InP or on silica- and silicon-based materials. As noted already, silica-based devices are more matched to the core of the ber and represent low coupling and propagation loss. However, due to low index contrast, they have larger dimensions and a larger bend radius, making them unsuitable for a higher level of integration, let alone integration with active devices. InP-based devices, on the other hand, can accommodate more components on a single chip. Though they offer more coupling and propagation losses, the smaller size more than compensates for the losses to make the overall loss per chip lower. They are not suitable for circuits of lower complexity, however. Some useful references for InP-based devices may be found in the paper by Smit and van Dam [7], who give a comprehensive review of the PHASAR-based devices.

In the present chapter, we shall rst discuss the structure and basic operating principle of the AWGs. An analysis of the grating will then follow, which will give the characteristic expressions for the device related to frequency response, insertion loss, BW, and cross-talk. The methods employed to remove polarization and temperature sensitivities will then be discussed. The present scenario of AWGs or WGRs grown on silica- and other silicon-based materials will then be presented. Finally another type of grating, the Eschelle grating, will be introduced.

12.2 Structure and Operation of AWGs

We shall rst describe the construction of an AWG and describe its basic working principle [8–12].

12.2.1 Structure and Working Principle

AWGs are the most important PHASAR device and are alternatively called WGRs. The principle of operation of the device may be understood with the help of Figure 12.1. The structure is symmetric, and either side could be input or output.

The input and output stages consist of star couplers called free space couplers (FSCs), also referred to as *slab regions*. The two FSCs are interconnected by the grating region consisting

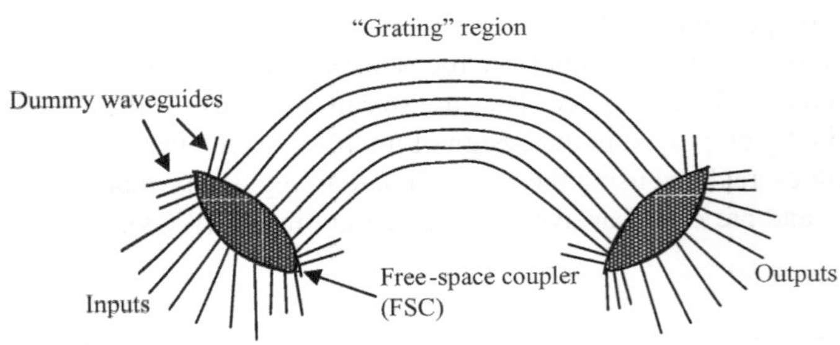

Figure 12.1 Structure of an arrayed waveguide grating (AWG).

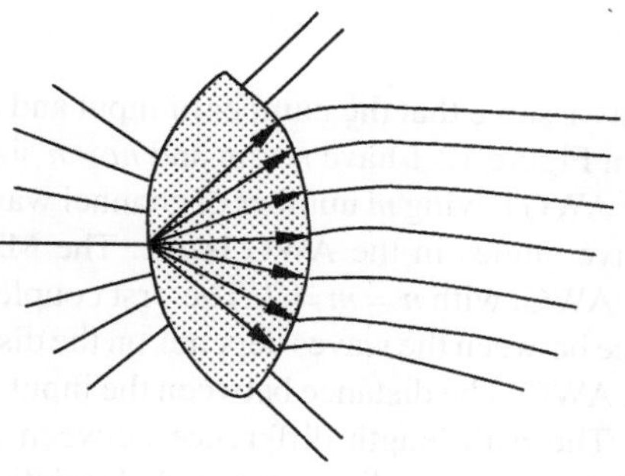

Figure 12.2 *Coupling of a signal to the input of all waveguides in the AWG via FSC.*

of an array of channel waveguides having different lengths and having no coupling with one another. When a single-mode signal is fed to one of the input waveguides, it couples to a large number of modes in FSC and then couples to the arrayed waveguides, as shown in Figure 12.2. The power coupled from any particular input to each of the arrayed waveguide is equal, but due to difference in lengths, the light in each waveguide will differ in phase.

The length of each channel guide differs from its neighbors by a xed amount ΔL. This results in further phase difference between the signals when they reach the output star coupler. Due to interference, a signal at a particular wavelength is reinforced at a particular output guide and is extinguished at all other guides. The reinforcement at a particular output guide and destructive interference at other guides are illustrated in Figure 12.3. The particular output guide, in which a given wavelength reinforces, depends on both the location of the input guide and the particular wavelength. Thus N wavelengths on a particular input are spatially separated onto N-separated output guides [10].

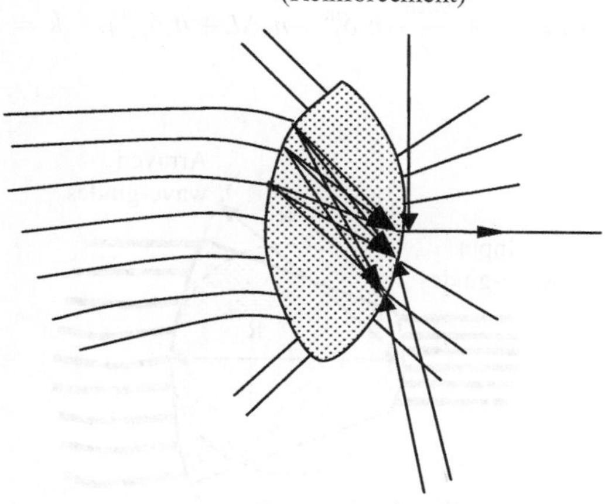

Figure 12.3 *Constructive interference of the signals at a particular output waveguide and destructive interference at all other waveguides.*

12.2.2 Analysis

Following [9], let us assume that the number of input and output waveguides of the AWG is n. Also, let FSCs in Figure 12.1 have $n \times m$ and $m \times n$ size, respectively, and each of them is connected to the AWG having m number of channel waveguides. The difference in length between consecutive guides in the AWG is ΔL. The MZI discussed in Section 12.1 is a special case of the AWG, with $n = m = 2$. The rst coupler splits the input into m parts, and the phase difference between the waves depends on the distance from input waveguide to the particular guide in AWG. The distance between the input guide i and the guide k in the array is denoted by r_{ik}^{in}. The path length difference between arrayed guide k and $k-1$ is ΔL. Similarly, the differences in the distance traveled (relative to any one of the array and any one of the output guides) between AWG k and output guide j are denoted by r_{kj}^{out}. Using these notations, the relative phases of signals from input i to output waveguide j passing through different paths between them are

$$\phi_{ijk} = \frac{2\pi}{\lambda}(n_s r_{ik}^{\text{in}} + n_c k \Delta L + n_s r_{kj}^{\text{out}}), \quad k = 1, \ldots m. \tag{12.1}$$

where n_s is the RI of input and output FSCs, and n_c is the RI in the AWG. Only the signal with wavelength λ from input guide i traversing through channel k in AWG will add up in phase in output guide j, only when the relative phase ϕ_{ijk} differs by an integral multiple of 2π.

The phase relationship is satis ed by a suitable construction of the FSCs and AWGs, known as the *Rowland circle construction* shown in Figure 12.4. The AWGs are located on the arc of the grating circle of radius R having a center at the end of the central input (output) waveguide.The input and output waveguides are located on the arc of a circle with diameter R known as the Rowland circle. The input and output couplers are constructed such that $r_{ik}^{\text{in}} = r_i^{\text{in}} + k\delta_i^{\text{in}}$ and $r_{kj}^{\text{out}} = r_j^{\text{out}} + k\delta_j^{\text{out}}$. The above equation then transforms to

$$\phi_{ijk} = \frac{2\pi}{\lambda}(n_s r_i^{\text{in}} + n_s r_j^{\text{out}}) + \frac{2\pi k}{\lambda}(n_s \delta_i^{\text{in}} + n_c \Delta L + n_s \delta_j^{\text{out}}), \quad k = 1, \ldots m. \tag{12.2}$$

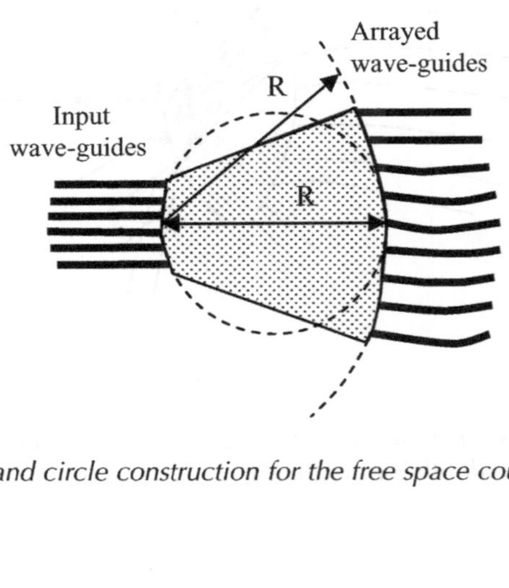

Figure 12.4 *The Rowland circle construction for the free space couplers used in the AWG.*

This equation suggests that the wavelength λ present in the *ith* input guide satisfying

$$n_s\delta_i^{\text{in}} + n_c\Delta L + n_s\delta_j^{\text{out}} = p\lambda, \tag{12.3}$$

where p is an integer, add in phase at output guide j.

When the AWG is used as a demultiplexer, all the wavelengths are present at the same input, say, input i. Therefore, if the above equation (12.3) is satis ed for λ_j for some integer p, then these wavelengths will be demultiplexed at the output. The spacing between demultiplexed wavelengths is determined primarily by δ_j^{out}. It may also be noted that the phase relationship may be satis ed by another wavelength λ'_j for the next integer $(p + 1)$, and both λ_j and λ'_j are demultiplexed at the same output j from input i. Thus the AWG has a periodic response and all the wavelengths must lie within one free spectral range (FSR). This situation is similar to what has been found in the Fabry–Perot lter discussed in Chapter 11.

In this way, the arrayed waveguide operates like a concave diffraction grating. The locations of the I–O waveguide ends and arrayed waveguide ends are based on the Rowland circle construction as is used for concave re ection-grating monochromator.

12.3 AWG Characteristics

12.3.1 Tuning and Free Spectral Range

We now consider the operation of the AWG in detail by using Figure 12.5. Following the notation given by Okamoto [11, 12], the input waveguide separation in the rst waveguide is D_1, the array waveguide separation is d_1, and the radius of curvature is f_1. We consider rst the general situation in which the parameters for the rst and second slab regions are different. Thus in the second slab region the output waveguide separation, the array waveguide separation, and the radius of curvature are denoted by D, d, and f, respectively. The position in the input waveguide is measured counterclockwise from the center. The light input at the x_1 position is radiated into the rst slab, and then all the guides in the array are

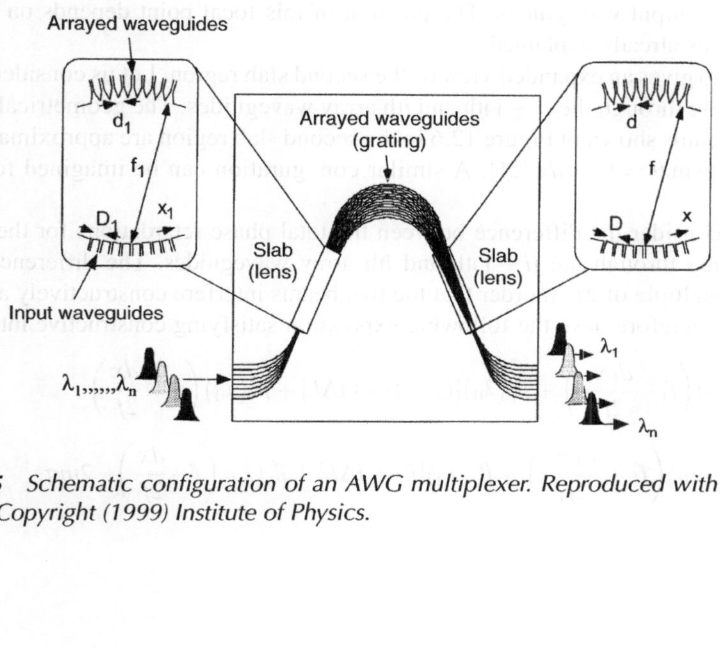

Figure 12.5 *Schematic configuration of an AWG multiplexer. Reproduced with permission from [12]. Copyright (1999) Institute of Physics.*

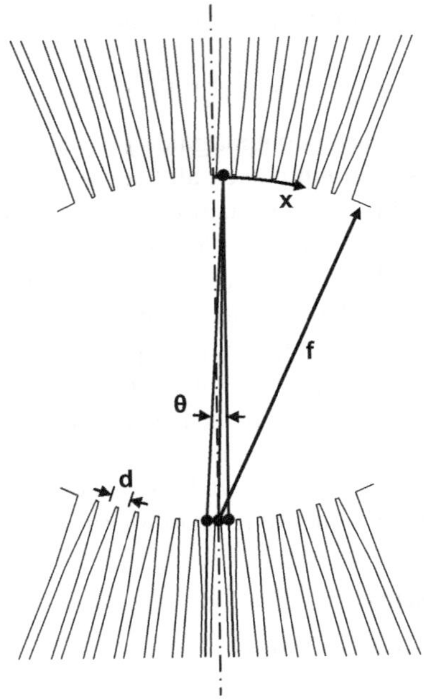

Figure 12.6 *Enlarged view of the second slab waveguide. Reproduced with permission from [12]. Copyright (1999) IEEE.*

excited. The excited field amplitude in ith guide of the array is a_i $(i = 1 - N)$, where N is the total number of waveguides. The amplitude profile for a_i is usually Gaussian in nature. After propagating through the arrayed waveguide, the light beams constructively interfere into one focal point x in the second slab, where x is again measured counterclockwise from the center of the output waveguides. The position of this focal point depends on the signal wavelength, as already explained.

Figure 12.6 gives an expanded view of the second slab region. Let us consider two light beams passing through the $(i-1)$th and ith array waveguides. The geometrical distances of the two beams shown in Figure 12.6 in the second slab region are approximately given as $f \pm (d/2)\sin\theta = f \pm (dx/2f)$. A similar configuration can be imagined for the first slab region.

We now consider the difference between the total phase retardations for the two light beams passing through the $(i-1)$th and ith array waveguides. The difference must be an integral multiple of 2π in order that the two beams interfere constructively at the focal point x. We therefore have the following expression satisfying constructive interference:

$$\begin{aligned} &\beta_s(\lambda_0)\left(f_1 - \frac{d_1 x_1}{2f_1}\right) + \beta_c(\lambda_0)[L_c + (i-1)\Delta L] + \beta_s(\lambda_0)\left(f + \frac{dx}{2f}\right) \\ &= \beta_s(\lambda_0)\left(f_1 + \frac{d_1 x_1}{2f_1}\right) + \beta_c(\lambda_0)[L_c + i\Delta L] + \beta_s(\lambda_0)\left(f - \frac{dx}{2f}\right) - 2m\pi, \end{aligned} \tag{12.4}$$

where β_s and β_c denote the propagation constants in the slab and array waveguide, respectively; m is an integer; λ_0 is the center wavelength of the WDM signal; and L_c is the minimum array waveguide length. Eliminating common terms in Eq. (12.4), we obtain

$$\beta_s(\lambda_0)\frac{d_1 x_1}{f_1} - \beta_s(\lambda_0)\frac{dx}{f} + \beta_c(\lambda_0)\Delta L = 2m\pi. \tag{12.5}$$

If now the condition $\beta_c(\lambda_0)\Delta L = 2m\pi$, or

$$\lambda_0 = \frac{n_c \Delta L}{m}, \tag{12.6}$$

is satis ed, where n_c is the RI of the array waveguide, the light input and output positions satisfy the condition

$$\frac{d_1 x_1}{f_1} = \frac{dx}{f}. \tag{12.7}$$

Equation (12.7) establishes the fact that the output position x is now uniquely determined when light is input at x_1. In Eq. (12.6), m represents the diffraction order. Usually the waveguide parameters for the rst and second slab regions are equal, thus making input and output distances equal ($x_1 = x$). Keeping the light input position x_1 xed, the dispersion of the focal position x with respect to wavelength λ may be obtained by differentiating Eq. (12.5) with respect to λ:

$$\frac{\Delta x}{\Delta \lambda} = -\frac{N_c f \Delta L}{n_s d \lambda_0}, \tag{12.8}$$

where n_s is the effective index of the slab region and N_c is the group index of the effective index n_c of the array waveguide ($N_c = n_c - \lambda dn_c/d\lambda$). Similarly, the dispersion of the input position x_1 with respect to the wavelength, for the xed light output position x, is given by

$$\frac{\Delta x_1}{\Delta \lambda} = \frac{N_c f_1 \Delta L}{n_s d_1 \lambda_0}. \tag{12.9}$$

The input and output waveguide separations are $|\Delta x_1| = D_1$ and $\Delta x = D$, respectively, when the channel spacing of the WDM signal is $\Delta\lambda$. Using these relations in Eqs. (12.8) and (12.9), the wavelength spacing in the output side for the xed input position x_1 is given by

$$\Delta\lambda_{\text{out}} = \frac{n_s d D \lambda_0}{N_c f \Delta L}. \tag{12.10}$$

Similarly the wavelength spacing in the input side for the xed light output position x is given by

$$\Delta\lambda_{\text{in}} = \frac{n_s d_1 D_1 \lambda_0}{N_c f_1 \Delta L}. \tag{12.11}$$

As mentioned, the waveguide parameters in the rst and second slab regions are usually the same. Thus $D_1 = D$, $d_1 = \mathrm{d}$, and $f_1 = f$, and the channel spacings as given by Eqs. (12.10) and (12.11) are the same. Denoting the spacing by $\Delta\lambda$, the path length difference may be expressed as

$$\Delta L = \frac{n_s d D \lambda_0}{N_c f \Delta\lambda}. \tag{12.12}$$

The free spatial range of the AWG is the spatial separation of the mth and $(m + 1)$th focused beam for the same wavelength and may be expressed by using Eq. (12.5) as

$$X_{FSR} = x_m - x_{m+1} = \frac{\lambda_0 f}{n_s d}. \tag{12.13}$$

The number of available wavelength channels N_{ch} is obtained by dividing X_{FSR} by the output waveguide separation as

$$N_{ch} = \frac{X_{FSR}}{D} = \frac{\lambda_0 f}{n_s d D}. \tag{12.14}$$

12.3.2 Frequency Response

If the wavelength is changed, the focal eld of the AWG moves along the receiver waveguides. The frequency response of different channels is governed by the overlap between the focused eld and the modal eld of the waveguide.We assume that both the elds are Gaussian. The overlap between two Gaussians is given by a simple function in Gaussian form as follows:

$$\Omega = \exp\left\{-\left(\frac{x}{w_0}\right)^2\right\}, \tag{12.15}$$

where x is the distance between the two elds and w_0 is their spot size. The distance x, which is proportional to the frequency shift δf from the pass frequency, is given by

$$x = \frac{\Delta x}{\Delta f}\delta f. \tag{12.16}$$

The ratio $\Delta x / \Delta f$ is the linear dispersion which is the product of angular dispersion and the focal length. The frequency response is thus given by

$$T(\delta f) = \exp\left\{-\left(\frac{\Delta x \delta f}{w_0 \Delta f}\right)^2\right\}. \tag{12.17}$$

It appears from this equation that the frequency response of the AWG is the same as that of a Gaussian bandpass lter. The full width at half maximum (FWHM) is therefore

$$\mathrm{FWHM} = \frac{2\sqrt{\ln 2}\, w_0 \Delta f}{\Delta x}. \tag{12.18}$$

12.3.3 Channel Cross-talk

There are different sources that contribute to the cross-talk. The coupling between receivers through the exponential tails of the eld distribution is an important source. The truncation of the eld due to nite width of the array aperture also plays its role. If the array waveguides are not strictly single moded, a rst-order mode excited near the junction of straight and curved parts of the waveguide propagates coherently through the arrays and creates ghost images. Since the propagation constants are different, the ghost image will be focused at a different location and may be coupled to an undesired receiver. Cross-talk can also occur by phase distortion due to coupling in the input and output sections of the array. Imperfections in the fabrication of the phased array may result in cross-talk.

12.3.4 Polarization Dependence

Waveguide birefringence leads to different propagation constants for the fundamental TE and TM modes. This in turn gives rise to a shift of Δf_{pol} of the spectral response with respect to each other. Light with different wavelengths in vacuo will be coupled into the same receiver waveguide, if the wavelengths of the fundamental TE and TM modes in the guide are equal. This implies

$$\lambda_{\mathrm{TM}}(f) = \frac{c}{f \cdot n_{\mathrm{TM}}(f)} = \lambda_{\mathrm{TE}}(f - \Delta f_{\mathrm{pol}}) = \frac{c}{f - \Delta f_{\mathrm{pol}} \cdot n_{\mathrm{TE}}(f - \Delta f_{\mathrm{pol}})}. \tag{12.19}$$

In Eq. (12.19), the subscripts TE and TM refer to the RIs for corresponding polarizations. Solving Eq. (12.19), we nd

$$\Delta f_{\mathrm{pol}} \approx f \cdot \frac{n_{\mathrm{TE}} - n_{\mathrm{TM}}}{n_{\mathrm{g-TE}}}, \tag{12.20}$$

where $n_{\mathrm{g-TE}}$ is the group index for TE polarization. For silica-based and other low-contrast waveguides, this value is quite small. A number of methods used to reduce polarization dependence are discussed in Section 12.4. The birefringence problem arising in silica-based waveguides is caused by strain induced due to different thermal expansion coef cients of silica and silicon. It can be reduced by using silica substrates instead of silicon substrates.

The origin of the birefringence has been explained, and expressions for the RI difference for TE- and TM-polarized waves have been derived by Janz [13] in terms of photo-elastic tensors and strain-optic tensors (see the "Problems" section of this chapter).

12.4 Methods for Improving Performance

This section discusses brie y the methods employed to make the grating insensitive to changes in polarization and temperature and to achieve at frequency response.

12.4.1 Flat Frequency Response

The dispersion of focal position x with respect to wavelength is almost constant. The transmission loss in each channel of the normal AWG increases monotonically from the center wavelength. Furthermore optical signals pass through several lters in the network, and therefore their frequency response also affects the overall response in the channel, making it still narrower. This necessitates the design of attened and broadened frequency response for AWG multiplexers.

Several approaches have been proposed to achieve a at frequency response for the passband of AWGs. A at-response AWG multiplexer with parabolic waveguide horns in the input waveguide with satisfactory performance has been reported in [14].

It has been con rmed that in order to obtain a at spectral response, a rectangular electric eld pro le must be produced at the focal plane, that is, the interface between the second slab and output waveguides. The electric eld pro le in the focal plane is, in turn, the Fourier transform of the eld in the output array aperture, that is, the interface between the array waveguide and the second slab. The needed electric eld pro le at the array output aperture is the sinc-type, which obeys a sin $(\xi)/\xi$ distribution, where ξ is measured along the array output aperture [15]. The sinc-type electric eld distribution has been measured in [16], and the distribution is achieved by introducing additional loss to each array waveguide.

12.4.2 Polarization Independence

Various authors employed different methods to grow AWGs that do not show polarization sensitivity. We shall discuss a few of them.

a. *Silica substrate*: Use of silica substrate instead of silicon substrate has been found to reduce the problem due to polarization [17].
b. *Halfwave plate*: Insertion of a halfwave plate in the middle of the phased array ef ciently overcomes the polarization problem. Light entering the array in a TE-polarized state will be converted into a TM-polarized state by the $\lambda/2$ plate and will travel the remaining half in that state. Similarly light entering in the TM-polarized state will travel the second half in the TE-polarized state. As a result, both the states will suffer the same amount of phase shift regardless of the birefringent property of the waveguides. This method was proposed by Takahashi *et al.* [18] and has been applied to a number of devices using polyimide half plates. This method is useful for low-contrast structures, but in semiconductors having a large numerical aperture, this method is not a practical solution.
c. *Dispersion compensation*: In this method a birefringence compensating waveguide of length δL with different birefringence is inserted in the phased array as shown in Figure 12.7. The compensating section is inserted in the ith waveguide. The phase transfer difference between the ith and $(i+1)$th waveguide is expressed as

$$\Delta\Phi = k_0[n_g\Delta L + \delta L(n_g - n_{gc})], \tag{12.21}$$

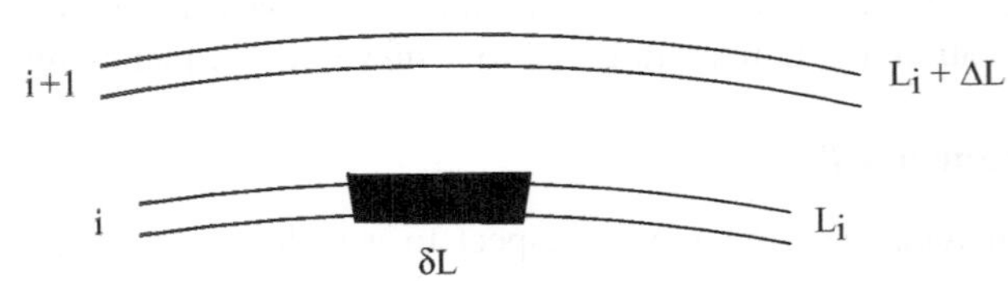

Figure 12.7 *Method of ensuring polarization independence by inserting a birefringent compensating section of length δL in the ith guide.*

where n_g and n_{gc} are the effective mode indices of the original waveguide and the compensation section. In order that $\Delta\Phi$ should be polarization independent, δL should be as follows:

$$\delta L = \Delta L/[(\Delta n_{gc}/\Delta n_g) - 1], \tag{12.22}$$

in which Δn_g and Δn_{gc} are the differences between the TE and TM values of n_g and n_{gc}, respectively.The whole array is made polarization independent by inserting a section of length δLin the rst guide, $2\delta L$in the second guide, and so on. The compensation section will then have a triangular shape having a maximum length of $N\delta L$, where N is the number of array waveguides. This method was also proposed by Takahashi *et al.* [19] and has been used by many workers for both silica- and InP-based waveguides.

d. *Order matching*: In another method, the value of FSR is made equal to polarization dispersion Δf_{pol}, so that the mth order beam for TE overlaps with the $(m-1)$th order beam of TM, which makes the response virtually polarization independent. The method has been used for InP-based guides [20].

12.4.3 Temperature Insensitivity

The RI of silica glass itself changes with temperature with a coef cient 1.1×10^{-5}/deg. This introduces a change in the passband wavelength $[d\lambda/dT = 1.2 \times 10^{-2}(\text{nm/deg})]$. The AWG multiplexer should be made temperature independent by using a heater or a Peltier cooler to stabilize the channel wavelength. This involves power consumption and adds to the complexity of circuits. The temperature-insensitive operation may be achieved by inserting a compensating groove lled with silicon adhesive having a negative thermal coef cient [21], as illustrated in Figure 12.8. Since the pass wavelength is given by $\lambda_0 = n_g\Delta L/m$, the optical path length difference $n_g\Delta L$ should be made insensitive to temperature. The groove should satisfy the following two conditions:

$$n_g\Delta L = n_g\Delta l + n_{\text{com}}\Delta l_{\text{com}}, \tag{12.23}$$

$$\frac{d(n_g\Delta L)}{dT} = \frac{dn_g}{dT}\Delta l + \frac{dn_{\text{com}}}{dT}\Delta l_{\text{com}}, \tag{12.24}$$

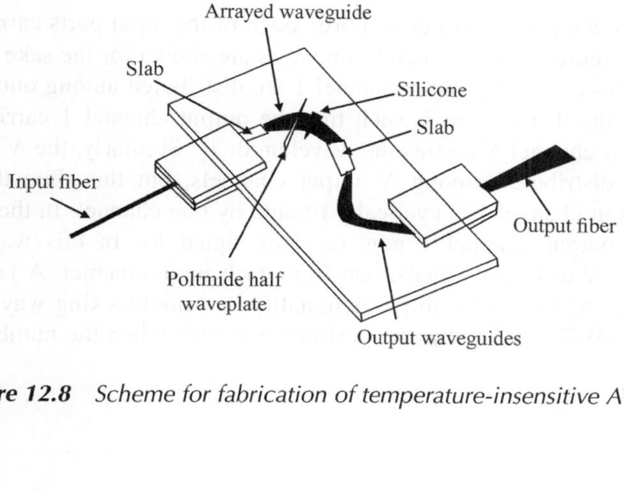

Figure 12.8 *Scheme for fabrication of temperature-insensitive AWG.*

where n_{com} is the RI of compensating liquid, and Δl and Δl_{com} are the path length differences in the waveguide and compensating region, respectively. The rst equation is a condition to satisfy for AWG speci cation, and the second is to ensure the athermal condition. As an example, the temperature sensitivity of silicone is $-37 \times 10^{-5}/°$, and therefore the path length difference of silicone should be $\Delta l_{com} = \Delta l/37$.

12.5 Applications of AWGs

AWGs perform four basic functions, namely, multiplexer, demultiplexer, passive waveguide router, and add-drop multiplexer used in WDM networks. In addition, they are also used in producing multiwavelength lasers and as tunable elements in photoreceivers. Some of these applications are now described in this section [8–11, 22].

12.5.1 Demultiplexers and Multiplexers

Figure 12.2 illustrates how an AWG can be used as a basic demultiplexer. The multichannel signals are fed to one of the input waveguides via a single ber. Signals having different wavelengths are focused onto different output waveguides. Separation of the different wavelength components, which is the basic function of the demux, is automatically performed by the AWG. In order to achieve multiplexing, different signals are fed to different input guides of the AWG, all of which are then focused into a single output waveguide as may be visualized if Figure 12.2 and Figure 12.3 are arranged in reverse sense. It may be mentioned at this stage that an integrated four-channel waveguide multiplexer was constructed by some workers by using three MZIs. However, when the number of input wavelengths increases, this type of design involves a larger number of MZIs leading to a complex, oversized structure. The use of a single AWG solves this problem. The number of wavelengths multiplexed currently by AWGs exceeds 1000.

12.5.2 Wavelength Routers

Wavelength routers are essential components in WDM networks. We describe here the functionality by using Figure 12.9 as an illustration for its operation and interconnectivity.

The WGR has *N* input and *N* output ports. Each of the input ports carries *N* different frequencies. In Figure 12.9 four such frequencies are shown for the sake of illustration. The *N* frequencies carried by input channel 1 are distributed among output channels 1 to *N*. However, the distribution is such that the output channel 1 carries wavelength λ_N and the output channel *N* carries the wavelength λ_1. Similarly, the *N* signals carried by input 2 are distributed among *N* output channels, but this time the wavelength received by output channel 1 is cyclically rotated by one channel. In the illustration of a 4×4 WGR, output channel 1 now receives signal b_3. In this way each output channel receives *N* different signals, one from each input channel. A large number of switches are needed to achieve this functionality in a nonblocking way. However, the use of a single AWG can achieve the performance even when the number of channels is quite large.

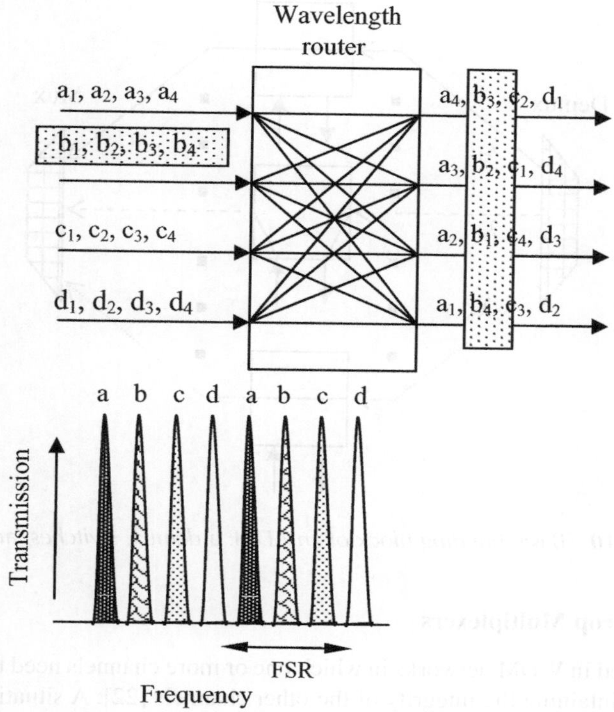

Figure 12.9 *Illustration of the principle of wavelength router used in the WDM network.*

In order to achieve the cyclical rotation of the input frequencies as mentioned in this chapter, the frequency response of the AWG should be periodic and the FSR should be equal to N times the channel spacing. Using Eq. (12.13), one obtains the following relation to be satis ed:

$$\Delta L = \frac{c}{N_g N_{ch} \Delta f_{ch}}, \tag{12.25}$$

where N_g is the group index of the waveguide mode, N_{ch} is the number of frequency channels, and Δf_{ch} is the channel spacing.

12.5.3 Multiwavelength Receivers and Transmitters

The straightforward method to achieve a multiwavelength receiver is to connect a demultiplexer with an array of photodiodes. However, in this respect the AWG demultiplexer formed on InP substrate proves to be more useful as it can be monolothically integrated with the InGaAs photodetectors.

A WGR was inserted in a cavity to extract multiple wavelengths from a laser [23]. However, the system uses InP-based devices. Hybrid integrated transmitters and receivers have been fabricated by using silica-based planar lightwave circuits [24].

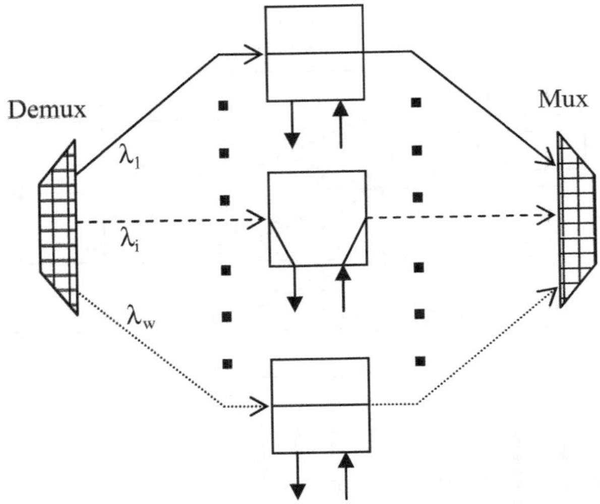

Figure 12.10 *Basic building block of an ADM: a demux, switches, and a mux.*

12.5.4 Add-Drop Multiplexers

ADMs are needed in WDM networks in which one or more channels need to be dropped or added while maintaining the integrity of the other channels [22]. A situation may arise in which only a single channel is to be demultiplexed, while the other channels remain unaffected. In such a case the speci c channel is diverted to one port and all other channels are sent to a different port. The device is then called an *add-drop filter*. In many situations, it is necessary to pick out one or more low-speed streams from a high-speed stream and likewise to add one or more low-speed streams to a high-speed stream. This function is performed by an add-drop multiplexer (ADM). ADM is an important component in the ber ring networks, which are expected to be deployed in the metropolitan-area market. One form of wavelength ADM is shown in Figure 12.10, which contains a demux, then a set of 2 × 2 switches, one switch for each wavelength, followed by a mux. When all the 2 × 2 switches are in the bar state, all the wavelengths ow through the ADM. However, if one of the switches is con gured in the cross-state by electronic control, then the signal on corresponding wavelength is dropped locally, and a new data stream can be added to the same switch on the same location. More than one wavelength can be dropped or added if corresponding processing capability is available.

The basic building blocks of an ADM as given in Figure 12.10 are a demux, a mux, and switches (or signal processors). It is necessary to precisely match the characteristics of the demux and mux.

12.5.5 Optical Cross-connects: Reconfigurable Wavelength Routers (WRs)

The function of a recon gurable WR is illustrated in Figure 12.11. It has *P* incoming and *P* outgoing bers. On each incoming ber there are *M* wavelengths, which are rst separated by a demux, the output of which is directed to an array of *M PxP* optical

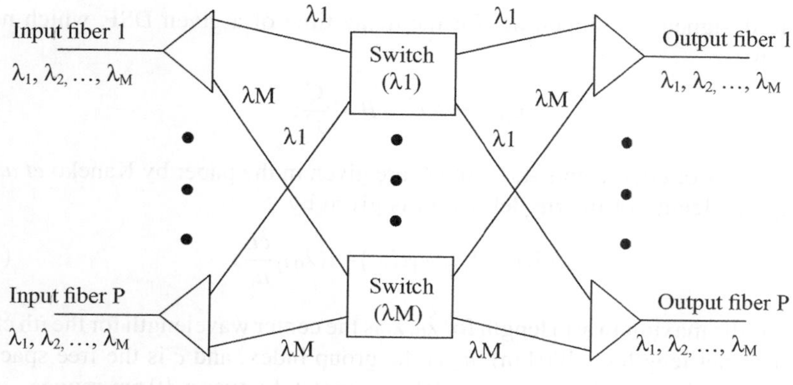

Figure 12.11 *Schematic of a reconfigurable optical cross-connect.*

switches between the demux and mux stages. All signals of the same wavelength are directed to the same switch. The switched signals are then directed to muxes which are connected to output fibers.

Networks using such switches provide additional control in setting up connections. The routing matrix can be reconfigured electronically on demand. These routers, also known as *wavelength-selective cross-connect* (WSXC) or *active switches*, are used in wide-area wavelength-routed networks. It may be made more versatile if a wavelength conversion is made just before entering the mux stage. They are then known as wavelength-interchangeable cross-connects (WIXCs).

The function of a reconfigurable wavelength-routing switch, alternately named the *wavelength-selective cross-connect* or *optical cross-connect*, has been briefly mentioned. It performs the same function that electronic switches perform in telephone networks. By using dynamic routing, a limited number of wavelengths can be used with the provision of wavelength reuse. The functional diagram of a reconfigurable WR is given in Figure 12.11.

12.5.6 Dispersion Equalizer

The transmission distance of optical fiber communications has been increased enormously with the development of erbium-doped fiber amplifier (EDFA). The maximum repeater span is now determined by the chromatic dispersion of the fiber. The use of dispersion-shifted fibers having minimum dispersion at 1.55 μm has reduced considerably the effect of dispersion, and the main contribution to dispersion comes from the second-order effects. There are different methods to compensate this effect, for example using a short length of fiber of opposite dispersion or using a Bragg grating. However, the method cannot meet the conflicting demand of wide bandwidth and large delay compensation capability. An elegant method to circumvent this problem by the use of AWGs is reported by Kaneko *et al.* [25]. Two AWGs are located such that they cross each other's slab waveguides and both of them have the same grating parameters (channel spacing = 25 GHz and FSR = 3200 GHz). The AWGs are connected to multiple delay arms much in the same way as AWG1 and AWG2 are connected to TO switches in ADMs. The delay arm lengths are designed to compensate for the delay time of a given DSF, which may be expressed as

$$\tau(\lambda) = A\lambda^2 + B + \frac{C}{\lambda^2}, \qquad (12.26)$$

where the values of coef cients A, B, and C are given in the paper by Kaneko *et al.* [25]. The waveguide length of the ith delay arm is given by

$$\Delta l(\lambda_i) = l_{\max} - [\tau(\lambda_i) - \tau(\lambda_0)]\frac{cL}{n_g}, \qquad (12.27)$$

where $l_{\max}$ is the maximum arm length for λ_0, λ_i is the center wavelength for the ith channel and L is the ber length (=400 km), n_g is the group index, and c is the free space light velocity. The device could compensate delay completely over a 10 nm range.

12.6 PHASAR-Based Devices on Different Materials

PHASAR-based devices, in particular AWGs, have been fabricated by using different materials systems that have been introduced in Section 12.1. We discuss in this section the characteristics of the devices grown on different materials systems.

12.6.1 Silica on Silicon

As already mentioned, devices realized with this materials system excessively outnumber the devices grown on other materials. The principal reason lies in the fact that a few large companies as well as several other groups use silica waveguides for their device fabrication [26–29]. All the devices and components discussed in this chapter have been realized using silica on silicon, and the athermal and polarization-insensitive characteristics are also demonstrated. A number of review articles have appeared in the literature describing the basic principles, device fabrication, characteristics, and applications of PHASAR-based devices, and the reader is referred to the articles and the long list of references given therein.

The characteristics of the devices have been discussed. The devices show polarization insensitivity when a half-wavelength polyimide plate is inserted in the guide regions. Furthermore, athermal behavior is ensured by the use of silicone resin in narrow optical grooves created in the arrayed waveguide region. A modi ed form of the groove to reduce the excess loss due to triangular design, the triangular groove is replaced by arrow-shaped stripes spaced at regular intervals. In another method, the strain-optic effect produced by a bimetallic stip below the guide compensates for the thermal effect.

One of the important issues in the design of AWGs is the spectral response, which is Gaussian in nature. In order to allow concatenation of many such devices and to suppress the loss deviation due to wavelength uctuation of light sources and AWGs, at and wide passband characteristics are required. A number of methods have been proposed to achieve the desired response curve. These include the *Y*-branch, sinc-functional eld, multimode interference, two-focal-point method, double-pass method, parabolic taper method, and uniform-loss cyclic frequency method [11].

The maximum number of channels gradually increased from 128 to 256 and then 400, and the frequency separation between the channels became reduced. A 1010-channel tandem

AWG lter that has only 10 GHz spacing has been reported. It is constructed by cascading a 1 THz spaced at-top 1×10 AWG and ten 10 GHz spaced 1×160 AWG secondary lters [30].

The reader is referred to the article by Janz [13] for details of AWG and other devices grown on the Si platform.

12.6.2 Silicon-on-Insulator

AWGs for use in WDM systems have been developed by using SOI as the guide material. The number of channels is, however, quite low; only four- and eight-channel devices have been reported in the literature [31]. An interesting feature of the waveguides is their weak polarization dependence. A TE–TM shift of only 0.4 nm was observed in the transmission characteristics of AWGs. Unlike silica, SOI lms do not have intrinsic stress, and this is believed to be the reason for polarization insensitivity.

12.6.3 Silicon Oxynitride

Reports on the growth and characterization of PHASAR-based devices using SiON as the guiding layer are not plentiful in the literature. de Ridder *et al.* [32] described the growth of waveguides and of some simpler devices. An ADM was fabricated by his group cascading a few asymmetric MZIs. Report on work on AWG comes from Schauwecker *et al.* [33]. They fabricated a ve-channel AWG with 5 nm channel spacing and obtained a chip insertion loss of 5 dB and a cross-talk level of −30 dB at 725 nm wavelength. Further advances in planar waveguide devices including ADMs are reviewed in the work of Worhoff *et al.* [34].

12.7 Echelle Grating

Diffractions gratings have been in use over decades to separate light into its different component wavelengths. In principle, gratings can be used as demultiplexers in WDM systems. A re ection grating has been used for demultiplexing WDM signals.

Due to dif culty in fabricating planar diffraction gratings, AWGs have so far enjoyed a monopoly in the WDM demultiplexing scenario. Three fundamental challenges in the fabrication of planar diffraction gratings must be overcome before these devices could be employed on a large scale in present DWDM communications systems. These are (1) fabrication of vertical grating facets in a waveguide structure, (2) reduction of polarization dependence of the grating diffraction ef ciency, and (3) elimination or compensation of the birefringence of the device.

Blaze is a tilt on the grooves of a re ection grating to concentrate the light at a higher angle, and therefore in orders $M > 0$. Thus light is concentrated in the dispersed orders. An echelle grating is an extreme example of a blazed grating. The blaze angle $\theta_B = (i + d)/2$ is set very high (to the order of 65°), where i and d are, respectively, the angles of incidence and diffraction.

In recent years, echelle gratings (EGs) have proven to be promising devices for demultiplexing DWDM light and potential competitors to the more advanced AWG devices. The EGs have some inherent advantages [13].

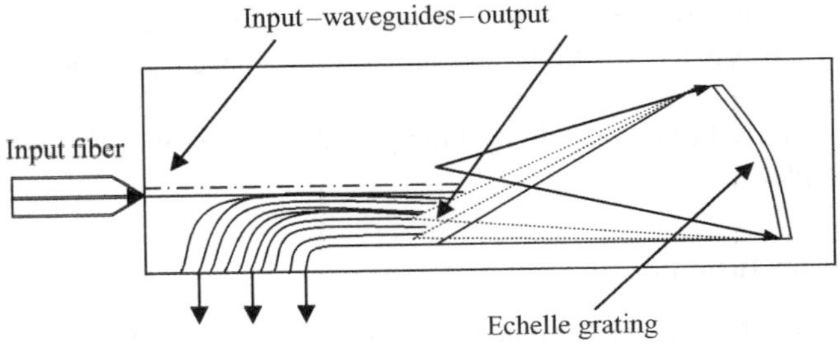

Figure 12.12 *Schematic diagram of an Echelle grating.*

The AWGs are usually realized with shallow-etched ridge waveguides. Grating-based devices, on the other hand, require deeply etched grating facets. The size of the phased array, however, is much larger than the grating demultiplexers. The number of waveguides in the AWG is much smaller than the number of teeth in the grating, and therefore the nesse is much smaller in AWGs. This limits the number of channels available over its free spectral range. The EC grating shown in Figure 12.12 has a folded beam path and does not need an array. Its size is several times smaller than the AWG for the same number of channels [13].

Problems

12.1 Derive an expression for the free spectral range in terms of frequency from Eq. (12.6).

12.2 Work out the steps to arrive at Eqs. (12.8) and (12.9) from Eq. (12.5).

12.3 The path length difference in a 32-channel AWG is 63 μm. Calculate the order of diffraction at 1.55 μm, assuming a value of 1.5 for the RI of the core. Also calculate the values of the free spectral range and channel spacings in nm and GHz.

12.4 Calculate the number of channels in an AWG when the arc length $f = 36.3$ mm and waveguide separations are 30 μm. Assume a value of 1.5 for the effective index of the slab.

12.5 A glass lm grown on Si substrate is under biaxial stress in the wafer. The stress-induced changes in RI for light polarized along the wafer plane is $\Delta n_{TE} = -(C_{11} + C_{12})\sigma$ and normal to the plane is $\Delta n_{TM} = -2C_{12}\sigma$., where C's are components of sress-optic tensors and σ is the in-plane stress. Calculate the change in RI given $C_{11} = 0.065 \times 10^{-11}$/Pa, $C_{12} = 0.452 \times 10^{-11}$/Pa, and $\sigma = 300$ MPa.

12.6 Using the de nition of overlap as given by Eq. (10.80), and following the steps that lead to Eq. (10.89), the overlap for misaligned guides, derive Eq. (12.15).

12.7 Derive Eq. (12.18).

References

1. Pennings, E., Khoe, G.D., Smit, M.K., Staring, T. *et al.* (1996) Integrated optic versus micro-optic devices for ber optic telecommunication systems: a comparion. *JSTQE*, **2**(2), 151–164, ref. 9, sec.1.

2. Smit, M.K. (1988) New focussing and dispersive planar component based on an optical phase array. *Electron. Lett.*, **24**(7), 385–386.
3. Dragone, C. (1989) Ef cient N×N star couplers using Fourier optics. *J. Lightwave Technol.*, **7**(3), 479–489.
4. Vellekoop, A.R. and Smit, M.K. (1989) Low-loss planar optical polarization splitter with small dimensions. *Electron. Lett.*, **25**(15), 946–947.
5. Takahashi, H., Suzuki, S., Katoh, K., and Nishi, I. (1990) Arrayed waveguide grating for wavelength division multi/demultiplexer with nanometer resolution. *Electron. Lett.*, **26**, 87–88.
6. Dragone, C. (1991) An N×N optical multiplexer using a planar arrangement of two star couplers. *IEEE Photon. Technol. Lett.*, **3**, 812–815.
7. Smit, M.K. and van Dam, C. (1996) PHASAR-based WDM-devices: principles, design and applications. *IEEE J. Sel. Top. Quantum Electron.*, **2**(2), 236–250.
8. Agrawal, G.P. (2002) *Fiber Optic Communication Systems*, 3rd edn, John Wiley & Sons, Inc., Hoboken, NJ.
9. Ramaswami, R. and Sivarajan, K.N. (2002) *Optical Networks: A Practical Perspectives*, Elsevier, Amsterdam.
10. Dutton, H.J.R. (1998) *Understanding Optical Communication*, Prentice Hall PTR, Upper Saddle River, NJ.
11. Okamoto, K. (2000) *Fundamentals of Optical Waveguides*, Academic Press, San Diego, CA.
12. Okamoto, K. (1999) Recent progress of integrated optical planar lightwave circuits. *Opt. Quant. Electron.*, **31**, 107–129.
13. Janz, S. (2004) Si based waveguide technology for wavelength division multiplexing, in *Si Photonics*(eds L. Pavesiand D.J. Lockwood), Springer, Berlin, pp. 323–361.
14. Okamoto, K. and Sugita, A. (1996) Flat spectral response arrayed waveguide grating multiplexer with parabolic waveguide horns. *Electron. Lett.*, **32**, 1661–1662.
15. Okamoto, K. and Yamada, H. (1995) Arrayed waveguide grating multiplexer with at frequency response. *Opt. Lett.*, **20**, 43–45.
16. Takada, K., Inoue, Y., Yamada, H., and Horiguchi, M. (1994) Measurement of phase error distributions in silica-based arrayed-waveguide grating multiplexers by using Fourirt transform spectroscopy. *Electron. Lett.*, **30**, 1671–1672.
17. Suzuki, S., Sumida, S., Inoue, Y. *et al.* (1997) Polarization insensitive arrayed waveguide gratings using dopant-rich silica based glass with thermal expansion adjusted to Si substrate. *Electron. Lett.*, **33**, 1173–1174.
18. Takahashi, H., Hibino, Y., and Nishi, I. (1992) Polarization insensitive arrayed waveguide grating wavelength multiplexer on silicon. *Opt. Lett.*, **17**, 499–501.
19. Takahashi, H., Hibino, Y., Ohmori, Y., and Kawachi, M. (1993) Polarization insensitive arrayed-waveguide wavelength multiplexer with birefringence compensating lm. *IEEE Photon. Technol. Lett.*, **5**, 707–709.
20. Spiekman, L.H. *et al.* (1996) Design and realization of polarization independent phased array wavelength demultiplexers using different array orders for TE and TM. *J. Lightwave Technol.*, **14**, 991–995.
21. Inoue, Y., Kaneko, A., Hanawa, F. *et al.* (1997) Athermal silica-based arrayed-waveguide grating multiplexer. *Electron. Lett.*, **33**, 1945–1946.
22. Mukherjee, B. (2000) WDM optical communication networks: progress and challenges. *IEEE J. Sel. Areas Commun.*, **18**, 1810–1824.
23. Zirngibl, M., Joyner, C.H., Doerr, C.R. *et al.* (1996) An 18-channel multifrequency laser. *IEEE Photon. Technol. Lett.*, **8**, 870–872.
24. Ikushima, I., Himi, S., Hamaguchi, T. *et al.* (1995) High-performance compact optical WDM transceiver module for passive double star subscriber systems. *J. Lightwave Technol.*, **13**, 517–524.

25. Kaneko, A., Goh, T., Yamada, H. *et al.* (1999) Design and application of Si based passive lightwave circuits. *IEEE J. Sel. Top. Quant. Electron.*, **5**, 1227–1236.
26. Li, Y.P. and Henry, C.H. (1996) Silica based optical integrated circuits. *IEE Proc. Optoelectron.*, **143**, 263–280.
27. Kawachi, M. (1996) Recent progress in silica-based planar lightwave circuits on silicon. *IEE Proc. – Optoelectron.*, **143**, 257–262.
28. Miya, T. (2000) Silica-based planar lightwave circuits: passive and thermally active devices. *IEEE J. Sel. Top. Quant. Electron.*, **6**, 38–45.
29. Hibino, Y. (2002) Recent advances in high-density and large-scale AWG multi/demultiplexers with higher index-contrast silica-based PLCs. *IEEE J. Sel. Top. Quant. Electron.*, **8**, 1090–1101.
30. Takada, K., Abe, M., Shibata, T., and Okamoto, K. (2002) A 25-GHz-spaced 1080-channel tandem multi/demultiplexer covering the S-, C-, and L-bands using arrayed-waveguide grating with Gaussian passbands as primary lter. *IEEE Photon. Technol. Lett.*, **14**, 648–650.
31. Jalali, B., Yegnanarayanan, S., Yoon, T. *et al.* (1998) Advances in silicon-on-insulator optoelectronics. *IEEE J. Sel. Top. Quantum Electron.*, **4**, 938–947.
32. de Ridder, R.M., Worhoff, K., Driessen, A. *et al.* (1998) Silicon oxynitride planar waveguiding structures for application in optical communication. *IEEE J. Sel. Top. Quantum Electron.*, **4**(6), 930–937.
33. Schauwecker, B., Przyrembel, G., Kuhlow, B., and Radehaus, C. (2000) Small size silicon oxynitride AWG demultiplexer operating around 725nm. *IEEE Photon. Technol. Lett.*, **12**(12), 1645–1646.
34. Worhoff, K., Lamdeck, P.V., Klunder, D.J.W. *et al.* (2002) Planar waveguide devices on a silicon compatible technology platform, in *Integrated Optoelectronics: Proc. 1st Intl. Symp* (eds M.J. Deen, D. Misra, and J. Ruzyllo), Electrochemical Society, Pennington, NJ, pp. 279–312.

13

Fabrication Techniques and Materials Systems

13.1 Introduction

Many planar lightwave circuits using optical waveguides and other devices are made by using silicon as the substrate. It is to be noted that planar and channel guide integrated optics are both analogous to very large-scale integrated (VLSI) electronics. In both circuits and systems, the main aim is to fabricate a large number of miniaturized interconnected devices on a single substrate. The processing of planar lightguides is very similar to the fabrication of VLSI devices, and both methods employ similar equipment. There are a number of advantages to using silicon. For example, the material silicon is obtained in large wafers, having diameters as large as 300 mm in 2005. Silicon is cheap, and thanks to decades of research, development, and manufacturing of microelectronics technology, Si-based lightwave circuits possess the same advantages, such as low cost, high reliability, and ability to be batch-fabricated. Also, the developed photonic devices may be readily integrated with electronic components grown on the same substrate.

In this chapter, we discuss some of the commonly used methods for fabrication of optical waveguides in silicon. Since the number of devices formed is quite large, the fabrication methods are widely different. We shall outline the most commonly used methods and briefly describe some of the special methods.

As noted already, techniques to fabricate Si-based lightwave circuits rely heavily on present-day electronics technology. We shall start from the growth process for the basic material and wafers needed for fabrication. Brief descriptions of epitaxy, doping methods like diffusion and ion implantation, photolithography, and etching and metallization processes will first be given. Basically, two different structures are employed to grow waveguides: silica-on-silicon and silicon-on-insulator (SOI). The methods of growing these multilayers will then be examined. The fabrication of the waveguide structures will be described next.

Silicon Photonics: Fundamentals and Devices, First Edition. M. Jamal Deen and P. K. Basu.

Other materials and structures employed, in addition to silica and SOI, are either variants of these two, or slightly different. This chapter will also describe the corresponding fabrication processes. At the end, a comparison of different materials systems used to fabricate lightwave circuits will be made.

The technology described in this chapter is also used to grow active devices, like light sources, detectors, and modulators. Although optical fibers and fiber-based devices are not described in this book, some of the technologies used in fiber production are also employed in producing silica-based planar lightwave circuits. We shall include a discussion of such methods and related instruments.

13.2 Planar Processing

The starting point in planar processing is a suitably oriented silicon wafer, polished flat and cleaned, to be used as the substrate. Thin layers of the same or different materials are then deposited on this substrate, and the desired pattern is then delineated on the grown layer. The deposited layer may be either amorphous or crystalline, and, depending on its nature, different techniques are employed to grow them. In an alternative scheme, the surface property of the substrate can be modified by diffusion or ion implantation of different impurities. To define a pattern on the top layer, different etching techniques are employed. The pattern definition in a specified region on the surface may be accomplished by a process known as *lithography*, which itself involves the deposition and removal of materials.

The dimensions involved in most optoelectronic devices are of the order of microns; nearly all the processing steps are carried out in a clean room, in which filtered air is circulated to remove dust particles. The rooms are kept at constant temperature and humidity to ensure process reproducibility and reliability. Similarly, the services to the room (e.g., water and gases) are also highly purified and workers inside wear special clothing to keep the contaminants to the minimal level. These precautions greatly improve the yield and reliability of the devices.

After completion of the process steps, the devices are packaged and sealed and electrical contacts are made to different terminals of the manufactured devices.

13.3 Substrate Growth and Preparation

The typical semiconducting substrate material is either silicon or compound semiconductors like GaAs or InP. These are crystalline materials in the form of large cylindrical ingots grown through a phase change from one of the three phases (vapor, liquid or amorphous solid) into a crystalline solid phase. The liquid-to-crystalline solid-phase transition is the most common method. In this case, a small seed crystal of the material to be grown is allowed to touch the surface of the melt of the material and is pulled slowly out of the melt, as a result of which each successive plane of atoms attaches to the surface of the seed crystal in the correct relative position and orientation. We illustrate this method of growth by the diagram shown in Figure 13.1, in which a Si single crystal is grown by the Czochralsky method.

In the Czochralsky method, the silicon boule is reshaped after recovery from the crystal puller. The ends are removed by sawing together with flawed regions, and the remainder is put to the shape of a cylindrical ingot on a lathe. By using X-ray diffraction, the crystal axes

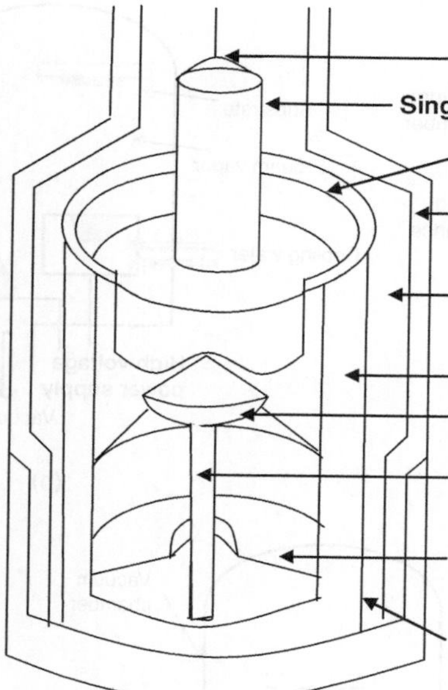

Figure 13.1 *The Czochralsky method of growing single-crystal silicon.*

are identified and orientational flats are ground on the surface so as to identify the axes at later times. The boule is then sawed into thin wafers by using diamond-impregnated saw blades. After separation the wafers are lapped flat, and then chemically etched to a depth of about 10 μm (or more for GaAs) to remove the crystalline defects introduced by cutting and shaping operations. Finally, one or both surfaces are polished to a mirror finish.

13.3.1 Deposition and Growth of Materials

13.3.1.1 Deposition

Vacuum evaporation: Several methods exist for depositing insulating or metallic films on another material or substrate. The simplest method is vacuum evaporation in which both the substrate and coating materials are placed in a vacuum chamber, separated from each other. The coating material is then heated to its melting point so that it evaporates and sticks to the surface of the substrate material. The purpose of the vacuum environment is to avoid a collision of evaporating atoms with unwanted atoms of gas. Molecules of the evaporating atoms travel at high speeds, but the distance they travel depends on the pressure of the environment in which they are traveling. The mean free path, which is the average distance the evaporating molecule travels between collisions, is an important parameter in the design of vacuum deposition systems. Typically, the lower the pressure in the deposition system, the better it would be for film deposition and the purer the film would be.

Several methods may be employed to melt the evaporant. In the resistive heating method shown in Figure 13.2a, the evaporant is placed in a metallic heater (e.g., W) formed in the shape of a coil or in the form of a boat (e.g., Mo or graphite). By sending direct current (DC)

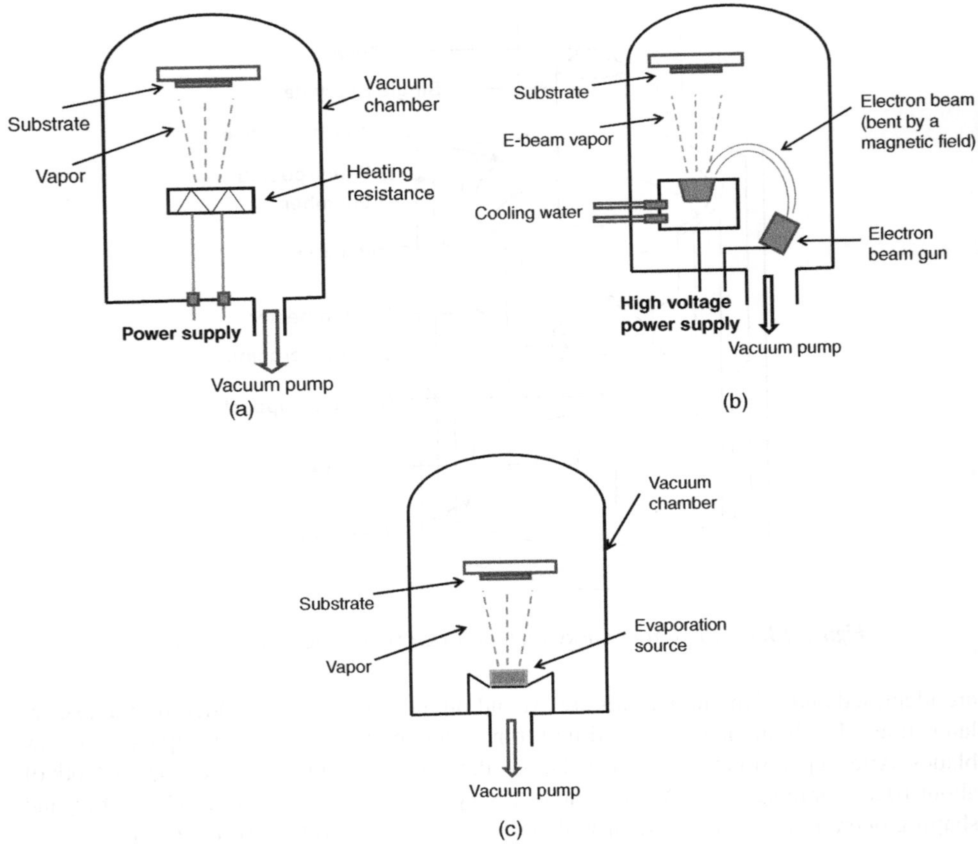

Figure 13.2 *Deposition using (a) resistive heating; and (b) e-beam heating. (c) Schematic diagram of a vacuum evaporation system.*

through the coil or boat, heat energy is provided to the evaporant. Many metals are successfully evaporated by using resistive heating.

Example 13.1

Consider that a film is deposited using vacuum evaporation in which the source is regarded as a point source. If the substrate is a smooth glass slide 10×10 cm, what should be the source to substrate separation if the metal film is to be uniformly thick to within 5%?

Here, we should consider that the point source is exactly below the middle of the substrate. From the center of the substrate, the furthest point is half a diagonal away. This distance $l = (10\sqrt{2})/2 = 5\sqrt{2}$ cm. Now, assuming that the distance from the point source to the center of the substrate is D, it can be shown that the thickness distribution t normalized to the thickness at the center of the substrate $t(0)$ can be given by

$$\frac{t}{t(0)} = \frac{1}{\left(1 + \frac{l^2}{D^2}\right)^{3/2}}$$

Since we want the film thickness uniformity to be within 5%, then $\frac{t}{t(0)} = 0.95$. Substituting for $l = 5\sqrt{2}$ cm and then calculating D result in $D = \sim 38$ cm. That is, the point source should be ~40 cm below the center of the glass substrate.

Electron beam evaporation: For some metals, Ti for example, the high melting point and reactivity at high melting temperatures lead to the failure of the filament or contamination of the film by impurities. Also some insulators have very high melting points and low thermal conductivity. In all such situations, an alternative method of heating, the electron beam (e-beam) heating, is preferred. A schematic representation of an e-beam system is shown in Figure 13.2b.

Here, a hot-wire filament acts as a thermionic electron emitter. The emitted electrons are accelerated by a charged aperture along the vertical direction, the field performing additionally as a collimator of the e-beam. A magnetic field applied perpendicular to the direction of the e-beam allows one to bend the beam in the form of an arc and to fall on the water-cooled crucible containing the evaporant. The material is now heated by the kinetic energy of the e-beam, rather than by direct contact with the filament in resistive heating, thus eliminating the problems mentioned in the paragraph above.

A schematic diagram of a vacuum evaporator is given in Figure 13.2c. The enclosure is a glass bell or metallic jar, evacuated down to a pressure of about 10^{-6} torr. The boat or coil is screwed to the metal posts through which DC current is passed. The sample is held face down over the evaporant with the help of a support frame and a sample holder. The atoms of the evaporant strike the surfaces of both the sample and a monitor quartz crystal. The coating changes the resonant frequency of the crystal connected to an oscillator circuit, from which the thickness of the evaporated film may be determined.

Sputtering: An entirely different method of coating, both on metals and insulators, is sputter deposition. Here the substrate is placed much closer to the coating material, which is usually a large sheet known as a *target*, as shown in Figure 13.3. Instead of working at high

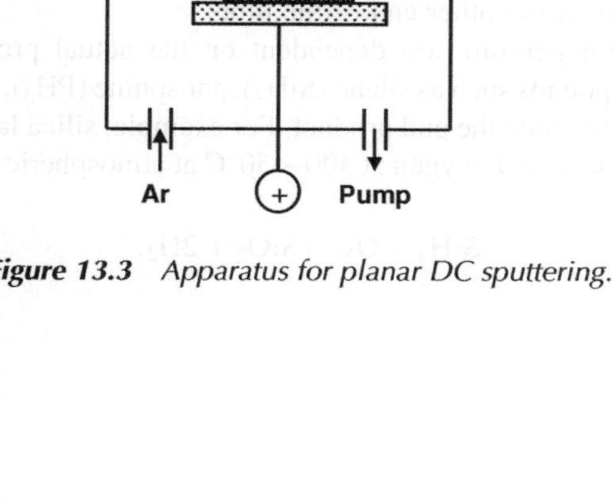

Figure 13.3 *Apparatus for planar DC sputtering.*

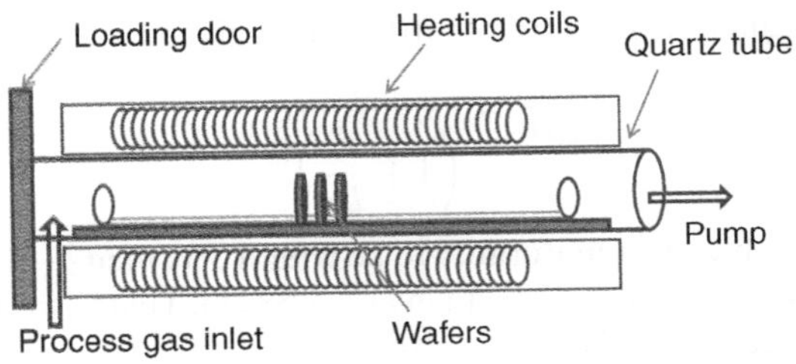

Figure 13.4 *Schematic drawing of a horizontal-tube hot-wall reactor, consisting of a quartz tube with an external resistive heater.*

vacuum, an inert gas like Ar is injected into the chamber at a pressure of 10^{-3}–10^{-2} torr. By using a radiofrequency (RF) electric field, a plasma combination of ions and electrons is created in the space between target and substrate. The target is made the cathode so that Ar^+ ions strike it and liberate atoms from it. The atoms are then attracted by the substrate, the anode, and are deposited onto the surface. The ejection of atoms from the target is known as *sputtering*. The sputtered atoms move randomly in all directions, but a sizable fraction of the atoms adheres to the substrate. Also, the sputtered atoms suffer random collisions with Ar ions and strike the surface from all directions. Thus the technique can be used to coat nonplanar substrates. The Ar atoms are inert, and so no further chemical reaction takes place.

Instead of radiofrequency, a DC field may be used to induce sputtering. However, if the target is a dielectric, then there is a build-up of surface charge and the process stops. As a result, DC sputter deposition is suitable for metal targets. The substrate is constantly bombarded by electrons and is thereby heated up, unless water cooling is used. A permanent magnet may be used to bend the trajectory of electrons, disallowing them to strike the substrate. The sputtering rate is dependent on the target material and is usually low ($\approx$0.01 μm/min for glassy materials). As the thickness of guiding layer is a few microns, it takes quite some time to deposit the required thickness of the material. Nevertheless, it is one of the first techniques used to grow successfully step-index planar waveguides using SiO_2 and Al_2O_3 films.

Chemical vapor deposition (*CVD*): In this process a hot mixture of gas is passed over the substrate in a furnace. A chemical reaction takes place on the surface, and consequently a film of correct composition is gradually deposited on the surface. Figure 13.4 shows a horizontal-tube hot-wall reactor, consisting of a quartz tube with an external resistive heater. The substrate is placed on a quartz support loaded into the tube with a removable end cap. The process gases are introduced into the tube from one end, are heated, and then complete the reaction on the heated surface. The by-products of the reaction and the unused gases are removed to the exhaust at the other end of the tube.

The pressure and temperature are dependent on the actual process. In many cases hydrogen-related compounds such as silane (SiH_4), phosphine (PH_3), and arsine (AsH_3) are thermally oxidized to produce the end product. For example, silica layers are deposited by heating a mixture of silane and oxygen at 400–450°C at atmospheric pressure according to the following reaction:

$$SiH_4 + O_2 \rightarrow SiO_2 + 2H_2. \qquad (13.1)$$

Similarly, silicon nitride may be produced from dichlorosilane and ammonia at about 750°C or from silane and ammonia at about 900°C following

$$3SiCl_2H_2 + 4NH_3 \rightarrow Si_3N_4 + 6HCl + 6H_2, \tag{13.2}$$

or

$$3SiH_4 + 4NH_3 \rightarrow Si_3N_4 + 12H_2. \tag{13.3}$$

By using the CVD process, it is possible to deposit doped silica glass used in planar waveguide circuits. The corresponding reactions are as follows:

$$SiH_4 + 2O_2 \rightarrow SiO_2 + 2H_2O. \tag{13.4}$$

$$2PH_3 + 4O_2 \rightarrow P_2O_5 + 3H_2O. \tag{13.5}$$

By changing the process gas flows, the index changes needed to form the planar guide structure may be achieved.

There exist a number of variations of the basic CVD process. For example, SiO_2 may also be produced by decomposing the metal-organic compound TEOS (tetraethyl orthosilicate, Si $(OC_2H_5)_4$) at a low pressure and a relatively high temperature (650–700°C). Low-pressure CVD (LPCVD) is used for this process. Another important CVD process, known as *metal-organic CVD* (MOCVD), is now commonly used to fabricate heterojunctions involving compound semiconductors, and it will be described later in this chapter.

In practice, the CVD process is complicated. Since the temperature, concentration, chemical composition, and velocity of gas mixtures can vary considerably along the length and diameter of the tube, it is difficult to ensure uniform film thickness and stoichiometry. Furthermore, the gases are toxic, explosive, or corrosive, so very special care is needed in constructing the furnace and the systems for delivering and exhausting the gases.

An electrically excited gas plasma, rather than heat, may supply the energy needed for decomposition of the gases. This variant in CVD is known as PECVD. Figure 13.5 shows a schematic diagram of a PECVD reactor. The substrate is placed on a heated susceptor, which serves as one of a pair of RF electrodes. The plasma is then created mostly by an inert gas and the process gases are introduced into the reactor. Silicon dioxide films may be grown by reacting silane and nitrous oxide in an argon plasma. PECVD processes are now used to fabricate waveguides on silicon substrate.

13.3.2 Epitaxial Growth

In all the processes discussed in Sections 13.3 and 13.3.1, the deposited films are amorphous in nature. A different process, named *epitaxy*, is needed to grow a crystalline film on a crystalline substrate. Such epitaxial layers are needed to fabricate light emitters and detectors as well as passive waveguide devices. If the grown material has the same chemical composition as the substrate, the process is known as *homoepitaxy*. In most of the compound semiconductor devices, both active and passive, the grown layer is of different composition than the substrate. The process is then heteroepitaxy. Heterojunctions made with $Ga_{1-x}Al_xAs$ and GaAs, or with $In_{1-x}Ga_xAs_{1-y}P_y$ and InP, are common examples of heteroepitaxial materials systems to grow lightguides. For Si-based systems, the heterojunctions are formed by using Si and its alloys, like $Si_{1-x}Ge_x$ and $Si_{1-x-y}Ge_xC_y$.

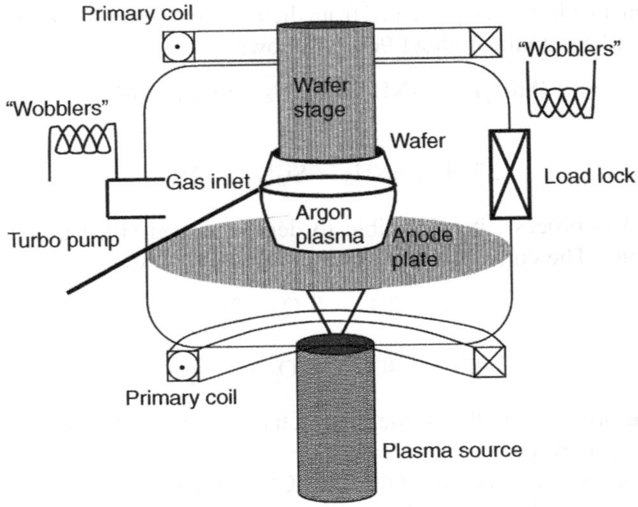

Figure 13.5 *Schematic drawing of a PECVD reactor.*

The epitaxial processes may be based on vapours or liquids, and epitaxial processes may be performed at temperatures well below the melting point of the substrate. Processes based on liquids are known as *liquid-phase epitaxy* (LPE). The growth is based on the principle that the melting point of a mixture is lower than that of their stoichiometric counterparts.

13.3.3 Molecular Beam Epitaxy (MBE)

Molecular beam epitaxy (MBE) is a process by which an ultrathin layer of a semiconductor may be grown on a substrate with very sharp or abrupt interfaces. The technique was first applied to grow heterostructures involving the ternary alloy $Al_xGa_{1-x}As$ on GaAs, or the quaternary alloy $In_{1-x}Ga_xAs_yP_{1-y}$ on InP, which are widely used to fabricate lasers, photodetectors, and other photonic components commonly used in long-distance communications. The technique has also been developed to grow Si-based heterostructures. In this subsection, we describe the basic principles involved in MBE.

In ultra-high vacuum (UHV) (10^{-9} mm Hg or even lower) a beam of atoms or, more generally, a beam of molecules is directed toward a crystalline substrate such that the atoms or molecules stick to the substrate's surface, forming a new layer of the deposited material. MBE not only deposits materials, as is done by conventional evaporation techniques, but also ensures the controlled epitaxial growth of a new layer using the very low rates of the impinging atoms, migration on the surface, and subsequent surface reactions. Simply stated, every atom reaching the surface of the heated substrate has enough time to migrate around and find its place to build up a new crystal lattice.

The typical construction of an MBE reactor is shown in Figure 13.6. At the bottom of the figure, there are the effusion cells to provide the molecular beam for either the bulk constituents or the dopants. These cells can be thermal evaporation cells (Knudsen cells), cells for gaseous media, or plasma sources. In front of each of the cells is a shutter, or a plate

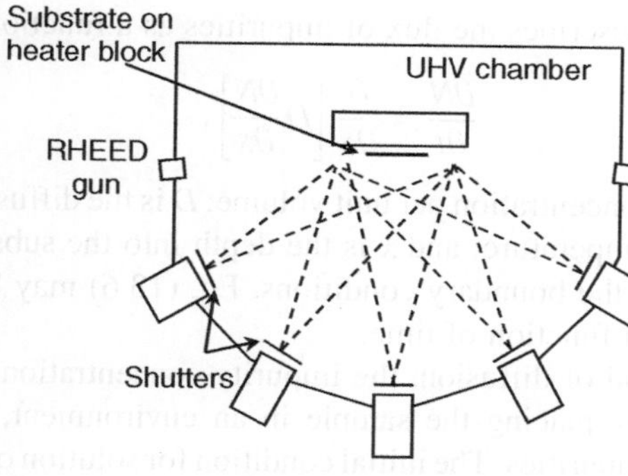

Figure 13.6 *Schematic diagram of an MBE reactor. The RHEED (reflection high energy electron diffraction) system is used for surface characterization during the MBE growth of crystalline layers.*

which could be brought into the beam for "switching" the beam on or off. Opposite to the cells and the shutters is the substrate, which is mounted on a substrate holder that may be heated and rotated.

The whole MBE system is in an ultra-high vacuum environment to guarantee the formation of a molecular "beam" which ensures purity and, therefore, the quality of the grown layer. Without an ultra-high vacuum, the atoms or molecules leaving the effusion cells will be scattered at residual gas molecules and not form a beam directed toward substrate. Often there are instruments for in situ analysis like reflection high-energy electron diffraction (RHEED) attached to the growth chamber. A cryopanel around the sample and the cells absorbs residual gases and provides a clean substrate environment.

13.4 Material Modification

There are different methods of modifying the properties of a semiconductor near its surface. The two widely used techniques are diffusion and ion implantation. We briefly describe the principles involved in these processes.

13.4.1 Diffusion

In this process, the substrate may be placed in intimate contact with the dopant material which may be a solid, liquid, or gas. Both the dopant and the substrate are heated to temperatures as high as 800–1000°C, because at higher temperatures the atoms in both of them are considerably more mobile than at room temperature. However, the substrate atoms are more fixed to lattice sites and the dopant atoms migrate to any of the vacancies existing in the substrate material. The dopant atoms may also occupy interstitial sites.

Diffusion of impurities occurs in all materials, whenever a concentration gradient of impurities and random thermal motion to drive the process exists. The following one-dimensional model describes the flux of impurities as a function of time and depth:

$$\frac{\partial N}{\partial t} = \frac{\partial}{\partial x}\left[D\frac{\partial N}{\partial x}\right], \tag{13.6}$$

where N is the impurity concentration per unit volume; D is the diffusion coefficient, which may depend on N and temperature; and x is the depth into the substrate, measured from the surface. By knowing the boundary conditions, Eq. (13.6) may be solved to yield the concentration profile as a function of time.

In one common method of diffusion, the impurity concentration at the surface is held constant. This is done by placing the sample in an environment, in which there is an inexhaustible supply of impurities. The initial condition for solution of the above equation is $N(x, t=0)=0$, for all x. The boundary conditions are

1. $N(x=0)=N_s$: the impurity concentration per unit volume at the surface; and
2. $N=0$, for large x.

The solution takes the form:

$$N(x,t) = N_s\text{erfc}\left[\frac{x}{2\sqrt{Dt}}\right]. \tag{13.7}$$

If the background or substrate doping concentration is N_B per unit volume, then the depth x_j to which the impurity atoms will diffuse is given by

$$x_j = \left(2\sqrt{Dt}\right)erfc^{-1}\left(\frac{N_B}{N_s}\right).$$

After depositing a fixed amount of impurities, Q atoms per unit surface area, onto the sample, it is removed to another chamber for processing at high temperature. In this process, called the *drive-in diffusion*, the solution becomes

$$N(x,t) = \frac{Q}{\sqrt{\pi Dt}}\exp\left[-\left\{\frac{x}{2\sqrt{Dt}}\right\}^2\right]. \tag{13.8}$$

In Eqs. (13.7) and (13.8), $\sqrt{Dt}$ is termed the characteristic length of the diffusion process.

Example 13.2

A boron (p-type dopant) predeposition is carried out at 950°C for 30 minutes. Using $D_P = 1.6 \times 10^{-3}$ μm²/hour, find the characteristic length of this process.

The characteristic length is given by $\sqrt{Dt}$. Therefore,

$$\sqrt{Dt} = \sqrt{\left(1.6\times10^{-3}\frac{\mu m^2}{hour}\right)\times 0.5\,hour}$$

$$= \sqrt{8\times10^{-4}\,\mu m^2} = 28\,nm$$

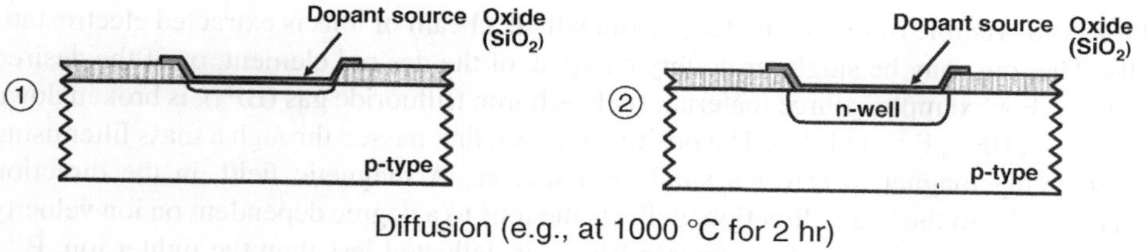

Figure 13.7 *Schematic representation of a diffusion process. (Left) Dopant deposition; and (right) diffusion of dopants into the p-substrate to create the n-well.*

Example 13.3

A boron predeposition is performed on a wafer with doping 5×10^{15} cm^{-3} for 30 minutes at 950°C. The boron surface concentration is 4.5×10^{20} cm^{-3} and $D_P = 1.6 \times 10^{-3}$ μm^2/hour. Find the surface concentration of predeposited boron atoms.

The total atoms per unit area Q, that is predeposited in time t_p, are given by

$$Q(t_p) = \frac{2N_s}{\sqrt{\pi}}\sqrt{D_p t_p}$$

where D_p and t_p are the diffusion coefficient and time of the predeposited species. Therefore

$$Q(t_p) = \frac{2N_s}{\sqrt{\pi}}\sqrt{D_p t_p} = \frac{2(4.5 \times 10^{20}\ cm^{-3})}{\sqrt{\pi}}\sqrt{(1.6 \times 10^{-11}\ cm^2/hr)0.5\ hr}$$

$$= 1.43 \times 10^{15}\ cm^{-2}$$

For small impurity concentrations, the change in refractive index (RI) due to impurities is proportional to the impurity concentration, so that

$$n - n_0 = KN, \tag{13.9}$$

where n_0 and n are, respectively, the RI in the absence and in the presence of impurities, and K is a proportionality constant. A schematic representation of the diffusion process to create an n-well in silicon is shown in Figure 13.7. After the dopant atoms are deposited on the surface, the substrate is heated to allow the n-type dopants to diffuse into the p-type substrate to crease the n-well.

An alternative to diffusion is ion exchange. This is suitable for fabrication of glass waveguides. Usually soda–lime glass is dipped into a hot silver nitrate melt. The Ag ions in the melt exchange positions with Na ions in the glass, and consequently the RI increases in regions of significant Ag concentrations.

13.4.2 Ion Implantation

As a final example of a technique for material modification, we may mention ion implantation, which is mainly used for doping semiconductors. The process is carried out in high vacuum, and the equipment needed is both sophisticated and expensive. The source of ions is a crucible of molten material from which a beam of ions is extracted electrostatically. The ions may be singly or doubly charged, of the desired element, or of the desired impurity. For example, source material, such as boron trifluoride gas (BF_3), is broken down into BF_2^+, BF^+, F^+, and B^+. The positive ions are first passed through a mass filter using an analyzing magnet to select a single ion species. A magnetic field, in the direction perpendicular to the beam direction, deflects the ions to a degree dependent on ion velocity and mass. The heavier ion, for example BF_2^+, is deflected less than the lighter ion, B^+. The ions are then accelerated by a high voltage (e.g., a few hundreds of kV) so that the final energy of the ions determines the depth of penetration into the target, that is, the silicon wafer. The process chamber houses the wafer to be implanted. Usually the beam is scanned relative to the wafer, in order to ensure uniform exposure to the ion beam. The beam may be stationary, and the wafer position is changed mechanically. Alternatively, the wafer is kept stationary, but the beam is scanned electrostatically. The ions suffer several collisions with the atoms of the target and finally stop. The substrate atoms are disturbed considerably so that an annealing step is needed after implantation.

The ion dose Q is the number of ions implanted per unit area (i.e., ions per cm^2) and is related to the beam current I and the time of implant t by

$$Q = \frac{It}{qnA}, \tag{13.10}$$

where q is the electronic charge, n is the charge state of the ion (e.g., for B^+, $n = 1$), and A is the area of the target.

Although each implanted ion impinges on the target with the same energy, the subsequent slowing-down process is randomly dependent on the interaction of the ions with the free electrons and the atoms in the target. The distribution of implanted ions is often described by a Gaussian distribution as follows:

$$N(x) = \frac{Q}{\sqrt{(2\pi)}\sigma} \exp\left\{\frac{-(x-r_p)^2}{2\sigma^2}\right\}, \tag{13.11}$$

where Q is the density of implanted ions per unit surface area, defined by $Q = \int_0^\infty N(x)dx$; r_p is the projected (or mean) range of the ions; and σ is the standard deviation of the projected range, also sometimes called the *straggle* (half width), and is a measure of the spread of the distribution. The peak concentration occurs at $x = r_\text{p}$ and is given by $N_\text{p} = N(r_\text{p}) = Q/(\sqrt{2\pi}\sigma) \approx 0.4\, Q/\sigma$. The projected range depends on several factors: the mass and charge of ion species, the substrate material, and the acceleration voltage. In practice the maximum practical voltage ($\approx$1 MV) limits the value of r_p to about 1 μm. The distribution of implanted ions for three different voltages is shown in Figure 13.8.

As shown in Figure 13.8, the penetration is deeper for higher accelerating voltage, but the distribution becomes broader and weaker at the same time. By suitably combining different implants with different acceleration voltages, doses, and other combinations, like diffusion and implantation, a suitable profile may be synthesized.

Collision of implanted ions with the host atoms, say Si, causes displacement of the host atoms from their lattice sites, creating vacancies and interstitials. As mentioned in this chapter, an annealing step (T > 900°C) will reconstruct the lattice to its crystalline condition, as well as place the dopant ions onto the lattice sites, thereby doping the

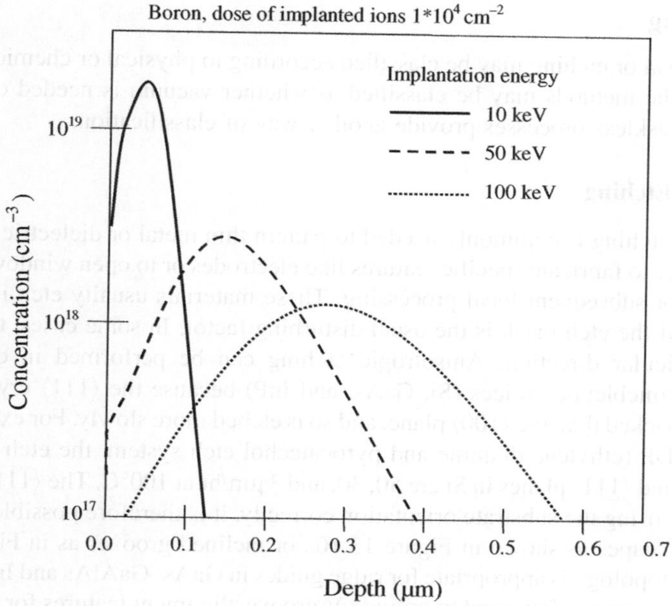

Figure 13.8 *Distribution of ions in Si caused by ion implantation for three different accelerating energies.*

material. The anneal also causes the dopant atoms to diffuse more into the substrate. This however, makes the dopant distribution broader. This *drive-in* process is done at a higher temperature (>1000°C) for many minutes. If it is required to retain the as-implanted profile, a *rapid thermal annealing* step for a few seconds places the dopants at the substitutional sites.

Ion implantation is quite attractive for direct fabrication of the waveguides. The use of ion implantation in producing SOI structures will be described in Section 13.7.2.1.

A schematic representation of the ion implantation process to create an *n*-well in silicon is shown in Figure 13.9. After ion implantation, the dopants should be activated using a high temperature, for example 800°C as shown. Note that the creation of the *n*-well by implantation results in less rounded edges compared to the diffusion, which is an isotropic process.

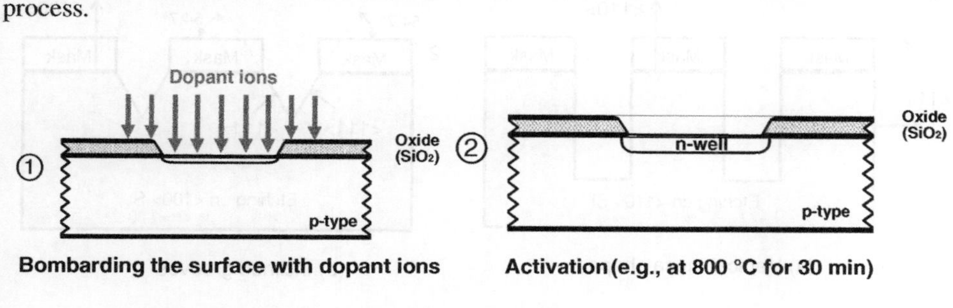

Figure 13.9 *Schematic representation of ion implantation process.*

13.5 Etching

Material removal or etching may be classified according to physical or chemical methods. Additionally, the methods may be classified as whether vacuum is needed or not. Also, masked and maskless processes provide another way of classification.

13.5.1 Wet Etching

Wet chemical etching is commonly needed to pattern thin metal or dielectric films on the substrate, either to fabricate specific features like electrodes or to open windows in surface mask layers for subsequent local processing. These materials usually etch isotropically; the undercut of the etch mask is the usual disturbing factor. In some cases, the etch may follow a particular direction. Anisotropic etching can be performed in crystals with diamond and zincblende lattices (Si, GaAs, and InP) because the $\langle 111 \rangle$ crystal plane is more closely packed than the $\langle 100 \rangle$ plane, and so is etched more slowly. For example, using the Water : EDP (ethylene diamine and pyrocatechol etch system, the etch rates for the $\langle 100 \rangle$, $\langle 110 \rangle$, and $\langle 111 \rangle$ planes in Si are 50, 30, and 3 μm/hr at 100°C. The $\langle 111 \rangle$ etch rate is slower. By choosing the substrate orientation correctly, it is therefore possible to fabricate vertical mesa shapes as shown in Figure 13.10a or inclined grooves as in Figure 13.10b.

The former topology is appropriate for ridge guides in GaAs–GaAlAs and InP–InGaAsP, while the latter is most often used to provide V-groove alignment features for optical fibers in Si. Suitable etch mixtures for a number of important amorphous and crystalline materials are summarized in Table 13.1.

13.5.2 Dry Etching

In modern microfabrication technology, wet etching is being replaced by different vacuum-etching techniques, all known as *dry-etching processes*. These processes require sophisticated

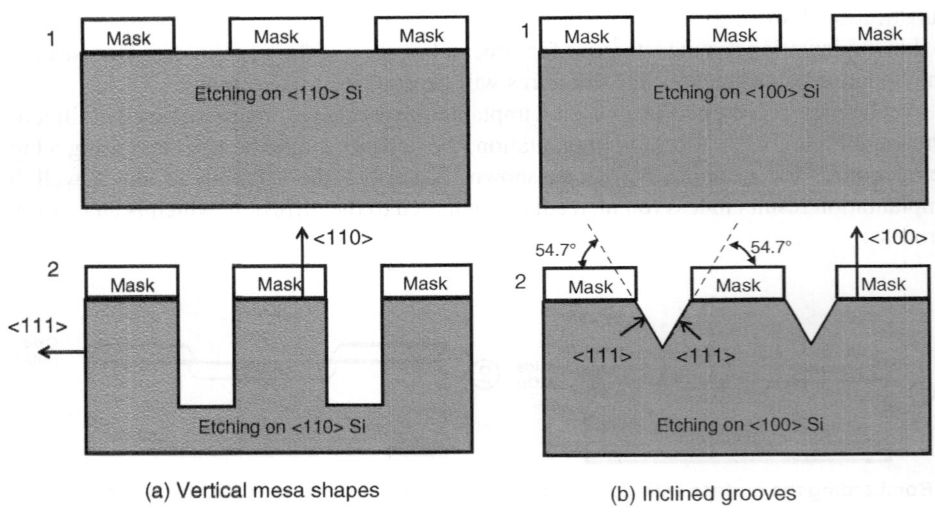

Figure 13.10 *Vertical mesa shapes and inclined grooves formed by wet etching.*

Table 13.1 *Wet etch mixtures for a variety of materials used in optoelectronic device fabrication*

Material	Etch system	Comments
Al	$HNO_3/CH_3COOH/H_3PO_4/H_2O$	Used for electrode patterning
Au	$KI/I_2/H_2O$	Used for electrode patterning
Ti	$HF/HNO_3/H_2O$	Used to pattern Ti prior to diffusion into $LiNbO_3$
SiO_2	$HF/H_2O/NH_4F$	Used to open windows in SiO_2 surface mask
GaAs	$H_2SO_4/H_2O_2/H_2O$	Used to etch rib guides on ⟨100⟩ substrates
InP	HCl/H_3PO_4	Has a negligible effect on InGaAsP
InGaAsP	$H_2SO_4/H_2O_2/H_2O$	Has a negligible effect on InP
Si	$KOH/H_2O/C_3H_7OH$	Used to etch V-grooves in ⟨100⟩ substrates

and costly equipment. However, the advantages of greatly improved process control and a high degree of selectivity and anisotropy in the etching outweigh the disadvantages.

The methods used in dry etching are purely physical or a combination of physical and chemical processes. The simplest physical method is sputtering, in which atoms are ejected from the surface by ion bombardment. RF sputter etching is a variation of RF sputter deposition, in which the electrodes are reversed. The substrate is now bombarded by positive ions rather than the target. This technique may be used for cleaning the surface of the substrate prior to deposition, but it may also be used as a method of etching. Since the ions strike the substrate from different angles, etching is roughly isotropic.

Another physical method is ion beam milling shown in Figure 13.11. This is similar to DC sputter etching (the electrode connections are now reversed in a DC sputter coater). A

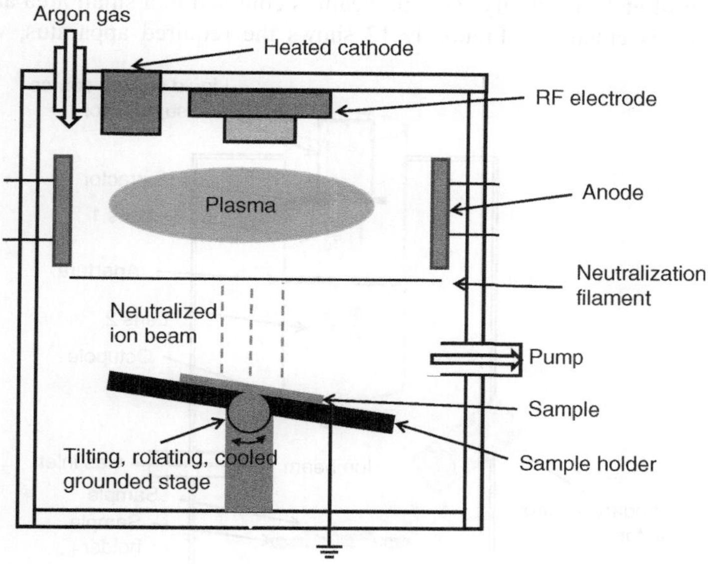

Figure 13.11 *Ion beam milling apparatus. Reproduced with permission from [1]. Copyright (1996) Institute of Physics.*

different arrangement is employed for plasma generation. The inert gas (e.g., Ar) is inserted into the far end of the evacuated chamber. The argon atoms are then ionized by electrons generated by a thermionic process. A magnetic field enhances the interaction between the electrons and the plasma. By using a biased grid, the Ar^+ ions are extracted from the plasma and are accelerated toward the target. The atoms are then sputtered off the substrate by this beam in the normal way. Use of a neutralizer filament, which itself is a thermionic emitter and injects electrons into the beam, prevents the accumulation of positive charge on the substrate. In contrast to RF sputter etching, the ions are incident from a defined direction, so that the etching is anisotropic. However, the phenomenon of ion bounce, in which ions recoil in random directions, erodes the walls of any deep feature. The substrate can be rotated to ensure even etching, and enhanced etch rates in most materials occur from machining at a tilt.

Some needs arise when a particular process is applied locally. For example, a small area on the surface may require sputter etching or ion milling. For this purpose, a layer of mask material is first deposited over the substrate and openings are made in the regions where etching is to be performed. Therefore, the mask material should have lower sputter yield than the substrate. Unfortunately, in most purely physical etching techniques, there is not much difference between the sputter yields of different materials. The edges are attacked even faster. This limits the depth of etched features, and prevents the fabrication of anything other than the sloping sidewalls. The fabrication of deep-etched features with walls accurately orthogonal to the substrate, as might be required in a waveguide mirror, is therefore extremely difficult.

13.5.3 Maskless Etching

In focused ion beam micromachining, the masks are not needed and it provides a solution to the problem of etch selectivity. The ion beam is confined to a small area and thereby local sputtering is enhanced. Figure 13.12 shows the required apparatus, which is a

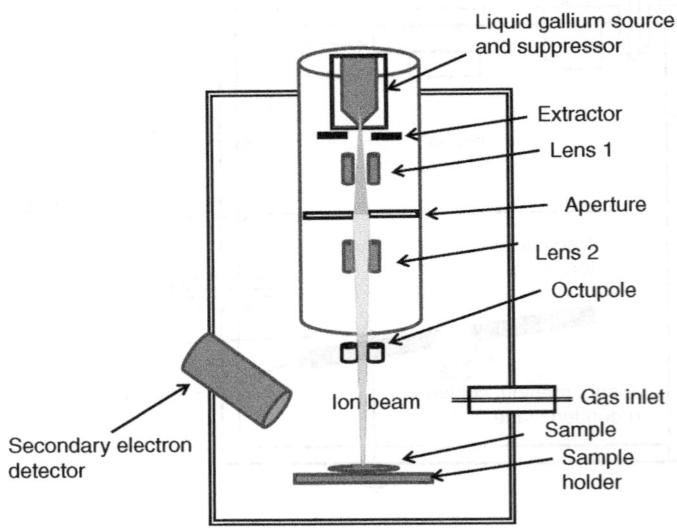

Figure 13.12 *Focused ion beam micromachining.*

modified version of a scanning ion microscope. Instead of having a plasma as the source of ions, a liquid-metal ion source (LIMS) acts as a high brightness near the point emitter. The ion source is a capillary tube filled with a metal of low melting point (e.g., In or Ga) surrounded by a resistive heater. A small aperture, raised to a negative potential, is placed just beyond the tip of the capillary. The metal is liquefied on heating, and the molten surface at the end of the capillary is distorted into a shape known as a *Taylor cone* due to the large electric field. At the apex of the cone the field is strong enough to overcome the surface forces and extract a large number of ions. The tip of the cone thus acts as a near-point ion source. The ions, after emerging, pass through the aperture as an expanding beam and refocus by electrostatic lenses to a point further down the column. Any portion of the substrate placed here will have a highly directed flux of ions and will eject atoms from the surface by sputtering. If the substrate is insulating, a metal layer connected to it leaks away the charges deposited on it.

By confining an area of the substrate that is exposed to the ions to a small spot, the sputtering is localized. The beam may also be deflected or blanked under computer control. The area to be machined is first identified by a suitable program, so that a pattern is etched into the substrate. This may be aligned to other features on the substrate prior to etching by using the equipment in an alternative ion microscope mode. In this case a smaller ion flux is used, and the beam is raster scanned across the entire substrate. Variations in the number of secondary electrons, also generated by the sputtering process, emitted during the scan are then used to build up a picture of the surface, so that the substrate can be moved to a desired coordinate origin by an x–y stage. In this alignment stage, the substrate is also eroded, albeit to a very small extent. Normally the machining is performed in a number of passes, using a raster scan of the addressed area. Due to the absence of an erodible mask, the etched features may be extremely deep (tens of microns) with near-vertical walls.

Focused ion beam micromachining may be used to cut very deep features. The technique has been used for the fabrication of a number of components in optoelectronics, such as mirrors. However, it suffers from two major disadvantages. First, being a serial process, the throughput is low and is therefore inappropriate for the definition of, say, ridge guides, where most of the substrate areas must be etched. Second, it shares the disadvantage of other physical processes; there is no distinction between the substrate and the sputtered material, which may be redeposited elsewhere on the surface.

13.5.4 Reactive Etching

Both the problems discussed in Section 13.5.3 are eliminated in reactive ion etching (RIE), a combined physical-chemical process requiring a mask layer for local feature definition. It is similar to sputter etching. However, the inert gas is now replaced by a reactive molecular gas. This gas is decomposed by the plasma, and the by-products react with the substrate material and form volatile compounds at low temperatures. The common etchant for silicon is CF_4 or CCl_2F_2. A schematic drawing of a basic RIE system is shown in Figure 13.13.

An analogous process based on ion milling with a reactive gas also exists; this is known as reactive ion beam etching (RIBE). In each case, the use of a mask restores the parallelism of the process, making the etching of large areas feasible. The selectivity between the mask and substrate is introduced by tailoring the etchant to attack mainly the latter. Etching is normally highly anisotropic, especially with crystalline substrates. This virtually eliminates

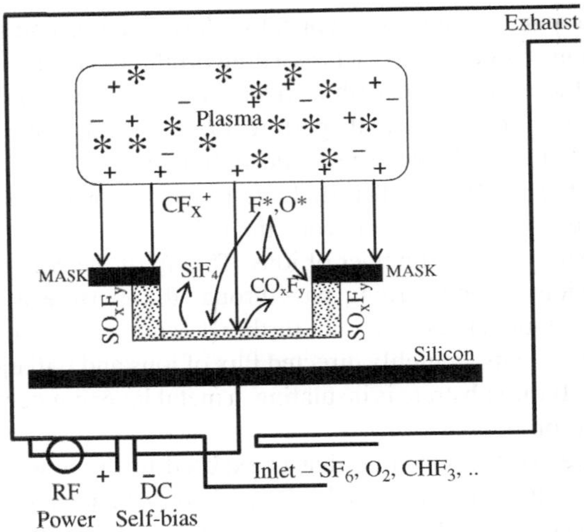

Figure 13.13 Basic reactive ion etching (RIE) system (after [1]). Note that *, + and – denote radicals, positive and negative ions.

mask erosion and undercut, allowing deep features to be etched easily. A final advantage of this process is that the etched material may be pumped away avoiding redeposition, although the pumping system must be specially protected against each gas.

13.6 Lithography

The method of forming an epitaxial layer has been described in Section 13.3.2. A planar waveguide structure may be realized by this process. However, to fabricate a 3D waveguide, the layer should be etched to form a rib or ridge structure. The width of the rib waveguide is primarily determined by the lithographic definitions. The same process is also needed to define patterns for other active and passive devices.

The process of pattern transfer is known as *lithography*. Basically it is a two-step process. In the first stage a mask, normally a glass plate containing an opaque pattern in chromium metal, is made. In the second stage the features on this master are copied onto the substrate by a form of printing.

13.6.1 Mask Fabrication

The mask fabrication process by using e-beam lithography is described here. First, the design dimensions of the device are specified, and its salient features are assigned to one of a number of mask levels.

13.6.2 Pattern Transfer

In the second stage, the pattern is transferred from the mask to the substrate using an optical printing step known as *photolithography*. It is based on the properties of light-sensitive organic materials or photoresists (PRs), whose behavior is qualitatively similar to the

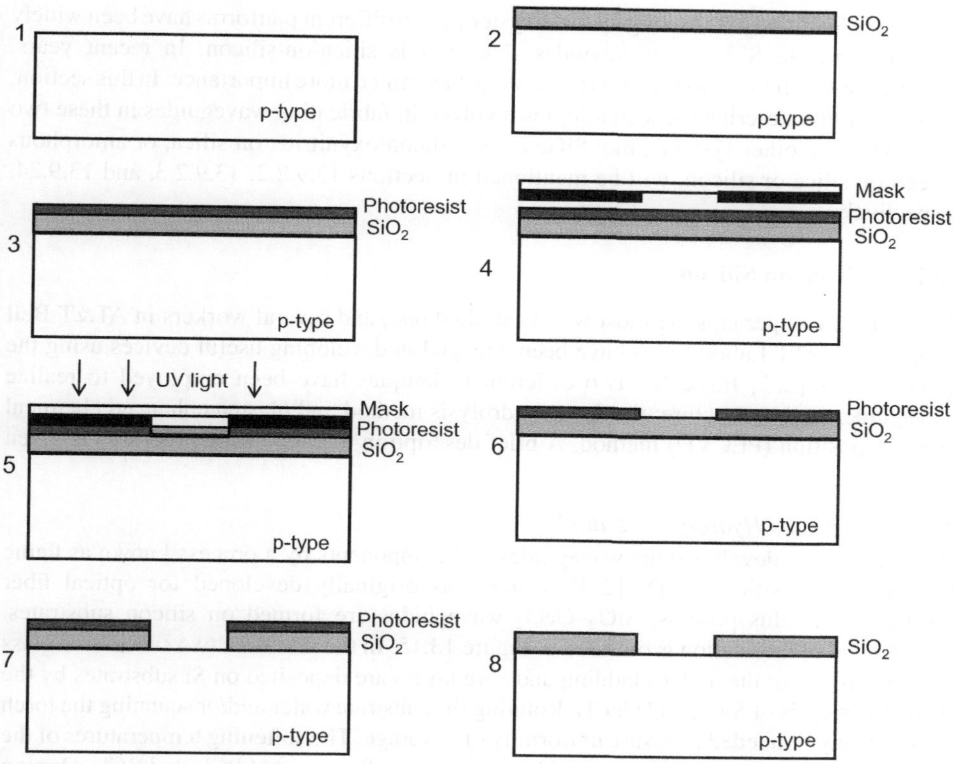

Figure 13.14 *Basic steps (eight) in photolithography for patterning silicon dioxide.*

electron resists that have been described. Both negative and positive optical resists exist. In the former, exposure to ultraviolet (UV) light polymerizes the material, and in the latter, cross-links are broken by the exposure. Figure 13.14 shows a typical process for subtractive pattern transfer using positive resist, with the mask. Exposure is carried out in a machine known as a *mask aligner*; this contains accurate translation stages, a high-power microscope, and a UV exposure source. First the relative position of the mask and substrate are carefully adjusted, so that the mask pattern is aligned with existing alignment marks and features on the substrate. The two are then clamped together by evacuating the space between them, and the resist is exposed by the flood beam. After removal from the mask aligner, the resist is developed. Negative resists generally require organic solvents, while alkaline aqueous solutions are used for positive resist. The result is a pattern of resist on the film layer; this may be transferred to the film itself by etching, followed by removal of the remaining PR.

13.7 Fabrication of Waveguides

A number of materials systems based on silicon have been used for realizing integrated optical waveguides. A description of the various materials systems used by researchers will be given in Section 13.9. Among all these systems, two different platforms have been widely used to fabricate Si-based waveguides. The first is silica-on-silicon. In recent years, however, SOI (silicon-on-insulator) technology has gained more importance. In this section, we shall briefly describe the technologies involved in fabricating waveguides in these two platforms. The other systems, like SiGe on Si, silicon oxynitride on silica, or amorphous silicon on silica or silicon, will be mentioned in Sections 13.9.2.2, 13.9.2.3, and 13.9.24, respectively.

13.7.1 Silica on Silicon

This structure is perhaps the most widely studied one, and several workers in AT&T Bell Labs and in NTT Laboratories have been engaged in developing useful devices using the grown layers [2, 3] Basically, two different techniques have been employed to realize waveguides in this structure: the flame hydrolysis method and plasma enhanced chemical vapor deposition (PECVD) method. A brief description of each of the processes is given below.

13.7.1.1 Flame Hydrolysis Method

The NTT group developed the waveguides and components by a process known as flame hydrolysis deposition (FHD) [2, 3], which was originally developed for optical fiber fabrication. In this process, SiO_2–GeO_2 waveguides are formed on silicon substrates. The method of fabrication is outlined in Figure 13.15. In the first step, two successive glass particle layers for the under-cladding and core layers are deposited on Si substrates by the flame hydrolysis of $SiCl_4$ and $GeCl_4$. Rotating the substrate wafer and/or scanning the torch position may be needed to ensure uniformity of coverage. The softening temperatures of the synthesized glass particles are reduced by adding a small amount of PCl_3 and BCl_3. Doping with Ti, Ge, P, N, B, and As may be done by including appropriate compounds containing these elements in the burned mixture. The next step is consolidation in which the silicon wafers with the deposited porous glass layers are heated to 1200–1350°C in an electric furnace. The core ridges of the waveguides are then defined by photolithography and RIE. The ridges are finally covered by a thick over-cladding layer of SiO_2 formed by the FHD

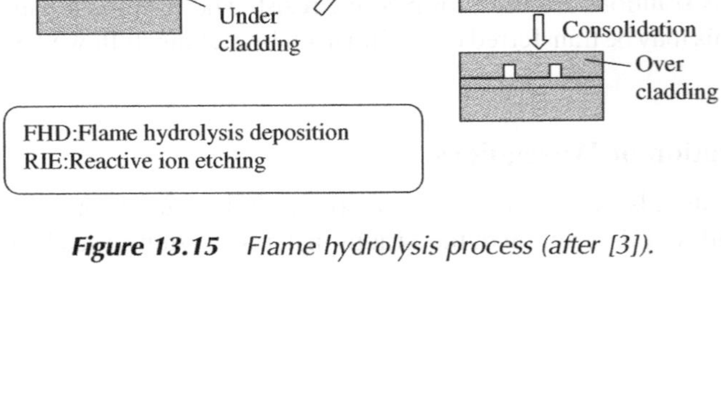

Figure 13.15 *Flame hydrolysis process (after [3]).*

process. In this method, the thickness of the films may be made over a wide range, and RI variation also may be achieved.

13.7.1.2 Plasma-Enhanced Chemical Vapor Deposition (PECVD)

The plasma-enhanced chemical vapor deposition (PECVD) technique has become more popular in recent times to grow silica layers on silicon [5]. The growth rate is several hundred nanometers per minute, with excellent control over the uniformity of RI and layer thickness.

In PECVD, a plasma is created between two parallel electrodes using a gas mixture of SiH_4, N_2O, and an inert carrier gas like He, Ar, or N_2. The substrate is placed on a heated susceptor, which serves as one of the pair of RF electrodes. The wafer itself is heated to a temperature of about 400°C. The gases react on the surface of silicon to form a layer of SiO_2. The RI of the glass and other material properties are modified by incorporating dopants during deposition by adding appropriate precursor gases to the gas mixture. By varying the ratio of the precursor gases, layers of silicon oxynitride (SiO_xN_y) with compositions and RIs ranging from SiO_2 ($n = 1.46$) to Si_3N_4 ($n = 2.05$) may be deposited. Silicon oxynitride has attracted considerable attention from scientists and engineers due to its high RI contrast with silica. PECVD is already a standard tool in the microelectronics industry, and commercial reactors intended for the semiconductor industry are easily adapted for manufacture of glass waveguides. PECVD reactors can also work on a large scale, as the number of wafers processed at a time to deposit silica layers may be tens or even hundreds.

The quality of as-deposited oxide by the PECVD process is not as good as those produced by the FHD process. The glass network in PECVD growth contains significant concentrations of H and N, thereby increasing the absorption. The N–H and Si–H bonds present give rise to strong absorption lines around 1480 and 1510 nm. These lines are particularly strong in silicon oxynitride (SiON). The broad absorption features increase the insertion loss and channel non-uniformity of the demultiplexers in the C band (1525–1565 nm). The glass network is not complete, and the voids and defects therein contribute to the scattering loss. With continuing progress, water is absorbed in the network and the material properties change with time. PECVD grown glass must therefore be annealed at high temperatures (about 1000°C) to release the trapped hydrogen in glass. The waveguides grown by PECVD, after annealing, show excellent optical properties comparable to those exhibited by guides grown with the FHD process.

13.7.2 Formation of Waveguides Using Silicon-on-Insulator

In recent years, a vast majority of silicon photonic devices have been being fabricated using the SOI platform. SOI structures consist of a film of single-crystalline Si separated by a layer of SiO_2 from the bulk substrate. The top silicon layer, which has a RI of approximately 3.5, acts as the guiding layer. The silica layer beneath the top silicon layer acts as the lower cladding layer. Silica has a RI = 1.46, and therefore mode confinement in silicon is better than in silica guides. Sometimes, another oxide layer of silica is grown on top of the silicon layer to convert the guide into a symmetric one.

In the following, we describe the commonly employed techniques to realize the SOI structures. Note that the same techniques are employed to grow SOI, which is then used to fabricate complementary metal oxide semiconductor (CMOS) devices on SOI. Fabrication, characterization, and devices have been described in [6].

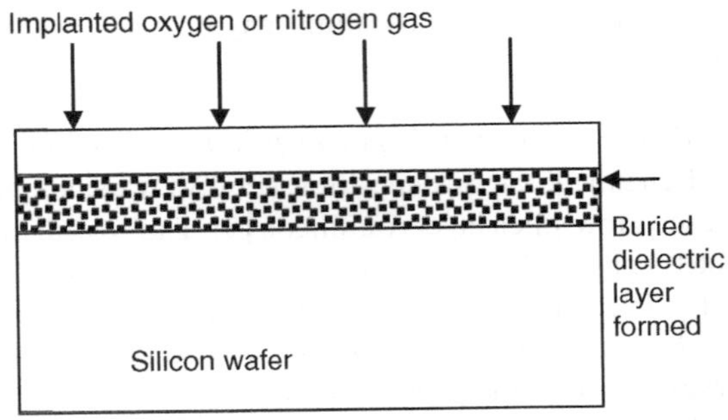

Figure 13.16 *Illustration of the SIMOX process.*

13.7.2.1 Separation by Implanted Oxygen (SIMOX) Process

The separation by implanted oxygen (SIMOX) process has been the most popular technique to fabricate a large volume of SOI material. This method is illustrated in Figure 13.16.

In essence, a large number of oxygen ions are introduced beneath a silicon layer by ion implantation. As stated above, implantation of a large concentration of ion species requires a large dose of ions. The total dose required in the SIMOX process is usually $>10^{18}\,cm^{-2}$. To prevent the growth of an amorphous silicon overlayer, the silicon substrate is maintained at a temperature of approximately 600°C during implantation. It might be of interest to note that to form the source and drain contacts in typical CMOS transistors, the dose is typically not greater than $10^{16}\,cm^{-2}$. The oxygen ions are implanted into crystalline silicon at an energy level of ~200 keV. This energy determines the depth of SiO_2 and the thickness of the Si overlayer. Also, as mentioned in Section 13.4.2, with low doses ($<10^{16}\,cm^{-2}$), the implanted dopant profile is approximately Gaussian. With increasing dose, the peak concentration of O^+ ions saturates and equals the concentration found in stoichiometric SiO_2. Further increase in implantation makes the oxygen profile flatter, forming a buried continuous layer of SiO_2. The next step, after implantation, is annealing at approximately 1300°C for several hours. This produces a uniform buried layer of SiO_2 having sharp interfaces with the upper and lower layers of silicon.

The silicon waveguide formed by ion implantation may have quite a large number of dislocations. However, the process technology has been improved vastly to decrease the number drastically to 10^2–$10^3\,cm^{-2}$. The propagation of electromagnetic waves depends on the dislocation density. Another detrimental effect is the interface roughness between the guiding and cladding layers, as discussed in Section 10.7.1.3.

The depth and thickness of the buried SiO_2 layer depend on the implantation energy. For a 200 keV implant, the thickness of the buried oxide and the silicon overlayer are about 0.5 and 0.2 μm, respectively. The thickness of the overlayer may be increased by several microns by epitaxial regrowth.

13.7.2.2 Bond and Etch-back Silicon-on-Insulator (BESOI) Process

The bond and etch-back SOI (BESOI) structure is a result of the strong bond developed between two hydrophilic surfaces (such as SiO_2) brought into intimate contact. The production of BESOI proceeds in three steps: (1) oxidation of two wafers to be bonded,

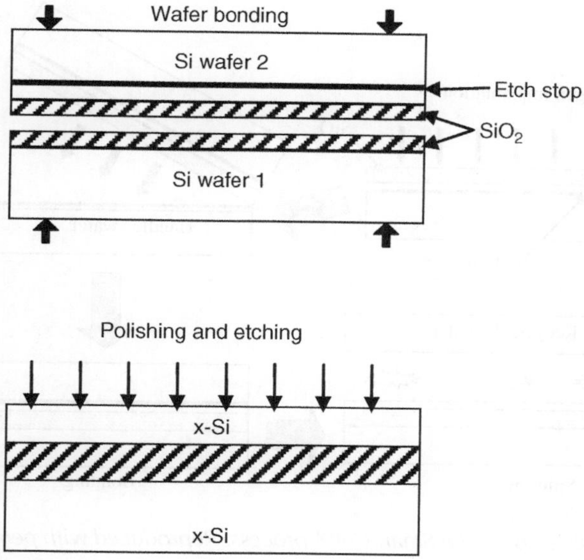

Figure 13.17 *A schematic representation of the BESOI process.*

(2) formation of the chemical bond, and (3) thinning of one of the wafers. Figure 13.17 illustrates the processes involved.

The two oxidized surfaces are brought into contact at room temperature when the initial bond is formed. Subsequent thermal processing at higher temperatures (>1100°C) enhances the bond strength.

Thinning of the wafer is usually accomplished by chemical mechanical polishing (CMP). The silicon surface to be polished is brought into contact with a rotating pad, in which a chemically reactive slurry containing the abrasive material, like alumina and glycerin, is put. The top silicon layer is thinned in this process to the desired thickness.

An improvement in the SOI thickness uniformity has been introduced by an etch stop technique described in [7]. In this process, one of the layers is doped heavily by acceptors before the bonding is made. After bonding, a highly selective etch that does not attack the p-type layer is used, and it removes the portion that is not p-doped. The etching eventually stops when the heavily doped p-layer is reached. Thus, the required depth of the silicon overlayer may be achieved.

13.7.2.3 Smart Cut Process

The smart-cut process is a promising newer addition [8] to the technology suite. The commercial version is known as the Smart Cut™ process. The four basic steps involved in the process using hydrogen as the implanted ions are shown in Figure 13.18. A "seed" wafer, from which a layer of Si is to be removed, is oxidized to the desired thickness. This oxide becomes the buried oxide (or BOX) after bonding. In the next step, hydrogen is implanted into silicon through the oxide with a typical dose $>5 \times 10^{16}\ \mathrm{cm}^{-2}$. After implantation, both the seed and handle wafers are carefully cleaned to eliminate any particle and surface contaminants and to make both surfaces hydrophilic. Wafer pairs are aligned and contacted

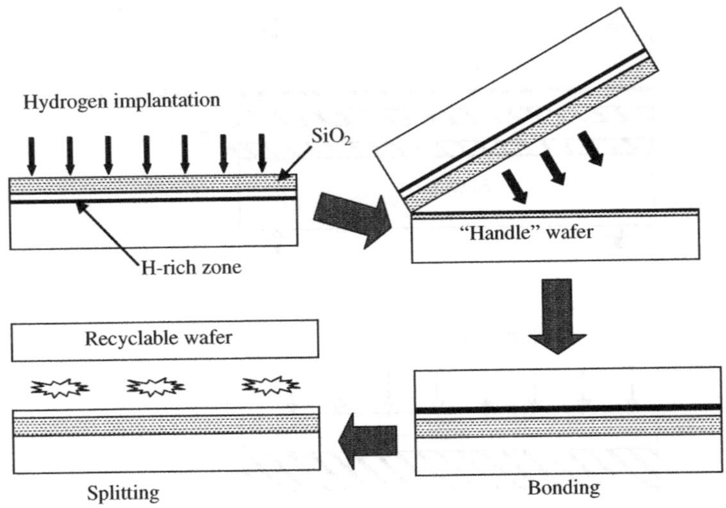

Figure 13.18 Basic steps in the Smart Cut™ process. Reproduced with permission from [16]. Copyright (1997) American Institute of Physics.

so that the fusion wave can propagate across the entire interface. Next, a batch of bonded wafer pairs is heated in a furnace to a temperature of 400–600°C. The wafers at this temperature split along the hydrogen implanted plane. The as-split wafer has a mean roughness of a few nanometers. A light touch polish is given to the handle wafer. The seed wafer is used again by repolishing its surface.

13.8 Grating Formation Process

A large volume of work has been performed in imprinting a good grating structure in the core of the optical fibers used for telecommunication. The methods are then applied in achieving grating structures in planar waveguide circuit (PLC) waveguides. As has been described, the grating formation means the creation of a RI modulation in the core. The amplitude of the RI change depends on the photosensitivity of the material under consideration. In this subsection, we shall first consider the available methods of control of photosensitivity in the core of the guide, mainly considering silica as the core material. After that, the techniques of imprinting the RI modulation will be discussed. Related activities on other materials like SiON or SOI will be discussed at the appropriate places.

13.8.1 Photosensitivity of Glass

The photosensitivity of germanium-doped silica fiber, when using argon ion laser emitting at 488 nm, was discovered and reported in [9]. In [10], it was found that a strong RI change occurs when a germanium-doped fiber is irradiated with UV light in the 240–250 nm wavelength range, around which an absorption peak due to a germania-related defect occurs. Since then, there have been considerable efforts in understanding and increasing the photosensitivity of fibers. Initially, it was thought that a core having a high concentration of germanium doping or a reduced oxidizing condition was necessary to increase the photosensitivity. Recently, other methods like hydrogen loading, flame brushing, and boron codoping have also been used for enhancing photosensitivity. In addition, exposure with a 193 nm ArF excimer laser has proven to be an efficient method of increasing photosensitivity.

In the hydrogen-loading scheme, hydrogen molecules are diffused into the core of the fiber at high pressures and temperatures. Ultraviolet irradiation after hydrogenation can produce an index change as high as 0.01. The method has also been used to increase the photosensitivity of planar silica waveguides (e.g., in [11]). In the flame-brushing technique, the region of the optical waveguide to be photosensitized is brushed repeatedly by a flame fueled with hydrogen and a small amount of oxygen [12]. At about 1700°C, hydrogen diffuses into the core and, by reacting with germanosilica glass, produces the germanium–oxygen defect center. As a result, a strong absorption band at 240 nm is created, rendering the core highly photosensitive. Boron, as a codopant in germanosilica fiber, has been found to increase the photosensitivity. The fiber with boron incorporation shows a higher photosensitivity than the fiber with higher germanium concentration but no boron doping.

In the three methods mentioned above, an enhancement of photosensitivity occurs due to absorption of UV light in the 240–250 nm band. Recently, it has been demonstrated that UV radiation at 193 nm from an ArF excimer laser can inscribe gratings that show stronger reflectivity than those inscribed by 240 nm radiation [13]. An advantage of this method lies in the fact that by using an intense ArF excimer laser light, a sizable UV-induced birefringence is produced, which compensates the natural waveguide birefringence of PLCs arising out of thermal expansion mismatch [14]. Though the core layers used in all the above-mentioned works are germanosilica glass, there are also reports on improved photosensitivity of P-doped glass [15]. Interested readers may find further details for fiber-based Bragg gratings in review articles [16, 17] and a book [18] that address this subject.

13.8.2 Grating Formation

Several methods are employed to imprint Bragg gratings in the core of the fiber or in a waveguide in PLCs. We shall briefly discuss two methods: the interferometric method and the phase mask technique.

In the interferometric technique first demonstrated in [10], an incoming UV light is split into two beams that are then allowed to recombine to form an interference pattern in the core of the guide. The resulting fringe pattern induces RI modulation in the core. One of the interferometric techniques, the amplitude splitting type, will be described here. In this method, the UV-writing laser light is split into two equal intensity beams that, after traversing unequal path lengths, recombine to form an interference fringe pattern on the core of the fiber. The scheme is illustrated in Figure 13.19. The interfering beams are focused on the fiber by using a cylindrical lens. The Bragg grating period Λ is equal to the period of the interference fringe pattern [16] and is given by

$$\Lambda = \frac{\lambda_w}{2 \sin \phi}, \qquad (13.12)$$

where λ_w is the UV laser wavelength and ϕ is the half angle between the two interfering UV beams. The grating period may be controlled by changing the laser wavelength or the angle ϕ.

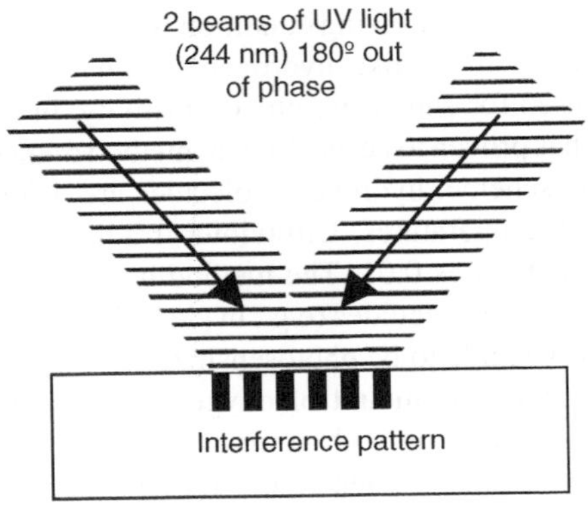

Figure 13.19 Amplitude splitting interferometric technique for the formation of Bragg grating. Reproduced with permission from [25]. Copyright (1996) IEEE.

In the phase mask technique, a phase mask is employed to spatially modulate the UV writing beam. Masks are produced holographically or by electron beam lithography. Phase masks are produced on a high-quality silica flat that is transparent to UV light. When UV light falls on it, the zero-order diffraction beam is suppressed, but the diffracted plus and minus first-order beams are maximized. A near-field fringe pattern is produced by these two first-order diffracted beams. The period of the fringe is half of that in the phase mask. Figure 13.20 shows schematically the phase mask method of writing the Bragg grating in the core of the fiber. KrF excimer lasers are usually employed to write the grating pattern. The same technique may be used to inscribe gratings on the planar waveguide.

The two processes described in this subsection may produce most of the patterns in integrated optics. However, the resolution of the copy process is limited by diffraction to approximately the wavelength of the incoherent UV exposure source. Therefore, these methods are not suitable for making reflection gratings, especially in semiconductor

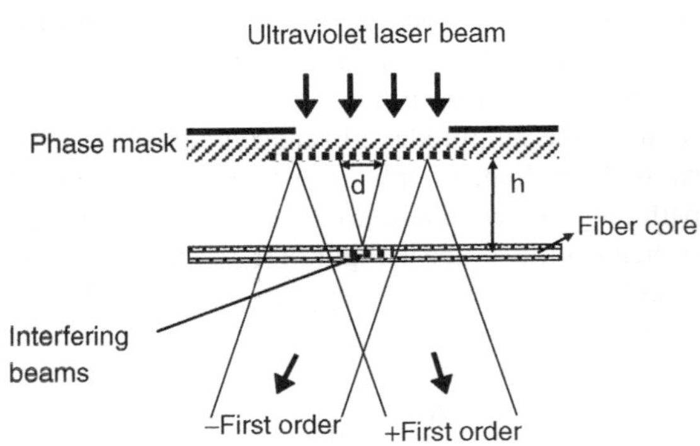

Figure 13.20 Schematic of the phase mask technique for writing Bragg gratings. The positive and negative first-order beams interfere at the core of the fiber placed at a distance h from the phase mask (after [16]).

materials. For this purpose, single-step processes are generally employed. There are two common techniques. In the first, the substrate is coated by e-resist onto which the grating pattern is directly written by e-beam lithography. After development, the pattern may be transferred to the substrate by anisotropic wet etching. A first-order grating for a DFB laser (period $\approx 0.23\ \mu m$) may be etched into InGaAsP in this way.

In the second method, a UV laser is used as a maskless exposure source for the photoresist. As noted earlier, the local distribution of irradiance resulting from two coherent plane waves that intersect at an angle is a periodic fringe pattern. This may be used to record a grating pattern directly in the PR, in a technique known as *holographic exposure*. In a particularly simple setup, which requires a single recording beam, a prism is used. The geometry is arranged such that the second beam necessary for the exposure is generated by total internal reflection of the input wave at the end face of the prism. Both beams then pass directly into the PR layer without reflection, since the prism is contacted onto the resist-coated substrate using an index-matching liquid.

13.9 Materials Systems for Waveguide Formation

Earlier chapters of this volume described the general principle of guiding light in a planar structure, and the earlier sections of the present chapter discussed the techniques used in forming planar waveguides and Bragg gratings. The basic requirement for having light guiding is to introduce a RI variation along one or two directions. Different techniques employed to bring about the changes have been outlined there. The other characteristics of the waveguide, that is, loss and various sources contributing to it, have been mentioned. The treatment in the earlier sections is general and applicable to most of the optical integrated circuits (OICs), irrespective of the substrate, be it $LiNbO_3$, III–V compounds, or Si.

In the present section, we shall consider waveguiding light in structures grown with Si as the substrate. After giving a general survey of the various methods, we shall consider specific materials systems consisting of Si and related materials that guide light. Then, the methods of fabrication of the waveguides will be introduced. The chapter is concluded by presenting a comparative study of the systems considered.

13.9.1 General Considerations

The structures considered in this section are thin or epitaxial layers grown on silicon substrates. Furthermore, the thin layers that serve as the guide and are considered in this work are all derived from silicon. It may be mentioned that layers of III–V compound semiconductors have been grown on Si, and active and passive planar lightwave circuits (PLCs) are realized by using these direct-gap materials. We shall not consider such PLCs in this section.

The different layers derived from Si are

1. silicon epilayers or doped Si,
2. SiGe and SiGeC alloy layers,
3. silica layers,
4. Si oxynitride layers, and
5. amorphous silicon layers.

Of these, the silica-on-silicon has been studied most extensively and the following discussion will emphasize the published work in this area. Recently, SOI structures are being widely studied by a number of workers. Depending on the RI of the guiding layer with respect to that of the cladding layer, one may expect weak or strong guiding. Each type of structure has its own merits and limitations. For example, since all these devices are to be coupled with the core of input and output optical fibers, a close match between the RIs of the fiber core and guiding layer is needed to reduce the coupling loss. Silica guides are well suited for this purpose. However, the RI step in such structures is quite small, so that the permissible minimum bend radius of the devices becomes large, leading to an increase in the overall size of the device. In many cases, a suitable compromise is made between several design parameters, or the problem is solved by proper engineering.

We shall, in the following subsections, consider the various guiding layers derived from and grown upon silicon substrates. In all these cases, a brief outline of the fabrication technique will be first given. The RI profile along the depth of the structure and horizontal direction will then be described. The loss characteristics of the waveguides are then be presented, and an account of research activity for achieving better performance will be given. Some basic planar devices like couplers, directional couplers (DCs), simple switches, and their characteristic features will be discussed in this section. More complex devices like arrayed waveguide gratings (AWGs) or reconfigurable switches are discussed briefly in Section 12.6.

13.9.2 Characteristics of Guides and Simple Planar Components

In this subsection, we shall discuss the processes of growing different waveguiding layers, mentioning the different characteristics of the waveguides, like material loss, coupling loss, and polarization sensitivity. After discussing the general properties, a brief description of the different passive components realized by the guides formed will be presented.

13.9.2.1 Silicon-on-Silicon

Growth Processes and Waveguide Formation. The basic requirement of optical application of any material is its transparency at $\lambda = 1.3$ and 1.55 μm. The absorption coefficient of Si (and other popular semiconductors), shown in Figure 8.3, indicate that the fundamental absorption edge occurs at ~1.2 μm. Therefore, the loss in these two wavelengths is quite small, less than 0.1 dB/cm, provided that the free carrier density is not appreciably large. When silicon is used as a waveguide, a RI change must occur in the vertical direction. The first IO waveguide in epitaxial Si was proposed and fabricated in [19] using a rib structure. The schematic diagram of the structure is shown in Figure 13.21 in which different dimensions and doping levels in the substrate and epilayer are indicated. The RI change in the vertical direction is achieved via a change in doping level (or free-carrier density) in the epilayer and the substrate.

As already discussed, free carriers can induce changes in absorption and also a change in RI, Δn, in bulk Si. From Eq. (4.113) one may notice that the RI decreases with increasing doping level. The epilayer used in the work mentioned was more lightly doped than the substrate in order to increase the RI and to confine light within it.

Epilayers with thicknesses in the range $7\,\mu\text{m} < H < 43\,\mu\text{m}$ with doping levels $\approx 10^{14}\,\text{cm}^{-3}$ using a higher doping level ($\approx 10^{18}$–$10^{19}\,\text{cm}^{-3}$) for the substrate were studied in [19]. The losses in the structure were due to free carrier absorption and evanescent field

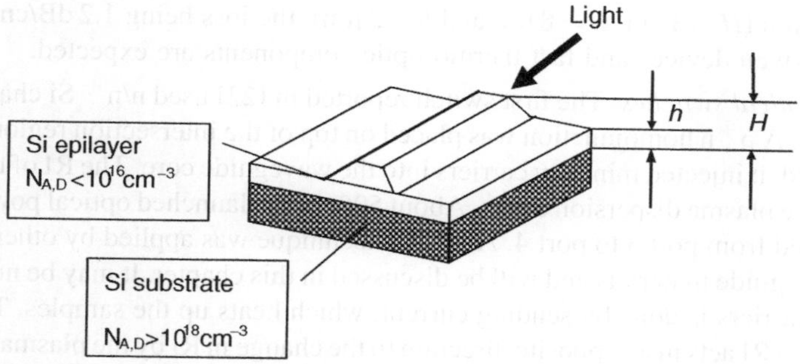

Figure 13.21 *Waveguide structure on epitaxial silicon (after [25]).*

in the substrate and were of high magnitude ≈5–13 dB/cm in slab waveguides and ≈15–20 dB/cm in rib waveguides. The substrate absorption is mainly responsible for the high loss. The remedy to this was suggested in [20], where it was shown that for a sufficiently thick waveguide layer H, the loss α due to fundamental mode is

$$\alpha = \frac{\lambda^3 \Delta\alpha}{8\pi H^3 \sqrt{(2n_0\Delta n)^3}}. \tag{13.13}$$

The symbols Δn and $\Delta\alpha$ denote, respectively, the RI change and the absorption coefficient, and both are proportional to the doping level in the layer under consideration. The expression points out the advantages of using a thick waveguide layer and a larger change in RI between the guiding and cladding layers. The authors used a single-mode guide in n/n$^+$ Si in which $H = 20\,\mu$m, $h = 13.4\,\mu$m, and $w = 12\,\mu$m (width at the top of the waveguide); the doping levels were $N_D = 2\times10^{15}\,\text{cm}^{-3}$ and $1\times10^{19}\,\text{cm}^{-3}$, respectively, in the epilayer and the substrate. The waveguide losses were 1.2 dB/cm and 1.5 dB/cm at $\lambda = 1.3$ and 1.55 μm, respectively, for both transverse electric (TE) and transverse magnetic (TM) polarizations. Further improvement may be expected with still higher substrate doping levels.

Equation (13.13) suggests that a thicker epilayer or a larger substrate doping may reduce the loss in the waveguide. The first alternative has already yielded the desired results. A higher substrate doping increases loss but improves radiation confinement also in a lossless epilayer. The authors in [21] were able to prove both theoretically and experimentally that under certain conditions the confinement effect prevails resulting in a global reduction of the propagation loss. They used a lightly doped substrate, on top of which was grown an ion-implanted (dose $5\times10^{15}\,\text{cm}^{-3}$ of As) cladding layer. It was followed by a CVD growth of the waveguide Si layer. Standard wet etching was employed to define the optical channel. The loss characteristics were compared with a non-implanted structure, and a 7 dB/cm improvement was observed for a width of 4 μm.

The advantages of this method lie in (a) an independent choice of substrate doping; (b) a relatively small thickness of epilayer, thus eliminating the time-consuming and expensive growth of a thick layer that is incompatible with standard microelectronics processing; and (c) a high thermal conductivity compared to the SOI waveguides. An optimized structure can be realized using a higher implantation dose ($10^{16}\,\text{cm}^{-2}$) for cladding layer and a larger rib cross-section ($H = 8\ \mu m$, $w = 8\ \mu m$ and $h = 2\ \mu m$), the loss being 1.2 dB/cm. A host of electrically driven devices and fast thermo-optic components are expected.

Devices. *Optical switches*: The first switch reported in [22] used n/n^+ Si channel guides crossing at 2°. A p^+n homojunction was placed on top of the intersection region and, when forward biased, it injected minority carriers into the waveguide core. The RI of the core was changed by the plasma dispersion effect. About 50% of the launched optical power into port 1 was switched from port 3 to port 4. A similar technique was applied by other workers to different waveguide materials and will be discussed in this chapter. It may be noted that the injection of carriers is done by sending current, which heats up the samples. The thermo-optic change in RI acts in an opposite direction to the change in RI by the plasma effect. Also of note is the fact that the electro-optic (EO) effect is absent in silicon. Therefore, EO modulators or switches cannot be realized in these structures.

A thermo-optical modulator was developed in [23] by using Si waveguides. The waveguiding planar structure was lightly doped epitaxial silicon layer grown on a heavily doped substrate with a lower RI. A cladding layer of poly-Si covers the whole structure and works as a heater when current flows in it. The guide structure was defined by anisotropic plasma etching, and the air–silicon surfaces acted as mirrors of a Fabry–Perot cavity. The maximum modulation depth reported was 60% with an electric energy of 0.2 μJ. The measured 3 dB BW was 700 kHz, and it is expected that it may provide useful functions in the fiber-in-the-local-loop application, where the data rate is not very high.

A tunable Fabry–Perot etalon filter has recently been fabricated on the silicon wafer [24] by depositing dielectric mirrors on a double-sided polished silicon wafer. The high-reflectivity mirrors were produced by depositing three $\lambda/4$ pairs of Si_3N_4 and SiO_2. The reflectivity of the mirrors was 66%, the FSR was 110 GHz, and the 3 dB BW of the filter was 14 GHz. The device was used in filtering the unwanted frequency components (chirp) of a directly modulated DFB laser, as well as relaxation oscillations at the edges of the modulating pulse.

13.9.2.2 SiGe and SiGeC Alloy Layers on Si

A review of the waveguiding action in these systems is given in [25], based on their own work in the area. This section will essentially report on their observations, and at the end, some recent results will be cited.

Refractive Index. The change in RI in Si–Si structure is only about 10^{-2}. An effective way to increase the RI of the epilayer is to incorporate Ge in Si. At $\lambda = 1.3\ \mu m$, the RI of pure Si is 3.5, whereas that for pure Ge is 4.3. If one assumes a linear relationship, the RI of a $Si_{1-x}Ge_x$ alloy with a small amount of Ge content x may be written as

$$n_{SiGe} = n_{Si} + x(n_{Ge} - n_{Si}) = 3.5 + x(0.8) \qquad (13.14)$$

This relationship applies to cases where a small amount of Ge is interdiffused into a Si layer.

In recent years, improved growth techniques such as MBE have given birth to a new kind of epitaxy on Si, the growth of a $Si_{1-x}Ge_x$ epilayer on Si substrate. Considering the lattice constants of Ge (5.43 Å) and of Si (5.14 Å), there is a 4% lattice mismatch, which is quite high. The growth of SiGe epilayers on Si necessarily introduces threading dislocations at the heterointerface. If, however, the epilayer thickness is kept below a critical value, the growth of a pseudomorphic layer is favored because of energy consideration. Since the lattice constants of the epilayer and the substrate are now equal in this growth, the epilayer is biaxially strained and its band structure becomes modified. The decrease of the band gap in a strained alloy layer occurs more rapidly with increasing Ge content than in an unstrained bulk SiGe alloy material. The experimentally found RI in a strained layer may be empirically expressed as

$$n_{SiGe} = n_{Si} + 0.3x + 0.32x^2. \tag{13.15}$$

The use of a SiGe strained layer with low x on Si gives rise to a weakly guiding structure. Large values of x are generally avoided because the reduced band gap of the alloy leads to a larger absorption at $\lambda = 1.3\ \mu m$ [26] and the value of x is kept below 0.2.

The waveguide structures using SiGe as the material are classified into two groups: (1) a grown $Si_{1-x}Ge_x$ heterostructure, and (2) an interdiffused SiGe channel waveguide.

Growth and Waveguide Formation. *SiGe–Si Heterostructures.* MBE or CVD has been employed so far to grow SiGe heterostructures on undoped or lightly doped Si substrates. To avoid the occurrence of dislocations, the grown layer thickness is kept below the critical layer thickness by using low Ge content. The layer thickness is $\approx 1\ \mu m$ for a Ge content $x < 10\%$ [27] and [28].

The authors in [29] used a MBE-grown strained $Si_{0.99}Ge_{0.01}$ layer with a rib thickness of $H = 3.45\ \mu m$. They used an anisotropic etch using buffered KOH solution to define the rib, after which the thickness of the SiGe layer remained up to $h = 2.15\ \mu m$ (see Figure 13.21 for notations). The width of the rib was varied in the range of $2\ \mu m < w < 12\ \mu m$. All waveguides were single moded along vertical direction, whereas single modedness was observed for a horizontal direction for $w < 9\ \mu m$ at $\lambda = 1.3\ \mu m$. The index enhancement was estimated to be $\Delta n = 3.8 \times 10^{-3}$ which agrees with Eq. (6.14) for $x = 0.01$. The waveguide losses were $\alpha = 3$–5 dB/cm at $\lambda = 1.3\ \mu m$. More recent measurements report a low value of less than 1 dB/cm.

In [30, 31], the authors used the CVD technique to grow strained and unstrained SiGe layers with x varying from 0.01 to 0.18 and thickness ranging from 1 to 10 μm. In their later study, the SiGe layer was covered by a Si layer to yield a buried guide. Their reported loss values are quite high, with the lowest value being 3.2 dB/cm for TE and 1.9 dB/cm for TM polarization. The 10 μm thick $Si_{0.9}Ge_{0.1}$ layer showed multimode operation. Drastically reduced losses have been achieved in [32], by using a buried waveguide with the top Si layer 3 μm thick and the guiding alloy layer, which is 6.5 μm thick, having only 1.2% of Ge. The authors used RIE to delineate the structures, and the loss was only 0.6 dB/cm. In addition, the waveguides offered low coupling loss to fibers.

Germanium Indiffused SiGe Waveguides. Germanium can simply be indiffused into commercially available Si by using standard technology like high vacuum deposition, liftoff processing, and diffusion. The process avoids expensive techniques like MBE and CVD. An evaporated alloy stripe of SiGe, obtained from simultaneous e-beam evaporation of Si and Ge, acts as the solid source for diffusion. The use of alloy increases the melting point from 937°C of pure Ge to about 1200°C. A SiO_2 overlay is sputtered on SiGe before diffusion. The RI profile of the indiffused channel may be expressed as follows:

$$n(x,y) = \frac{\Delta n_{max}}{2}\left[\operatorname{erf}\left(\frac{\frac{w}{2}-y}{d}\right)^2 + \operatorname{erf}\left(\frac{\frac{w}{2}+y}{d}\right)^2\right]\exp(x/d)^2. \tag{13.16}$$

Here Δn_{max} is the maximum deviation in RI; x and y represent, respectively, the vertical and horizontal directions; and $d = 2\surd(Dt)$, with D being the diffusion coefficient. Values of the maximum RI change depend on the Ge content, and typically $\Delta n_{max} \approx 10^{-1}$ are obtained.

The waveguide losses have been reduced to about 0.3 dB/cm [33] at both 1.3 and 1.55 μm, and the loss has been seen to be polarization independent. The horizontal and vertical spot sizes were measured and compared with a spot size of 9 μm of a single-mode fiber. The mismatch losses are estimated to be only 1 dB/facet. The corresponding figure for III–V waveguides are 6–10 dB/facet, indicating a higher coupling loss compared to indiffused waveguides in silicon.

In a more recent work [34], a method of fabrication of SiGe waveguides on Si using VLSI processing techniques was described. The SiGe layers were grown on Si by the UHV–CVD technique. As the fabrication is conducted at high temperatures, dislocation formation at the heterojunction seems unavoidable. The authors used a strain-compensating mask layer of SiO_2 and Si_3N_4 to help compensate the strain of the SiGe layer. By suitably choosing the thickness and deposition temperatures of the masking layer combination, the resulting strain was tailored so that it opposed the strain inherent in the SiGe layer during high-temperature oxidation. The strain-absorbing layer reduced the dislocation formation at high temperatures. They used local oxidation of silicon (LOCOS) processing to fabricate a dual-channel directional coupler that demultiplexed 1.3 and 1.55 μm wavelength signals. Recent progress in this area made by this group has been reviewed in [35].

Use of SiGeC. As mentioned already, the large lattice mismatch between Ge and Si is a major limitation in the growth of SiGe–Si heterostructures. The SiGe layer must be kept below a critical layer thickness to achieve high-quality pseudomorphic growth. It has been suggested and found experimentally that a small amount of substitutional C can counterbalance the strain caused by Ge [36]. The waveguiding property of $Si_{1-x-y}Ge_xC_y$ has been studied in [37]. Such waveguiding structure has been used in a SiGeC p–i–n waveguide photodetector working in the 1.3–1.55 μm wavelength range (see [38] for fabrication and other details).

Devices. In order to fabricate input–output (IO) components, the waveguides should be bent and the bend losses depend on the amount of Ge indiffused into Si. The bend losses have been investigated in [39] as a function of Ge content, and minimum values of bend radii were specified. The authors examined the performance of directional couplers (DCs) with Ge-indiffused guides in the above-cited work and also in [40] and observed nearly polarization-independent behavior at both 1.3 and 1.55 μm.

Modulators and switches have been realized by exploiting the plasma dispersion effect in a p–i–n diode Mach–Zehnder interferometer (MZI) having Ge-indiffused guides. The static modulator characteristic depended on the injected current. Its nonsinusoidal character indicated the presence of the thermo-optic (TO) effect also. Detailed measurements in [40], however, proved the role of the plasma effect also.

In [41], a fabricated SiGe–Si bifurcation optical active switch based on a plasma dispersion effect was described. The central rib waveguide supports two transverse modes coupled with single-moded input and output waveguides of width w. The index step in the vertical direction is obtained by a SiGe layer grown on Si. A n^+ p region located at the center of the coupling region injects carriers. When a fundamental mode A propagates to the central guide of width $2w$, through input guide 1, it is turned into fundamental mode B and at the same time excites a first-order mode C. The difference between the propagation constants of the two modes is a constant $\Delta\beta$. When $\Delta\beta L_c = \pi$, where L_c is the physical length of the coupling section, the modes B and C are out of phase in branch 3, and the whole power appears at arm 4. By injecting carriers, an additional difference of propagation constant $\Delta\beta'$ may be induced and when $(\Delta\beta' + \Delta\beta)L_c = 2\pi$, the light will pass through port 3.

Details of the fabrication process are given in [41]. The device reached the switching at 36 mA of injection current. The cross talk was −11 dB, and the insertion loss was 3.8 dB. The measured response time (fall time of the pulse from 90% to 10%) was <100 ns.

13.9.2.3 Silicon Oxynitride

Silicon oxynitride (SiO_xN_y, or SiON for short) is an interesting material that has been increasingly applied in various optical integrated circuits (OICs) over the first decade of the 21st century. The material has low absorption loss in the visible and near-infrared regions. The RI can be changed over a wide range from 1.45 (SiO_2) to 2.0 (Si_3N_4). This large flexibility in choosing the RI favors its use in OICs. In addition, Si substrates' standard Si integrated circuit (IC) processing steps can be used, thus leading to low-cost mass production.

Growth of Layers and Waveguides. Basically, two methods have been employed to deposit thin films of SiON on SiO_2. They are (a) plasma-enhanced chemical vapor deposition (PECVD) and (b) low-pressure CVD (LPCVD). [42]. PECVD layers are deposited from 2% SiH_4–N_2, N_2O, and NH_3 using a PECVD machine, while LPCVD layers are grown from SiH_2Cl_2, O_2, and NH_3. The RI and deposition rate of the films depend on the flow ratio of N_2O–SiH_4 in PECVD and on O_2 flow in the LPCVD process. From these two processes, a wide variation of the RI is achieved. For both types of grown layers, the optical loss in slab waveguide is below 0.2 dB/cm for the visible wavelength range, but it increases considerably at 1.55 μm. The enhancement is due to absorption by vibrational overtones of N–H and Si–H bonds. The value has been decreased by post-deposition annealing and the value for low-index PECVD SiON has been reduced to less than 0.2 dB/cm at 1.55 μm [43–45].

In order to fabricate waveguide structures for communication devices with low channel, coupling, and bend losses and polarization insensitivity, a simulation study was conducted in [42]. The optimized structure is shown in fig. 2 of [42]. A birefringence compensating layer (BCL) is to be introduced in the structure for compensating polarization dependence introduced by material birefringence. The core in the structure is PECVD-grown SiON, the BCL was LPCVD-grown Si_3N_4 with thickness of the BCL $d = 26$ nm, and the cladding layer was PECVD SiO_2. In order to reduce the mismatch loss due to fiber-to-chip coupling, the waveguide was tapered horizontally, predicting a loss of 0.2 dB/facet.

The actual structure is grown on a thermally oxidized ⟨100⟩ Si wafer. The BCL film was grown using LPCVD, and the cladding layers were grown by PECVD. Details of the fabrication process may be found in [42]. The same group and the collaborators reported the growth of SiON-based waveguides and devices like arrayed waveguide gratings (AWGs) by using similar techniques [46, 47]. In [48], the authors also deposited SiON core, and SiO_2 upper and lower cladding layers, by using the PECVD process with silane, ammonia, nitrous oxide, helium, and nitrogen. RIE was used to define the waveguides. And, in [49], the growth of a SiON waveguide–based AWG demux operating around 725 nm was reported.

Very recently, the authors in [50] reported fabrication of Si_3N_4–SiO_2 waveguides by low-pressure chemical vapor deposition (LPCVD) within a CMOS fabrication pilot line. They measured the propagation losses for different waveguide geometries (channel and rib loaded) near 1550 nm, as a function of polarization, width of the waveguide, and light wavelength. A maximum thickness of a single layer of Si_3N_4 of 250 nm is allowed by the large stress between SiON and SiO_2. This small thickness translates into significant propagation losses at 1544 nm of about 4.5 dB/cm because of the poor optical mode confinement factor. The authors in [50] showed that strain release and control are possible by using multilayer waveguides by alternating SiON and SiO_2 layers. Due to an improved optical mode confinement factor and the good quality of the interfaces in the waveguide, propagation losses of about 1.5 dB/cm have been demonstrated in the multilayered structure.

Devices. Several basic photonic passive devices were grown by the University of Twente group (see [42]). Now, we describe some of these basic devices.

A MZI filter was realized by the group in [42] using two unequal branches of SiON waveguides, having a $3 \times 3\,\mu m^2$ core, but without the birefringence compensating layer (BCL). The path length difference was 820 μm, and the bend radii were 7 mm. The tuning was accomplished by the thermo-optic (TO) effect, for which a 200 nm thick, 20 μm wide, and 4.5 mm long gold layer was deposited on top of a silica cladding layer above one branch of MZI. The crosstalk (TE) was −20 dB for cross points and −10 dB for bar points. The heating power was 0.8 W for tuning the device over a full free spectral range (FSR).

An EO modulator was realized by employing a resonant coupling between a guided mode in SiON with a lossy surface plasmon mode bound to a thin silver film with an electro-optic polymer cladding. An adiabatic TE–TM mode splitter was designed and realized by using an asymmetric Y-junction.

13.9.2.4 Silicon-on-Insulator (SOI)

Silicon-on-insulator (SOI) technology has been investigated as an alternative to the earlier ones – Si on Si, SiGe–Si, and SiON. The waveguiding action in these structures occurs due to index change between silicon and the insulator. Although sapphire was used as the insulating material in some studies, most of the work reported so far uses SiO_2. The OICs using SOI technology (sometimes referred to as SOI photonic integrated circuits, or SOIPICs) are compatible with SOI-based CMOS circuits. The RI change is large between Si ($n = 3.45$) and silica ($n = 1.45$). Therefore, SOI seems to be a promising candidate for future low-cost PICs.

Growth and Waveguide Formation. SOI structures for OICs are fabricated using several different technologies, such as

a. zone-melting recrystallization SOI (ZMR-SOI),
b. separation by implantation of oxygen SOI (SIMOX-SOI),
c. bond and etch-back SOI (BESOI), and
d. the smart-cut method.

The first method (ZMR-SOI) gives a large value of material loss, and therefore will not be considered. In the SIMOX process [51–53], a high dose ($1.8 \times 10^{18}\,cm^{-2}$) of O-implantation is performed on Si wafers at an energy of about 200 keV. This produces a buried layer of SiO_2 in the Si channel. After anneal, the top Si layer is made thicker by epitaxy and the rib structure is formed by RIE. The waveguiding along the vertical direction is due to high RI contrast in the air–Si–SiO_2 layers, while horizontally, the guiding occurs due to patterning of the etched rib. In BESOI, a silicon wafer is first oxidized using dry or wet oxidation followed by a hydrophilic bonding to a Si wafer and then heating. The first wafer is then thinned and polished by mechanical and mechanical-chemical processes to the desired thickness. The smart-cut process is a promising newer addition [8] and comprises four steps, as described in Section 13.7.2.

The material losses in the SOI waveguides follow a $1/H^3$ law, as in Eq. (13.13) and loss values of about 0.1 dB/cm for a top-layer thickness of 11 μm were reported. The single-mode behavior of the waveguides has been the subject of investigation because, at first sight, it may appear that single-mode behavior is difficult to attain in view of the large RI change and thick guiding layer. However, the analysis in [54] showed that for a certain rib dimension, a single-mode guide exists even if the slab guide is multimoded. For waveguides with $h/H > 0.5$ (see Figure 13.21), no higher order vertical modes exist. For single-mode operation, the rib width should satisfy

$$w \leq 0.3H_{eff} + h_{eff}/\sqrt{1-(h_{eff}/H_{eff})^2} \tag{13.17}$$

where $H_{eff} \approx H$ and $h_{eff} \approx h$, denote the effective heights of the central and slab regions, respectively.

Various authors reported values of propagation loss in SOI waveguides prepared in different ways. A planar guide by direct wafer bonding resulting in a thick silicon layer which allows straightforward coupling to fibers was reported in [55]. The lowest propagation loss for the BESOI structure reported was 0.27 dB/cm for a guiding layer thickness of 7.1 μm at 1.523 μm wavelength. Also, in [56], the authors reported 0.14 dB/cm for optical loss of SIMOX waveguides having a guiding layer thickness of 6 μm at the same wavelength and for TE polarization. In [57], the authors achieved 0.1 dB/cm loss for SOI single-mode rib waveguide. The loss characteristics in unibond SOI waveguides were measured in [58], and the reported value of 0.15 dB/cm is among the lowest loss for SOI reported. The loss value is comparable to that of pure Si, allowing OICs to be fabricated at low cost.

A view of the waveguides grown using SOI platform is given in Figure 13.22. The process steps involved in fabricating waveguides are shown in Figure 13.23.

Devices. A compact view of the PICs grown on SOI (SOIPIC) has been given in [59]. We discuss in this subsection the characteristic features of a few of them.

Fabricated bends and Y-junctions by the SIMOX method were reported in [60]. The bend radius was made 10 mm or more to reduce loss, and losses of 1.82 dB (TE) and 3.51 dB (TM) were found at 1.523 μm. The first directional coupler, fabricated in [61] using the BESOI method along with RIE, showed an excess loss of 1.9 dB at 1.55 μm. Using the same technology, the authors also realized an asymmetric MZ filter with a channel spacing of 4 nm (FSR 8 nm) with a cross-talk of −18 dB [62].

Modulators in SOI technology were fabricated that used the thermo-optic effect and free-carrier induced plasma effects and, in [63], modulators using a MZI were first demonstrated. These authors evaporated NiCr on one arm and used it as a heater. The modulation depth was 40% at a power of 60 mW, and the modulation BW was only 10 kHz. Better performance was achieved in [64].

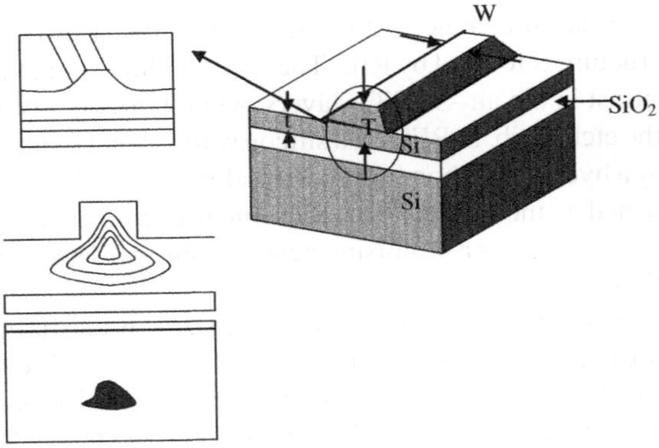

Figure 13.22 A view of an SOI-based waveguide and the computed mode field pattern (after [59]).

Free carriers injected into a material change both the RI and absorption coefficient. In [65], the authors reported on injecting carriers in one arm of MZI to achieve modulation. In [66], a Fabry–Perot etalon-based intensity modulator, in which the RI of the silicon waveguide on silica was changed by injecting carriers in a p–i–n structure, was constructed. The buried layer of SiO_2 served as the bottom mirror, and the top mirror was formed by air–polysilicon–silica–Si multilayers. In [67], an improved design was presented that used an elongated p–i–n structure grown on SOI that formed one arm of MZI. The p+ region was grown on top of the rib, and two n^+ regions were grown at the edges of the rib. The schematic diagram of the structure is given in Figure 4.11. Bias applied

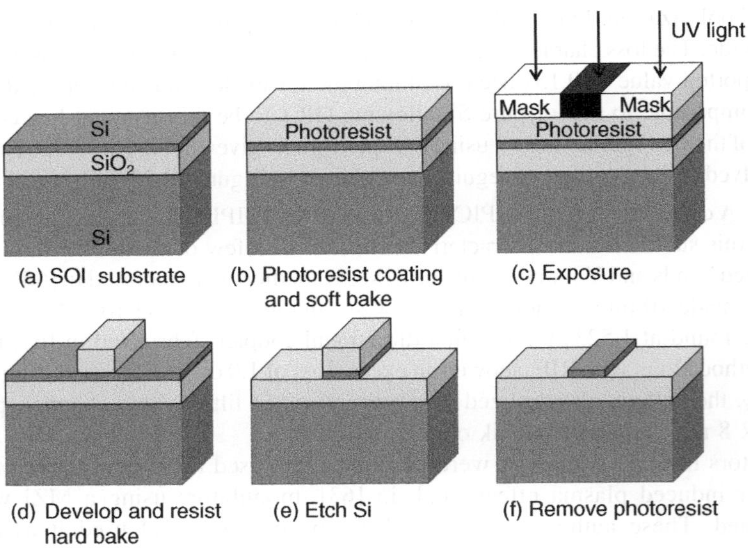

Figure 13.23 Fabrication steps for SOI ridge waveguides.

across the p^+–n^+ structure injected carriers. They obtained a switching voltage $V_\pi =$ 1.7 V, at a current density of 175 A/mm^2 leading to a power of 11.9 mW. The figure of merit, that is, the induced phase shift per volt per mm, was 212°/V/mm. A MZI modulator using the carrier effect was also developed in [68].

The change in absorption by free carriers was utilized in [69] to fabricate an absorption modulator that combines a rib guide and a p–i–n diode.

13.9.2.5 Amorphous Silicon

In comparison to the other materials systems discussed in this chapter, less work has been reported for amorphous silicon–based passive and active guided wave devices for use in telecommunications wavelength bands. A review of the work done in this area is given in [70], where mostly the group's own work was described. The following discussion is essentially based on their publication.

Growth and Waveguide Formation. The main advantages of the amorphous silicon (a-Si)–based technology lie in the low cost and low temperatures involved in the fabrication process. Thin films of a-Si may be deposited by physical deposition techniques, like sputtering or evaporation, which take place at temperatures below 300°C. Unfortunately, however, the density of states in the forbidden gap of sputtered a-Si is quite high and the introduction of dopants has virtually no control over the Fermi level position. The absorption tail extends below the energy gap, so that the absorption coefficient is quite high (10^3 cm^{-1}) at 1.3 eV. These materials are not useful for waveguiding.

Silicon thin films grown by the PECVD process show low absorption coefficients near 1.3 and 1.55 μm. The films are grown by decomposition of SiH_4. Then, by mixing it with other gases like CH_4, CO_2, and NO_2 during the plasma reaction, it is possible to obtain semiconductor alloys with wider band gaps. If GeH_4 is added, the gap is lowered. Two different layers, a-Si:H and a-SiC:H, have been studied in detail. The absorption spectra of both the layers have been measured over a wide range of wavelengths. The value for a-Si:H is quite low, $\approx$0.1 cm^{-1} at 1.3 μm, and even lower at 1.55 μm, while that for a-SiC:H is slightly higher but less than 1 cm^{-1}. This sets a lower limit of about 0.4 dB/cm to the loss for the waveguide at 1.3 μm.

Waveguides: The waveguides fabricated consisted of an a-Si:H or a-SiC:H heterostructure grown on a crystalline silicon (c-Si) substrate. A heavily doped c-Si substrate is first cleaned and then loaded in the deposition system, where the a-SiC cladding layer was deposited on silicon by RF-assisted decomposition of SiH_4 and CH_4. The a-Si:H layer was then grown by decomposing SiH_4. Details of the deposition process may be found in [71]. After photolithographic patterning of the planar layers, the rib structure was defined by plasma etching. The loss characteristics were evaluated as a function of waveguide width, and the minimum loss of 0.7 dB/cm was found for a 15 μm wide waveguide. Also, it was found that the loss increases with decreasing width.

Devices. An interferometric modulator has been constructed by using the PECVD technique. The device utilized the thermo-optic effect. It was shown that a-Si:H has a thermo-optic coefficient $\partial n/\partial T = 2.1 \times 10^{-4}$ K^{-1} at 1.55 μm, which compares favorably with the value for c-Si. The RI change in one of the arms of the interferometric modulator was produced by heating. A current pulse was applied to a resistive tungsten film lying on the top of the rib. In another structure, films of ZnO, deposited by sputtering, were used as the

Table 13.2 Characteristics of waveguides fabricated by the NTT Photonics Laboratory

	Low Δ	Middle Δ	High Δ	Super-high Δ
Refractive index difference Δ (%)	0.25	0.45	0.75	1.5
Core size (μm)	8 × 8	7 × 7	6 × 6	4.5 × 4.5
Propagation loss (dB/cm)	<0.01	0.02	0.04	0.07
Fiber coupling loss (dB/point)	<0.1	0.1	0.5	2.0 (SMF) 0.4 (DSF)
Minimum bending radius (mm)	25	10	5	2
Application field	Small-scale PLC	Medium-scale PLC	Large-scale PLC	Very large scale and high density PLC

under and top cladding layers. The modulators showed a depth of modulation of about 60%. Furthermore, a value of 10 μs for the maximum switching time was measured.

13.9.2.6 Silica Layers on Silicon

This structure is perhaps the most widely studied one, and several workers at AT&T Bell Labs and at NTT Laboratories have been engaged in developing useful devices using the grown layers.

Growth of Layers and Waveguides. The NTT group developed the waveguides and components by a process known as flame hydrolysis deposition (FHD) [2, 3], which was originally developed for optical fiber fabrication. In this process, SiO_2–GeO_2 waveguides are formed on silicon substrates. The method of fabrication is outlined in Section 13.7.1 and illustrated by Figure 13.15. In this method, the thickness of the films may be made over a wide range, and usable RI variation also may be achieved. Some of the characteristics of the waveguides are listed in Table 13.2.

As may be seen from the Table 13.2, both propagation loss and the bending loss depend on the value of the refractive index difference Δ. The propagation loss (which is a sum of scattering and absorption losses) increases with increasing Δ or doping concentration. However, increasing the Δ value results in a decrease in the bending loss.

The CVD process has been employed by Lucent Technologies, Bell Labs, and other companies to fabricate waveguides. The waveguide is made from three layers: base, core, and cladding. The RI of the cladding is chosen to be nearly equal to that of the base layer. The increase in the core layer RI is achieved by P doping, and a value of $\Delta \approx 0.60$–0.70% is obtained. The base layer is made of undoped silica. The cladding layers are doped with P and B. The addition of B reduces the RI increase due to P, and thus the RIs of base and cladding may be made approximately equal.

A schematic representation of a waveguide was previously shown in Figure 13.23. In such a waveguide, the base layer can be formed by either CVD or by high-pressure steam oxidation of the silicon substrate. The subsequent annealing densifies the glass. The core layer is then deposited and annealed. Photolithography and RIE are then employed to define the core layers. After this, the cladding layers are deposited in several steps.

The processes are time consuming, and often take a couple of days. Details of the fabrication process may be found in [4].

The propagation losses of different waveguides have been measured for P-doped cores having $\Delta = 0.70\%$. The loss values are 4.72, 1.22, and 0.85 dB/m for radii of 10, 20 and 30 mm, respectively, at a wavelength of 1.55 μm. The bend radius and component size have been measured. It was observed that for medium Δ waveguides, bend radii $\geq$10 mm are required. For high Δ waveguides, ($\Delta \geq 1.2\%$), the bend radius drops down to 2–3 mm. Another significant contribution to loss comes from fiber-to-waveguide coupling. This loss was minimized by using a mode taper of height 6.7 μm and width of 9 μm with P-doped guides ($\Delta \approx 0.63\%$), and an insertion loss of 0.3 dB was found.

An alternative way of forming the waveguide by ion implantation of 5 MeV Si, P, and Ge ions in fused silica or silica on silicon substrate has been suggested and tested in [72]. The cladding and core layers are made by flame hydrolysis deposition. The core layer was then doped with B, P, and either 3% or 7% Ge. The channel waveguides are then formed by ion implantation of P, B, and Ge at an energy of 5 MeV and a dose of $1 \times 10^{15}\ \text{cm}^{-2}$. After anneal, the channels are delineated by photolithography.

The waveguide loss coefficient was measured as a function of anneal temperature for Si, Ge, and P implantations. For the As implanted guides, the value of α was in the range of 0.9–1.0 dB/cm. The Ge implanted guides showed a minimum value of 0.10–0.15 dB/cm after annealing at 500°C.

Devices. A large number of planar lightwave circuits (PLCs) have already been developed by using the guides grown on silica-on-silicon. The PLCs may be divided, according to [2], into four generations. The first generation includes simple Y-branches and directional couplers, which are used as optical splitters and taps. The most important device in the second generation is the Mach–Zehnder interferometer and ring resonators. The third generation includes $N \times N$ star couplers and arrayed waveguide grating multiplexers. The fourth generation includes optical analogues of digital filters. As was done in earlier subsections of this chapter, we shall discuss the components appearing at the first and second generations here.

Y-coupler units can be integrated to form $1 \times N$ splitters. They are used in distributing optical signals from a trunk line to branch lines in optical subscribers' networks. The family includes not only 1×8 splitters, but also 1×16, 1×32, and 2×8, 16, or 32 splitters. Cascading N stages of Y-branches give a 1×2^N power splitter. In the fabrication process, the minimum width of the gap between two waveguides is limited by the photolithographic resolution and the chemical etching process. The sudden opening of a gap results in an extra insertion loss. A 0.13 dB extra insertion loss per Y-branch fabricated with a low $\Delta(0.3)$ waveguide was reported in [73]. As the value of Δ increases, the loss associated with a sudden opening of the gap also increases. In a recent work [74], the realization of 1×32 splitters having a chip size of $6 \times 24\ \text{mm}^2$ using 0.4%-Δ waveguides with a 15 mm bending radius was reported. Wavelength-insensitive couplers [75] are optical couplers in which the coupling ratios do not depend on wavelengths. They are installed in monitoring systems in optical networks, in which the transmission signal is at 1.55 μm, while the monitoring wavelength is 1.65 μm.

A number of components have been grown by using CVD-grown silica-on-silicon guides. As noted already, a Mach–Zehnder interferometer (MZI) usually consists of two 3 dB couplers connected by paths of unequal lengths. It is, however, difficult to fabricate accurate 3 dB couplers, which are normally directional couplers. The equal splitting depends on coupling length, which in turn depends sensitively on the size and composition of the waveguide, and so it is hard to control. The coupling length is a function of wavelength, and hence an equal division of power occurs over a narrow spectral range. In the work reported in [76], a broadband operation was achieved by the use of an adiabatic design. In an adiabatic coupler, for example, the two waveguides are not identical and have different widths and separation along the direction of propagation. The waveguides are tapered slowly, and the two guides have a small separation without merging with each other.

The requirements for adiabatic design are discussed in [77]. In [76], the fabrication of Y-branches, and adiabatic 3 dB couplers that gave nearly even splitting of power and showed broadband wavelength performance was reported. The excess insertion loss of the Y-branches was about 0.4 dB and that of 3-dB coupler was 0.1–0.2 dB. The Y-branches were then used to make Mach–Zehnder (MZ) transmission filters working at 1.55 μm. A MZ power combiner combining two channels at 1.55 μm and 1.48 μm wavelengths was realized by combining a Y-branch and 3 dB couplers. The combiner showed almost no polarization sensitivity.

A noteworthy feature about silica-on-silicon waveguides is that strain develops in the silica layer, and as a result the silica layer shows birefringence. The TE and TM polarized waves therefore travel with different phase velocities due to the difference in RIs. Furthermore, the optical modes shift radially outward while traversing a bend, and consequently have a greater path length and phase. The path lengths are different for TE and TM polarizations. There are several methods to eliminate the effect of polarization. In [78], the authors inserted a thin patch of Si_3N_4 under the longer waveguide core. TM polarized light induces charge on the surface of the nitride layer, which in turn produces a field opposing the TM field. Consequently the RI due to the nitride layer is larger for TE than TM. The phase difference due to this RI difference compensates the phase differences occurring due to the path length difference between the two arms and the RI difference in the bend. In a different method [79], a 60-μm deep groove was etched along the side of the waveguide, the groove relieving the strain in the silica layer. In another method, a 6 μm thick amorphous silicon layer, which has strain opposite that of silica, was deposited on top of the structure [80]. In [81], the authors employed a $\lambda/4$ plate inside an arrayed waveguide gratings (AWG) multiplexer to rotate the polarization by 90°.

Results on other types of lightwave components including Fourier filters were reported in [4]. The Fourier filter is a generalization of the MZI. It consists of a chain of N ($N > 2$) optical couplers of different coupling ratios linked by $N - 1$ differential delays of different lengths.

13.9.3 A Comparative Study of Materials Systems

The key features of all the materials systems discussed in Section 13.9.2 are summarized in Table 13.3, the wavelength of interest being 1.55 μm.

The first three materials systems have been compared in [82]. Also, in [83], a table summarizing the properties of materials applied in integrated optics was provided. In addition to following the arguments given in [82, 83], we shall include in our discussion our

Table 13.3 *Properties of some key optical materials systems used in PLCs at 1.55 μm*

Material	Propagation loss (dB/cm)	Fiber coupling loss (dB/chip)	Refractive index	Index contrast range in waveguide	Birefringence	TO coefficient dn/dT (K^{-1})	Maximum modulation frequency
Silica	0.1	0.5	1.44–1.47	0–1.5%	10^{-4}–10^{-2}	10^{-5}	1 kHz (TO)
Si-on-insulator (SOI)	0.1	1.0	3.48	70% (fixed)	Same	1.8×10^{-4}	1 kHz(TO)
SiON	0.1	1.0	SiO_2:1.44–1.47 Si_3N_4:2.0	0–30% (with Si_3N_4 core)	10^{-3}–5×10^{-6}	10^{-5}	1 kHz (TO)
SiGe on Si	0.3	2.0	3.48				90 kHz
Si on Si	4.0	?	3.48–3.475	3%	As in Si	10^{-5}	700 kHz
Amorphous Si	0.7		a-Si core: 3.5 a-SiC: 2.9–3.4			2.1×10^{-4}	100 kHz

observations regarding the relative merits and demerits of all the materials systems discussed so far in the above subsections.

By far, the most widely used technology is silica-on-silicon. Mainly, two different methods, CVD and FHD, are employed to grow thin layers of silica on a silicon substrate. Both these processes are lengthy, and it requires a few to several days to grow a few tens of micron-thick layers at high temperatures. The deposited layers have a high level of stress which may lead to a bending of wafers. As a result, there may be misalignment between the waveguides on a chip and the fibers in a fiber array unit. The problem may be minimized by growing a layer on the backside of the wafer, but the growth time increases. Even when the wafer-bending problem is alleviated or minimized, the stress is not altogether eliminated. The stress-induced polarization birefringence and stress-induced scattering loss pose additional problems.

The birefringence problem has been tackled with a number of methods, such as etching grooves, adding a thin birefringence-compensating layer, or, in the case of interferometric devices, inserting a Half-wave plate at an appropriate position of the device. However, these methods introduce additional complexity and also increase the cost. The RIE used to define the rib structure introduces a rough surface that contributes to the scattering loss. The highest RI contrast is only 1.5%, so that the minimum value of the bend radius is quite large. The yield is poor and, in large interferometric devices, it does not exceed 10%. The core of the guide is perfectly index-matched with the core of the fiber, and hence the coupling loss is somewhat lower. The waveguide formation process is fully compatible with IC processing. However, due to a smaller RI change, the overall thickness of the waveguide is currently ~50 μm for the production process, which is too high compared to the overall layer thickness of ~10 μm in CMOS circuits.

The silicon-on-insulator technology has been developed over the last few years and is supposed to replace the silica-based technology. The layers are grown by SIMOX, BESOI, or smart-cut methods. The starting substrate is a silicon wafer with a buried or grown silica layer. The core rib is patterned in the top silicon layer. An over-cladding silica layer is to be grown on silicon. The growth allows faster turnaround time and higher yield. The high index contrast between silicon and silica allows only rib structure for the guide. Although high index contrast allows one to achieve the sharpest bends in current integrated technology, the coupling loss to the fiber becomes large due to the large RI mismatch between the cores of the waveguide and of the fiber. A large Fresnel reflection loss of the order of 1.5 dB per chip (0.75 dB per facet) occurs. An antireflection coating or an adiabatic waveguide taper is necessary to reduce this loss, adding to the process complexity. The technology is fully compatible with the CMOS technology, as the thickness of the total guide structure is about 10 μm, almost equal to the total thickness of epilayer and other layers in CMOS devices.

Silicon oxynitride is a relatively new planar waveguide technology. The cladding layer is SiO_2 with RI = 1.45, and the core layer is tunable between SiO_2 and Si_3N_4, the RI being maximum (2) for the nitride compound. The adjustable index contrast may be as large as 30%. LPCVD or PECVD techniques are used to grow different layers, and the required growth time is on the order of days. The waveguide structure is rib, the minimum bend radius is small, but fiber coupling losses and polarization dependence are higher. To reduce the birefringence, a birefringence-compensating layer (BCL) is grown beneath the core rib. Also, a mode size transformer is needed to reduce the mismatch loss of the fiber-to-chip coupling.

Silicon waveguides grown on silicon are the first reported integrated optical waveguide system using silicon-based materials. The RI change between the guide layer and substrate is obtained by introducing a dopant variation. Initially the waveguide loss was in the higher level. However, recent work with a high-dose ion-implanted cladding layer reported a significant reduction of loss. Still the loss is quite high due to substrate radiation loss. The advantages of all silicon guides lie in the potential for monolithic integration between optoelectronics and electronic devices, suitability for high-speed, low-driving-power, electronically controlled devices; and availability of sophisticated, controllable and low-cost fabrication processes, as well as the simple method of end facet preparation (scribe and cleave). The all-Si (SOS) structures are easier to grow than the SOI or SiGe–Si heterostructures; the SOIPICs are compatible with only SOI-based VLSI circuits. Most importantly, the higher thermal conductivity of silicon compared to that in silica may prove to be useful in fabricating thermo-optic (TO) or free-carrier based switches and modulators. However, there is a limit to the reduction of RI imposed by the higher limit of doping density.

SiGe and more recently SiGeC have been used as the guiding materials. The layers are now grown by CVD or MBE. The RI contrast is theoretically high due to higher RI in Ge. However, a large concentration of Ge in the alloy layer leads to higher strain and the occurrence of misfit dislocations. Therefore, the concentration of Ge never exceeds 20%. The waveguide and coupling losses have been reduced, but still the values are higher than the values in silica, SOI, or SiON guides. Incorporation of C in SiGe alloy has been found to relieve strain, and some waveguide structures incorporating photodetectors have been fabricated. However, complete details of the propagation and coupling loss and reports describing OICs are not available.

Some work on amorphous Si as a waveguide material has been reported in the literature. The confinement occurs in a-Si:H layer, while the cladding layer is a-SiC:H. The main advantages of the structure are in the inexpensive growth of the devices, the high optical confinement, and the high thermal conductivity of the a-Si and a-SiC layers. In the absence of a large amount of data, a fruitful comparison between the OICs using a-Si and other materials systems cannot be undertaken at this stage.

Problems

Note that in some of these problems, you will need to consult an *erfc* graph or solve numerically.

13.1 A boron predeposition is performed on a wafer with uniform doping of $5 \times 10^{15}\,\text{cm}^{-3}$ for 30 minutes at 950°C. Find the resulting junction depth. Given that at 950°C, for boron, the surface concentration N_s is $4.5 \times 10^{20}\,\text{cm}^{-3}$ and $D_P = 1.6 \times 10^{-3}\,\propto\text{m}^2/\text{hour}$.

13.2 A boron predeposition is performed on a wafer with doping $5 \times 10^{15}\,\text{cm}^{-3}$ for 30 minutes at 950°C. The boron surface concentration is $4.5 \times 10^{20}\,\text{cm}^{-3}$ and $D_P = 1.6 \times 10^{-3}\,\propto\text{m}^2/\text{hour}$. Find the resulting junction depth, that is, the depth that the boron penetrates into the silicon substrate.

13.3 A p-type silicon wafer is uniformly doped with boron to a concentration of $10^{16}\,\text{cm}^{-3}$. This wafer is then subjected to a phosphorus diffusion step at a constant surface concentration of $10^{18}\,cm^{-3}$. If the diffusion coefficient of phosphorus at 1250°C is $7.5 \times 10^{-2}\,cm^2$/second, how much time is required to produce a junction depth of 5 ∝m?

13.4 A p–n diode is formed by diffusing boron (p-type dopant) into a n-substrate of doping $10^{16}\,cm^{-3}$. The surface concentration is kept constant at $10^{20}\,cm^{-3}$ to make the p-type part of the diode heavily doped. If the diffusion temperature is kept constant at 1050°C, calculate the time required to form the p-junction at a depth of 1 ∝m, given that the diffusion coefficient of boron at 1050°C is 10^{-5} ∝m²/second.

13.5 A two-step boron diffusion is performed into n-type silicon that is uniformly doped with $10^{16}\,cm^{-3}$ phosphorus atoms per unit volume. The predeposition is carried out at 950°C for 15 minutes with a surface concentration of $10^{21}\,cm^{-3}$, and the drive-in is performed at 1100°C. How long should the drive-in last for a junction depth of 3 ∝m? And what is the impurity concentration on the surface after the drive-in diffusion? Given the following information, at 950°C, $D = 10^{-6}$ ∝m²/second, and at 1100°C, $D = 2.5 \times 10^{-5}$ ∝m²/second. For a two-step diffusion process,

$$N(x, t_d) = \frac{2N_s}{\pi}\sqrt{\frac{D_p t_p}{D_d t_d}}\exp\left(\frac{-x^2}{4D_d t_d}\right),$$

provides a good approximation of the actual measured profiles provided that $D_p t_p \ll D_d t_d$. Also, the subscripts p and d refer to the predeposition and drive-in processes.

13.6 A boron predeposition is formed on an n-type silicon that is uniformly doped with phosphorus atoms at $5 \times 10^{15}\,cm^{-3}$. The predeposition is carried out at 950°C for 30 minutes, and the measured junction depth is 0.2 ∝m. The wafer is now subjected to a drive-in diffusion in nitrogen atmosphere for 2 hours at 1150°C. The surface concentration per unit area is $1.43 \times 10^{15}\,cm^{-2}$. Find the resulting junction depth.

13.7 The article by D.H. Lee and J.W. Mayer, "Ion-implanted Semiconductor Devices," *Proceedings of the IEEE*, vol. 62, pp. 1242–1255 (December 1974), provides a listing of the range and straggle for boron, phosphorus, and arsenic ions implanted into silicon for a range of energies. Using the information in this reference, determine the range and straggle of phosphorus, boron, and arsenic ions implanted into silicon using an ion implantation beam at 50 keV.

References

1. Jansen, H., Gardeniers, H., de Boer, M. *et al.* (1996) A survey on the reactive ion etching of silicon in microtechnology. *J. Micromech. and Microeng.*, **6**(1), 14–28.
2. Kawachi, M.(October 1996) Recent progress in silica-based planar lightwave circuits on silicon. *IEE Proc. – Optoelectron.*, **143**(5), 257–262.
3. Miya, T. (2000) Silica-based planar lightwave circuits: passive and thermally active devices. *IEEE J. Sel. Top. Quantum. Electron.*, **6**(1), 38–45.
4. Li, Y.P. and Henry, C.H. (1996) Silica based optical integrated circuits. *IEE Proc. Optoelectron.*, **143**(5), 263–280.
5. Janz, S. (2004) *Silicon-based Waveguide Technology for Wavelength Division Multiplexing, Silicon Photonics*(eds L. Pavesiand D.J. Lockwood), Springer Verlag, Berlin, pp. 323–360.

6. Celler, G.K. and Cristoloveanu, S. (2003) Frontiers of Si-on-insulator. *J. Appl. Phys.*, **93**(9), 4955–4578.
7. Maszara, W.P., Goetz, G., Caviglia, A., and McKitterick, J.B. (1988) Bonding of silicon wafers for silicon-on-insulator. *J. Appl. Phys.*, **64**, 4943–4950.
8. Bruel, M. (1995) Silicon on insulator material technology. *Electron. Lett.*, **31**(14), 1201–1202.
9. Hill., K.O., Fuji, Y., Johnson, D.C., and Kawasaki, B.S. (1978) Photosensitivity in optical wave guides: application to reflection filter fabrication. *Appl. Phys. Lett.*, **32**(10), 647–649.
10. Meltz, G., Morey, W.W., and Glenn, W.H. (1989) Formation of Bragg gratings in optical fibers by a transverse holographic method. *Opt. Lett.*, **14**(15), 823–825.
11. Maxwell, G.D. and Ainslie, B.J. (1995) Demonstration of a directly written directional coupler using UV-induced photosensitivity in a planar silica waveguide. *Electron. Lett.*, **31** (2), 95–96.
12. Bilodeau, F., Malo, B., Albert, J. *et al.* (1993) Photosensitization of optical fiber and silica-on-silicon/silica waveguides. *Opt. Lett.*, **18**, 953–955.
13. Dyer, P.E., Farley, R.J., Giedl, R. *et al.* (1994) High reflectivity fiber gratings produced by incubated damage using a 193 nm ArF laser. *Electron. Lett.*, **30**(11), 860–862.
14. Albert, J., Bilodeau, F., Johnson, D.C. *et al.* (1998) Polarization-independent strong Bragg gratings in planar lightwave circuits. *Electron. Lett.*, **34**(5), 485–486.
15. Kohnke, G.E., Henry, C.H., Laskowski, E.J. *et al.* (1996) Silica based Mach-Zehnder add-drop filter fabricated with UV induced gratings. *Electron. Lett.*, **32**(17), 1579–1580.
16. Orthonos, A. (1997) Fiber Bragg gratings. *Rev. Sci. Instrum.*, **68**, 4309–4341.
17. Giles, C.R. (1997) Lightwave applications of fiber Bragg gratings. *J. Lightwave Technol.*, **15**(8), 1391–1404.
18. Kashyap, R. (1999) *Fiber Bragg Gratings*, Academic Press, New York.
19. Soref, R.A. and Lorenzo, J.P. (1986) All silicon active and passive guided wave components for λ=1.3 and 1.6 μm. *IEEE J. Quantum Electron.*, **QE-22**, 873–879.
20. Splett, A. and Petermann, K. (1994) Low loss single-mode optical waveguides with large cross section in standard epitaxial silicon. *IEEE Photon. Technol. Lett.*, **6**(3), 425–427.
21. Cocorullo, G., Della Corte, F.G., Iodice, M. *et al.* (1998) Silicon-on-silicon rib waveguides with a high-confining ion-implanted lower cladding. *IEEE J. Sel. Top. Quant. Electron.*, **4**(6), 983–989.
22. Lorenzo, J.P. and Soref, R.A. (1987) 1.3 μm electro-optic silicon switch. *Appl. Phys. Lett.*, **51**(1), 6–8.
23. Cocorullo, G., Iodice, M., Rendina, I., and Sarro, P.M.(Apr 1995) Silicon thermo-optical micro-modulator with 700 kHz-3 dB bandwidth. *IEEE Photon. Technol. Lett.*, **7**(4), 363–365.
24. Niemi, T., Uusimaa, M., Tamela, S. *et al.* (Jan 2001) Tunable silicon etalon for simultaneous spectral filtering and wavelength monitoring of a DWDM transmitter. *IEEE Photn. Technol. Lett.*, **13**(1), 58–60.
25. Schuppert, B., Schmidtchen, J., Splett, A. *et al.* (Oct 1996) Integrated optics in silicon and SiGe-heterostructures. *J. Lightwave Technol.*, **14**(10), 2311–2323.
26. Naval, L., Jalali, B., Gomelsky, L., and Liu, J.M. (1996) Optimization of $Si_{1-x}Ge_x$/Si waveguide photodetectors operating at λ=1.3 μm. *J. Lightwave Technol.*, **14**(5), 787–797.
27. People, R. (1986) Physics and applications of Ge_xSi_{1-x}/Si strained layer heterostructures. *IEEE J. Quant. Electron.*, **22**, 1696–1710.
28. Bean, J.C. (1986) Recent developments in the strained layer epitaxy of germanium-silicon alloys. *J. Vac. Sci. Technol. B*, **4**(6), 1427–1429.
29. Splett, A., Schmidtchen, J., Schuppert, B. *et al.* (1990) Low loss optical ridge waveguides in a strained GeSi epitaxial layer grown on silicon. *Electron. Lett.*, **26**, 1035–1036.
30. Soref, R.A., Namavar, F., and Lorenzo, J.P. (1990) Optical waveguiding in a single crystal layer of germanium silicon grown on silicon. *Opt. Lett.*, **15**, 270–272.

31. Namavar, F. and Soref, R.A. (1991) Optical waveguiding in GexSi1-x/Si heterostructures. *J. Appl. Phys.*, **70**, 3370–3372.
32. Pesarcik, S.F., Treyz, G.V., Iyer, S.S., and Halbout, J.M. (1992) Silicon germanium optical waveguides with 0.5 dB/cm losses for single mode fiber optic systems. *Electron. Lett.*, **28**, 159–160.
33. Schmidtchen, J., Schuppert, B., Splett, A., and Petermann, K. (1992) Germanium diffused waveguides in silicon for $\lambda = 1.3$ μm and $\lambda = 1.55$ μm with losses below 0.5 dB/cm. *IEEE Photon. Technol. Lett.*, **4**, 875–877.
34. Pearson, M.R.T., Jessop, P.E., Bruce, D.M. *et al.* (Mar 2001) Fabrication of SiGe optical waveguides using VLSI processing techniques. *J. Lightwave Technol.*, **19**(3), 363–370.
35. Jessop, P.E., Pearson, M.R.T., Janz, S., and Xu, D-X. (2002) Silicon-based integrated optical devices, *Integrated Optoelectronics; Proc. 1st Intl. Conf*, **2002–4**(eds M.J. Deen *et al.*.), Electrochemical Society, Pennington, NJ, pp. 333–344.
36. Mi, J., Warren, P., Letourneau, P. *et al.* (1995) High quality $Si_{1-x-y}Ge_xC_y$ epitaxial layers grown on (100) Si by rapid thermal chemical vapor deposition using methylsilane. *Appl. Phys. Lett.*, **67**, 3915–3917.
37. Soref, R.A., Atzmon, Z., Shaapur, F., and Robinson, M. (1996) Infrared waveguiding in $Si_{1-x-y}Ge_xC_y$ upon silicon. *Opt. Lett.*, **21**, 345–347.
38. Huang, F.Y., Sakamoto, K., Wang, K.L. *et al.* (Feb 1997) Epitaxial SiGeC waveguide photodetector grown on Si substrate with response in the 1.3–1.55 μm wavelength range. *IEEE Photon. Technol. Lett.*, **9**(2), 229–231.
39. Schmidtchen, J., Schuppert, B., and Petermann, K. (1994) Passive Integrated optical waveguide structures by Ge-diffusion in silicon. *J. Lightwave Technol.*, **12**, 842–848.
40. Schmidtchen, J., Schuppert, B., Splett, A., and Petermann, K. (1992) Ge diffused passive optical waveguide structures in silicon. *Tech. Dig. Integr. Photon. Res., Paper TUD4, Opt. Soc. Amer., Washington, DC, OSI, Wash DC, vol. 7, pp. 122–123.*
41. Gao, Y., Li, G.Z., Liu, X.D. *et al.* (1995) SiGe/Si bifurcation optical active switch based on plasma dispersion effect. *Electron. Lett.*, **31**(20), 1740–1741.
42. de Ridder, R.M., Worhoff, K., Driessen, A. *et al.* (1998) Silicon oxynitride planar waveguiding structures for application in optical communication. *IEEE J. Sel. Top. Quant. Electron.*, **4**(6), 930–937.
43. Albers, H. (1996) Proceedings of the 1996 National Sensor Conference, Delft.
44. Worhoff, K., Lambeck, P.V., Albers, H. *et al.* (1997) Optimization of LPCVD silicon oxynitride growth to large refractive index homogeneity and layer thickness uniformity. *Proc. SPIE*, **3099**, 257–268.
45. Albers, H., Hilderink, L.T.H., Szilagyi, E. *et al.* (2000) Reduction of hydrogen induced losses in PECVD SiO_xN_y optical waveguides in the near infrared. *Proceedings of IEEE-LEOS, 1995, San Francisco, CA*, pp. 88–89.
46. Offerin, B.J., Bona, G.L., Horst, F. *et al.* (1999) Wavelength tunable optical add-after-drop filter with flat passband for WDM networks. *IEEE Photon. Technol. Lett.*, **11**(2), 239–241.
47. Roeloffzen, C.G.H., Horst, F., Offrein, B.J. *et al.* (2000) Tunable passband flattened 1-from-16 binary-tree structured add-after-drop multiplexer using SiON waveguide technology. *IEEE Photon. Technol. Lett.*, **12**(9), 1201–1203.
48. Sun, Z.J., McGreer, K.A., and Broughton, J.N. (1998) Demultiplexer with 120 channels and 0.29-nm channel spacing. *IEEE Photon. Technol. Lett.*, **10**(1), 90–92.
49. Schauwecker, B., Przyrembel, G., Kuhlow, B., and Radehaus, C. (2000) Small size silicon oxynitride AWG demultiplexer operating around 725 nm. *IEEE Photon. Technol. Lett.*, **12**(12), 1645–1646.
50. Melchiorri, M. *et al.* (2005) Propagation losses of silicon nitride waveguides in the near-infrared range. *Appl. Phys. Lett.*, **86**, 121111-1–121111-3.

51. Kurdi, B.N. and Hall, D.G. (1988) Optical waveguides in oxygen implanted buried-oxide silicon-on-insulator structures. *Opt. Lett.*, **13**, 175–177.
52. Schmidtchen, J., Splett, A., Schuppert, B., and Petermann, K. (1991) Low loss singlemode optical waveguides with large cross-section in silicon-on-insulator. *Electron. Lett.*, **27**(16), 1486–1488.
53. Rickman, A.G. and Reed, G.T. (1994) Silicon-on-insulator optical rib waveguides: loss, mode characteristics, bends and y-junctions. *IEEE Proc. Optoelectron.*, **141**(6), 391–393.
54. Soref, R.A., Schmidtchen, J., and Petermann, K. (1991) Large single-mode rib waveguides in GeSi-Si and Si-on-SiO_2. *IEEE J. Quantum. Electron.*, **27**(8), 1971–1974.
55. Reed, G.T., Jinhua, L., Tang, C.K. *et al.* (1992) Silicon on insulator optical waveguides formed by direct wafer bonding. *Mater. Sci. Eng.*, **B15**, 156–159.
56. Rickman, A., Reed, G.T., Weiss, B.L., and Namavar, F. (1992) Low loss planar optical waveguides fabricated in SIMOX materials. *IEEE Photon. Technol. Lett.*, **4**(6), 633–635.
57. Fischer, U., Zinke, T., Kropp, J-R. *et al.* (1996) 0.1 dB/cm waveguide losses in single-mode SOI rib waveguides. *IEEE Photon. Technol. Lett.*, **8**(5), 647–648.
58. Ang, T.W., Reed, G.T., Vonsovici, A. *et al.* (1999) 0.15 dB/cm loss in unibond SOI waveguides. *Electron. Lett.*, **35**(12), 977–978.
59. Jalali, B., Yegnanarayanan, S., Yoon, T. *et al.* (1998) Advances in silicon-on-insulator optoelectronics. *IEEE J. Sel. Top. Quant. Electron.*, **4**(6), 938–947.
60. Weiss, B.L. and Reed, G.T. (1991) The transmission properties of optical waveguides in SIMOX structures. *Opt. Quant. Electron.*, **23**, 1061–1065.
61. Trinh, P.D., Yegnanarayanan, S., and Jalali, B. (1995) Integrated optical directional couplers in silicon-on-insulator. *Electron. Lett.*, **31**(24), 2097–2098.
62. Trinh, P.D., Yegnanarayanan, S., and Jalali, B. (1996) Guided wave optical circuits in silicon-on-insulator technology. *Technical Digital Integrated Photonics Research Conference, Boston, MA, vol. 6*, pp. 273–277.
63. Treyz, G.V. (1991) Silicon Mach-Zehnder waveguide interferometers operating at 1.3 μm. *Electron. Lett.*, **27**, 118–120.
64. Schuppert, B., Schmidtchen, J., Splett, A. *et al.* (October 1996) Integrated optics in silicon and SiGe-heterostructures. *J. Lightwave Technol.*, **14**(10), 2311–2323.
65. Treyz, G.V., May, P.G., and Halbout, J.-M. (1991) Silicon Mach-Zehnder waveguide-interferometers based on the plasma dispersion effect. *Appl. Phys. Lett.*, **59**(7), 771–773.
66. Xiao, X., Sturm, J.C., Goel, K.K., and Schwartz, P.V.(Mar 1991) Fabry-Perot optical intensity modulator at 1.3 μm in silicon. *IEEE Photon. Technol. Lett.*, **3**(3), 230–231.
67. Tang, C.K. and Reed, G.T. (1995) Highly efficient optical phase modulator in SOI waveguides. *Electron. Lett.*, **31**(6), 451–452.
68. Zhao, C.Z., Li, G.Z., Liu, E.K. *et al.* (1995) Silicon on insulator Mach-Zehnder waveguide interferometers operating at 1.3 μm. *Appl. Phys. Lett.*, **67**(17), 2448–2449.
69. Treyz, G.V., May, P.G., and Halbout, J.M. (1991) Silicon optical modulators at 1.3 μm based on free-carrier absorption. *IEEE Elect. Devices Lett.*, **12**, 276–278.
70. Cocorullo, G., Della Corte, F.G., De Rosa, R. *et al.* (Nov/Dec 1998) Amorphous silicon based guided-wave passive and active devices for silicon integrated optoelectronics. *IEEE J. Sel. Top. Quantum Electron.*, **4**(6), 997–1002.
71. Cocorullo, G., Della Corte, F.G., Rendina, I. *et al.* (1996) Amorphous silicon waveguides and light modulators for integrated photonics realized by low-temperature plasma-enhanced chemical-vapor deposition. *Opt. Lett.*, **21**(24), 2002–2004.
72. Leech, P.W., Ridgway, M., and Faith, M. (1996) Channel waveguides formed in fused silica and silica-on-silicon by Si, P and Ge ion implantation. *IEEE Proc. – Optoelectron.*, **143**(5), 281–286.
73. Hibino, Y., Hanawa, F., Nakagome, H. *et al.* (1995) High reliability optical splitters composed of silica-based planar lightwave circuits. *J. Lightwave Technol.*, **13**(8), 1728–1735.

74. Hida, Y., Fukumitsu, T., Hanawa, F. *et al.* (1998) Highly compact silica-based PLC-type 1 × 32 splitters using 127 μm-spacing output and 0.4%-Δ waveguide. *Electron. Lett.*, **34**(1), 75–76.
75. Jinguji, K., Takato, N., Sugita, A., and Kawachi, M. (1990) Mach-Zehnder type interferometer type optical waveguide coupler with wavelength-flattened coupling ratio. *Electron. Lett.*, **26**(17), 1326–1327.
76. Adar, R., Henry, C.H., Kazarinov, R.F. *et al.* (1992) Adiabatic 3-dB couplers, filters, and multiplexers made with silica waveguides on silicon. *J. Lightwave Technol.*, **10**(1), 46–50.
77. Shani, Y., Henry, C.H., Kistler, R.C. *et al.* (1991) Integrated optic adiabatic devices on silicon. *IEEE J. Quant. Electron.*, **27**(3), 556–566.
78. Yaffe, H.H., Henry, C.H., Kazarinov, R.F., and Milbrodt, M.A. (1994) Polarization-independent silica-on-silicon Mach-Zehnder interferometers. *J. Lightwave Technol.*, **12**(1), 64–67.
79. Inoue, Y., Katoh, K., and Kawachi, M. (1992) Polarization sensitivity of a silica waveguide thermooptic phase shifter for planar lightwave circuits. *IEEE Photon. Technol. Lett.*, **4**(1), 36–38.
80. Kawachi, M., Takato, N., Jinguji, K., and Sugiita, A. (1989) *Proceedings of CLEO'89, TuJ17, Baltimore, MD.*
81. Takahashi, H., Hibino, Y., and Nishi, I. (1992) Polarization-insensitive arrayed-waveguide grating wavelength multiplexer on silicon. *Opt. Lett.*, **17**(7), 499–501.
82. Eldada, L.(July 2001) Advances in telecom and datacom optical networks. *Opt. Eng.*, **40**(7), 1165–1178.
83. Worhoff, K. *et al.* (2002) Planar waveguide devices on a silicon compatible technology platform, *Proceedings of the 1st Intl. Conf. Integrated Optoelectronics*(eds M.J. Deen *et al.*.), Electrochemical Society Proceedings Series, vol. **2002-4**, Electrochemical Society, Pennington, NJ, pp. 279–312.

Suggested Reading

Chapman, B.N. (1980) *Glow Discharge Processes: Sputtering and Plasma Etching*, John Wiley & Sons, Inc., New York.

Cho, A. (1994) *Molecular Beam Epitaxy*, American Institute of Physics, College Park, MD.

Coburn, J.W. (1982) *Plasma Etching and Reactive Ion Etching*, American Institute of Physics, College Park, MD.

Deen, M.J., Misra, D., and Ruzyllo, J. (eds) (2002) *Integrated Optoelectronics (First International Symposium), Electrochemical Society Proceedings Series*, vol. **2002–4**, Electrochemical Society, Pennington, NJ, p. 436.

Doering, R. and Nishi, Y. (2007) *Handbook of Semiconductor Manufacturing Technology*, CRC Press, Boca Raton, FL.

Ghandi, S.K. (1983) *VLSI Fabrication Principles: Silicon and Gallium Arsenide*, John Wiley & Sons, Inc., New York.

Glaser, A.B. and Subak-Sharpe, G.E. (1979) *Integrated Circuit Engineering – Design, Fabrication and Applications*, Addison Wesley Publishing, Reading, MA.

Grym, J. (ed.) (2010) *Semiconductor Technologies*, InTech Publisher, Rijeka, Croatia.

Jackson, K.A. and Schröter, W. (2000) *Handbook of Semiconductor Technology (2 Volume Set)*, Wiley-VCH, Weinheim, Germany.

Kawachi, M. (1994) Silica waveguides on silicon and their application to integrated-optic components. *Optical and Quantum Electronics*, **22**, 391–416.

Lockwood, D. and Pavesi, L. (2011) *Silicon Photonics I and II*, Springer Verlag, Berlin.

Maissel, L.I.and Glang, R. (eds) (1970) *Handbook of Thin Film Technology*, McGraw-Hill, New York.

Ohring, M. (1992) *The Materials Science of Thin Films*, Academic Press, New York.

Panish, M.B. and Temkin, H. (1993) *Gas Source Molecular Beam Epitaxy: Growth and Properties of Phosphorus Containing III-V Heterostructures*, Springer-Verlag, Berlin.

Park, H., Fang, A.W., Liang, D. *et al.* (2008) *Review Article – Photonic Integration on the Hybrid Silicon Evanescent Device Platform, Advances in Optical Technologies*, Hindawi Publishing Corporation, Cairo.

Plummer, J.D., Deal, M.D., and Griffin, P.B. (2000) *Silicon VLSI Technology: Fundamentals, Practice and Modeling*, Prentice Hall, Upper Saddle River, NJ.

Shul, R.J. and Pearton, S.J. (2000) *Handbook of Advanced Plasma Processing Techniques*, Springer Verlag, Berlin.

Smith, D.L. (1995) *Thin-Film Deposition: Principles and Practice*, McGraw Hill, New York.

Soref, R. (2006) The past, present, and future of silicon photonics. *IEEE J. Sel. Top. Quant.*, **12**(6), 1678–1687.

Sze, S.M. (1988) *VLSI Technology*, 2nd edn, McGraw-Hill, New York.

Sze, S.M. (2002) *Semiconductor Devices: Physics and Technology*, 2nd edn, John Wiley & Sons, Inc., New York.

Wasa, K. and Hayakawa, S. (1992) *Handbook of Sputter Deposition Technology: Principles, Technology, and Applications*, Noyes Publications, Bracknell, UK.

Wasa, K., Kitabatake, M., and Adachi, H. (2004) *Thin Film Materials Technology: Sputtering of Compound Materials*, Springer Verlag, Berlin.

Appendix A: k.p Method

A.1 Band Theory: Bloch Functions

A solid consists of many atoms and electrons. The total energy of the system is therefore the sum of the kinetic energies of all the nuclei and electrons, the potential energy due to nuclear forces, the potential energies of electrons in the field of nuclei, the potential energy due to electron–electron interactions, and the magnetic energy associated with the spin and the orbit. The total Hamiltonian of the system may be constructed accordingly. The formidable problem of solving the resultant Schrödinger equation is bypassed by introducing several approximations. Since the motion of nuclei is sluggish, the electrons instantaneously adjust their motion to that of the ions. The total wavefunction is then written as a wavefunction for ions $\phi(R)$ and that for all electrons $\psi(r, R)$ instantaneously dependent on all ionic positions **R**. An approximation, known as the *adiabatic approximation*, is introduced to decouple the Schrödinger equation into a purely ionic and a purely electronic equation, which are expressed, respectively, as

$$H_L\phi(\mathbf{R}) = E_L\phi(\mathbf{R}) \tag{A.1}$$

and

$$H_e\psi(\mathbf{r}, \mathbf{R}) = E_e\psi(r, R) \tag{A.2}$$

where **r** denotes the electronic coordinates [1, 2].

The electron potential energy is due to electron–electron and electron–ion interactions. If a suitable average is found for the first, a constant repulsive contribution can be added to the electron energy and then each electron becomes independent. The one-electron Schrödinger equation then takes the form

$$H_{ei}\psi_i(\mathbf{r}_i, \mathbf{R}) = E_{ei}\psi_i(\mathbf{r}_i, \mathbf{R}) \tag{A.3}$$

where

$$H_{ei} = \frac{\mathbf{p}_i^2}{2m_0} + \sum_i V(\mathbf{r}_i, \mathbf{R}_i) \tag{A.4}$$

Silicon Photonics: Fundamentals and Devices, First Edition. M. Jamal Deen and P. K. Basu.

and $\boldsymbol{p}_i$ is the momentum of the ith electron. The Hamiltonian still depends on the fluctuating position of the ion. In the next approximation, the ions are assumed to lie in their equilibrium position and the effect of ionic vibration is taken as a perturbation. Thus the problem is reduced to solving the equation

$$\frac{\mathbf{p}^2}{2m_0}+\sum_i V(\mathbf{r}-\mathbf{R}_{10})\psi(\mathbf{r})=E\psi(\mathbf{r}) \tag{A.5}$$

The ionic potential V is periodic and the eigenfunctions are Bloch functions expressed as

$$\psi_{n\mathbf{k}}(\mathbf{r})=U_{nk}(r)\exp\left(j\mathbf{k}.\mathbf{r}\right), \tag{A.6}$$

where the cell periodic part $U(\mathbf{r})$ obeys the relation

$$U_{nk}(\mathbf{r}+\mathbf{R})=U_{nk}(\mathbf{r}). \tag{A.7}$$

In the above equations, $\mathbf{R}$ is a vector of the Bravais lattice, n denotes the band index, and $\boldsymbol{k}$ is a wave vector of the electron in the first Brillouin zone. From Eqs. (A.6) and (A.7)

$$\psi_{nk}(\mathbf{r}+\mathbf{R})=\psi_{nk}(\mathbf{r})\exp\left(j\mathbf{k}.\mathbf{R}\right)$$

The Bloch functions are eigenfunctions of the one-electron Schrödinger equation and therefore they are orthogonal to one another. Thus

$$\int \psi_{n'k'}\psi_{nk}d^3r=\delta_{n',n}\delta_{k'k} \tag{A.8}$$

The wavefunctions are also normalized over the volume V of the crystal and therefore

$$\psi_{nk}=V^{-1/2}U_{nk}(r)\exp\left(j\mathbf{k}.\mathbf{r}\right) \tag{A.9}$$

A.2 The k.p Perturbation Theory Neglecting Spin

Complete knowledge of the band structure of a semiconductor requires that the full E–$\boldsymbol{k}$ dispersion relation be known completely. The Schrödinger equation should therefore be solved completely. This is a rather difficult task since the form of the periodic potential $V(\mathbf{r})$ must be specified. Fortunately, in most descriptions of the electron and hole properties in semiconductors, the mostly populated electron and hole states lie within a fraction of an eV from the band edges. Thus, if the wavefunctions and energies of the carriers are known at the band extrema, then perturbation methods may be applied to find out the wavefunctions and energies at other points in the Brillouin zone, leading to knowledge of the E–$\boldsymbol{k}$ relationship. The method, known as ***k.p** perturbation theory*, is most widely used in the study of transport and optical processes in common semiconductors [3–9].

The $\boldsymbol{k}.\boldsymbol{p}$ perturbation theory is based on the fact that the cell periodic part U_k of the electrons, for any value of k but different bands, forms a complete set. Let us consider the wavefunctions for the electrons having a value $\boldsymbol{k}$ near the minima in the nth band. For simplicity we assume that the minima are located at $\boldsymbol{k}=0$. The theory is applicable also when the minima are located at $\boldsymbol{k}=\boldsymbol{k}_0$. The wavefunction is given by

$$\psi=U_{nk}(\mathbf{r})\exp(j\mathbf{k}.\mathbf{r})=\left[\sum_m c_m U_{m0}(\mathbf{r})\right]\exp\left(j\mathbf{k}.\mathbf{r}\right), \tag{A.10}$$

since U_{m0} forms a complete orthonormal set. Using this form of ψ in the Schrödinger equation, one obtains

$$\left[-\frac{\hbar^2}{2m_0}\nabla^2+\frac{\hbar^2}{m_0}\mathbf{k}.\mathbf{p}+\frac{\hbar^2k^2}{2m_0}+V(\mathbf{r})\right]U_{nk}(\mathbf{r})=E_n(\mathbf{k})U_{nk}(\mathbf{r}). \tag{A.11}$$

However, U_{m0} is the wavefunction for $\mathbf{k}=0$ in the nth band satisfying the equation

$$\left[-\frac{\hbar^2}{2m_0}\nabla^2+V(\mathbf{r})\right]U_{m0}(\mathbf{r})=E_m(0)U_{m0}(\mathbf{r}). \tag{A.12}$$

We now put Eq. (A.10) in (A.11) and use (A.12) to obtain

$$\sum c_m\left[E_m(0)+\frac{\hbar^2}{2m_0}k^2+\frac{\hbar^2}{m_0}\mathbf{k}.\mathbf{p}\right]U_{m0}(\mathbf{r})=\sum_{\mathbf{m}} c_mE_n(\mathbf{k})U_{m0}(\mathbf{r}). \tag{A.13}$$

Multiplying both sides of Eq. (A.13) by $U^*_{l0}(\mathbf{r})$ and integrating over a volume of a unit cell (V_c), the following set of linear homogeneous equations is obtained:

$$c_l\left[E_n(\mathbf{k})-E_l(0)-\frac{\hbar^2}{2m_0}k^2\right]-\sum c_m\frac{\hbar}{m_0}(\mathbf{k}.\mathbf{p}_{lm})=0, \tag{A.14}$$

where

$$\mathbf{p}_{lm}=\int_{V_c}U^*_{l0}(\mathbf{r})\,\mathbf{p}\,U_{m0}(\mathbf{r})d^3r. \tag{A.15}$$

By giving l successive integer values, one obtains the full set of equations.

In the general case, the set of equations has a nontrivial solution if the determinant of the coefficients c_l is zero. This condition gives the energy eigenvalues $E_n(\mathbf{k})$ in terms of the quantities $E_m(0)$ and p_{lm}. The relative values of the expansion coefficients c_m are then obtained by using the values of $E_n(\mathbf{k})$. The absolute values of c_m are obtained by imposing normalization conditions on ψ. The accuracy of the calculation is increased if many such coefficients are included. For practical reasons we need to limit ourselves to a few bands. The bands of greatest interest in common semiconductors are conduction (C), heavy hole (HH), light hole (LH), and split-off (SO) bands. Each of these four bands has two spin components, so there are altogether eight bands. Depending on the problem at hand and the degree of accuracy required, we may use some or all of these eight bands. We present below the results by using different approximations regarding the number of bands.

A.2.1 Single-Electron Band

Let us assume that U_{nk} is determined mostly by U_{n0} and the contributions from other bands are smaller. In other words, we assume that $c_m(m\neq n)\ll c_n$. Then

$$\psi=\left[c_nU_{n0}(\mathbf{r})+\sum c_mU_{m0}(\mathbf{r})\right]\exp\,(j\mathbf{k}.\mathbf{r}). \tag{A.16}$$

Since ψ is normalized, $\sum_m|c_m|^2=1$; but we have assumed $c_m\ll c_n$. It follows, therefore, that $c_n\approx 1$.

To solve Eq. (A.16), stationary perturbation theory is applied. First, neglect c_m $(m \neq n)$ in comparison to c_n in the nth equation. The result is

$$E_n(\mathbf{k}) = E_n(0) + \frac{\hbar^2 k^2}{2m_0} + \frac{\hbar}{m_0}\mathbf{k}.\mathbf{p_{nn}}. \tag{A.17}$$

Next put $c_n = 1$ and obtain from Eq. (A.14)

$$c_m \approx \frac{\hbar}{m_0}\frac{\mathbf{k}.\mathbf{p}_{mn}}{E_n(0) - E_m(0)}c_n, \tag{A.18}$$

neglecting the ***k.p*** term in the denominator. If this expression for c_m is now used in Eq. (A.14), a second-order approximation results and one obtains

$$E_n(\mathbf{k}) = E_n(0) + \frac{\hbar^2 k^2}{2m_0} + \frac{\hbar}{m_0}\mathbf{k}.\mathbf{p}_{nn} + \sum_{m \neq n}\left(\frac{\hbar}{m_0}\right)^2 \frac{|\mathbf{k}.\mathbf{p}_{nn}|^2}{E_n(0) - E_m(0)}. \tag{A.19}$$

Since the extrema occur at $\boldsymbol{k} = 0$, $\boldsymbol{p}_{\rm nn} = 0$. Therefore, by choosing a proper coordinate system, one may write

$$E_n(\mathbf{k}) = E_n(0) + \frac{\hbar^2 k^2}{2m_i},$$

where, from Eq. (A.19),

$$\frac{1}{m_i} = \frac{1}{m_0} + \frac{2}{m_0{}^2}\sum_m \frac{|\mathbf{i}.\mathbf{p}_{\rm nm}|^2}{E_n(0) - E_m(0)}, \tag{A.20}$$

where $\mathbf{i}$ is a unit vector along the ith coordinate axis. The above equation predicts a parabolic *E*–***k*** relation.

The analysis presented in this subsection may be improved by combining with the band under consideration other bands close to it and treating the effects of the additional bands as small perturbations.

A.2.2 Four Bands

We now consider four bands: the C, HH, LH, and SO bands. The valence bands are triply degenerate, and the C-band minima and H-band maxima occur at $\boldsymbol{k} = 0$. For the present, the spin–orbit interaction is neglected. We denote the cell periodic parts of the conduction band by U_c and those of the three valence bands by U_{v1}, U_{v2}, U_{v3}. Also the symbols E_c and E_v are used to denote, respectively, the energy for conduction band minima and valence band maxima. We may write the wavefunction for any ***k*** as

$$\psi = (a_k U_c + b_k U_{v1} + c_k U_{v3} + d_k U_{v2})\exp(j\mathbf{k}.\mathbf{r}), \tag{A.21}$$

in accordance with Eq. (A.10). Using the symbol $E' = E - \hbar^2 k^2/2m_0$, the linear homogeneous equations are

$$\begin{aligned}
&a_k(E' - E_c) - (\hbar/m_0)\mathbf{k}.(b_k\mathbf{p}_{cv1} + d_k\mathbf{p}_{cv2} + c_k\mathbf{p}_{cv3}) = 0\\
&-a_k(\hbar/m_0)\mathbf{k}.\mathbf{p_{cv1}} + b_k(E' - E_v) - (\hbar/m_0)\mathbf{k}.(d_k\mathbf{p}_{v1v2} + c_k\mathbf{p}_{v1v3}) = 0\\
&-a_k(\hbar/m_0)\mathbf{k}.\mathbf{p_{cv2}} - b_k(\hbar/m_0)\mathbf{k}.\mathbf{p}_{v1v2} - d_k(E' - E_v) - (\hbar/m_0)\mathbf{k}. = 0\\
&-a_k(\hbar/m_0)\mathbf{k}.\mathbf{p_{cv3}} - b_k(\hbar/m_0)\mathbf{k}.\mathbf{p}_{v1v3} - d_k(\hbar/m_0)\mathbf{k}. + c_k(E' - E_v) = 0
\end{aligned} \tag{A.22}$$

The quantities p_{cv1}, p_{v1v2}, and so on are defined in Eq. (A.15). The matrix elements may be evaluated once the Us are known. Since U_c is an atomic s-like function and the U_v' s are p-like functions, the $\boldsymbol{k}.\boldsymbol{p}_{cv1}$ term may be expressed as

$$h_1 = \frac{\hbar^2}{m_0 j}\int U_t^*\left(k_x\frac{\partial}{\partial x}+k_y\frac{\partial}{\partial y}+k_z\frac{\partial}{\partial z}\right) U_m d^3r \quad t, m = s, x, y, z. \tag{A.23}$$

Since $(\partial/\partial x)U_s$ is an odd function of x, the matrix element $\int U_j^*\,(\partial/\partial x)U_s\,dx$ is nonzero only when $j = x$. The same is true for $\int U_s^*\,(\partial/\partial x)U_j dx$. The function $(\partial/\partial x)\,U_y$ is odd in both x and y. Thus

$$\iint U_j^*\frac{\partial}{\partial x}U_y dxdy = 0, \quad j = s, x, y, z.$$

The only nonvanishing matrix elements are the ones defined by the following expressions:

$$P = \frac{\hbar^2}{m_0}\int U_j\frac{\partial}{\partial j}U_s d^3r = -\frac{\hbar^2}{m_0}\int U_s\frac{\partial}{\partial j}U_j d^3r, \quad j = x, y, z. \tag{A.24}$$

In the following, we shall assume that $\boldsymbol{k}$ is parallel to the z direction. Then we may rewrite the four homogeneous equations (A.22) as

$$\begin{aligned} a_k(E'-E_c)-c_k Pk &= 0,\\ b_k(E'-E_v) &= 0,\\ -a_k Pkc_k(E'-E_v) &= 0,\\ d_k(E'-E_v) &= 0. \end{aligned} \tag{A.25}$$

The energy eigenvalues are thus given by

$$E' = E_c, E_V \text{ and } (E'-E_c)(E'-E_v)-P^2k^2 = 0. \tag{A.26}$$

These equations give the dispersion relations when the conduction bands and valence bands are strongly coupled. Denoting the energy gap by $E_g = E_c - E_v$, we may write Eq. (A.26) as

$$(E'-E_c)(E'-E_c+E_g)-P^2k^2 = 0.$$

When E' tends to E_c, we may neglect E'–E_c in comparison to E_g and write

$$(E'-E_c)E_g = P^2k^2. \tag{A.27}$$

If the band edge effective mass is denoted by m_{e0}, we obtain

$$P^2 = \left(\frac{E-E_c-\hbar^2k^2}{2m_0}\right)\frac{E_g}{k^2} = \hbar^2\left(\frac{1}{m_{e0}}-\frac{1}{m_0}\right)\frac{E_g}{2}. \tag{A.28}$$

A.3 Spin–Orbit Interaction

The electron is a fermion with spin $^1/_2$ in units of $\hbar$. In classical mechanics, a point particle rotating about an axis has an angular momentum $\mathbf{L} = \mathbf{r} \times \mathbf{p}$. In the quantum picture, the angular momentum of a point particle is quantized and the intrinsic value of the momentum is called the *spin*. There is strong interaction between the spin and orbital motion of the electrons. This spin–orbit coupling may be calculated for isolated atoms; however, it is difficult to do so in crystals.

A.3.1 Spin–Orbit Interaction Term

To calculate the interaction a general form of spin–orbit interaction is assumed with a fitting parameter that is adjusted to fit experimentally observed effects. The total Hamiltonian in the presence of spin–orbit interaction is written as $H = H_0 + H_{so}$, where H_0 is the Hamiltonian without interaction and H_{so} is the spin–orbit interaction written as

$$H_{so} = \lambda \mathbf{L} \cdot \mathbf{S}. \tag{A.29}$$

Here $\boldsymbol{L}$ represents the operator for orbital angular momentum, $\boldsymbol{S}$ is the operator for spin angular momentum, and λ is treated as a constant. The total angular momentum $\boldsymbol{J}$ may be expressed as

$$\mathbf{J}^2 = (\mathbf{L} + \mathbf{S})^2 = L^2 + S^2 + 2\mathbf{L} \cdot \mathbf{S}. \tag{A.30a}$$

Thus

$$\langle \mathbf{L} \cdot \mathbf{S} \rangle = (1/2)\langle J^2 - L^2 - S^2 \rangle = \frac{\hbar^2}{2}[j(j+1) - l(l+1) - s(s+1)], \tag{A.30b}$$

where j, l, and s are the quantum numbers for the operators $\boldsymbol{J}$, $\boldsymbol{L}$, and $\boldsymbol{S}$, respectively. However, to calculate the spin–orbit interaction energy, one needs the pure angular momentum states to which Eq. (A.30) is applicable. One should note that states like $|X\alpha\rangle$ are mixed states, with the symbol α denoting the spin-up state. To illustrate this statement, we express $|X\rangle$ in terms of pure angular momentum states, that is,

$$\begin{aligned} |X\rangle &= \frac{1}{\sqrt{2}}(-\phi_{1,1} + \phi_{1,-1}) \\ |Y\rangle &= \frac{j}{\sqrt{2}}(\phi_{1,1} + \phi_{1,-1}) \\ |Z\rangle &= \phi_{1,0} \end{aligned} \tag{A.31}$$

The $\phi_{i,j}$'s are pure angular momentum states, and the expressions for the lower eigenstates are

$$\begin{aligned} \phi_{1,\pm 1} &= Y_{1,\pm 1}(\theta, \phi) = \mp \sqrt{\frac{3}{8\pi}} \sin\theta \exp(\pm j\phi) \\ \phi_{1,0} &= Y_{1,0}(\theta, \phi) = \sqrt{\frac{3}{4\pi}} \cos\theta \end{aligned}$$

The $\phi_{i,j}$'s are eigenfunctions of L^2 and L_z. The respective quantum numbers are $l = I$ and $l_z = j$. For example, $L^2\phi_{1,-1} = \hbar^2(1)(1+1)\phi_{1,-1} = 2\hbar^2\phi 1, -1$ and $L_z\phi_{1,-1} = -1\hbar\phi_{1,-1}$.

Equation (A.31) is modified if spin is included; for example, the spin-up state $p_x = |X\alpha\rangle$ is expressed as

$$|X\alpha\rangle = \frac{1}{\sqrt{2}}(-\phi_{1,1} + \phi_{1,-1})\alpha.$$

This formulation is still in terms of mixed states. To decompose the mixed states into states of pure angular momentum, the spin and orbital angular momentum must be added to obtain the total angular momentum states. The standard Clebsch–Gordan (CG) technique is employed for this addition. The following six equations are obtained as a result:

$$\begin{aligned}
\phi_{3/2,3/2} &= \phi_{1,1}\alpha = (-1/\sqrt{2})|(X + jY)\alpha\rangle \\
\phi_{3/2,1/2} &= \frac{1}{\sqrt{3}}\phi_{1,1}\beta + \frac{2}{\sqrt{6}}\phi_{1,0}\alpha = \frac{-1}{\sqrt{6}}[|(X + jY)\beta\rangle - |2Z\alpha\rangle] \\
\phi_{3/2,-1/2} &= \frac{1}{6}\phi_{1,0}\beta + \frac{1}{\sqrt{3}}\phi_{1,-1}\alpha = \frac{1}{\sqrt{6}}[|(X - jY)\alpha\rangle + |2Z\beta\rangle] \\
\phi_{3/2,-3/2} &= \phi_{1,-1}\beta = (1/\sqrt{2})|(X - jY)\beta\rangle \\
\phi_{1/2,1/2} &= \frac{-1}{\sqrt{3}}\phi_{1,0}\alpha + \frac{2}{\sqrt{6}}\phi_{1,1}\beta = \frac{-1}{\sqrt{3}}[|(X + jY)\beta\rangle + |Z\alpha\rangle] \\
\phi_{1/2,-1/2} &= \frac{-2}{\sqrt{6}}\phi_{1,-1}\alpha + \frac{1}{\sqrt{3}}\phi_{1,0}\beta = \frac{-1}{\sqrt{3}}[|(X - jY)\alpha\rangle - |Z\beta\rangle]
\end{aligned} \tag{A.32}$$

These six equations are inverted to find states like $\phi_{1,0}$, and from the resultant equations one gets states like $|X\alpha\rangle$, and so on.

$$\begin{aligned}
|X\alpha\rangle &= \frac{1}{\sqrt{2}}\left[-\phi_{3/2,3/2} + \frac{1}{\sqrt{3}}\phi_{3/2,-1/2} - \sqrt{\frac{2}{3}}\phi_{1/2,-1/2}\right] \\
|X\beta\rangle &= \frac{1}{\sqrt{2}}\left[-\frac{1}{\sqrt{3}}\phi_{3/2,1/2} - \frac{2}{\sqrt{3}}\phi_{1/2,1/2} - \phi_{3/2,-3/2}\right] \\
|Y\alpha\rangle &= \frac{j}{\sqrt{2}}\left[\phi_{3/2,3/2} + \frac{1}{\sqrt{3}}\phi_{3/2,-1/2} - \sqrt{\frac{2}{3}}\phi_{1/2,-1/2}\right] \\
|Y\beta\rangle &= \frac{j}{\sqrt{2}}\left[\frac{1}{\sqrt{3}}\phi_{3/2,1/2} + \frac{2}{\sqrt{3}}\phi_{1/2,1/2} + \phi_{3/2,-3/2}\right] \\
|Z\alpha\rangle &= \sqrt{\frac{2}{3}}\phi_{3/2,1/2} - \frac{1}{\sqrt{3}}\phi_{1/2,1/2} \\
|Z\beta\rangle &= \sqrt{\frac{2}{3}}\phi_{3/2,-1/2} + \frac{1}{\sqrt{3}}\phi_{1/2,-1/2}.
\end{aligned} \tag{A.33}$$

The phases used in the above expressions for ϕ_{j,m_j} in terms of $|X\alpha\rangle, \ldots .|Z\beta\rangle$ are obtained in the standard derivation of Clebsch–Gordan coefficients. The overall phase of a state is arbitrary and has no effect on the physical predictions. The convention used by Luttinger and Kohn [10] is in widespread use and will be used here. The states are expressed in terms of CG states as

$$\begin{aligned}
\phi_{3/2,3/2}(\mathrm{LK}) &= -\phi_{3/2,3/2}(\mathrm{CG}) = \frac{1}{\sqrt{2}}|(X+jY)\alpha\rangle \\
\phi_{3/2,1/2}(\mathrm{LK}) &= -j\phi_{3/2,1/2}(\mathrm{CG}) \\
\phi_{3/2,-1/2}(\mathrm{LK}) &= \phi_{3/2,-1/2}(\mathrm{CG}) \\
\phi_{3/2,-3/2}(\mathrm{LK}) &= j\phi_{3/2,-3/2}(\mathrm{CG}) \\
\phi_{1/2,1/2}(\mathrm{LK}) &= -\phi_{1/2,1/2}(\mathrm{CG}) \\
\phi_{1/2,-1/2}(\mathrm{LK}) &= j\phi_{1/2,-1/2}(\mathrm{CG})
\end{aligned} \tag{A.34}$$

The spin–orbit Hamiltonian may be calculated now with the above states. The interaction is

$$H_{so} = \frac{\lambda\hbar^2}{2}[j(j+1) - l(l+1) - s(s+1)] \tag{A.35}$$

For p-type electron orbitals $l=1$ and $s=1/2$, j is given by the first subscript of ϕ in Eq. (A.33). Many terms become zero as the pure states are orthogonal. We conclude that only the following terms are nonzero:

$$\begin{aligned}
&\langle X\alpha|H_{so}|Y\alpha\rangle = \langle Y\alpha|H_{so}|Z\beta\rangle = \langle Y\beta|H_{so}|Z\alpha\rangle = -j\frac{\Delta}{3} \\
&\langle X\alpha|H_{so}|Z\beta\rangle = \frac{\Delta}{3};\ \langle X\beta|H_{so}|Z\alpha\rangle = -\frac{\Delta}{3};\ \langle X\beta|H_{so}|Y\beta\rangle = j\frac{\Delta}{3},
\end{aligned} \tag{A.36}$$

where Δ is a parameter known as *spin–orbit splitting* given by $\Delta = \Delta_{so} = 3\lambda\hbar^2/2$.

A.3.2 Conduction Band Energy

The calculation of energy levels in the conduction band by including the spin–orbit interaction is easier, and therefore we consider it first. As mentioned previously, we are interested in four basic vectors, the $|S\rangle$ state for the conduction band and the $|X\rangle.|Y\rangle, |Z\rangle$ states for the valence bands. There are four coefficients, $a_k \ldots .d_k$ as in Eq. (A.21), needed to describe a state. With the inclusion of spin, the number of basis vectors to be considered becomes eight. The secular equation containing the coefficients and basis vectors involves an 8×8 matrix. It turns out that if the basis vectors are arranged in the following manner,

$$|S\alpha\rangle.|(X+jY)\beta\rangle, |Z\alpha\rangle, |(X-jY)\beta\rangle, |S\beta\rangle, |(X-jY)\alpha\rangle, |Z\rangle, \text{ and } |-(X+jY)\alpha\rangle,$$

the matrix may be written in the form

$$\begin{bmatrix} H & 0 \\ 0 & H \end{bmatrix},$$

where H is a 4×4 matrix. Using the matrix elements between different states and the earlier elements obtained without the spin–orbit interactions, we may express the different matrix elements in terms of P in the following form:

$$\begin{array}{ccccc} & |S\alpha\rangle & |(X+jY)\beta\rangle & |Z\alpha\rangle & |(X-jY)\beta\rangle \\ |S\alpha\rangle & E_s-E' & 0 & -jkP & 0 \\ |(X+jY)\beta\rangle & 0 & E_p-E'-\Delta/3 & \sqrt{2}\Delta/3 & 0 \\ |Z\alpha\rangle & jkP & \sqrt{2}\Delta/3 & E_p-E' & 0 \\ |(Xxx-jY)\beta\rangle & 0 & 0 & 0 & E_p-E'+\Delta/3 \end{array} \tag{A.37}$$

To simplify the calculation, we choose the $\boldsymbol{k}$-vector in (A.37) along the z-direction. Furthermore, to account for the shift of band energies due to spin–orbit interaction, notations E_p and E_s are used. The difference in signs in $E_p-E'\pm\Delta/3$ is due to the fact that $L_zS_z|(X+jY)\beta\rangle = -|(X+jY)\beta\rangle$ while $L_zS_z|(X-jY)\beta\rangle = |(X-jY)\beta\rangle$. Expanding (A.37) one obtains

$$E' = E_p + \Delta/3 \tag{A.38a}$$

$$\left(E'-E_p+\frac{2\Delta}{3}\right)\left(E'-E_p-\frac{\Delta}{3}\right)(E'-E_s)-k^2P^2\left(E'-E_p+\frac{\Delta}{3}\right)=0 \tag{A.38b}$$

For small values of k^2, the cubic equation can easily be solved by treating the term k^2P^2 as a small perturbation. This yields

$$E_1' = E_s + \frac{k^2P^2(E_s-E_p+\Delta/3)}{(E_s-E_p+2\Delta/3)(E_s-E_p-\Delta/3)}. \tag{A.38c}$$

Let $E_v = E_p+\Delta/3 = 0, E_c = E_s = E_{g0}$ and $E_c-E_v = E_{g0}$, the direct gap. We then rewrite Eq. (A.38a) as

$$E'(E'-E_{g0})(E'+\Delta)-k^2P^2(E'+2\Delta/3) = 0. \tag{A.39}$$

Taking $E_c = E_{g0}$ in the first approximation,

$$E_c(\mathbf{k}) = E_{g0} + \frac{\hbar^2k^2}{2m_0} + \frac{k^2P^2}{3}\left[\frac{2}{E_{g0}} + \frac{1}{E_{g0}+\Delta}\right]. \tag{A.40a}$$

We also obtain, by putting $E_v = E_p+\Delta/3 = 0$,

$$E_{v1}(\mathbf{k}) = \frac{\hbar^2k^2}{2m_0} \tag{A.40b}$$

Taking $E = 0$,

$$E_{v2} = \frac{\hbar^2k^2}{2m_0} - \frac{2k^2P^2}{3E_{g0}}. \tag{A.40c}$$

Finally, taking $E' = -\Delta$ as a first approximation,

$$E_{v3} = -\Delta + \frac{\hbar^2k^2}{2m_0} - \frac{k^2P^2}{3(E_{g0}+\Delta)}. \tag{A.40d}$$

In Eq. (A.40a), E_c is the energy of the conduction band electrons, while E_{v1}, E_{v2}, and E_{v3}, are, respectively, the energies of the three valence bands. From Eq. (A.40a), we may define the band edge effective mass for conduction band electrons by writing

$$E_c = E_{g0} + \hbar^2 k^2 / 2m_{e0}.$$

It then follows that

$$\frac{1}{m_{e0}} = \frac{1}{m_0} + \frac{2P^2}{3\hbar^2}\left(\frac{2}{E_{g0}} + \frac{1}{E_{g0} + \Delta}\right). \tag{A.41}$$

The momentum matrix element, P, which is central to all calculation of transition probabilities from the valence band to the conduction band, may be expressed in terms of m_{e0} as

$$P^2 = \frac{\hbar^2}{2m_{e0}} \frac{E_{g0}(E_{g0} + \Delta)}{E_{go} + 2\Delta/3} \frac{m_0 - m_{e0}}{m_0}. \tag{A.42}$$

A.3.3 Valence Band Energies

The earlier treatment of the dispersion relation cannot explain the properties of electrons in the valence band. Referring to Eq. (A.40a), one notices that the energy of the electrons increases with k, which however is opposite to what is observed experimentally. To treat the valence band properties correctly, degenerate perturbation theory is needed [11].

Let the state of the electron in the lth band at $k = 0$ be degenerate, having f-fold degeneracy. It follows from the theory of perturbation of degenerate states that the second-order corrections $\Delta E^{(2)}$ due to the $(\hbar/m_0)(\mathbf{k}.\mathbf{p})$ perturbation are the roots of the secular equation

$$\left[\left(\frac{\hbar^2}{m_0}\right)^2 \sum_{n,s}{}' \frac{\langle l, r'|\mathbf{k}.\mathbf{p}|n, s\rangle \langle n, s|\mathbf{k}.\mathbf{p}|l, r\rangle}{E_l(0) - E_n(0)} - \Delta E^{(2)} \delta_{rr'}\right] = 0, \tag{A.43}$$

where the primed summation sign indicates that the summation is over all $n \neq l$ and over s. $|l, r\rangle$ and $|l, r'\rangle$ are the unperturbed f-fold degenerate wavefunctions (r, $r' = 1,2,\ldots f$) satisfying Eq. (A.43) for the energy eigenvalue $E_l(0)$. The $|n, s\rangle$'s are the wavefunctions for energy level $E_n(0)$. The order of the determinant of the secular equation is equal to the degree of degeneracy of the level E_l (0). In the present situation, the degenerate eigenstates at $k = 0$ are the three $l = 1$, $m_l = 0$, and ± 1 states. To the second order in perturbation, the energy in the nth band is

$$E_n(k) = E_n(0) + \frac{\hbar^2 k^2}{2m_0} + \Delta E_n^{(2)}(k), \tag{A.44}$$

where $\Delta E_n^{(2)}$ is given by Eq. (A.43) above. Hence we have the set of three equations

$$\sum_{i=1}^{3}\left[\sum_{m}{}' \frac{\langle j|H_l|m\rangle\langle m|H_l|l\rangle}{E_l(0)-E_m(0)}+\left\{E_l+\frac{\hbar^2k^2}{2m_0}-E_n(k)\right\}\delta_{j,l}\right]\langle n|k\rangle=0. \tag{A.45}$$

Nontrivial solutions of this set of N coupled homogeneous equations occur only if

$$\det[|H|-E_n(k)\mathbf{I}]=0,$$

where $\mathbf{I}$ is the identity matrix and H is a 3×3 matrix whose elements are

$$H_{jl}=\left[E_l+\frac{\hbar^2k^2}{2m_0}\right]\delta_{jl}+\sum_{m}{}'\frac{\langle j|H_l|m\rangle\langle m|H_l|l\rangle}{E_l(0)-E_m(0)}.$$

The calculation of the matrix element is first made by ignoring the spin–orbit interaction for the present. We take the basis sets as $|X\rangle.|Y\rangle$ and $|Z\rangle$. Then

$$H_{11}=\langle X|H|X\rangle=E_1(0)+\frac{\hbar^2k^2}{2m_0}+\sum_{m}{}'\frac{|\langle X|H_1|m\rangle|^2}{E_l(0)-E_m(0)}. \tag{A.46}$$

Since $|X\rangle$ is proportional to $xf(\mathbf{r})$, we may verify that

$$\frac{m_0^2}{\hbar^2}|\langle X|H_l|m\rangle|^2=|\langle X|p_x|m\rangle|^2k_x^2+|\langle X|p_y|m\rangle|^2k_y^2+|\langle X|p_z|m\rangle|^2k_z^2.$$

Thus we may write

$$H_{11}=E_1+\sum_{j=x,y,z}\left[\frac{\hbar^2}{2m_0}+\frac{\hbar^2}{m_0^2}\sum{}'\frac{|\langle X|p_j|m\rangle|^2}{E_l-E_m}\right]k_j^2.$$

Due to symmetry at $k=0$,

$$|\langle X|p_y|m\rangle|^2=|\langle X|p_z|m\rangle|^2.$$

Therefore,

$$H_{11}=E_1+Ak_x^2+B(k_y^2+k_z^2), \tag{A.47a}$$

where

$$A=\frac{\hbar^2}{2m_0}+\frac{\hbar^2}{m_0{}^2}\sum_{j}{}'\frac{|\langle X|p_x|j\rangle|^2}{E_l-E_j}, \tag{A.47b}$$

$$B=\frac{\hbar^2}{2m_0}+\frac{\hbar^2}{m_0^2}\sum_{j}{}'\frac{|\langle X|p_y|j\rangle|^2}{E_l-E_j}. \tag{A.47c}$$

The remaining matrix elements can be evaluated in a similar way to give the following Hamiltonian matrix,

$$H = \begin{bmatrix} E_1 + Ak_x^2 + B(k_y^2 + k_z^2) & Ck_xk_y & Ck_xk_z \\ Ck_xk_y & E_1 + Ak_y^2 + B(k_x^2 + k_z^2) & Ck_yk_z \\ Ck_xk_z & Ck_yk_z & E_1 + Ak_z^2 + B(k_x^2 + k_y^2) \end{bmatrix} \tag{A.48}$$

where

$$C = \frac{\hbar^2}{m_0^2} \sum_j{}' \frac{\langle X|p_x|j\rangle\langle j|p_y|Y\rangle + \langle X|p_y|j\rangle\langle j|p_x|Y\rangle}{E_l - E_j}. \tag{A.49}$$

Let us now consider the effect of spin. As noted earlier in Eq. (A.30) for $j = 1/2$, $\langle \mathbf{L.S}\rangle = -\hbar^2$, and for $j = 3/2$, $\langle \mathbf{L.S}\rangle = +\hbar^2/2$. Thus the states are split by an amount Δ proportional to $(3/2)\hbar^2$, the doubly degenerate state with $j = 3/2$ moving up by $\Delta/3$ and the single $j = 1/2$ state moving down by $2\Delta/3$. Since in the designation $|ls; jm_j\rangle$, $l = 1$ and $s = 1/2$, we shall use only the symbol $|jm_j\rangle$ to denote the states. As noted in this appendix, the Hamiltonian for the spin–orbit coupling is diagonalized if the states are chosen according to Eq. (A.32).

We shall treat the $j = 3/2$ states and $j = 1/2$ states separately, since the splitting energy is large. The Hamiltonian matrix $|H|$ now becomes a 4×4 matrix for $j = 3/2$ states and a 2×2 matrix for $j = 1/2$ states. We may evaluate the matrix elements for the 4×4 matrix using Eq. (A.47). Thus

$$\begin{aligned} H_{11} &= \langle 3/2, 3/2|H|3/2, 3/2\rangle = (1/2)\langle (X + jY)\alpha|H|(X + jY)\alpha\rangle \\ &= (1/2)[\langle X\alpha|H|X\alpha\rangle + \langle Y\alpha|H|Y\alpha\rangle + j\langle X\alpha|H|Y\alpha\rangle - j\langle Y\alpha|H|X\alpha\rangle] \\ &= E_1 + \frac{A}{2}(k_x{}^2 + k_y{}^2) + \frac{B}{2}(k_x{}^2 + k_y{}^2 + 2k_z{}^2) \\ &= H_{44} \end{aligned} \tag{A.50}$$

where the symbols 1, 2, 3, and 4 are used in the order in which the states are written in Eq. (A.48). Similarly,

$$H_{12} = \frac{1}{2\sqrt{3}}\langle (X + jY)\alpha|H|[(X + jY)\beta - 2Z\alpha]\rangle = \frac{1}{\sqrt{3}}(H_{xz} - jH_{yz}).$$

Instead of computing matrix elements $\langle n|p|m\rangle$ from first principles, one replaces them with experimentally determined parameters called *Luttinger parameters*, defined as

$$\gamma_1 = -2m_0(A + 2B)/3\hbar^2, \quad \gamma_2 = -m_0(A - B)/3\hbar^2, \quad \gamma_3 = -m_0C/3\hbar^2. \tag{A.51}$$

In terms of Luttinger parameters,

$$H_{11} = E_1 - \frac{\hbar^2 k_z^2}{2m_0}(\gamma_1 - 2\gamma_2) - \frac{\hbar^2(k_x^2 + k_y^2)}{2m_0}(\gamma_1 + \gamma_2) \tag{A.52}$$

Since in measurements the parameters conform to holes, and since the hole energy is positive, we write $H_{11} = -H_{hh}$ and the zero energy reference is $E_1 = 0$. Repeating the above calculation for the other matrix elements, we obtain for the Luttinger Hamiltonian,

$$H = \begin{bmatrix} H_{hh} & -c & -b & 0 \\ -c^* & H_{lh} & 0 & b \\ -b^* & 0 & H_{lh} & -c \\ 0 & b^* & -c^* & H_{hh} \end{bmatrix} \tag{A.53}$$

where

$$\begin{aligned} H_{lh} &= \frac{\hbar^2 k_z^2}{2m_0}(\gamma_1 + 2\gamma_2) - \frac{\hbar^2 (k_x^2 + k_y^2)}{2m_0}(\gamma_1 - \gamma_2) \\ c &= \frac{\sqrt{3}\hbar^2}{2m_0}\left[\gamma_2 (k_x^2 - k_y^2) - 2j\gamma_3 k_x k_y\right] \\ b &= \frac{\sqrt{3}\hbar^2}{m_0}\gamma_3 k_z (k_x - jk_y). \end{aligned} \tag{A.54}$$

In the vicinity of $k = 0$, one may use the axial approximation, where γ_2 and γ_3 are replaced by an effective Luttinger parameter,

$$\bar{\gamma} = (1/2)(\gamma_2 + \gamma_3) \tag{A.55}$$

The function c is then expressed as

$$c \cong \frac{\sqrt{3}\hbar^2 \bar{\gamma}}{2m_0}(k_x - jk_y)^2 \tag{A.56}$$

The dispersion relation for valence band holes may be written as

$$\begin{aligned} E_v &= -\frac{\hbar^2}{2m_0}\left[Ak^2 \pm \left\{B^2 k^4 + C^2 (k_x^2 k_y^2 + k_x^2 k_z^2 + k_y^2 k_z^2\right\}^{1/2}\right] \\ &= \frac{\hbar^2}{2m_0}\left[-\gamma_1 k^2 \pm \left\{4\gamma_2^2 k^4 + 12(\gamma_3^2 - \gamma_2^2)(k_x^2 k_y^2 + k_x^2 k_z^2 + k_y^2 k_z^2\right\}^{1/2}\right] \end{aligned} \tag{A.57}$$

Introducing the spherical polar coordinate system with the polar axis along the z-direction, we obtain

$$E_v = \frac{\hbar^2 k^2}{2m_0}\left[A \pm (B^2 + C^2/5)^{1/2}\right].$$

This enables us to define effective masses for heavy and light holes as

$$m_{hh} = \frac{m_0}{A - (B^2 + C^2/5)^{1/2}} \tag{A.58a}$$

$$m_{lh} = \frac{m_0}{A + (B^2 + C^2/5)^{1/2}} \tag{A.58b}$$

Example A.1

Using the values of γ's in Table A.1, the band edge effective masses for Ge are $m_{hh} = 0.33\, m_0$ and $m_{lh} = 0.04\, m_0$.; the values for Si are $m_{hh} = 0.56\, m_0$ and $m_{lh} = 0.16\, m_0$. For GaAs, the values are $m_{hh} = 0.059\, m_0$ and $m_{lh} = 0.08\, m_0$.

Table A.1 Band structure parameters for Ge, Si, and GaAs

	m_e/m_0	γ_1	γ_2	γ_3
Ge	1.58/0.082	13.25	4.20	5.56
Si	0.916/0.191	4.26	0.34	1.45
GaAs	0.067	6.8	2.1	2.9

A.3.4 Momentum Matrix Elements

It follows from Eq. (A.41) that the conduction band effective mass is expressed in terms of P and is related to p_{cv}, the momentum matrix element. The momentum matrix element also appears in the calculation of the optical absorption coefficient or the recombination rate in semiconductors. The conduction band edge state for a direct-gap semiconductor has been found to have s-type symmetry and is denoted by $|S\alpha\rangle$ and $|-S\beta\rangle$. The valence band states are written in terms of angular momentum spin representation in Eq. (A.32).

From symmetry, we find that only the matrix elements of the form

$$\langle X|p_x|S\rangle = \langle Y|p_y|S\rangle = \langle Z|p_z|S\rangle$$

are nonzero. The nonvanishing matrix elements are

$$\langle \pm 3/2|p_x|\pm S\rangle = (1/\sqrt{2})\langle X|p_x|S\rangle, \quad \langle \pm 1/2|p_x|\mp S\rangle = (1/\sqrt{6})\langle X|p_x|S\rangle$$
$$\langle \pm 3/2|p_x|\pm S\rangle = (2/\sqrt{6})\langle X|p_x|S\rangle$$

One may define a quantity:

$$E_p = \frac{2}{m_0}|\langle X|p_x|S\rangle|^2 = \frac{2}{m_0}p_{cv}^2$$

The values of E_p for different semiconductors are remarkably close to ~25 eV.

A.4 Quantum Wells

The subband structures for electrons and holes have been calculated in Chapter 3 by using simple theory. However, for refined calculation, complete knowledge of the *E*–***k*** dispersion relation is needed in order to explain the experimental results. Here we shall give the outline of the theory for valence band states in a quantum well.

A.4.1 Subband Structures for Holes

The degenerate nature of the valence bands prompts us to employ the multiband effective mass approximation. The Hamiltonian is written as

$$\sum_{v'} [H_{vv'}(\mathbf{k}) + V(z)\delta_{vv'}]\phi_m^{v'} = E_m\phi_m^{v}. \tag{A.59}$$

The complete wavefunction for the valence band hole in the mth subband in the νth valence band $\varphi_{mk}(\mathbf{r})$ is written in terms of the envelope function ϕ_m^ν as

$$\varphi_{mk}(r) = \sum_{\nu=1}^{4} \phi_m^\nu(\mathbf{k}, z) \exp(j\mathbf{k}.\mathbf{r})\, U^\nu(\mathbf{r}). \tag{A.60}$$

We have assumed as before that the conduction band is decoupled from the valence bands and only the two top valence bands are considered. We therefore consider four eigenstates: $|1\rangle = |3/2, 3/2\rangle, |2\rangle = |3/2, -1/2\rangle, |3\rangle = |3/2, 1/2\rangle, |4\rangle = |3/2, -3/2\rangle$.

The Luttinger Hamiltonians for the hole states are given by Eq. (A.53) and the coefficients are given in Eq. (A.54).

For a rectangular QW, $H_{\nu\nu'}(\mathbf{k})$ given in the matrix form described here should be replaced by $H_{\nu\nu'}(\mathbf{k}, -j\partial/\partial z)$, where $\mathbf{k}$ is now the in-plane wave vector. The simple solution for $\mathbf{k} = 0$ has been worked out in Chapter 3 (see Eq. (3.57)).

The in-plane effective masses for holes in different subbands have been calculated by different authors using various degrees of approximations. We assume that the band gap as well as the separation between the heavy-hole and split-off bands are large, so that the Hamiltonian matrix is treated as a 4×4 matrix as before. The eigenvalues are obtained by solving the secular determinant of the 4×4 matrix, and the expression is given by (A.57).

The upper and lower signs correspond, respectively, to heavy- and light-hole bands. The character of the bands becomes increasingly mixed for higher values of $\mathbf{k}$. The effective mass in the xy plane depends, in general, on the direction, and the magnitude of the anisotropy is determined by the difference in γ_2 and γ_3. In many cases the difference is small and it is justified to take a spherical average. It can be shown that the average $\left\langle k_x^2k_y^2 + k_y^2k_z^2 + k_z^2k_x^2 \right\rangle \cong k^4/5$. Putting this in Eq. (A.57), we obtain

$$E = \frac{\hbar^2}{2m_0}(-\gamma_1 \pm 2\bar{\gamma})k^2 \tag{A.61}$$

where

$$\bar{\gamma}^2 = (2\gamma_2^2 + 3\gamma_3^2)/5. \tag{A.62}$$

Each of the four eigenfunctions is of the form:

$$\varphi = [A|3/2, 3/2\rangle + B|3/2, -1/2\rangle + C|3/2, 1/2\rangle + D|3/2, -3/2\rangle] \exp(j\,\mathbf{k}.\mathbf{r}) \tag{A.63}$$

The four eigenfunctions may be expressed as column matrices.

In the practical situation when the barrier height is finite, a parameter $\bar{\gamma} = (\gamma_2 + \gamma_3)/2$ is introduced. The 4×4 Hamiltonian matrix is transformed to a new matrix $\bar{H}$ by using a unitary matrix U such that [12]

$$\bar{H} = UHU^+ = \begin{bmatrix} H^U & 0 \\ 0 & H^L \end{bmatrix}, \tag{A.64}$$

where

$$H^U = \begin{bmatrix} H_{hh} & R \\ R^* & H_{hh} \end{bmatrix} \quad \text{and} \quad H^L = \begin{bmatrix} H_{lh} & R \\ R^* & H_{lh} \end{bmatrix} \tag{A.65}$$

$$R = |c| - j|b|$$

The upper and lower blocks are then decoupled. We write now the upper- and lower-block envelope functions as

$$\varphi_{mk}^{U}(\mathbf{r}) = \sum_{\nu=1,2} g_m^{(\nu)}(\mathbf{k}, z)\exp(j\mathbf{k}.\mathbf{r})|\nu\rangle \tag{A.66a}$$

and

$$\varphi_{mk}^{L}(\mathbf{r}) = \sum_{\nu=2,4} g_m^{(\nu)}(\mathbf{k}, z)\exp(j\mathbf{k}.\mathbf{r})|\nu\rangle, \tag{A.66b}$$

where $\{|\nu\rangle\}$ denotes the transformed basis set, and the envelope functions satisfy

$$\sum_{\nu'=1,2}\left[H_{\nu\nu'}^{L}\left(\mathbf{k}.-j\frac{\partial}{\partial z}\right) + V(z)\delta_{\nu\nu'}\right]g_m^{(\nu'+2)}(\mathbf{k}, z) = E_m^{L}(\mathbf{k})g_m^{(\nu+2)}(\mathbf{k}, z). \tag{A.67}$$

A.4.2 Subband Structures for Strained Ge on GeSiSn

The theoretical calculation by Chang and Chuang [13] is somewhat along the lines shown, but includes the strain effect. The upper- and lower-block Hamiltonians are expressed as

$$H_U = \begin{bmatrix} V_v(z)-P-Q & \bar{R} \\ \dot{R}^{+} & V_v(z)-P+Q+\Delta Q \end{bmatrix} \tag{A.68a}$$

$$H_L = \begin{bmatrix} V_v(z)-P+Q+\Delta Q & \bar{R} \\ \bar{R}^{+} & V_v(z)-P-Q \end{bmatrix} \tag{A.68b}$$

$$P = -\frac{\hbar^2}{2m_0}\frac{\partial}{\partial z}\gamma_1\frac{\partial}{\partial z} + \frac{\gamma_1\hbar^2 k_t^2}{2m_0} + P_\varepsilon$$

$$P_\varepsilon = -a_v(\varepsilon_{xx} + \varepsilon_{yy} + \varepsilon_{zz})$$

$$Q = \frac{\hbar^2}{2m_0}\frac{\partial}{\partial z}\gamma_2\frac{\partial}{\partial z} + \frac{\gamma_2\hbar^2 k_t^2}{2m_0} + Q_\varepsilon$$

$$Q_\varepsilon = -\frac{b_v}{2}(\varepsilon_{xx} + \varepsilon_{yy} + \varepsilon_{zz})$$

$$\Delta Q = \frac{1}{2}\left[\sqrt{(\Delta + Q_\varepsilon)^2 + 8Q_\varepsilon^2} - (\Delta + Q_\varepsilon)\right]$$

$$\bar{R} = \frac{\sqrt{3}\hbar^2}{2m_0}\left[\left(\frac{\gamma_2+\gamma_3}{2}\right)k_t^2 - k_t\left(\frac{\partial}{\partial z}\gamma_3 + \gamma_3\frac{\partial}{\partial z}\right)\right]$$

$$\bar{R}^{+} = \frac{\sqrt{3}\hbar^2}{2m_0}\left[\left(\frac{\gamma_2+\gamma_3}{2}\right)k_t^2 + k_t\left(\frac{\partial}{\partial z}\gamma_3 + \gamma_3\frac{\partial}{\partial z}\right)\right].$$

The symbols have their usual meanings.

In a QW grown along the [100] plane, all four L valleys are equivalent. The Hamiltonian for the [111] L valley may be written as

$$H_L^{[111]}\left(k_t, k_z=-j\frac{\partial}{\partial z}\right)-\frac{\hbar^2}{2}\frac{\partial}{\partial z}\left(\frac{1}{3m_l}+\frac{2}{3m_t}\right)\frac{\partial}{\partial z}-j\frac{\sqrt{2}\hbar^2 k_1}{6}$$

$$\times\left[\begin{array}{l}\frac{\partial}{\partial z}\left(\frac{1}{m_l}-\frac{1}{m_t}\right)+\left(\frac{1}{m_l}-\frac{1}{m_t}\right)\frac{\partial}{\partial z}+\left(\frac{2}{3m_l}+\frac{1}{3m_t}\right)\frac{\hbar^2 k_1^2}{2}\\ +\frac{\hbar^2 k_2^2}{2m_t}+V^{[111]}(z)+V_\varepsilon^{[111]}(z)\end{array}\right]$$

$$V_\varepsilon^{[111]}(z)=a_L(\varepsilon_{xx}+\varepsilon_{yy}+\varepsilon_{zz})$$
$$k_1=\frac{1}{\sqrt{2}}\left(k_x+k_y-\frac{2\pi}{a}\right)$$
$$k_2=\frac{1}{\sqrt{2}}(-k_x+k_y) \tag{A.69}$$

A.4.3 Expression for Gain

In order to calculate the gain spectra, one needs to evaluate the squared momentum matrix elements for transverse electric (TE) and transverse magnetic (TM) configurations. We quote here the results for the TE momentum matrix element as given in Chang and Chuang.

$$M_{nm}^{U,TE}(k_t)=\frac{3}{2}\left|\int_{-\infty}^{\infty}dz\phi^*{}_{n,k_t=0}(z)M_b g_m^{(1)}(k_t,z)\right|^2+\frac{1}{2}\left|\int_{-\infty}^{\infty}dz\phi^*{}_{n,k_t=0}(z)M_b g_m^{(2)}(k_t,z)\right|^2$$

$$M_{nm}^{L,TE}(k_t)=\frac{3}{2}\left|\int_{-\infty}^{\infty}dz\phi^*{}_{n,k_t=0}(z)M_b g_m^{(4)}(k_t,z)\right|^2+\frac{1}{2}\left|\int_{-\infty}^{\infty}dz\phi^*{}_{n,k_t=0}(z)M_b g_m^{(3)}(k_t,z)\right|^2$$

$$M_b^2=\frac{|\langle S|p_x|X\rangle|^2}{3}=\frac{m_0}{6}E_p. \tag{A.70}$$

where M_b^2 is the bulk momentum matrix element squared and E_p is the corresponding energy parameter. $g_m^{(1)}(k_t,z)$ and $g_m^{(2)}(k_t,z)$ are the eigencomponents of the upper Hamiltonian H_U; $g_m^{(3)}(k_t,z)$ and $g_m^{(4)}(k_t,z)$ are the eigencomponents of the lower Hamiltonian H_L, and ϕ is the eigenfunction for the Γvalley. Similar equations may be obtained for TM waves [13].

References

1. Madelung, O. (1978) *Introduction to Solid State Theory*, Springer, Berlin.
2. Ridley, B.K. (1982) *Quantum Processes in Semiconductors*, Oxford Univ. Press, Oxford.

3. Basu, P.K. (2003) *Theory of Optical Processes in Semiconductors: Bulk and Microstructures*, Clarendon Press, Oxford.
4. Singh, J. (2003) *Electronic and Optoelectronic Properties of Semiconductor Structures*, Cambridge Univ. Press, Cambridge, New York.
5. Nag, B.R. (1980) *Electron Transport in Compound Semiconductors*, Springer, Berlin.
6. Wang, S. (1989) *Fundamentals of Semiconductor theory and Device Physics*, Prentice Hall, Englewood Cliffs, NJ.
7. Anselm, A.I. (1981) *Introduction to Semiconductor Theory*(translation: M.M. Samokkvalov), Mir Publishers, Moscow.
8. Kane, E.O. (1957) Band structure of indium antimonide. *J. Phys. Chem. Solids*, **1**, 249–261.
9. Chuang, S.L. (1995) *Physics of Optoelectronic Devices*, John Wiley & Sons, New York.
10. Luttinger, J.M. and Kohn, W. (1955) Motion of electrons and holes in perturbed periodic fields. *Phys. Rev.*, **97**, 869–883.
11. Landau, L.D. and Lifshitz, E.M. (1977) *Quantum Mechanics: Non-relativistic Theory*, Pergamon, Oxford.
12. Broido, D.A. and Sham, L.J. (1985) Effective masses of holes at GaAs–AlGaAs heterojunctions. *Phys. Rev. B*, **31**, 888–892.
13. Chang, S-W. and Chuang, S.L. (2007) Theory of optical gain of Ge–$Si_xGe_ySn_{1-x-y}$ quantum well lasers. *IEEE J. Quantum Electron.*, **43**, 249–256.

Appendix B: Values of Parameters

Table B.1 *Values of parameters*

Parameter	Unit	Si	Ge	C	Sn
Atomic number		14	32	6	
Direct band gap: $E_{g\Gamma}$	eV [1]	4.185	0.7985	6.5	−0.413
Indirect band gap: E_{gL}	eV [1]	1.65	0.664		0.092
Indirect band gap: E_{gX}	eV [1]			9.2	
Average valence band energy: ΔE_{vav}	eV [2]	−0.47 [2]	0		0.69 [3, 4]
Spin-orbit splitting: Δ	eV [1]	0.044	0.29		0.8
E_p	eV [1]	21.6	26.3		24.0
Longitudinal mass: m_l		0.915			
Transverse mass: m_t		0.19			
Electron mass in Γ valley: m_Γ		0.528	0.038		0.058
Luttinger parameter: γ_1	[1]	4.22	13.38		−15
Luttinger parameter: γ_2	[1]	0.39	4.24		−11.45
Luttinger parameter: γ_3	[1]	1.44	5.69		−8.55
Lattice constant: a	Å [1]	5.4307	5.6573	3.567	6.4892
Deformation potential: a_c	eV	1.98	−8.38	−16.8	
a_L	eV				
a_v	eV				
b_v	eV	−1.5	−2.2	−4.67	
Stress tensor: C_{11}	$\times 10^{11}$ Pa	1.675	1.315	10.76	
Stress tensor: C_{12}	$\times 10^{11}$ Pa	0.65	0.494	1.25	
Deformation potential for electrons for uniaxial stress: Ξ_u	eV	9.16	9.42		
$\Xi_d + (1/3)\,\Xi_u$	eV	1.5	−2.34		
Relative permittivity: ε_r	(msw)	11.9	16.2		24.0

Table B.2 *Parameters for calculation of strain in SiGeSn*

	Ge	C	Si	Sn
a(Å)	5.657	3.567	5.431	6.489
$E_{dir}(E_0)$(eV)	0.805	6.5	3.37	
E_{indir}(eV)	0.664	9.2	1.16	
Δ_0(eV)	0.30	0.0	0.044	
$C_{11}(10^{11}Pa)$	1.285	10.76	1.66	
$C_{12}(10^{11}Pa)$	0.483	1.25	0.639	
$b = b_1 + 2b_2$(eV)	−2.86	−4.67	−2.2	
a_v(eV)	1.26	−13.9	2.46	
a_c(eV)	−8.38	−16.8	1.98	
$\Xi_d + (1/3)\Xi_u$(eV)	−2.34		1.5	

References

1. Madelung, O., Schultz, M., and Weiss, H. (eds) (1982) *Physics of Group IV Elements and III-V Compounds*, vol. **17a,** 1st edn, Springer-Verlag, NY.
2. Reiger, M.M. and Vogl, P. (1993) Electronic band parameters in strained Si_{1-x} alloys on $Si_{1-y}Ge_y$ substrates. *Phys. Rev. B*, **48**, 14276–14287.
3. Jaros, M. (1988) Simple analytic model for heterojunction band offsets. *Phys. Rev. B*, **39**, 7112–7114.
4. Menendez, J. and Kouvetakis, J.J. (2004) Type I Ge/$Ge_{1-x-y}Si_xSn_y$ strained layer heterostructures with a direct Ge bandgap. *Appl. Phys. Lett.*, **85**, 1175–1177.

Index

Note: Page numbers in *italics* refer to Figures; those in **bold** to Tables

Silicon Photonics: Fundamentals and Devices, First Edition. M. Jamal Deen and P. K. Basu.

Printed and bound by CPI Group (UK) Ltd, Croydon, CR0 4YY
16/09/2026
14963390-0005